Early Events in Monocot Evolution

Tracing the evolution of one of the most ancient major branches of flowering plants, this is a wide-ranging survey of state-of-the-art research on the early clades of the monocot phylogenetic tree. It explores a series of broad but linked themes, providing for the first time a detailed and coherent view of the taxa of the early monocot lineages, how they diversified and their importance in monocots as a whole.

Featuring contributions from leaders in the field, the chapters trace the evolution of the monocots from largely aquatic ancestors. Topics covered include the rapidly advancing field of monocot fossils, aquatic adaptations in pollen and anther structure and pollination strategies, and floral developmental morphology. The book also presents a new phylogenetic tree of early monocots based on sequence data from 17 plastid regions, and a review of monocot phylogeny as a whole, placing in an evolutionary context a plant group of major ecological, economic and horticultural importance.

Abstracts and key words for each chapter are available for download at www.cambridge.org/9781107012769.

PAUL WILKIN is Lilioid and Alismatid Monocots and Ferns Team Leader in the Herbarium, Library, Art and Archives Directorate of the Royal Botanic Gardens, Kew. His main research foci are systematics of Dioscoreales (yams and their allies) and dracaenoids (dragon trees and mother-in-law's tongues), lilioid monocots widely used in human diet and horticulture, with taxa of high conservation and ecological importance. He is principal investigator of the eMonocot Biodiversity Informatics Project.

SIMON J. MAYO is an Honorary Research Associate at the Royal Botanic Gardens, Kew. Since 1977 he has worked on the systematics and phylogeny of the Araceae, the largest plant family of the early-divergent clades in monocots. He has been active in post-graduate teaching in Brazilian universities since 1988, focussing on monocot families and especially on the Araceae.

The Systematics Association
Special Volume Series

SERIES EDITOR

DAVID J. GOWER

Department of Life Sciences, The Natural History Museum, London, UK

The Systematics Association promotes all aspects of systematic biology by organizing conferences and workshops on key themes in systematics, running annual lecture series, publishing books and a newsletter, and awarding grants in support of systematics research. Membership of the Association is open globally to professionals and amateurs with an interest in any branch of biology, including palaeobiology. Members are entitled to attend conferences at discounted rates, to apply for grants and to receive the newsletter and mailed information; they also receive a generous discount on the purchase of all volumes produced by the Association.

The first of the Systematics Association's publications *The New Systematics* (1940) was a classic work edited by its then-president Sir Julian Huxley. Since then, more than 70 volumes have been published, often in rapidly expanding areas of science where a modern synthesis is required.

The Association encourages researchers to organize symposia that result in multi-authored volumes. In 1997 the Association organized the first of its international Biennial Conferences. This and subsequent Biennial Conferences, which are designed to provide for systematists of all kinds, included themed symposia that resulted in further publications. The Association also publishes volumes that are not specifically linked to meetings, and encourages new publications (including textbooks) in a broad range of systematics topics.

More information about the Systematics Association and its publications can be found at our website: www.systass.org

Previous Systematics Association publications are listed after the index for this volume.

Systematics Association Special Volumes published by Cambridge University Press:

78. Climate Change, Ecology and Systematics (2011)
   Trevor R. Hodkinson, Michael B. Jones, Stephen Waldren and John A.N. Parnell
79. Biogeography of Microscopic Organisms: Is everything small everywhere? (2011)
   Diego Fontaneto
80. Flowers on the Tree of Life (2011)
   Livia Wanntorp and Louis Ronse De Craene
81. Evolution of Plant–Pollinator Relationships (2011)
   Sébastien Patiny
82. Biotic Evolution and Environmental Change in Southeast Asia (2012)
   David J. Gower, Kenneth G. Johnson, James E. Richardson, Brian R. Rosen, Lukas Rüber and Suzanne T. Williams

THE SYSTEMATICS ASSOCIATION SPECIAL

VOLUME 83

Early Events in Monocot Evolution

EDITED BY

PAUL WILKIN
and SIMON J. MAYO

Herbarium, Library, Art and Archives Directorate, Royal Botanic Gardens, Kew

THE
Systematics
ASSOCIATION

CAMBRIDGE
UNIVERSITY PRESS

CAMBRIDGE UNIVERSITY PRESS
Cambridge, New York, Melbourne, Madrid, Cape Town,
Singapore, São Paulo, Delhi, Mexico City

Cambridge University Press
The Edinburgh Building, Cambridge CB2 8RU, UK

Published in the United States of America by
Cambridge University Press, New York

www.cambridge.org
Information on this title: www.cambridge.org/9781107012769

First published 2013

Printed and bound in the United Kingdom by the MPG Books Group

A catalogue record for this publication is available from the British Library

Library of Congress Cataloguing in Publication Data

Early events in monocot evolution / edited by Paul Wilkin, Simon J. Mayo, Herbarium, Library, Art and
Archives Directorate, Royal Botanic Gardens, Kew.
    pages   cm. - (The Systematics Association special volume ; 83)
  Based on a conference held in London at the Linnean Society and the Royal Botanic Gardens, Kew,
July 20-22, 2010.
   ISBN 978-1-107-01276-9 (Hardback)
   1. Monocotyledons–Congresses. 2. Monocotyledons–Evolution–Congresses. 3. Angiosperms–
Congresses. 4. Angiosperms–Evolution–Congresses. 5. Plants–Congresses. I. Wilkin, Paul, editor
of compilation. II. Mayo, S. J., editor of compilation.
  QK495.A14E27 2013
  580–dc23

          2012046333

ISBN 978-1-107-01276-9 Hardback

Additional resources for this publication at
www.cambridge.org/9781107012769

Contents

Colour plate section appears between pages 180 and 181.

Contributors

DENIS BARABÉ Institut de Recherche en Biologie Végétale, Jardin Botanique de Montréal, Montréal, Canada

CRAIG F. BARRETT L. H. Bailey Hortorium and Department of Plant Biology, Cornell University, Ithaca, NY, USA

JOSEF BOGNER Botanischer Garten, München-Nymphenburg, München, Germany

MARK W. CHASE Jodrell Laboratory, Royal Botanic Gardens, Kew, Richmond, Surrey, UK

JAMES I. COHEN Department of Biology and Chemistry, Texas A&M International University, Laredo, TX, USA

NATALIE CUSIMANO Institute of Systematic Botany and Mycology, LMU München, München, Germany

JERROLD I. DAVIS L. H. Bailey Hortorium and Department of Plant Biology, Cornell University, Ithaca, NY, USA

MELVIN R. DUVALL Department of Biological Sciences, Northern Illinois University, DeKalb, IL, USA

CAROL A. FURNESS Micromorphology Section, Jodrell Laboratory, Royal Botanic Gardens, Kew, Richmond, Surrey, UK

THOMAS J. GIVNISH Department of Botany, University of Wisconsin-Madison, Madison, WI, USA

RAFAËL GOVAERTS Royal Botanic Gardens, Kew, Richmond, Surrey, UK

SEAN W. GRAHAM UBC Botanical Garden and Centre for Plant Research (Faculty of Land and Food Systems), and Department of Botany, University of British Columbia, Vancouver, Canada

WILLIAM J. D. ILES UBC Botanical Garden and Centre for Plant Research (Faculty of Land and Food Systems), and Department of Botany, University of British Columbia, Vancouver, Canada

F. ANDREW JONES Imperial College London, Silwood Park Campus, Ascot, Berkshire, UK

JIM LEEBENS-MACK Department of Plant Biology, University of Georgia, Athens GA, USA

DONALD H. LES Department of Ecology and Evolutionary Biology, University of Connecticut, CT, USA

SIMON J. MAYO Honorary Research Associate, Royal Botanic Gardens, Kew, Richmond, Surrey, UK

JOEL R. MCNEAL Department of Plant Biology, University of Georgia, Athens GA, USA.

RENATO MELLO-SILVA Instituto de Biociências, Universidade de São Paulo, São Paulo, Brazil

JIN MURATA Botanical Gardens, Graduate School of Science, The University of Tokyo, Bunkyo-ku, Tokyo, Japan

C. DAVID L. ORME Imperial College London, Silwood Park Campus, Ascot, Berkshire, UK.

GITTE PETERSEN Botanical Garden and Museum, Natural History Museum of Denmark, University of Copenhagen, Copenhagen, Denmark

J. CHRIS PIRES Division of Biological Sciences, Christopher S. Bond Life Sciences Center, University of Missouri, Columbia, MO, USA

MARGARITA V. REMIZOWA Department of Higher Plants, Faculty of Biology, Moscow State University, Moscow, Russia

PAULA J. RUDALL Jodrell Laboratory, Royal Botanic Gardens, Kew, Richmond, Surrey, UK

MARIA DAS GRAÇAS SAJO Departamento de Botânica, Instituto de Biociências, UNESP, Rio Claro, Brazil

VINCENT SAVOLAINEN Imperial College London, Silwood Park Campus, Ascot, Berkshire, UK

ROBERT W. SCOTLAND Department of Plant Sciences, South Parks Road, University of Oxford, Oxford, UK

OLE SEBERG Botanical Garden and Museum, Natural History Museum of Denmark, University of Copenhagen, Copenhagen, Denmark

SELENA Y. SMITH Department of Geological Sciences and Museum of Paleontology, University of Michigan, Ann Arbor, MI, USA

BENJAMIN SOBKOWIAK Imperial College London, Silwood Park Campus, Ascot, Berkshire, UK

DMITRY D. SOKOLOFF Department of Higher Plants, Faculty of Biology, Moscow State University, Moscow, Russia

DENNIS W. STEVENSON New York Botanical Garden, Bronx, New York, USA

NORIO TANAKA Tsukuba Botanical Garden, National Museum of Nature and Science, Tsukuba, Japan

NICHOLAS P. TIPPERY Department of Ecology and Evolutionary Biology, University of Connecticut, Storrs, CT, USA

KOICHI UEHARA Graduate School of Horticulture, Chiba University, Matsudo, Japan

PAUL WILKIN Herbarium, Royal Botanic Gardens, Kew, Richmond, Surrey, UK

Preface

The monocotyledons represent a major subgroup of flowering plants, with many species of economic and ecological significance. They include food crops such as cereal grasses, palms, bananas, taro, yams and onions, and also many plants important for ornamental horticulture, such as lilies, orchids, aroids, bromeliads and sedges, some of which are critically endangered in the wild. Since monocots form *c.* 20% of the angiosperms they are of significance to all those working with or studying the biology and evolution of flowering plants. There is now a community of at least 300 researchers pursuing monocot systematic and evolutionary research, as evidenced by attendance at the most recent of the quinquennial Monocots Conferences (Monocots IV) held in Copenhagen in 2008.

Recent new research on both extant and fossil monocots has significantly increased our knowledge of their biology and evolution. Despite these advances, several questions remain, especially regarding aspects of early-divergent monocots and higher-level ordinal relationships. At the Monocots IV Conference there was relatively little focus on the deep branches of the monocot tree or ordinal relationships, or the early-branching taxa themselves, particularly outside the major alismatid radiation, Araceae. Since 2008, new comparative data on monocot evolution and systematics have emerged on an unprecedented scale, addressing whole-genome phylogenetics and character trait evolution in relation to global patterns of dispersal and diversification.

The papers published here were originally presented at the three-day conference *Early Events in Monocot Evolution*, organized to bring together scientists with a special interest in early-divergent monocots and held in London (at the Linnean Society and the Royal Botanic Gardens Kew) on 20–22 July 2010. Although not intended to be equivalent to a quinquennial Monocots Conference, this meeting commemorated the first one, originally conceived by Paula Rudall and Simon Mayo and hosted at the Royal Botanic Gardens Kew in 1993 (Rudall et al. 1995). The 2010 meeting was also planned to coincide with the official retirement of Simon Mayo and to celebrate his career after 37 years working on Araceae and monocot systematics at Kew and in Brazil.

Although first planned to take place in April 2010, the conference had to be postponed until July due to the travel problems arising from the eruption of the volcano Eyjafjallajökull in Iceland, which unfortunately prevented the attendance in July of some monocot scientists. The meeting was nevertheless attended by over 80 participants, with oral presentations by 19 speakers and their collaborators from Brazil, Canada, Denmark, France, Germany, Japan, Mexico, Russia, UK and the USA, as well as posters by authors from a similarly diverse range of nations. The scientific programme had three main themes: monocot origins and relationships, including fossils, evolution and systematics of early-divergent monocots (including Acorales and Alismatales) and deep monocot relationships: early divergence of lilioid monocots, especially pandans and yams (Pandanales and Dioscoreales). Papers presented at the meeting but not published here include contributions by Dennis Stevenson on alismatids, Sabine von Mering and Joachim Kadereit on Juncaginaceae, Mark Chase, Lídia Cabrera and Gerardo Salazar on Araceae phylogeny, Marc Gibernau on floral character evolution in Araceae, and Ana Maria Giulietti Harley on the contribution of Brazilian researchers to monocot taxonomy and phylogenetic research, in recognition of Simon Mayo's contributions to Brazilian botany.

In the present volume the first few chapters focus on general topics. Iles et al. (Chapter 1) set the scene of early monocot phylogeny in a study of the Alismatales, the largest early clade of the monocot phylogeny. They use plastid sequence data in a new analysis which results in more robust support than obtained by previous studies for the key internal nodes of the alismatalean phylogenetic tree. Smith (Chapter 2) provides a fascinating survey of monocot fossils, an area which has advanced rapidly in recent years. She focusses especially on fossils known from Cretaceous times and describes modern techniques which greatly enhance data gathering from fossils. The clear need for continuing morphological research on extant monocots to complement fossil studies is highlighted. Sokoloff et al. (Chapter 3) target the evolution of syncarpy in early-divergent lineages of mono-cots and eudicots, a classic issue in angiosperm evolution, and conclude that intercarpellary fusion is ancestral for monocots. Furness (Chapter 4) surveys pollen and anther characters in the Alismatales and concludes that aquatic and semi-aquatic environments are likely to have had an important influence in their evolution in comparison to other early-branching monocot lineages. Jones et al. (Chapter 5) move the focus towards future prediction of monocot diversity in a macroecological study. They compare species richness in all accepted monocot genera with data on their geographical distribution and ecological (biome) associ-ations. Among other things they find differences between the major monocot orders in their diversification in relation to climate, and highlight the importance of understanding the role of niche conservatism in the response of species to environmental change.

The following chapters move the focus to more detailed studies of families and orders. Les and Tippery (Chapter 6) present a comprehensive and detailed review of molecular phylogenetic studies in the alismatid monocots, a taxon of particular interest since it includes the marine angiosperms and most water-pollinated species. They confirm the monophyly of the core elements of alismatids and note that some genera, including *Sagittaria*, remain poorly understood systematically. Tanaka et al. (Chapter 7) studied pollen and stigma morphology of Hydrocharitaceae and discuss these fascinating and unusual structures in relation to pollination mechanisms and molecular phylogeny. In a developmental study, Remizowa et al. (Chapter 8) investigated floral bract reduction in genera of early-divergent monocots (*Potamogeton*, *Tofieldia*, *Triglochin*) and propose the evolution of two different patterns of reduction, although both may occur within the same genus.

Three chapters follow on Araceae, the largest family of early-divergent monocots. Mayo et al. (Chapter 9) review molecular phylogenetic work in Araceae since 1995, and in Chapter 10 Mayo and Bogner discuss the interpretation of the first evolution-based classification of the family by A. Engler, tracing in its conception the influence of the orthogenetic ideas of C.W. Nägeli. Barabé (Chapter 11) reviews work on floral morphogenesis in Araceae in relation to recently published phylogenies, highlighting the great diversity of developmental features that have been observed to date.

The next two chapters shift attention to developmental studies on groups in later-emergent clades of the monocots. Scotland (Chapter 12) reviews an old controversy concerning the homological relationships of the corona of daffodil flowers and reports on work using ABC developmental genes to investigate the genetic basis of this structure. A study of the Velloziaceae by Sajo et al. (Chapter 13) focusses on the contribution of embryology to assessing relationships within the order Pandanales and reveals considerable complexity both at family and ordinal levels.

Finally, Davis et al. (Chapter 14) conclude the volume by presenting the general picture of the monocots as a whole. They using novel data from the ongoing monocot tree-of-life project (MonAToL) to address the origins of the diversity of the lilioid and commelinid monocot lineages, based on a set of 600 representative taxa.

Early Events in Monocot Evolution presents a range of papers which explore a series of broad but linked themes and provides for the first time a more detailed and coherent view of the taxa of the early monocot lineages, how they diversified and their importance in monocots as a whole. As with Monocot Conference volumes that have already appeared, we are confident that this volume will stimulate further research and discussion of early monocot plants as well as demonstrate the vitality and rapid progress of this area of scientific research.

It is a great pleasure to acknowledge the following people and organizations for their help and support in bringing this project to its conclusion: The Systematics

Association, the Royal Botanic Gardens Kew and the Linnean Society of London, provided the essential financial and infrastructural support for the meeting, the Systematics Association having sponsored the project from its inception. We are also very grateful to the Annals of Botany Company and the Bentham Moxon Trust at Kew for supporting the participation of various researchers. We are especially grateful to the following people for their enthusiastic and essential support in making the meeting so successful and enjoyable; from the Royal Botanic Gardens Kew: Bill Baker, Claire Carter, Karen Etheridge, Lauren Gardiner, Anna Haigh, Anne Morley-Smith, Kamil Rebacz, Paula Rudall, Dave Simpson, Laura Smith, Karen van der Vat, Odile Weber; from the Linnean Society of London: Gren Lucas, Ruth Temple, Vaughn Southgate, Claire Inman and Kate Longhurst and from the Systematics Association, David Gower, Alan Warren and Peter Olson.

For their essential help in producing this book we are most grateful to David Gower (Natural History Museum London), Abigail Jones, Dominic Lewis, Megan Waddington and Zewdi Tsegai from Cambridge University Press, and finally but especially, the many botanists who generously gave their time to review the manuscripts of the papers presented here.

Reference

Rudall, P.J., Cribb, P.J., Cutler D.F. and Humphries C.J. (1995). *Monocotyledons: Systematics and Evolution*. Richmond: Royal Botanic Gardens, Kew.

1

A well-supported phylogenetic framework for the monocot order Alismatales reveals multiple losses of the plastid NADH dehydrogenase complex and a strong long-branch effect

WILLIAM J. D. ILES, SELENA Y. SMITH AND SEAN W. GRAHAM

1.1 Introduction

The order Alismatales is a cosmopolitan and enormously diverse clade of monocotyledons, comprising ~4500 extant species in 13 families, as currently defined (Stevens, 2001+; Janssen and Bremer, 2004; APG III, 2009). Some of the oldest monocot fossils (late Barremian and early Albian; 125–112 Ma) have been assigned to this lineage (Friis et al., 2004, 2010), and most phylogenetic studies (e.g. Chase et al., 2006; Givnish et al., 2006, 2010; Graham et al., 2006) resolve Alismatales as the sister group of all monocots except *Acorus* (Acorales: Acoraceae). Refining our understanding of the phylogenetic backbone of Alismatales will therefore be important for understanding the early evolutionary history of the monocots.

The overall composition of Alismatales remained relatively constant until a recent expansion to include Araceae and Tofieldiaceae (e.g. Dahlgren and Clifford, 1982; Tomlinson, 1982; Les et al., 1997; APG I, 1998; APG II, 2003; APG III, 2009;

Early Events in Monocot Evolution, eds P. Wilkin and S. J. Mayo. Published by Cambridge University Press. © The Systematics Association 2013.

Chase, 2004). This shift reflects substantial molecular systematic evidence (e.g. Duvall et al., 1993; Chase et al., 1995, 2000, 2006; Tamura et al., 2004a; Givnish et al., 2006; Graham et al., 2006) for a close relationship between Araceae, Tofieldiaceae and a clade of 'core alismatid' families that corresponds approximately to the order Helobiae (Engler, 1892) and subclass Alismatidae (Cronquist, 1988). Les and Tippery (Chapter 6, this volume) favour a narrower definition of the clade (as Alismatidae, with two orders, and excluding Araceae and Tofieldiaceae), but we find the broader circumscription of the order more appealing, because it underlines the evolutionary links among these diverse lineages. *Acorus* has also sometimes been recovered within Alismatales (e.g. Davis et al., 2004, 2006), but this placement may reflect substantial rate elevation in several mitochondrial genes (Petersen et al., 2006a, 2006b; Mower et al. 2007; Cuenca et al., 2010). There have been multiple morphological and molecular phylogenetic studies of individual families and major genera of Alismatales (e.g. Les et al., 1993, 1997, 2002a, 2002b, 2005, 2006, 2008, 2010; Tanaka et al., 1997, 2003; Waycott et al., 2002, 2006; Kato et al., 2003; Iida et al., 2004; Rothwell et al., 2004; Tamura et al., 2004b, 2010; Keener, 2005; Lehtonen, 2006, 2009; Lindqvist et al., 2006; Jacobson and Hedrén, 2007; Wang et al., 2007; Cabrera et al., 2008; Lehtonen and Myllys, 2008; Zhang et al., 2008; Ito et al., 2010; von Mering and Kadereit, 2010; Azuma and Tobe, 2011; Cusimano et al., 2011). However, only a few studies (e.g. Les et al., 1997) have surveyed the broad phylogenetic backbone of the order.

Les et al. (1997) provided the most comprehensive study of higher-order relationships in Alismatales. They sampled the plastid gene *rbc*L for exemplar taxa representing all families except Tofieldiaceae, and most of the genera except in Araceae. In addition to improving our knowledge of phylogenetic relationships in the order, and refining family-level circumscriptions, they were interested in reconstructing the evolution of characters that may be associated with hydrophilous (water-mediated) pollination. The core alismatid families are mostly fully aquatic (Les et al., 1997), and semi- to fully aquatic plants are also found in Araceae and Tofieldiaceae, consistent with an aquatic or semi-aquatic habit for the most recent common ancestor of the monocots (e.g. Chase, 2004; note that *Acorus* is also semi-aquatic). If so, terrestrial species in the order (i.e. most Araceae, some Tofieldiaceae) would therefore represent subsequent reversions in habit. The order encompasses all major aquatic life forms (i.e. emergent, floating-leaved, free-floating and submersed; Sculthorpe, 1967), and includes the only fully marine angiosperms, the seagrasses, a life form that evolved several times in the order (Les et al., 1997). Morphological features linked to hydrophily and an aquatic habit are expected to have an unusually high level of homoplasy, which may have contributed to the fluidity of earlier family-level classification schemes based on morphology (see Les and Haynes, 1995, Les et al., 1997).

Les et al. (1997) reconstructed the overall phylogenetic backbone of the order using a single plastid gene, and recovered multiple poorly to moderately supported branches underpinning the higher-order relationships. The monophyly and extent of several families were also unclear (this latter uncertainty was partly accommodated in the APG classification systems by the expanded circumscription of several families).

A few studies have revisited their *rbc*L data set, either alone or in combination with morphology (Chen et al., 2004a, 2004b; Li and Zhou, 2009), but no subsequent studies have sampled the order broadly using additional genes, with the exception of a suite of papers focussed primarily on mitochondrial gene evolution (Petersen et al., 2006b; Cuenca et al., 2010). Here we substantially expand the number of plastid genes sampled from exemplar species that represent the broad phylogenetic backbone of the order. Our major goal is to re-examine and further refine the overall backbone of Alismatales phylogeny recovered by Les et al. (1997) by considering more plastid data per taxon. This general approach has proved to be effective for the inference of broad-scale monocot phylogeny (e.g. Graham et al., 2006; Saarela et al., 2008; Givnish et al., 2010; Saarela and Graham, 2010). We confirm much of the broad phylogenetic backbone recovered by Les et al., (1997), with some notable exceptions. We also obtain substantially improved branch support in many cases. However, we demonstrate that too limited taxon sampling can lead to spurious inference of some local relationships when using plastid genes, which may be a consequence of elevated rates of evolution in a subset of regions examined. Finally, we document and characterize multiple independent losses of plastid genes that code for two subunits of the plastid NADH dehydrogenase chlororespiratory complex.

1.2 Materials and methods

1.2.1 Taxon sampling

Our main analyses focus on a set of 92 exemplar (representative) species comprising 31 species from Alismatales, 49 other monocots and 12 other angiosperms. We expanded taxon sampling in Alismatales by 26 species compared to our most recent broad study of monocot phylogeny (Saarela and Graham, 2010), and included all currently recognized families in the order (Appendix). Our overall taxon sampling for Alismatales is generally less dense than Les et al. (1997), but the included lineages constitute a highly representative subsample of the broad backbone of Alismatales phylogeny. As far as possible we included multiple representatives per family and targeted species within families that span their deepest phylogenetic splits, at least as defined in Les et al. (1997). We included the south-eastern Australian endemic *Maundia triglochinoides* because of a recent report that it lies outside Juncaginaceae (von Mering and Kadereit, 2010), rendering that family paraphyletic as currently circumscribed (Les et al., 1997). Our most complete generic sampling in the order is in Tofieldiaceae, with four of its five genera included (only *Isidrogalvia* is not sampled).

Outside Alismatales we excluded some taxa that were included previously (Graham and Olmstead, 2000; Graham et al., 2006; Saarela et al., 2007, 2008; Saarela and Graham, 2010) to facilitate maximum likelihood analysis, but our taxon sampling is broadly representative of Petrosaviidae (Cantino et al., 2007; this name was coined for the large clade that encompasses all monocots except *Acorus* and Alismatales). We also included new sequences for exemplar species from each of the following families: Acoraceae (Acorales), Bromeliaceae (Poales), Nartheciaceae (Dioscoreales), Nymphaeaceae (Nymphaeales), Orchidaceae (Asparagales), Philesiaceae and Rhipogonaceae (Liliales); see Appendix for details.

1.2.2 Gene sampling

We extracted total genomic DNAs from silica-gel dried leaf material (Appendix) using standard protocols (Doyle and Doyle, 1987; Graham and Olmstead, 2000), or by using a DNeasy Plant Mini Kit (Qiagen Inc, Valencia, California, USA) for recalcitrant material. Some DNAs were provided by the Royal Botanic Gardens, Kew. In several cases we included sequences from GenBank (*Nuphar, Phalaenopsis*; several eudicots) or from other workers (*Vallisneria*; Appendix). In total we sampled 17 plastid genes and associated noncoding regions (omitting several noncoding regions from analysis, see below). These genes are involved in several different plastid functions: photosynthesis (*atp*B, *psb*B, *psb*C, *psb*D, *psb*E, *psb*F, *psb*H, *psb*J, *psb*L, *psb*N, *psb*T, *rbc*L), chlororespiration (*ndh*B, *ndh*F) and protein translation (*rpl*2, *rps*7, 3'-*rps*12). Our sample included the following multigene clusters: *psb*B-*psb*T-*psb*N-*psb*H (which we refer to here as *psb*BTNH), *psb*E-*psb*F-*psb*L-*psb*J (= *psb*EFLJ), and 3'-*rps*12-*rps*7-*ndh*B-*trn*L(CAA). We surveyed these regions using amplification and sequencing protocols noted in Graham and Olmstead (2000) and Saarela et al. (2008), and designed several modified primers for the *psb*BTNH region: modB60F (5'-CATACAGCTTTAGTTGCTGGTTGG), modB64R (5'-GGGATCAGGGATATTTCCAGCAAG), mod65R (5'-GGAAATGTTTCAAAAAAAG-TAGGC A) and modB71R (5'- CCCGGCGCCACTTTACCATATTC).

1.2.3 Data assembly

We carried out base-calling and contig assembly using Sequencher 4.2.2 (Gene Codes Corp., Ann Arbor, Michigan, USA), determining gene boundaries using tobacco and *Ginkgo* reference sequences (Saarela et al., 2008). We added the new sequences to an existing alignment (Saarela et al., 2008), which we adjusted manually using Se-Al 1.0 (Rambaut, 1998) following criteria described in Graham et al. (2000). We coded gaps as missing data. The total aligned length is 23 903 bp, a large portion of which consists of 'offset' noncoding regions that are unique to individual taxa (for a justification of this approach see Saarela and Graham, 2010). For comparison, the unaligned sequence lengths for the newly determined sequences range from 11 009 bp for *Najas* to 15 560 bp for *Stratiotes*. We recovered all 17 gene regions from most species (Appendix). However, for a subset of taxa the *ndh* genes appear to be pseudogenes (i.e. their reading frames are interrupted by stop codons, out-of-phase indels, or both; see below). We recovered a probable *ndh*F pseudogene from *Amphibolis*, and *ndh*B pseudogenes from *Amphibolis*, *Najas*, *Posidonia* and *Thalassia* (partial sequences in several cases, see below). We could not retrieve *ndh*F for *Najas*, *Posidonia* and *Thalassia*, despite extensive attempts at amplification. The apparently pseudogenized *ndh* genes were generally straightforward to align, and so we included them in the analyses. However, a possible *ndh*F pseudogene sequence for *Vallisneria* was so divergent that it could not be aligned reliably, and other *ndh* genes were not recovered for this taxon (M. Moore, Oberlin College, Ohio; pers. comm.).

1.2.4 Phylogenetic analyses

We focussed on coding regions and several conservative noncoding regions from the plastid IR region for the main analysis, following Saarela et al. (2007) and Graham and Iles (2009); the included noncoding regions are intergenic spacers in the contiguous region spanning 3'-*rps*12, *rps*7, *ndh*B and *trn*L, and single introns in each of *rpl*2, 3'-*rps*12 and *ndh*B. We performed heuristic maximum parsimony (MP) searches using PAUP* (Swofford, 2002) with 100 random addition replicates and tree-bisection-reconnection branch swapping, and otherwise using default settings. We used RAxML version 7.2.6 (Stamatakis, 2006; Stamatakis et al., 2008) at the Bioportal website (www.bioportal.uio.no) to perform maximum likelihood (ML) analyses. jModelTest (Posada 2008) was employed to infer the optimal DNA substitution model using the AICc (the Akaike Information Criterion, correcting for sample size) considering the full matrix or subpartitions (see below) for Alismatales only. The GTR+Γ or GTR+Γ+I models were selected in all cases (GTR is the general-time-reversible model, the gamma distribution [Γ] accounts for among-site rate heterogeneity, and the 'I' parameter accommodates invariable sites). Previous analyses of the same gene set across monocots as a whole (e.g. Saarela et al., 2008; Saarela and Graham, 2010) favoured the GTR+Γ+I model. We omitted the I parameter here, as it may be adequately accounted for using the gamma distribution alone (Yang, 2006). We initiated the ML search from 104 random MP starting trees (multiples of eight are required on the Bioportal website), retaining the tree with the highest likelihood score across all searches. We also performed a partitioned ML analysis by distinguishing four partitions, one for each codon position and a separate one for the set of noncoding regions included here, but otherwise using the same settings and general DNA substitution model. We evaluated branch support using the nonparametric bootstrap (Felsenstein, 1985). We considered 500 (MP) or 104 (ML) bootstrap replicates using the search settings described above, but with 10 random addition replicates (MP) or a single random starting tree (ML) per bootstrap replicate. We use the terms 'weak', 'moderate,' and 'strong' to refer to bootstrap support values recovered in the ranges <70%, 70–89%, and ≥90% respectively (Graham et al., 1998).

In earlier unpublished analyses using fewer taxa in Alismatales we noticed that inferred phylogenetic relationships among three families (Alismataceae, Butomaceae and Hydrocharitaceae) depended strongly on the regions and phylogenetic criteria used, and sometimes conflicted with the main results reported here. To explore the possibility that this effect was related to taxon sampling, phylogenetic method or rate heterogeneity in plastid genes, we performed multiple ML and MP analyses for different gene and taxon subsamplings. Specifically, we ran ML and MP analyses on various subsets of the plastid genes, in addition to the full set of regions, using the search settings described above (although in some MP bootstrap analyses we set a MaxTrees limit of 1000 trees). We repeated these analyses for two different taxon densities in Alismatales, a 'reduced' taxon set of 11 exemplar taxa, vs. a 'dense' taxon set comprising all 31 exemplar species. We used two outgroups for these analyses: *Acorus calamus* (Acoraceae) and *Japonolirion osense* (Petrosaviaceae).

1.3 Results

1.3.1 The phylogenetic backbone of Alismatales inferred with 17 plastid genes

Outside Alismatales, the backbone relationships inferred from the full combined data set for 92 taxa are broadly similar to previous estimates using these genes (Fig 1.1, cf. Graham et al., 2006), and so we do not discuss them further here. A portion of the (unpartitioned) ML tree representing Alismatales is presented in Fig 1.2; considering four data partitions in ML analysis did not result in a substantially different topology (not shown: the latter scheme differed in one poorly supported branch inside Araceae). The MP analysis yielded a single most parsimonious tree (tree length = 26 194 steps) that is also highly congruent with the unpartitioned ML tree for Alismatales (not shown). Unpartitioned ML and MP bootstrap values are noted beside individual branches in Fig 1.2; partitioned ML values are indicated in Table 1.1. To facilitate comparisons across analyses and to other studies we have also tabulated support values from the various ML and MP analyses for a subset of branches (Table 1.1; labelled with letters in Fig 1.2); these correspond to interfamilial relationships in the order, in addition to two branches that contradict the monophyly of Juncaginaceae and Cymodoceaceae, respectively. These major backbone relationships in Alismatales are generally strongly supported (ML) or strongly to moderately supported (MP) by the 17-gene data (summarized in the second major column in Table 1.1).

Well-supported clades include Alismatales as a whole (branch a), the core alismatid clade (branch c), a 'petaloid' clade (branch d, comprising three core alismatid families; Les and Tippery, Chapter 6, this volume, refer to this subclade as Alismatales), a 'tepaloid' clade (branch f, comprising the remaining eight core alismatid families; Les and Tippery refer to this subclade as Potamogetonales) and most other branches (branches e and h-m). Core alismatids were distinguished as having either petaloid or tepaloid perianths by Posluszny and Charlton (1993); note that taxa lacking obvious perianths (e.g. *Halodule* and *Najas*) belong to both clades (Fig 1.2). In some cases MP bootstrap support values are marginally (10–20%) weaker than the corresponding ML values (i.e. branch g, which defines the first split in the tepaloid clade above its root node; branch i, which rejects monophyly of Juncaginaceae by placing *Maundia* as the sister group of five families in the tepaloid clade; branch j, for the clade comprising these five families). Two of these three are moderately strongly supported by MP (branches i, j), but all three have strong support (94–100%) from unpartitioned and partitioned ML analyses.

A branch that contradicts the monophyly of Cymodoceaceae here (branch n, which links *Ruppia*, Ruppiaceae, a monogeneric family, with one of the two sampled genera of Cymodoceaceae, *Halodule*; Fig 1.2) is only weakly to moderately supported by all three methods. All other families with multiple exemplar species are strongly supported as monophyletic at the taxon sampling here, and several families with denser sampling also have well-supported internal phylogenetic structure. Specifically, all three internal branches in Tofieldiaceae have strong support, including a placement of *Pleea* as the sister group of the remaining genera, and of *Harperocallis* as the sister group of *Tofieldia-Triantha*; two of four internal branches in Hydrocharitaceae are strongly supported, including a placement of *Stratiotes* as the sister group of other Hydrocharitaceae (Fig 1.2) at the current sparse taxon sampling for this family.

The only major relationship that is not well supported in Alismatales concerns the relative arrangement of its three major subclades: Araceae, Tofieldiaceae and the core clade of alismatid families. Branch b, recovered in the best ML trees here, depicts Araceae as the sister group of the core alismatid families (hence, Tofieldiaceae are the sister group to these two clades, as the order as a whole is also strongly supported). However, this arrangement receives relatively weak support from all three phylogenetic criteria (i.e. 62–66% support, Table 1.1). The two other

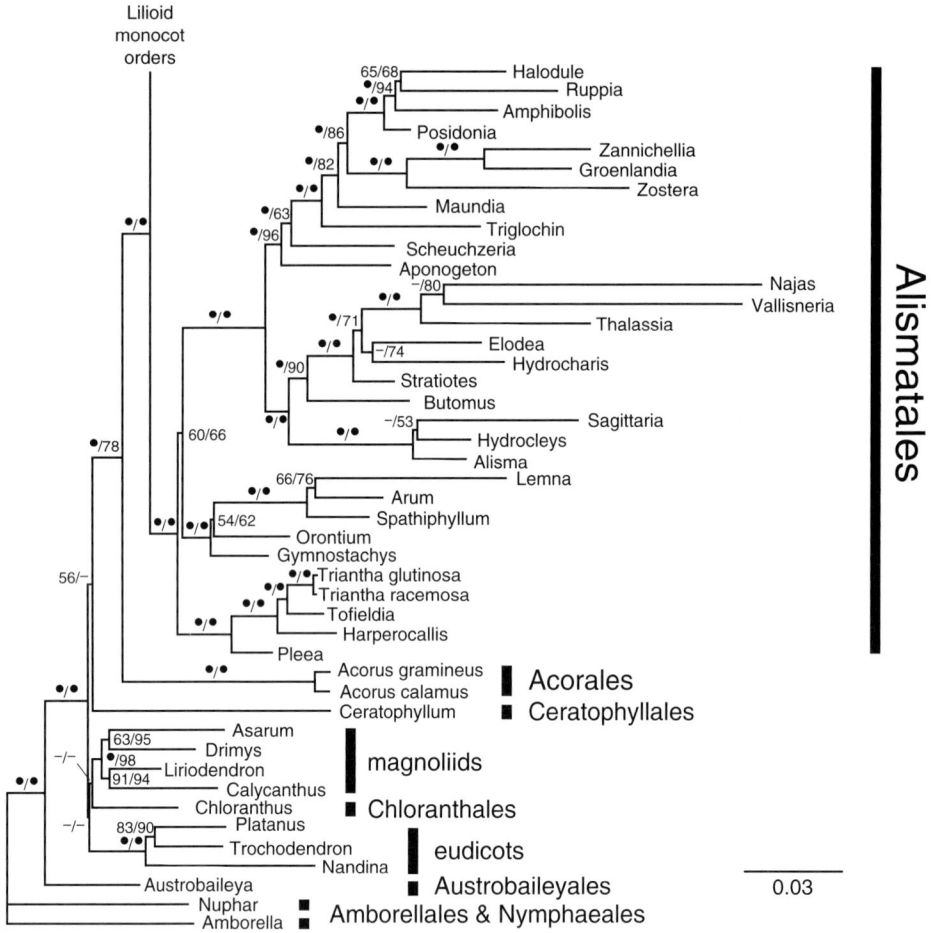

Fig 1.1 Higher-order monocot phylogeny inferred from 17 plastid genes and associated noncoding regions (see text for details). Several outgroup taxa are included. This is the best likelihood tree from an unpartititioned ML analysis (–lnL = 158 904.167). A portion of this tree is shown in magnified form in Fig 1.2. Support values based on bootstrap analyses are noted beside branches (left-hand value = unpartitioned ML, right-hand value = MP); filled circles indicate 100% bootstrap support, dashes <50% bootstrap support. Scale: substitutions per site. An expanded view of Alismatales is presented in Fig 1.2.

possible arrangements for these three clades have been recovered elsewhere with weak to strong support (compare column 2 with columns 4–7 in Table 1.1, which summarize relevant support values in Alismatales across several other studies). One of these alternative possibilities (Araceae sister to Tofieldiaceae; clade b3) has negligible support here, but the other (Tofieldiaceae sister to core alismatids; clade b2) has poor but non-negligible support (i.e. 31–40% of bootstrap replicates from the 17-gene data).

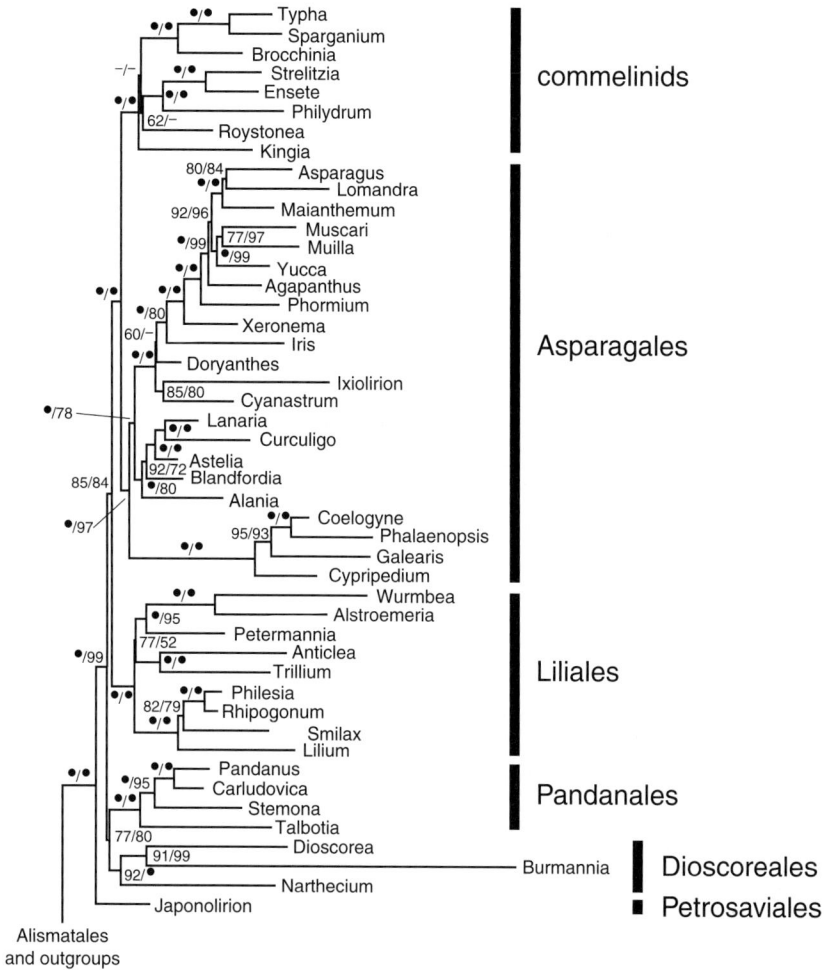

Fig 1.1 (*cont.*)

1.3.2 Effect of taxon density on branch support and conflict

Our reduced taxon set (for the full 17-gene set) has only 11 exemplar taxa from Alismatales, and so several major clades are no longer applicable when compared to the full taxon set (i.e. clades i-n; Fig 1.2 and Table 1.1). We did not run a partitioned ML analysis for this taxon set. The unpartitioned ML and MP bootstrap values are lower for several clades compared to the full taxon sampling (cf. columns 2 and 3 in Table 1.1). These reductions in support are more acute for MP than ML for two clades (i.e. for e and f). However, two clades saw marginal increases with fewer taxa sampled: clade b for ML and MP (this corresponds to Araceae + core alismatids), and clade g for MP only (this is a seven-family clade of core alismatids that corresponds to all of the tepaloid families except Aponogetonaceae; Fig 1.2).

Table 1.1 Comparison of the support values (bootstrap or jackknife) for interfamilial relationships (or for clades i and n, support for the relationships disrupting family monophyly). Author abbreviations: Les et al. (1997; their Fig 2) = L97; Chase et al. (2006; their Fig 2) = C06; Davis et al. (2006; their Fig 2) = D06; von Mering and Kadereit (2010; their Fig 3) = vM&K10. ML-p = partitioned ML; pt = plastid. A dash ('-') means support was not noted or assessed in the corresponding study; a '<' means the branch had <50% support (<70% in von Mering and Kadereit, 2010); 'na' = not applicable due to disrupted monophyly. Clade labels are depicted in Fig 1.2, except for those with a number (b2, b3, etc).

	Full taxon set[b]	Reduced taxon set[b]	L97	C06	D06	vM&K10
No. of exemplar species (Alismatales):	31 species	11 species	78 species	13 species	12 species	37 species
No. of genes:	17 pt genes	17 pt genes	1 gene[c]	7 genes[d]	4 genes[e]	1 gene[c]
Branch support determined using:	ML, ML-p (MP)	ML (MP)	MP	MP	MP	ML (MP)
Clade label and description[a]						
a Alismatales	100, 100 (100)	100 (99)	–	100	na	–(–)
b Araceae + core alismatids	60, 62 (66)	85 (82)	–	–	na	–(–)
b2 Tofieldiaceae + core alismatids	40, 34 (31)	15 (16)	–	99	na	–(–)
b3 Tofieldiaceae + Araceae	<5, <5 (<5)	<5, <5	–	–	na	72 (<)
c Core alismatid clade	100, 100 (100)	100 (100)	96	100	100	93 (95)
d Petaloid clade	100, 100 (100)	100 (100)	88	87	<	92 (96)
e Hydrocharitaceae + Butomaceae	100, 100 (90)	69 (<5)	31	–	<	<(<)
e2 Hydrocharitaceae + Alismataceae	<5, <5 (<5)	31 (99)	<	<	<	–(–)

f	Tepaloid clade	100, 100 (96)	77 (41)	77	–	–	77 (<)
g	Tepaloid clade excl. Aponogetonaceae	100, 94 (63)	92 (85)	–	–	–	– (–)
h	Tepaloid clade excl. Aponogetonaceae + Scheuchzeriaceae	100, 100 (100)	100 (100)	78	–	100	88 (87)
i	*Maundia* + Ruppiaceae/Cymodoceaceae + Posidoniaceae + Potamogetonaceae + Zosteraceae	100, 100 (82)	– (–)	–	–	–	<(71)
j	Ruppiaceae/Cymodoceaceae + Posidoniaceae + Potamogetonaceae + Zosteraceae	100, 100 (86)	– (–)	71	–	98	– (–)
k	Potamogetonaceae + Zosteraceae	100, 100 (100)	– (–)	100	100	–	99 (99)
l	Ruppiaceae/Cymodoceaceae + Posidoniaceae	100, 100 (100)	– (–)	40	–	–	77 (<)
m	Ruppiaceae/Cymodoceaceae	100, 91 (94)	– (–)	29	–	–	<(<)
n	*Ruppia* + *Halodule*	65, 73 (68)	– (–)	–	–	–	– (–)

[a] According to APG (2009); [b] 92 exemplar species here in the full taxon set (including 61 outgroups), vs. 13 exemplars in the reduced taxon set (including two outgroups); [c] *rbcL* only; [d] Four plastid genes, two nuclear genes, one mitochondrial gene; [e] Two plastid genes, two mitochondrial genes.

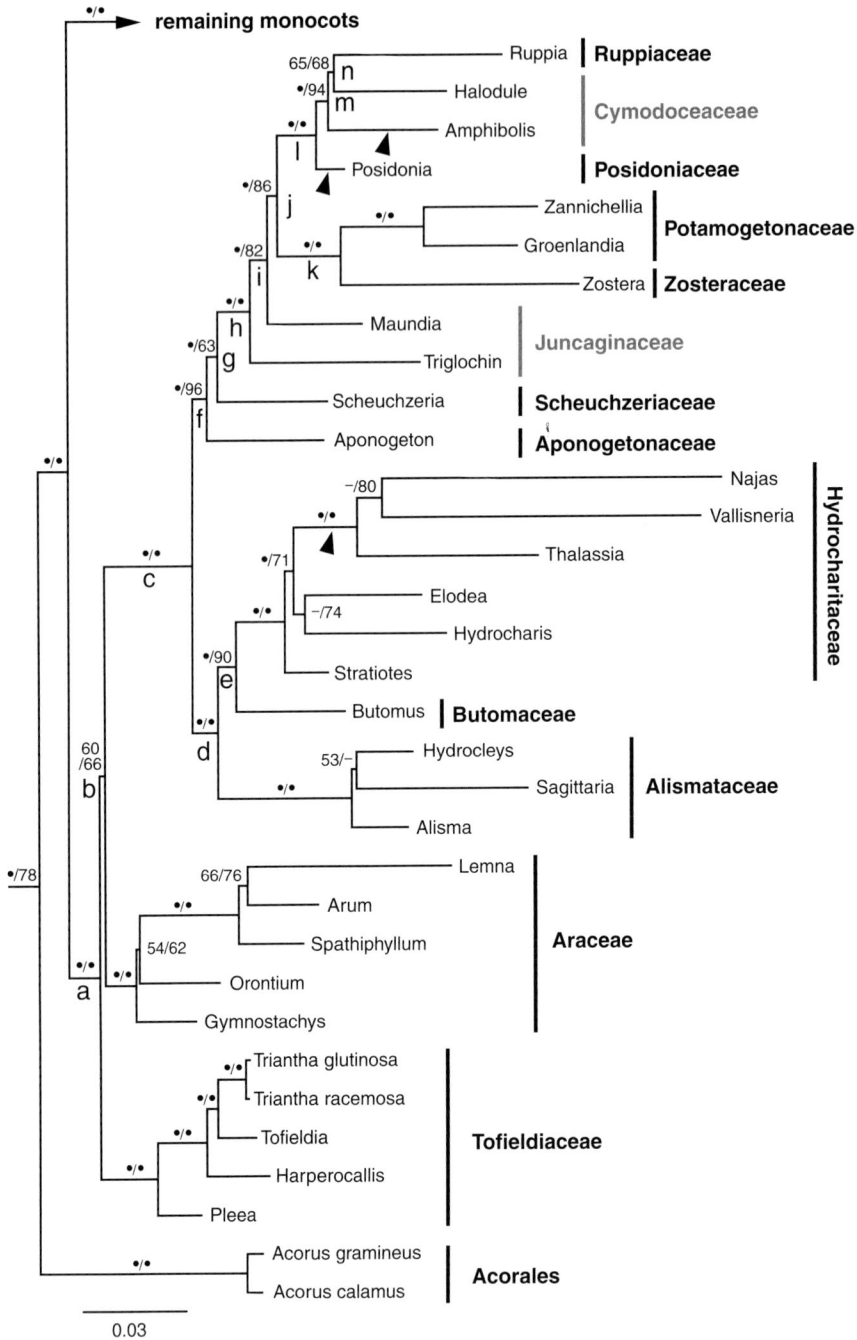

Fig 1.2 A portion of the monocot tree (Fig 1.1), focussed on Alismatales. Two families not resolved as monophyletic are noted in grey. Letter labels refer to clades noted in the text and Tables 1.1 and 1.2. Arrowheads indicate putatively independent losses of one or more *ndh* (NADH dehydrogenase subunit) loci. Scale and support values as for Fig 1.1.

The most surprising shift with the reduced taxon sampling was clade e, (Hydrocharitaceae + Butomaceae) which is weakly (ML) to negligibly (MP) supported (column 3 in Table 1.1); this lineage of petaloid alismatids was a strongly supported clade for the full taxon sampling (Fig 1.2; column 2 in Table 1.1). In contrast, a conflicting clade (clade e2; Hydrocharitaceae + Alismataceae) that previously had negligible (<5%) support (Table 1.1), receives non-negligible bootstrap support from ML (31%) and strong support from MP (99% support) with the reduced taxon sampling. We examined the conflicting signal for these two contrasting branches by performing bootstrap analyses on subsets of the full plastid data set that corresponds to individual genes or sets of genes (Table 1.2). We repeated these analyses for two different taxon densities (i.e. 11 vs. 31 ingroup taxa, the 'reduced' and 'dense' taxon samplings in Table 1.2), and by considering two optimality criteria (i.e. unpartitioned ML and MP). At the reduced taxon sampling, branch e, corresponding to Hydrocharitaceae + Butomaceae, is strongly supported by some data partitions for ML (e.g. coding regions only; single copy plastid genes; *ndh*B) and weakly supported by others (*psb*BTNH); no data partition supported this arrangement even moderately well for MP. In contrast, the conflicting relationship, branch e2 (corresponding to Hydrocharitaceae +Alismataceae) is moderately to strongly supported by multiple data partitions for MP at this weak taxon density, including some partitions that have strong support for the contrasting relationship for ML (e.g. the combined coding plastid regions).

Likelihood and parsimony analyses appear to converge to clade e at the dense taxon sampling (Table 1.2). This taxon sampling substantially breaks up the long terminal branches subtending *Najas* (Hydrocharitaceae) and *Sagittaria* (Alismataceae) in Fig 1.3A (note how both are divided for at least part of their length in Fig 1.2). One exception concerns the analysis of *rpl*2 alone, which recovers branch e2 for the reduced taxon sampling with strong support from ML and MP, but which, in contrast, has no well-supported relationship at the dense taxon sampling, for either phylogenetic method (Table 1.2). This gene has the greatest disparity of branch lengths for any data partition considered here (e.g. compare the relative lengths of the terminal branches subtending *Najas* and *Sagittaria* for the single-copy coding regions combined vs. *rpl*2 alone Fig 1.3A, B).

1.3.3 Parallel loss of *ndh* genes in the core alismatid clade

We predict that one or more plastid *ndh* loci have been lost independently in multiple lineages of core alismatids, based on the phylogenetic distribution of taxa that have accumulated stop codons in *ndh*B, *ndh*F or both (see arrowheads in Fig 1.2). These include a loss in the petaloid alismatids (perhaps in the common ancestor of *Najas*, *Thalassia* and *Vallisneria*, which comprise a well-supported subclade of Hydrocharitaceae), and two parallel losses in the tepaloid alismatid

Table 1.2 Comparison of MP and ML support values for contrasting relationships among the three petaloid families of Alismatales for different taxon densities and data partitions; A = Alismataceae; B = Butomaceae; H = Hydrocharitaceae; the '<' symbol corresponds to less than 50% bootstrap support. All listed plastid partitions lack noncoding regions unless noted.

	Reduced ingroup sampling[a]		Dense ingroup sampling[a]	
Clade recovered:	[H + B]	[H + A]	[H + B]	[H + A]
Clade label (Table 1.1)	e	e2	e	e2
Phylogenetic criterion:	ML (MP)	ML (MP)	ML (MP)	ML (MP)
Data partition				
All plastid genes (+IR noncoding[b])	69 (<)	<(99)	100 (91)	<(<)
Coding plastid regions only	100 (<)	<(92)	100 (87)	<(<)
Single copy plastid genes[c]	92 (<)	<(88)	100 (68)	<(<)
*atp*B	<(<)	<(<)	77 (61)	<(<)
*psb*BTNH	55 (<)	<(<)	<(62)	<(<)
*psb*DC	<(<)	69 (90)	54 (62)	<(<)
*psb*EFLJ	<(<)	<(<)	<(<)	<(<)
*rbc*L	<(<)	<(<)	54 (<)	<(<)
Entire IR region (incl. noncoding)	<(<)	73 (100)	100 (98)	<(<)
*ndh*B (IR gene, incl. intron)	92 (65)	<(<)	100 (97)	<(<)
*rpl*2 (IR gene, incl. intron)	<(<)	85 (93)	<(<)	<(<)

[a] The reduced ingroup sampling comprises 11 members of Alismatales vs. 31 species for the dense ingroup sampling; note that both samples include only two outgroups (*Acorus calamus* and *Japonolirion osense*); [b] Inverted repeat (IR) region = *rpl*2, 3′-*rps*12, *rps*7, *ndh*B, several intergenic spacer regions (between 3′-*rps*12 and *rps*7, *rps*7 and *ndh*B, and *ndh*B to *trn*L), and three introns (one each in *rpl*2, 3′-*rps*12 and *ndh*B); [c] 12 single copy loci (i.e. 10 *psb* genes, *atp*B and *rbc*L; *ndh*F was not retrieved here for *Najas*).

clade (one in Cymodoceaceae, one in Posidoniaceae). Although *Amphibolis* and *Posidonia* are close relatives (Fig 1.2), the former species is closely related to other sampled taxa (*Halodule* and *Ruppia*) that have retained open reading frames for *ndh*B and *ndh*F, supporting a convergent loss of function.

The internal stop codons in the *ndh*B locus are due to DNA substitutions in two cases (*Amphibolis, Najas*) and to indels in *Najas, Posidonia* and *Thalassia* (summarized in Fig 1.4). Reading-frame shifts resulting from these indels reveal

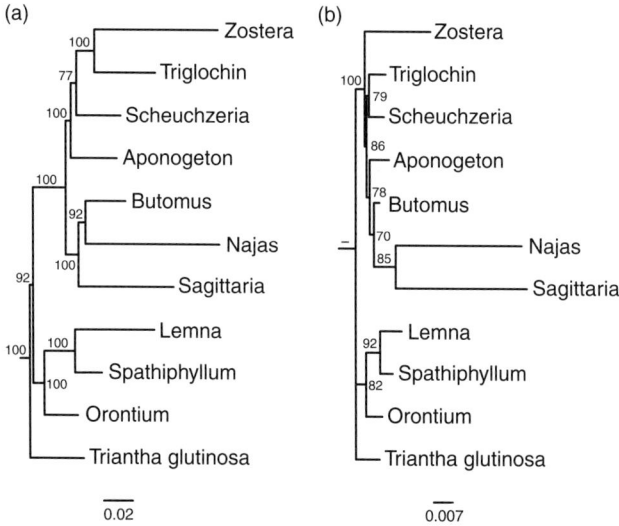

Fig 1.3 Likelihood trees inferred for a reduced taxon set for two plastid gene subsets, demonstrating contrasting patterns of rate heterogeneity. The taxon set includes 11 ingroup taxa (and two outgroups, *Acorus calamus* and *Japonolirion osense*, not shown here). A: Concatenated coding regions from the plastid single copy regions only; B: The *rpl*2 locus (two exons and an intron). Numbers adjacent to branches are ML bootstrap support values (dashes indicates <50% support; note the basal trichotomy in B). Scale: substitutions per site.

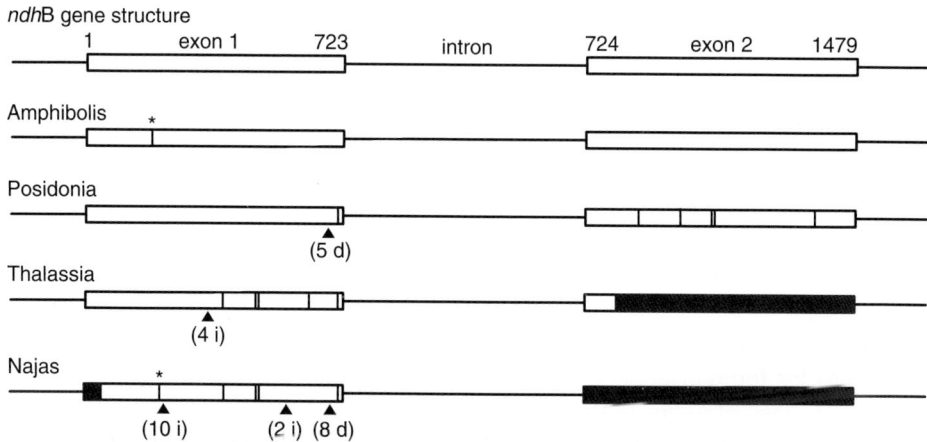

Fig 1.4 Putative pseudogenes of the plastid gene *ndh*B from *Amphibolis*, *Najas*, *Posidonia* and *Thalassia* (top panel; nucleotide positions for exon boundaries are noted relative to tobacco; the intron is not completely to scale across taxa). Vertical lines indicate predicted internal stop codons; those annotated with asterisks are inferred to be due to substitution events, whereas the remainder are a consequence of frameshifts following one or more nontriplet indel events noted below the first exon (size in bp noted; i = inferred insertion, d = inferred deletion). Shaded regions were not recovered.

otherwise out-of-frame stop codons that were present in their common ancestor. However, the relevant indels are not shared among them (Fig 1.4). We amplified only partial *ndh*B genes for two taxa (*Najas* and *Thalassia*) and were unable to retrieve another plastid-encoded NADH dehydrogenase subunit gene, *ndh*F, from three species (*Najas, Posidonia* and *Thalassia*). The *ndh*F locus recovered for *Amphibolis* also has multiple indels and stop codons. The putative *ndh*F sequence from *Vallisneria* recovered by Moore and colleagues (not included in analysis here) appears to be highly degraded.

1.4 Discussion

1.4.1 A refined estimate of the phylogenetic backbone of Alismatales

Our phylogenetic inferences based on 17 slowly evolving plastid genes and several associated noncoding regions are generally highly congruent with the backbone that was inferred by Les et al. (1997), who used more ingroup taxa but only a single gene. Only six of the internal branches that correspond to interfamilial relationships (summarized in Table 1.1) were moderately to strongly supported by Les et al., with a similar or smaller number of branches supported in other recent studies that have a moderately representative sampling of the backbone (12–37 exemplar species from the order sampled in Chase et al., 2006; Davis et al., 2006 and von Mering and Kadereit, 2010; see Table 1.1). In contrast, we recover strong support for 11 of 12 of these branches using ML bootstrap analysis (for partitioned and unpartitioned ML; branches a, c–m, ignoring several poorly supported alternative relationships noted in Table 1.1). MP bootstrap support for these branches was also generally comparable for the full taxon set, although there was only moderately strong MP bootstrap support for branches g, i and j.

Our new data provide consistently strong support for the division of the core alismatid families into 'petaloid' vs. 'tepaloid' clades, each comprising multiple families, and they also resolve several major branches in each case that until now have been neither well supported nor consistently resolved across studies. Butomaceae and Hydrocharitaceae are strongly supported as sister groups in the petaloid alismatids for the dense taxon sampling. In the tepaloid alismatids we find strong support for several major branches that have been recovered before but with only moderate to weak support (e.g. the clade comprising Ruppiaceae, Cymodoceaceae and Posidoniaceae). For the first time we resolve the deepest splits in the tepaloid alismatids with strong support: Aponogetonaceae and Scheuchzeriaceae are, respectively, the successive sister groups of the remaining families in this subclade.

The monophyly of most families of Alismatales as currently circumscribed (APG III, 2009) is also confirmed with strong bootstrap support here (Fig 1.2), within the limits of the current taxon sampling (an advance on previous studies in several cases). Cymodoceaceae may well be paraphyletic, although this result is not strongly supported here (Fig 1.2). We also confirm the finding of von Mering and Kadereit (2010) that the family Juncaginaceae is not monophyletic as currently construed, as the two genera that we sampled from this family, *Triglochin* and *Maundia*, are successive sister groups of five other families in the tepaloid clade (Fig 1.2). This paraphyletic arrangement has strong ML bootstrap support, and moderately strong MP bootstrap support (Fig 1.2, Table 1.1). *Maundia* may therefore deserve to be recognized as its own family Maundiaceae Nakai (see Takhtajan, 1997; von Mering and Kadereit, 2010). It may be premature to do so until the remaining genera of Juncaginaceae are included in studies using gene samples that are comparable to or larger than ours, in case further paraphyly is uncovered. However, the remainder of Juncaginaceae are monophyletic in von Mering and Kadereit (2010), although the corresponding clade was not well supported in their analyses. Les and Tippery (Chapter 6, this volume) recovered a clade that included all genera of Juncaginaceae (excluding *Maundia*), with poor support, in a study that had an expanded taxon sampling compared to the original *rbc*L-based study of Les et al. (1997). We are currently sampling additional taxa of Juncaginaceae s.s., and in preliminary analyses, we recover its monophyly with strong support (Iles and Graham, unpublished data).

APG III (2009) note that an alternative to recognizing an additional family (Maundiaceae) in rank-based classifications would be to lump multiple families from all or most of the tepaloid clade into one family; Aponogetonaceae and Scheuchzeriaceae might be included in the resulting family, as both are small, monogeneric families (see Backlund and Bremer, 1998). However defined, the resulting family would be referred to as Potamogetonaceae Bercht and J. Presl, as this name has priority (it corresponds to Potamogetonales as circumscribed by Les and Tippery, Chapter 6, this volume). It would be an extremely heterogeneous family, if recognized.

1.4.2 Impact of rate heterogeneity on phylogenetic inference in Alismatales

We observed substantial rate heterogeneity in the plastid data among different lineages of Alismatales (Fig 1.2), and the order as a whole clearly includes some of the longest branches of angiosperms considered here (Fig 1.1). Although not as extreme as the rate elevation observed in some mitochondrial genes (Mower et al., 2007), we worried that this may have a misleading impact on phylogenetic inference in at least some cases, due to long-branch artefacts (Felsenstein, 1978; Hendy and Penny, 1989). Long-branch problems can be minimized by dense taxon sampling (e.g. Hillis, 1998; Zwickl and Hillis, 2002; Hillis et al., 2003; Hedtke et al., 2006; Heath et al., 2008), but it is probably the case that additional taxon sampling here would be unlikely to break up the broad Alismatales backbone much further, as we sampled all families and used multiple exemplar species per family where this was feasible. Nonetheless, our examination of a less dense taxon sampling for these genes demonstrated that strongly supported but likely spurious findings are still possible with plastid data (Table 1.2): we know that at least one of the strongly conflicting arrangements of Alismataceae, Butomaceae and Hydrocharitaceae found using fewer taxa for different plastid data partitions or phylogenetic criteria must be incorrect. In this particular case we favour the arrangement that places Butomaceae as the sister group of Hydrocharitaceae, as this is what we see with the dense taxon sampling (Fig 1.2) and for most data partitions (Table 1.2) for parsimony and likelihood. Nonetheless, even model-based methods may be led astray by imperfectly modelled DNA substitution events (e.g. Matsen and Steel, 2007). This is a particular concern in Alismatales as there is fairly extensive heterogeneity in rate variation apparent among different plastid regions (e.g. Fig 1.3), although not as extreme as that observed for mitochondrial genes (see Petersen et al., 2006a, 2006b; Cuenca et al., 2010). Improved maximum likelihood models (e.g. ones that take better account of heterotachy) may help in these situations. We are also encouraged by how the denser taxon sampling led to convergence of ML and MP across most of the regions considered here, and apparently removed the strong conflict for all data partitions examined (Table 1.2). Les and Haynes (1995) suggest several morphological synapomorphies for a clade comprising only Butomaceae and Hydrocharitaceae, lending further support to this arrangement.

1.4.3 Loss of NADH dehydrogenase subunit genes

The plastid *ndh* genes encode protein subunits of the thylakoid NADH dehydrogenase complex, which is homologous at a very deep level of phylogeny with the mitochondrial NADH dehydrogenase/complex I (Shinozaki et al., 1986). The plastid complex is encoded by 11 plastid genes and additional nuclear-encoded subunits. Plastid *ndh* genes have been retained in most embryophytes and charophytes (Martín and Sabater, 2010), but appear to have been lost frequently in heterotrophic plants (e.g. for several parasitic plants, dePamphilis and Palmer, 1990; Stefanović and Olmstead, 2005) along with other plastid genes, apparently associated with a loss or reduction in photosynthetic capability. In monocots they have been inferred to be lost in some orchids (e.g. Neyland and Urbatsch, 1996; Chang et al., 2006), although the full extent of their loss in this mycoheterotrophic family is unclear. There are only a few putative losses in fully autotrophic plants (the suite of 11 *ndh* plastid-encoded genes is absent or pseudogenized in extant Pinaceae, Gnetales and *Erodium* of Geraniaceae; Braukmann et al., 2009; Blazier et al., 2011), and so the possibility of multiple apparently independent losses of *ndh*B and *ndh*F within a single order is surprising (note that mycoheterotrophy is thought to be lacking in the order as currently circumscribed; see Wang and Qiu, 2006).

While a complete understanding of the function of this complex is lacking, it has been implicated in chlororespiration, programmed cell death and protection against photo-oxidative stress during photosynthesis (see Martín and Sabater, 2010). The last may be important for understanding its evolutionary loss, as tobacco plants with experimentally induced nonfunctional *ndh* genes grow normally under optimal conditions, but not when environmentally stressed (Martín and Sabater, 2010). In Alismatales the losses could therefore be related to altered physiological constraints in an aquatic environment, such as reduced light stress in the subtidal zone (Martín and Sabater, 2010).

The phylogenetic distribution of *ndh* pseudogenes observed here suggests at least three independent losses in the order (Fig 1.2), but the various ways in which *ndh*B has become pseudogenized (Fig 1.4) is potentially consistent with different scenarios of loss (however, note that the lack of shared stop codons is not in itself indicative of separate losses, as a common initial loss of function may have occurred in some unsampled *ndh* subunit, with subsequent independent pseudogenization in the other subunits). At this point we can only predict loss of function based on presence of stop codons (and to some extent the general difficulty of retrieving *ndh* loci in these taxa), and make predictions of convergent loss based on phylogenetic distribution. It would therefore be useful to characterize whether other *ndh* subunit genes from the plastid genome are also pseudogenized, and to confirm loss of function of the whole complex using physiological or transcriptome-based methods. A denser taxonomic sampling would also be helpful for characterizing the extent and nature of independent *ndh*B and *ndh*F losses in the order.

1.4.4 Future work on the Alismatales backbone

Although we have refined our understanding of the broad phylogenetic backbone of Alismatales, additional work remains to be done. Several major branches within the order are very short according to the plastid data (e.g. branch b, subtending Araceae+core alismatids; major intrafamilial splits within Araceae, Alismataceae, Hydrocharitaceae), and the putative paraphyly of Cymodoceaceae needs further confirmation. Other clades may benefit from expanded taxon sampling for the current gene sampling. Within Tofieldiaceae, for example, the subtropical *Pleea* and *Harperocallis* form successive sister lineages to the predominately temperate and arctic *Tofieldia* and *Triantha*, which agrees closely with Tamura et al. (2004b, 2010) and Azuma and Tobe (2011). However, one South American montane genus, *Isidrogalvia*, was not included here. This genus may be the sister group of *Harperocallis* (Azuma and Tobe, 2011). Suggestions of a placement of *Isidrogalvia* within Nartheciaceae (Tamura et al., 2004b) appear to be the result of misidentification or contamination (Azuma and Tobe, 2011).

Some of the unresolved relationships here may profit from consideration of more genes per taxon (e.g. of the order of the whole plastid genome). However, the finding that an apparent strong long-branch effect seems to be ameliorated by dense taxon sampling suggests that future comparative genomic studies of the order should aim to include a density of taxa that meets or exceeds that used here. The phylogenetic framework that we inferred should be useful for other workers interested in exploring the evolution of morphological or other molecular characters in the order, facilitating our understanding of the origin and early evolution of the major clades of monocotyledons.

1.5 Acknowledgements

We thank J. Bogner, J. Bruinsma, S. Jacobs, D. Kolterman, M. Moore, G. Rothwell, R. Stockey, M. Waterway, W. Zomlefer and colleagues, the Royal Botanic Gardens, Kew, for plant material, J. Saarela, H. Rai and J. Zgurski for unpublished sequences (*Brocchinia*, *Narthecium*, *Philesia* and *Rhipogonum*), and several anonymous reviewers. This study was supported by a Natural Sciences and Engineering Research Council of Canada (NSERC) Discovery Grant to S.W.G., an NSERC graduate fellowship and an Alberta Ingenuity studentship to S.Y.S., and a graduate fellowship to W.J.D.I. from the University of British Columbia.

1.6 References

APG I (Angiosperm Phylogeny Group I). (1998). An ordinal classification for the families of flowering plants. *Annals of the Missouri Botanical Garden*, **85**, 531–553.

APG II (Angiosperm Phylogeny Group II). (2003). An update of the Angiosperm Phylogeny Group classification for the orders and families of flowering plants: APG II. *Botanical Journal of the Linnean Society*, **141**, 399–436.

APG III (Angiosperm Phylogeny Group III). (2009). An update of the Angiosperm Phylogeny Group classification for the orders and families of flowering plants: APG III. *Botanical Journal of the Linnean Society* **161**, 105–121.

Azuma, H. and Tobe, H. (2011). Molecular phylogenetic analyses of Tofieldiaceae (Alismatales): family circumscription and intergeneric relationships. *Journal of Plant Research*, **124**, 349–357.

Backlund, A. and Bremer, K. (1998). To be or not to be – principles of classification and monotypic plant families. *Taxon*, **47**, 391–400.

Blazier, J. C., Guisinger, M. M. and Jansen, R. K. (2011). Recent loss of plastid-encoded *ndh* genes within *Erodium* (Geraniaceae). *Plant Molecular Biology*, **76**, 263–272.

Braukmann, T. W. A., Kuzmina, M. and Stefanović, S. (2009). Loss of all plastid *ndh* genes in Gnetales and conifers: Extant and evolutionary significance for the seed plant phylogeny. *Current Genetics*, **55**, 323–337.

Cabrera, L. I., Salazar, G. A., Chase, M. W. et al. (2008). Phylogenetic relationships of aroids and duckweeds (Araceae) inferred from coding and noncoding plastid DNA. *American Journal of Botany*, **95**, 1153–1165.

Cantino, P. D., Doyle, D. A., Graham, S. W. et al. (2007). Towards a phylogenetic nomenclature of *Tracheophyta*. *Taxon*, **56**, 822–846.

Chang, C.-C., Lin, H.-C., Lin, I-P. et al. (2006). The chloroplast genome of *Phalaenopsis aphrodite* (Orchidaceae): comparative analysis of evolutionary rate with that of grasses and its phylogenetic implications. *Molecular Biology and Evolution*, **23**, 279–291.

Chase, M. W. (2004). Monocot relationships: an overview. *American Journal of Botany*, **91**, 1645–1655.

Chase, M. W., Duvall, M. R., Hills, H. G. et al. (1995). Molecular phylogenetics of Lilianae. In *Monocotyledons: Systematics and Evolution, Volume I*, ed. P. J. Rudall, P. J. Cribb, D. F. Cutler and C. J. Humphries. Kew: Royal Botanic Gardens, pp. 109–137.

Chase, M. W., Soltis, D. E., Soltis, P. S. et al. (2000). Higher-level systematics of the monocotyledons: an assessment of current knowledge and a new classification. In *Monocots: Systematics and Evolution*, ed. K. L. Wilson and D. A. Morrison. Melbourne: CSIRO, pp. 3–16.

Chase, M. W., Fay, M. F., Devey, D. S. et al. (2006). Multigene analyses of monocot relationships: a summary. In *Monocots: Comparative Biology and Evolution (excluding Poales)*, ed. J. T. Columbus, E. A. Friar, J. M. Porter, L. M. Prince and M. G. Simpson. Claremont, CA: Rancho Santa Ana Botanic Garden, pp. 63–75.

Chen, J.-M., Chen, D., Gituru, W. R., Wang, Q.-F. and Guo, Y.-H. (2004a). Evolution of apocarpy in Alismatidae using phylogenetic evidence from chloroplast *rbc*L gene sequence data. *Botanical Bulletin of Academia Sinica*, **45**, 33–40.

Chen, J.-M., Gituru, W. R. and Wang, Q.-F. (2004b). Evolution of aquatic life-forms in Alismatidae: phylogenetic estimation from chloroplast *rbc*L gene sequence data. *Israel Journal of Plant Sciences*, **52**, 323–329.

Cronquist, A. 1988. *The Evolution and Classification of Flowering Plants*. New York: New York Botanical Garden.

Cuenca, A., Petersen, G., Seberg, O., Davis, J. I and Stevenson, D. W. (2010). Are substitution rates and RNA editing correlated? *BMC Evolutionary Biology*, **10**, 349.

Cusimano, N., Bogner, J., Mayo, S. J. et al. (2011). Relationships within the Araceae: comparison of morphological patterns with molecular phylogenies. *American Journal of Botany*, **98**, 654–668.

Dahlgren, R. M. T. and Clifford, H. T. (1982). *The Monocotyledons: A Comparative Study*. London: Academic Press.

Davis, J. I, Stevenson, D. W., Petersen, G. et al. (2004). A phylogeny of the monocots, as inferred from *rbc*L and *atp*A sequence variation, and a comparison of methods for calculating jackknife and bootstrap values. *Systematic Botany*, **29**, 467–510.

Davis, J. I, Petersen, G., Seberg, O. et al. (2006). Are mitochondrial genes useful for the analysis of monocot relationships? *Taxon*, **55**, 857–870.

dePamphilis, C. W. and Palmer, J. D. (1990). Loss of photosynthetic and chlororespiratory genes from the plastid genome of a parasitic flowering plant. *Nature*, **384**, 337–339.

Doyle, J. J. and Doyle, J. L. (1987). A rapid DNA isolation procedure for small quantities of fresh leaf tissue. *Phytochemical Bulletin*, **19**, 11–15.

Duvall, M. R., Clegg, M. T., Chase, M. W. et al. (1993). Phylogenetic hypotheses for the monocotyledons constructed from *rbc*L sequence data. *Annals of the Missouri Botanical Gardens*, **80**, 607–619.

Engler, A. (1892). *Syllabus der Vorlesungen über spezielle und medizinisch-pharmaceutische Botanik*. Berlin: Bornträger.

Felsenstein, J. (1978). Cases in which parsimony or compatibility methods will be positively misleading. *Systematic Zoology*, **27**, 401–410.

Felsenstein, J. (1985). Confidence limits on phylogenies: an approach using the bootstrap. *Evolution*, **39**, 783–791.

Friis, E. M., Pedersen, K. R. and Crane, P. R. (2004). Araceae from the Early Cretaceous of Portugal: evidence on the emergence of monocotyledons. *Proceedings of the National Academy of Sciences of the United States of America*, **101**, 16565–16570.

Friis, E. M., Pedersen, K. R. and Crane, P. R. (2010). Diversity in obscurity: fossil flowers and the early history of angiosperms. *Philosophical Transactions of the Royal Society B: Biological Sciences*, **365**, 369–382.

Givnish, T. J., Pires, J. C., Graham, S. W. et al. (2006). Phylogenetic relationships of monocots based on the highly informative plastid gene *ndh*F: evidence for widespread concerted convergence. In *Monocots: Comparative Biology and Evolution (excluding Poales)*, ed. J. T. Columbus, E. A. Friar, J. M. Porter, L. M. Prince and M. G. Simpson. Claremont, CA: Rancho Santa Ana Botanic Garden, pp. 27–50.

Givnish, T. J., Ames, M., McNeal, J. R. et al. (2010). Assembling the tree of monocotyledons: plastome sequence phylogeny and evolution of Poales. *Annals of the Missouri Botanical Garden*, **97**, 584–616.

Graham, S. W. and Iles, W. J. D. (2009). Different gymnosperm outgroups have

(mostly) congruent signal regarding the root of flowering plant phylogeny. *American Journal of Botany*, **96**, 216–227.

Graham, S. W. and Olmstead, R. G. (2000). Utility of 17 chloroplast genes for inferring the phylogeny of the basal angiosperms. *American Journal of Botany*, **87**, 1712–1730.

Graham, S. W., Kohn, J. R., Morton, B. R., Eckenwalder, J. E. and Barrett, S. C. H. (1998). Phylogenetic congruence and discordance among one morphological and three molecular data sets from Pontederiaceae. *Systematic Biology*, **47**, 545–567.

Graham, S. W., Reeves, P. A., Burns, A. C. E. and Olmstead, R. G. (2000). Microstructural changes in noncoding chloroplast DNA: interpretation, evolution, and utility of indels and inversions in basal angiosperm phylogenetic inference. *International Journal of Plant Sciences*, **161** (Suppl. 6), S83–S96.

Graham, S. W., Zgurski, J. M., McPherson, M. A. et al. (2006). Robust inference of monocot deep phylogeny using an expanded multigene plastid data set. In *Monocots: Comparative Biology and Evolution (excluding Poales)*, ed. J. T. Columbus, E. A. Friar, J. M. Porter, L. M. Prince and M. G. Simpson. Claremont, CA: Rancho Santa Ana Botanic Garden, pp. 3–20.

Heath, T. A., Hedtke, S. M., and Hillis, D. M. (2008). Taxon sampling and the accuracy of phylogenetic analyses. *Journal of Systematics and Evolution*, **46**, 239–257.

Hedtke, S. M., Townsend, T. M. and Hillis, D. M. (2006). Resolution of phylogenetic conflict in large data sets by increased taxon sampling. *Systematic Biology*, **55**, 522–529.

Hendy, M. D. and Penny, D. (1989). A framework for the quantitative study of evolutionary trees. *Systematic Zoology*, **38**, 297–309.

Hillis, D. M. (1998). Taxonomic sampling, phylogenetic accuracy, and investigator bias. *Systematic Biology*, **47**, 3–8.

Hillis, D. M., Pollock, D. D., McGuire, J. A. and Zwickl, D. J. (2003). Is sparse taxon sampling a problem for phylogenetic inference? *Systematic Biology*, **52**, 124–126.

Iida, S., Kosuge, K. and Kadono, Y. (2004). Molecular phylogeny of Japanese *Potamogeton* species in light of noncoding chloroplast sequences. *Aquatic Botany*, **80**, 115–127.

Ito, Y., Ohi-Toma, T., Murata, J. and Tanaka, N. (2010). Hybridization and polyploidy of an aquatic plant, *Ruppia* (Ruppiaceae), inferred from plastid and nuclear DNA phylogenies. *American Journal of Botany*, **97**, 1156–1167.

Jacobson, A. and Hedrén, M. (2007). Phylogenetic relationships in *Alisma* (Alismataceae) based on RAPDs, and sequence data from ITS and *trnL*. *Plant Systematics and Evolution*, **265**, 27–44.

Janssen, T. and Bremer, K. (2004). The age of major monocot groups inferred from 800+ *rbc*L sequences. *Botanical Journal of the Linnean Society*, **146**, 385–398.

Kato, Y., Aioi, K., Omori, Y., Takahata, N. and Satta, Y. (2003). Phylogenetic analysis of *Zostera* species based on *rbc*L and *mat*K nucleotide sequences: implications for the origin and diversification of seagrasses in Japanese waters. *Genes and Genetic Systems*, **78**, 329–342.

Keener, B. R. (2005). *Molecular Systematics and Revision of the Aquatic Monocot Genus* Sagittaria *(Alismataceae). (Dissertation)*. Tuscaloosa, AL: University of Alabama.

Lehtonen, S. (2006). Phylogenetics of *Echinodorus* (Alismataceae) based on

morphological data. *Botanical Journal of the Linnean Society*, **150**, 291–305.

Lehtonen, S. (2009). Systematics of the Alismataceae – a morphological evaluation. *Aquatic Botany*, **91**, 279–290.

Lehtonen, S. and Myllys, L. (2008). Cladistic analysis of *Echinodorus* (Alismataceae): simultaneous analysis of molecular and morphological data. *Cladistics*, **24**, 218–239.

Les, D. H. and Haynes, R. R. (1995). Systematics of subclass Alismatidae: a synthetic approach. In *Monocotyledons: Systematics and Evolution, Volume II*, ed. P. J. Rudall, P. J. Cribb, D. F. Cutler and C. J. Humphries. Kew: Royal Botanic Gardens, pp. 353–377.

Les, D. H., Garvin, D. K. and Wimpee, C. F. (1993). Phylogenetic studies in the monocot subclass Alismatidae: evidence for a reappraisal of the aquatic order Najadales. *Molecular Phylogenetics and Evolution*, **2**, 304–314.

Les, D. H., Cleland, M. A. and Waycott, M. (1997). Phylogenetic studies in Alismatidae, II: evolution of marine angiosperms (seagrasses) and hydrophily. *Systematic Botany*, **22**, 443–463.

Les, D. H., Crawford, D. J., Landolt, E., Gabel, J. D. and Kimball, R. T. (2002a). Phylogeny and systematics of Lemnaceae, the duckweed family. *Systematic Botany*, **27**, 221–240.

Les, D. H., Moody, M. L., Jacobs, S. W. L. and Bayer, R. J. (2002b). Systematics of seagrasses (Zosteraceae) in Australia and New Zealand. *Systematic Botany*, **27**, 468–484.

Les, D. H., Moody, M. L. and Jacobs, S. W. L. (2005). Phylogeny and systematics of *Aponogeton* (Aponogetonaceae): the Australian species. *Systematic Botany*, **30**, 503–519.

Les, D. H., Moody, M. L. and Soros, C. L. (2006). A reappraisal of phylogenetic relationships in the monocotyledon family Hydrocharitaceae (Alismatidae). In *Monocots: Comparative Biology and Evolution (excluding Poales)*, ed. J. T. Columbus, E. A. Friar, J. M. Porter, L. M. Prince and M. G. Simpson. Claremont, CA: Rancho Santa Ana Botanic Garden, pp. 211–230.

Les, D. H., Jacobs, S. W. L., Tippery, N. P. et al. (2008). Systematics of *Vallisneria* (Hydrocharitaceae). *Systematic Botany*, **33**, 49–65.

Les, D. H., Sheldon, S. P. and Tippery, N. P. (2010). Hybridization in hydrophiles: natural interspecific hybrids in *Najas* (Hydrocharitaceae). *Systematic Botany*, **35**, 736–744.

Li, X. and Zhou, Z. (2009). Phylogenetic studies of the core Alismatales inferred from morphology and *rbc*L sequences. *Progress in Natural Science*, **19**, 931–945.

Lindqvist, C., De Laet, J., Haynes, R. R et al. (2006). Molecular phylogenetics of an aquatic plant lineage, Potamogetonaceae. *Cladistics*, **22**, 568–588.

Martín, M. and Sabater, B. (2010). Plastid *ndh* genes in plant evolution. *Plant Physiology and Biochemistry*, **48**, 636–645.

Matsen, F. A. and Steel, M. (2007). Phylogenetic mixtures on a single tree can mimic a tree of another topology. *Systematic Biology*, **56**, 767–775.

Mower, J. P., Touzet, P., Gummow, J. S., Delph, L. F. and Palmer, J. D. (2007). Extensive variation in synonymous substitution rates in mitochondrial genes of seed plants. *BMC Evolutionary Biology*, **7**, 135.

Neyland, R. and Urbatsch, L. E. (1996). Phylogeny of subfamily Epidendroideae (Orchidaceae) inferred from *ndh*F chloroplast gene sequences.

American Journal of Botany, **83**, 1195–1206.

Petersen, G., Seberg, O., Davis, J. I. et al. (2006a). Mitochondrial data in monocot phylogenetics. In *Monocots: Comparative Biology and Evolution (excluding Poales)*, ed. J. T. Columbus, E. A. Friar, J. M. Porter, L. M. Prince and M. G. Simpson. Claremont, CA: Rancho Santa Ana Botanic Garden, pp. 52–62.

Petersen, G., Seberg, O., Davis, J. I and Stevenson, D.W. (2006b). RNA editing and phylogenetic reconstruction in two monocot mitochondrial genes. *Taxon*, **55**, 871–886.

Posada, D. (2008). jModelTest: phylogenetic model averaging. *Molecular Biology and Evolution*, **25**, 1253–1256.

Posluszny, U. and Charlton, W. A. (1993). Evolution of the helobial flower. *Aquatic Botany*, **44**, 303–324.

Rambaut, A. (1998). *Se-Al (sequence alignment editor version 10 alpha 1)*. Oxford: Department of Zoology, University of Oxford.

Rothwell, G. W., Van Atta, M. R., Ballard, Jr. H. E. and Stockey R. A. (2004). Molecular phylogenetic relationships among Lemnaceae and Araceae using the chloroplast *trn*L-*trn*F intergenic spacer. *Molecular Phylogenetics and Evolution*, **30**, 378–385.

Saarela, J. M. and Graham, S. W. (2010). Inference of phylogenetic relationships among the subfamilies of grasses (Poaceae: Poales) using meso-scale exemplar-based sampling of the plastid genome. *Botany*, **88**, 65–84.

Saarela, J. M., Rai, H. S., Doyle, J. A. et al. (2007). Hydatellaceae identified as a new branch near the base of the angiosperm phylogenetic tree. *Nature*, **446**, 312–315.

Saarela, J. M., Prentis, P. J., Rai, H. S. and Graham, S. W. (2008). Phylogenetic relationships in the monocot order Commelinales, with a focus on Philydraceae. *Botany*, **86**, 719-731.

Sculthorpe, C. D. (1967). *The Biology of Aquatic Vascular Plants*. London: Edward Arnold.

Shinozaki, K., Ohme, M., Tanaka, M. et al. (1986). The complete nucleotide sequence of the tobacco chloroplast genome: its gene organization and expression. *EMBO Journal*, **5**, 2043-2049.

Stamatakis, A. (2006). RAxML-VI-HPC: maximum likelihood-based phylogenetic analyses with thousands of taxa and mixed models. *Bioinformatics*, **22**, 2688-2690.

Stamatakis, A., Hoover, P. and Rougemont, J. (2008). A rapid bootstrap algorithm for RAxML web servers. *Systematic Biology*, **57**, 758-771.

Stefanović, S. and Olmstead, R. G. (2005). Down the slippery slope: plastid genome evolution in Convolvulaceae. *Journal of Molecular Evolution*, **61**, 292-305.

Stevens, P. F. (2001+). *Angiosperm Phylogeny Website. Version 9, June 2008 [and more or less continuously updated since]*. http://www.mobot.org/mobot/research/apweb/.

Swofford, D. L. (2002). *PAUP*: Phylogenetic analysis using parsimony (*and other methods), version 4.0d100*. Sunderland, MA: Sinauer.

Takhtajan, A. (1997). *Diversity and Classification of Flowering Plants*. New York: Columbia University Press.

Tamura, M. N., Yamashita, J., Fuse S. and Haraguchi, M. (2004a). Molecular phylogeny of monocotyledons inferred from combined analysis of plastid *mat*K and *rbc*L gene sequences. *Journal of Plant Research*, **117**, 109-120.

Tamura, M. N., Fuse, S., Azuma, H. and Hasebe, M. (2004b). Biosystematic studies on the family Tofieldiaceae I. Phylogeny and circumscription of the family inferred from DNA sequences of *mat*K and *rbc*L. *Plant Biology*, **6**, 562-567.

Tamura, M. N., Azuma, H., Yamashita, J., Fuse, S. and Ishii, T. (2010). Biosystematic studies on the family Tofieldiaceae II. Phylogeny of species of *Tofieldia* and *Triantha* inferred from plastid and nuclear DNA sequences. *Acta Phytotaxonomica et Geobotanica*, **60**, 131-140.

Tanaka, N., Setoguchi, H. and Murata, J. (1997). Phylogeny of the family Hydrocharitaceae inferred from *rbc*L and *mat*K gene sequence data. *Journal of Plant Research*, **110**, 329-337.

Tanaka, N., Kuo, J., Omori, Y., Nakaoka, M. and Aioi, K. (2003). Phylogenetic relationships in the genera *Zostera* and *Heterozostera* (Zosteraceae) based on *mat*K sequence data. *Journal of Plant Research*, **116**, 273-279.

Thiers, B. *[continuously updated]. Index Herbariorum: A Global Directory of Public Herbaria and Associated Staff*. New York Botanical Garden's Virtual Herbarium. http://sweetgum.nybg.org/ih/

Tomlinson, P. B. (1982). *Anatomy of the Monocotyledons. VII Helobiae (Alismatidae)*. Oxford: Clarendon Press.

von Mering, S. and Kadereit, J. W. (2010). Phylogeny, systematics, and recircumscription of Juncaginaceae – a cosmopolitan wetland family. In *Diversity, Phylogeny, and Evolution in the Monocotyledons*, ed. O. Seberg, G. Petersen, A. S. Barfod and J. I Davis. Aarhus, Denmark: Aarhus University Press, pp. 55-79.

Wang, B. and Qiu, Y.-L. (2006). Phylogenetic distribution and evolution of mycorrhizas in land plants. *Mycorrhiza*, **16**, 299-363.

Wang, Q. D., Zhang, T. and Wang, J. B. (2007). Phylogenetic relationships and hybrid origin of *Potamogeton* species (Potamogetonaceae) distributed in China: insights from the nuclear ribosomal internal transcribed spacer sequence (ITS). *Plant Systematics and Evolution*, **267**, 65–78.

Waycott, M., Freshwater, D. W., York, R. A., Calladine, A. and Kenworthy, W. J. (2002). Evolutionary trends in the seagrass genus *Halophila* (Thouars): insights from molecular phylogeny. *Bulletin of Marine Science*, **71**, 1299–1308.

Waycott, M., Procaccini, G., Les, D. H. and Reusch, T. B. H. (2006). Seagrass evolution, ecology and conservation: a genetic perspective. In *Seagrasses: Biology, Ecology and Conservation*, ed. A. W. D. Larkum, R. J. Orth and C. M. Duarte. Dordrecht, Netherlands: Springer, pp. 25–50.

Yang, Z. (2006). *Computational Molecular Evolution*. Oxford: Oxford University Press.

Zhang, T., Wang, Q., Li, W., Cheng, Y. and Wang, J. (2008). Analysis of phylogenetic relationships of *Potamogeton* species in China based on chloroplast *trn*T-*trn*F sequences. *Aquatic Botany*, **89**, 34–42.

Zwickl, D. J. and Hillis, D. M. (2002). Increased taxon sampling greatly decreases phylogenetic error. *Systematic Biology*, **51**, 588–598.

1.7 Appendix

List of specimens and associated GenBank accessions used in this study. Taxon; collection information, herbarium (acronyms according to Thiers [continuously updated]); GenBank accessions for *atp*B, *ndh*F, *psb*B-T-N-H, *psb*D-C, *psb*E-F-L-J, *rbc*L, *rpl*2, 3′*rps*12-*rps*7-*ndh*B-*trnL*(CAA). Missing regions are indicated by 'N/A'. Sequence data of *Vallisneria spiralis* L. (Hydrocharitaceae) were provided by Mike Moore (Oberlin College, Ohio). An asterisk '*' indicates a sequence published previously. Whole plastid genomes were sampled from GenBank for the outgroup taxa *Nandina domestica* Thunb. (Berberidaceae, DQ923117), *Nuphar advena* Aiton (Nymphaeaceae, NC_008788), *Phalaenopsis aphrodite* Rchb. f. (Orchidaceae, NC_007499) and *Platanus occidentalis* L. (Platanaceae, NC_008335). Details of previously published Alismatales and outgroup taxa can be found in Graham and Olmstead (2000), Graham et al. (2000, 2006).

Acorales. Acoraceae. *Acorus gramineus* Sol. ex W. Aiton; G. A. Rothwell & Williams s. n., ALTA; HQ901511, HQ901538, HQ901404, HQ901484, HQ901457, HQ901561, HQ901430, HQ901377. **Alismatales**. Alismataceae. *Alisma triviale* Pursh; S. Y. Smith 47, ALTA; HQ901513, HQ901541, HQ901405, HQ901486, HQ901459, HQ901563, HQ901432, HQ901380. *Hydrocleys martii* Seub.; R. A. Stockey & G. W. Rothwell 86, no voucher, Botanischer Garten München-Nymphenburg living collection; HQ901514, HQ901542, HQ901406, HQ901487, HQ901460, HQ901564, HQ901433, HQ901381. Aponogetonaceae. *Aponogeton distachyos* L. f.; R. A. Stockey &

G. W. Rothwell 6, no voucher, Botanischer Garten München-Nymphenburg living collection; HQ901529, HQ901552, HQ901421, HQ901502, HQ901475, HQ901579, HQ901448, HQ901395. Araceae. *Arum italicum* Mill.; W. J. D. Iles 2010–001, UBC; HQ901533, HQ901556, HQ901425, HQ901506, HQ901479, HQ901583, HQ901452, HQ901399. *Gymnostachys anceps* R. Br.; M. W. Chase 3841, K; HQ901532, HQ901555, HQ901424, HQ901505, HQ901478, HQ901582, HQ901451, HQ901398. *Lemna trisulca* L.; R. A. Stockey & G. W. Rothwell 82, ALTA; HQ901530, HQ901553, HQ901422, HQ901503, HQ901476, HQ901580, HQ901449, HQ901396. *Orontium aquaticum* L.; R. A. Stockey & G. W. Rothwell 40, ALTA; HQ901531, HQ901554, HQ901423, HQ901504, HQ901477, HQ901581, HQ901450, HQ901397. Cymodoceaceae. *Amphibolis griffithii* (J. M. Black) Hartog; Hopper 8539, KPBG; HQ901524, HQ901548, HQ901416, HQ901497, HQ901470, HQ901574, HQ901443, HQ901390. *Halodule wrightii* Asch.; D. A. Kolterman & I. López 1003, ALTA; HQ901525, HQ901549, HQ901417, HQ901498, HQ901471, HQ901575, HQ901444, HQ901391. Hydrocharitaceae. *Elodea canadensis* Michx.; S. Y. Smith 55, ALTA; HQ901516, HQ901544, HQ901408, HQ901489, HQ901462, HQ901566, HQ901435, HQ901383. *Hydrocharis morsus-ranae* L.; S. Y. Smith 51, ALTA; HQ901517, HQ901545, HQ901409, HQ901490, HQ901463, HQ901567, HQ901436, HQ901384. *Najas flexilis* (Willd.) Rostk. & W. L. E. Schmidt; S. Y. Smith 30, ALTA; HQ901519, N/A, HQ901411, HQ901492, HQ901465, HQ901569, HQ901438, HQ901590. *Stratiotes aloides* L.; Bogner s. n., ALTA; HQ901515, HQ901543, HQ901407, HQ901488, HQ901461, HQ901565, HQ901434, HQ901382. *Thalassia testudinum* Banks & Sol. ex K. D. Koenig; D. A. Kolterman & I. López 1001, ALTA; HQ901518, N/A, HQ901410, HQ901491, HQ901464, HQ901568, HQ901437, HQ901385. Juncaginaceae. *Maundia triglochinoides* F. Meull.; L. Stanberg & G. Sainty LS 80, NSW; HQ901527, HQ901551, HQ901419, HQ901500, HQ901473, HQ901577, HQ901446, HQ901393. *Triglochin maritima* L.; M. Buzgo 1011, K DNA 10463; HQ901528, AF546998*, HQ901420, HQ901501, HQ901474, HQ901578, HQ901447, HQ901394. Posidoniaceae. *Posidonia australis* Hook. f.; M. van Keulen s. n., ALTA; HQ901523, N/A, HQ901415, HQ901496, HQ901469, HQ901573, HQ901442, HQ901389. Potamogetonaceae. *Groenlandia densa* (L.) Fourr.; Bogner s. n., ALTA; HQ901521, HQ901546, HQ901413, HQ901494, HQ901467, HQ901571, HQ901440, HQ901387. *Zannichellia palustris* L.; Bruinsma s. n., UBC; HQ901522, HQ901547, HQ901414, HQ901495, HQ901468, HQ901572, HQ901441, HQ901388. Ruppiaceae. *Ruppia maritima* L.; D. A. Kolterman, G. J. Breckon, J. Vélez-Gavilán & A. R. Lewis 1005, ALTA; HQ901526, HQ901550, HQ901418, HQ901499, HQ901472, HQ901576, HQ901445, HQ901392. Tofieldiaceae. *Harperocallis flava* McDaniel; M. W. Chase 306, NCU; HQ901536, HQ901559, HQ901428, HQ901509, HQ901482, HQ901586, HQ901455, HQ901402. *Pleea tenuifolia* Michx.; W. Zomlefer 789, GA; HQ901537, HQ901560, HQ901429, HQ901510, HQ901483, HQ901587, HQ901456, HQ901403. *Tofieldia coccinea* Richardson; M. J. Waterway 2006–241, UBC; HQ901535, HQ901558, HQ901427, HQ901508,

HQ901481, HQ901585, HQ901454, HQ901401. *Triantha racemosa* (Walter) Small; W. Zomlefer 801, GA; HQ901534, HQ901557, HQ901426, HQ901507, HQ901480, HQ901584, HQ901453, HQ901400. Zosteraceae. *Zostera angustifolia* (Hornem.) Rchb.; M. W. Chase 2795 W2, K; HQ901520, AF547022*, HQ901412, HQ901493, HQ901466, HQ901570, HQ901439, HQ901386. **Dioscoreales**. Nartheciaceae. *Narthecium ossifragum* L.; R. A. Stockey & G. A. Rothwell 59, ALTA; AY147597, AY147763, AY147503, AY147642, AY147550, AY149348, AY147689, AY147454. **Liliales**. Philesiaceae. *Philesia magellanica* J. F. Gmel.; M. W. Chase 545, K; AY465551, AY465656, AY465578 & AY465744, AY465682, AY465605, AY465707, AY465734, AY465633. Rhipogonaceae. *Rhipogonum elseyanum* F. Muell.; M. W. Chase 187, NCU; AY465553, AY465658, AY465580 & AY465745, AY465684, AY465607, AY465709, AY465736, AY465635. **Poales**. Bromeliaceae. *Brocchinia micrantha* (Baker) Mez; no voucher, U. Wisconsin Botany Greenhouse living collection; EU832849, EU832884, EU832899, EU832915, EU832935, EU832951, EU832964, EU832867 & EU832982.

2

The fossil record of noncommelinid monocots

SELENA Y. SMITH

2.1 Introduction

The fossil record is an invaluable source of information for biologists, providing data on novel features, past diversity, evolution and phylogeny of groups, biogeography and response to past disturbances such as climate change. Continuing progress has been made in the recognition of many new fossil monocot taxa that has consequently expanded our understanding of the monocot fossil record. While the monocot fossil record has been reviewed by Daghlian (1981), Collinson et al. (1993), Herendeen and Crane (1995), Gandolfo et al. (2000), Greenwood and Conran (2000), Stockey (2006) and Friis et al. (2011), and some records are documented in the Paleobiology Database (http://www.paleodb.org), new finds continue to advance our knowledge of the fossil record and further work on phylogenetic relationships and taxonomy (e.g. APG II, 2003; Chase et al., 2006; Graham et al., 2006; APG III, 2009) refines our ability to place them in an evolutionary context.

The earliest records of monocots recognized to date are fossils from the Early Cretaceous (Aptian-Albian) of Portugal fossils and appear to belong to Araceae. These include remains of pollen of *Mayoa* (Friis et al., 2004), although Hofmann and Zetter (2010) question the affinities with Araceae, and small araceous inflorescences and flowers (Friis et al., 2010). The monocot fossil record shows that both early-divergent (noncommelinid; e.g. Alismatales: Alismataceae, Araceae, and seagrasses; Pandanales: Triuridaceae, Pandanaceae) and more derived (commelinid)

Early Events in Monocot Evolution, eds P. Wilkin and S. J. Mayo. Published by Cambridge University Press. © The Systematics Association 2013.

monocot orders (e.g. Zingiberales, Poales and Arecales) are present beginning in the Cretaceous. However, many monocot families appear much later in the fossil record. This could be due to biases of the fossil record, biases in recognizing the fossils and/or evolutionary processes (e.g. evolution of a plant group for which there is little/no fossil record). Where does our understanding of the early evolution of monocots lie today, and what should monocot (paleo)botanists concentrate on? For the purposes of this chapter I will focus on the early-divergent monocot lineages, i.e. the noncommelinids: Acorales, Alismatales, Dioscoreales, Petrosaviales, Pandanales, Liliales and Asparagales, and their fossil record through the Cretaceous (145.5–65.5 million years ago), Paleogene (65.5–23.0 million years ago) and Neogene (23.0–2.6 million years ago). Family-level nomenclature follows APG III (2009).

2.1.1 Nature of the fossil record

Monocots, like all plants, can be preserved in a variety of ways by the fossilization process. Plants may be preserved as carbon films or impressions on rock (compression/impression fossils); charcoalified; mummified; permineralizations (with 3D and cellular preservation); or unaltered material (e.g. phytoliths and pollen grains). Unlike animals, plants shed various organs throughout life that can end up in the fossil record. Ideally, whole plants are reconstructed, representing both the vegetative and reproductive organs that form one natural, biological species. In practice this is difficult to accomplish. Extensive and careful collection of material may result in finding material in organic or biological connection, such as leaves and fruits attached to a branch or pollen in a flower (Figure 2.1). Repeated co-occurrence and exclusive associations of organs attributed to the same genus/family may be taken as a whole-plant concept.

Monocots have a relatively poor fossil record compared to other groups (Daghlian, 1981; Herendeen and Crane, 1995; Gandolfo et al., 2000, 2009; Stockey, 2006; Smith et al., 2010; Friis et al., 2011). This is related to two main factors: (1) the lower preservation potential of monocots and (2) that more comparative data from modern taxa is needed to identify synapomorphies useful for study of fossils.

The lower preservation potential of monocots is due in part to the fact that most are insect-pollinated herbaceous plants (Herendeen and Crane, 1995) that do not produce as much biomass as nonmonocot angiosperms and are not as likely to be preserved. Construction of the plants themselves also has an effect: organs with lignified tissue (e.g. wood, and some fruits and seeds) will make their way through more taphonomic filters than softer tissues. As many monocots are herbaceous and small, they are less likely to be preserved. Many monocots are not deciduous like woody trees and they produce fewer organs, but have a tendency to retain those organs on the plant, both factors that reduce the probability of entering the fossil record to start with. Also, some habitats are associated with a lower

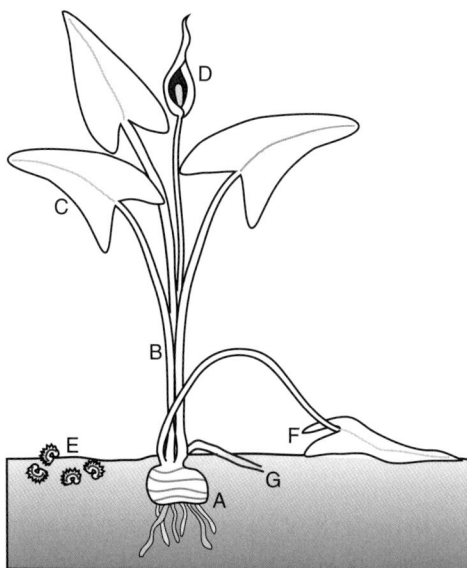

Fig 2.1 Generalized monocot illustrating fossilization potential of different organs. Whole plants are rare; other organs that may be preserved are underground stems and roots (A); petioles (B); leaf lamina (C); inflorescences/infructescences (D) which may have *in situ* seeds and/or pollen; dispersed fruits and seeds (E). Leaves often remain attached to the parent plant as they senesce and rot rather than abscising (F, G). Fruits, seeds, pollen and leaves are more commonly preserved and easily identified.

preservation potential; those near quiet bodies of water are likely to preserve fossils since the water carries sediment that buries the plants and protects them from degradation. Plants from higher-altitude environments and those further away from depositional basins will be less likely to be preserved. Thus, monocots generally have a lower preservation potential than nonmonocot angiosperms.

Another bias is introduced with collection methods. Just as with living plants, much more work has been done in the northern hemisphere, which leaves us with fewer data on southern-hemisphere and equatorial fossil sites (Greenwood and Conran, 2000; Gandolfo et al., 2009). Further studies in these areas will rectify this paucity of data.

Once the fossils are collected, they can be difficult to identify. The effects of taphonomy may alter their appearance; the style of preservation determines what type of information is preserved. Isolated organs provide a glimpse of an extinct plant, but it is more difficult to account for small variations resulting from ontogeny or local ecology (e.g. sun vs. shade leaves; underwater vs. emergent leaves), which may inflate diversity estimates. Loss of outer layers or preservation of internal casts produces morphologies that are not readily apparent from modern material, for example while looking at herbarium specimens. The lack of 'search images' (morphologies that can be visually matched with a fossil) provides difficulties in assessing affinities of fossil plants. Depending on the style of preservation there may also be an absence of or difficulty in recognizing identifying features. If the specimen is preserved in three dimensions, it is possible to extract both internal structural information and external morphology. This can be done by using the peel technique (for permineralized specimens) or X-ray techniques such as CT (computed tomography) and SRXTM (synchrotron radiation X-ray tomographic microscopy). Such data provide insights into the morphology of a fossil under study, allow the creation of potential fossils and virtual dissections of fruits and seeds, and enable comparisons between material preserved in different ways (e.g. compression vs. permineralization), all of which can help us interpret the monocot fossil record. 3D techniques have the potential to refine the monocot fossil record and provide important data for monocot evolutionary biology in understanding spatial and temporal patterns of paleodiversity.

2.2 The fossil record of monocots

2.2.1 Enigmatic Cretaceous monocots

Some fossils have been confidently identified as monocots, but are not placed in modern families. Other fossils that were suggested to have monocotyledonous affinities are either contested (and therefore in need of reinvestigation) or have since been ascribed to nonmonocot groups. Doyle et al. (2008) reinvestigated affinities of *Acaciaephyllum*, *Liliacidites*, the *Similipollis/Anacostia* plant and the *Pennipollis* plant using a phylogenetic framework. They confirmed the monocot identity of *Acaciaephyllum* leaves that were described from the Cretaceous Potomac group, USA and of *Liliacidites* pollen (defined as boat-shaped, reticulate, monosulcate and having finer sculpture at the ends of the grain). '*Liliacidites*' *minutus* pollen was found associated with *Virginianthus* flowers, but the pollen does not show the sculptural grading that is typical of *Liliacidites*, and phylogenetic analysis by Doyle et al. (2008) placed this fossil in Laurales instead. Doyle et al. (2008) also argue against a monocot affinity of *Similipollis/Anacostia* (suggested to be in the Austrobaileyales) and the *Pennipollis* plant (suggested to be Chloranthaceae). Fragmentary monocot leaves from the latest Albian-early Cenomanian described by Pole (1999) from Australia are also known, with monocotyledonous stomatal structure, but the affinities of these leaves are not clear (Pole, 1999; Doyle et al., 2008).

Sanmiguella, a palm-like plant from the Triassic of Colorado (Brown, 1956), is too poorly preserved to reliably classify it as an angiosperm (see Daghlian, 1981 for more extensive discussion). *Caricopsis*, likewise, was reported as an early Cretaceous record of Cyperaceae by Samylina (1968), but is not accepted as a monocot either (Daghlian, 1981; Friis and Crepet 1987). *Klitzschophyllites* is a genus that was erected for orbiculate leaves with trifurcate stems (Mohr and Rydin, 2002; Mohr et al., 2006), but the structure is quite different to that of typical monocotyledonous leaves (Doyle et al., 2008). Although it was recently suggested to be more similar to basal eudicots (e.g. Ranunculales; Gomez et al., 2009), the leaf venation and margins of *Klitzschophyllites* differ from those of ranunculaleans as well, and further studies are needed to determine its exact affinities.

2.2.2 Cretaceous commelinids

It is important to note that commelinids have also been present since the Cretaceous (Figure 2.2). Palms are one of the more conspicuous fossil monocots of the Late Cretaceous (summarized by Harley, 2006; Dransfield et al., 2008). Fossils assignable to Zingiberales are also found (summarized by Rodriguez-de la Rosa and Cevallos-Ferriz, 1994). These gingers and palms are interesting because they indicate the presence of more tropical conditions in mid and high latitudes. Other groups that are ecologically significant today do not have a strong presence in the Cretaceous. Grasses (Poaceae) are known only from phytoliths found in dinosaur coprolites (Prasad et al., 2005), but sedges (Cyperaceae) are not known until the Paleocene (Smith et al., 2010). The appearance of both early-divergent and derived (commelinid) monocots in the Cretaceous suggests a rapid evolutionary or ecological radiation.

2.2.3 Acorales

Although Acorales are sister to all other monocots, their fossil record does not extend back to the Cretaceous (Figure 2.2); the oldest record is of *Acorites heeri*, a spadix from the Eocene of North America (Crepet, 1978). *Acorus spitsbergensis* was recently described from the Eocene of Spitsbergen (Budantsev and Golovneva, 2009). Fruits and seeds of *Acorus* are also found from the Quaternary of the Soviet Union (Katz et al., 1965). *Acoropsis eximia* (syn. *Acoropsis minor*), an inflorescence from Eocene Baltic amber, was found to belong to Araceae, subfamily Monsteroideae, tribe Monstereae (Bogner, 1976). Several fossils attributed to Acoraceae have been re-examined and found not to belong to this group (Bogner, 2001; Wilde et al., 2005). The lack of fossil record is somewhat puzzling since this rhizomatous helophyte can form dense stands and grows in environments with good preservation potential, including rivers, streams, ponds, lakes, and swamps (Bogner and Mayo, 1998). The seeds are small, *c.* 4 mm, with a parenchymatous integument (Buell, 1935); these may be difficult to preserve or recognize in the fossil record. *Acorus* leaves may not be recognized due to lack of diagnostic characters.

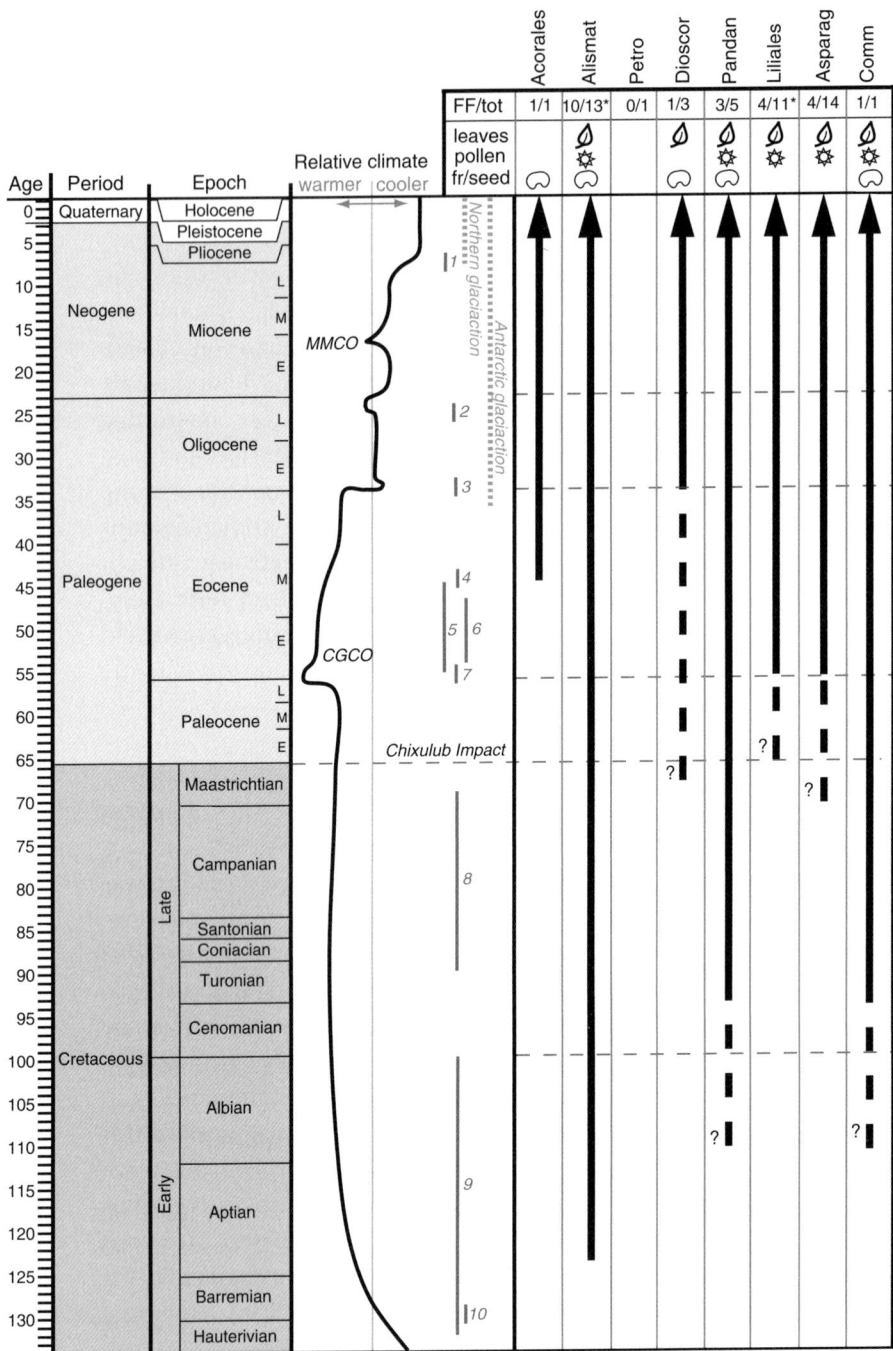

Fig 2.2 Ranges of monocot orders over geologic time. Ages (left) are in millions of years. Generalized climate curve and events based on Zachos et al. (2001), Veizer et al. (1999) and Boucot et al. (2004). Dashed range lines represent uncertain records. Alismat: Alismatales; Asparag: Asparagales; CGCO: Cenozoic Global Climatic Optimum; Comm: commelinids; Dioscor: Dioscoreales; E: early; FF/tot: numbers of families within the order with a fossil record over the total number of families (* = includes *incertae sedis*); L: late; M: middle; MMCO: Middle Miocene Climatic Optimum; Pandan: Pandanales; Petro: Petrosaviales; 1: C4 grass expansion; 2: North Andean uplift; 3: decline of broad-leaf forests and appearance of grasslands; 4: Central Andean uplift starts; 5: Alpine-Pyreneean orogeny; 6: Laramide orogeny (North America); 7: India–Asia contact; North Atlantic rifting; 8: Separation of Australia and Antarctica; 9: splitting of South America and Africa; 10: first fossil evidence of angiosperms.

2.2.4 Alismatales

The fossil record of Alismatales was most recently summarized in detail by Stockey (2006). Most families of this order have a fossil record, probably attributable to the fact that many are aquatics or semi-aquatics and thus grow close to environments with a high preservation potential. There is no known fossil record for Juncaginaceae and Scheuchzeriaceae (Collinson et al., 1993; Herendeen and Crane, 1995). Butomaceae seeds have been identified from Oligocene and younger strata in Europe (Mai, 1985; Collinson et al., 1993), but there are no other records.

The Alismataceae are recognized from several fossils dating from the Late Cretaceous onwards. Various records have been previously rejected or are based on poorly preserved and incomplete material, meaning they can only be assigned tentatively to the family (Daghlian, 1981; Stockey, 2006); these include the Cretaceous leaves of *Alismaphyllum cretaceum* (Berry, 1925a), leaves of *Alismaphyllites grandifolius* (Brown, 1962) and fruits of *Sagittaria megaspermum* (Brown, 1962) from the Paleocene of North America. Fossils that are more confidently placed in Alismataceae include a few extinct genera of leaves, including several species of *Haemanthophyllum* Budanstev, from the Cretaceous to Miocene of Europe, Russia and North America (Golovneva, 1997; Riley and Stockey, 2004) and *Cardstonia tolmanii* from the Cretaceous (Late Campanian-Early Maastrichtian) of Alberta, Canada (Riley and Stockey, 2004). *Cardstonia* and *Haemanthophyllum* are similar, but detailed comparison by Riley and Stockey (2004) demonstrates that *Cardstonia* lacks primary veins that merge with the leaf margin, which is found in *Haemonthophyllum* and many other Alismataceae, and that *Cardstonia* is most similar to extant *Limnocharis*. However, it is difficult to make confident taxonomic determinations from only partial leaves and some of these described fragments may represent other species or even families such as Aponogetonaceae (Golovneva, 1997; Riley and Stockey, 2004; Stockey, 2006). There is also a permineralized petiole of *Heleophyton helobiaeoides* from the Middle Eocene Princeton Chert (British Columbia, Canada) whose vascular bundle structure is like that of *Butomus*, while their arrangement is most like *Sagittaria* and *Echinodorus* (Erwin and Stockey, 1989).

Alismataceous fruits and seeds are represented in the fossil record by several extant and extinct genera. The first reliable carpological records date back to the Eocene London Clay (Chandler, 1964; Haggard and Tiffney, 1997). Fossil fruits and seeds are largely found from the Miocene and Pliocene of Europe and include the genera *Alisma, Baldellia, Caldesia, Luronium, Ranalisma, Sagittaria* and *Damasonium* (discussed by Haggard and Tiffney, 1997). *Alismaticarpum* is an extinct genus with winged fruits from the Oligocene of England (Collinson, 1983), and *Sagisma* is an extinct genus of wingless fruits with a recurved seed from the Oligocene and Miocene of western Siberia, with characteristics of *Caldesia, Sagittaria* and *Limnophyton* (Haggard and Tiffney, 1997). There are also many other fruits suggested to have affinities with Alismataceae but that have not been studied in detail (Haggard and Tiffney, 1997; Paleobiology Database, 2010). The possibly alismataceous flowers and fruits from the Eocene Princeton Chert discussed by Stockey (2006) are not monocots, but have been shown to be *Saururus* (Saururaceae, Piperales; Smith, 2006; Smith and Stockey, 2007). The pollen record of Alismataceae was considered largely unreliable and in need of further study by Stockey (2006).

Several leaves have been suggested to have affinities with Aponogetonaceae (Collinson et al., 1993; Herendeen and Crane, 1995; Riley and Stockey, 2004; Stockey, 2006). *Aponogeton* leaves have been described from the Oligocene of Kazakhstan (Zhilin, 1974; Pneva, 1988), but at least some of these fossils are now included in *Haemanthophyllum* (Golovneva, 1997; Riley and Stockey, 2004) and should be reinvestigated to confirm their affinities. The similarities between Alismataceae, Aponogetonaceae and Potamogetonaceae suggests a need to closely study leaf characters in Alismatales (Riley and Stockey, 2004) and highlights the potential difficulty of confidently identifying fragmentary material. *Aponogeton* leaves and reproductive material was reported from the Cretaceous of South America by Selling (1974); however, it was not described in detail, nor was it figured and as such this cannot be considered a reliable record without further investigations. Recent work has found *Aponogeton*-type pollen (sulcate reticulate pollen grains with characteristic small echinae at the muri) from the Cretaceous (Campanian; Eagle Formation) of Wyoming and Middle Eocene Princeton Chert of British Columbia (R. Zetter, pers. comm.). The lack of other fossil reproductive material is not surprising, as both the fruit and seed coats decay rapidly in nature (van Bruggen, 1998), which would leave little material to enter the fossil record.

Fossil Araceae are known from the Cretaceous onwards, including the oldest fossil monocots from the Aptian-Albian of Portugal (Friis et al., 2004, 2010). The aroid fossil record includes pollen, fruits, seeds, leaves, stems and whole plants (Mayo et al., 1997; Wilde et al., 2005; Stockey, 2006; Hesse and Zetter, 2007; Bogner, 2009), and although it is described as sparse by some authors (Herrera et al., 2008) it is actually a relatively diverse record compared to other noncommelinid monocot families. While more precise affinities are still unclear for a few fossils, such as leaves of *Petrocardium* from the Paleocene of Colombia (Herrera et al., 2008) and Middle Eocene leaves from Eckfeld, Germany (Wilde and Frankenhauser, 1998), most can be placed in one of the subfamilies Aroideae, Lasioideae, Lemnoideae, Monsteroideae, Orontioideae or Pothoideae (following the classification of Cabrera et al., 2008). It is worth noting also that aroid spadices have sometimes been incorrectly identified for other plant structures, for example, galls (D. Erwin, pers. comm.), twigs, or nonmonocot inflorescences (Wilde et al., 2005). In addition, Araceae are a large family with a range of diversity of leaves that make it difficult to thoroughly compare venation characters. Also, there are many putative aroid fossils that have since been removed from the family (either because they were misidentified or are too incomplete to be confident in their identification) or are in need of re-examination (see Mayo et al., 1997; Wilde et al., 2005 for details). Floating aquatic plants of *Cobbania corrugata* have been described from the Cretaceous (Campanian) to Eocene and were interpreted as having closest similarities to the subfamily Aroideae (Stockey et al., 2007). Fruits and seeds from the Late Cretaceous of the Russian Far East were found associated with *Cobbania* foliage, and described as *Cobbanicarpites amurensis* (Krassilov and Kodrul, 2009). *Pistia*, another floating aquatic aroid, is known from fruits and seeds of Oligocene and Miocene age in Eurasia (Kvaček and Bogner, 2008). Paleocene leaves from Colombia, *Montrichardia aquatica*, were also considered to be in Aroideae (tribe Montrichardieae) (Herrera et al., 2008), as are *Caladiosoma* leaves from the Eocene of Germany (Wilde et al., 2005) and Neogene of Trinidad (Berry, 1925b), and *Nitophyllites* from the Paleocene and Eocene of Eurasia and North America (Wilde et al., 2005). Friis et al. (2010) discussed Early Cretaceous inflorescences (Araceae sp. A) from the Vila Verde 2 locality in Portugal and interpreted it as showing similarities to Aroideae. Other material from the Vila Verde 2 showed similarities to Pothoideae (Friis et al., 2010). Subfamily Orontioideae is represented in the Cretaceous to Eocene of North America and Europe by a Cretaceous infructescence from Alberta, Canada (*Albertarum pueri*, Bogner et al., 2005) and leaves of *Lysichiton*, *Symplocarpus* and *Orontium* (Kvaček and Herman, 2004, 2005; Bogner et al., 2007), although venation patterns of the extinct species differ from the modern genera and further studies may refer some of this back to extinct genera; without the whole plant, it is generally better to be conservative and not place fossil leaves in extant genera. Lemnoideae are also known from the Cretaceous to Miocene of North America, Europe and East Asia from whole plants, seeds and pollen. Whole plants of *Limnobiophyllum* have been found, which offer chances to connect leaves with reproductive structures like *Pandaniidites* pollen (Kvaček, 1995; Stockey et al., 1997). A sterile plant of *Lemna cestmirii* is known from the Miocene of central Europe (Kvaček, 2003) and dispersed seeds of *Lemnospermum* occur in the Paleogene to Miocene of Europe and North America (Wilde et al., 2005; Bogner, 2009). Lemnoid fruits from the Maastrichtian of Mongolia (Krassilov and Makulbekov, 1995) are similar to duckweed fruits in some aspects, but have some characters that are unknown in Lemnoideae and therefore should be excluded from Araceae (J. Bogner, pers. comm.). Lasioideae are first recognized from the Late Cretaceous of Siberia on the basis of pollen (Hofmann and Zetter, 2010). The first macrofossils of subfamily Lasioideae are from the Middle Eocene of British Columbia, Canada, represented by seeds of *Keratosperma* (Cevallos-Ferriz and Stockey, 1988; Smith and Stockey, 2003; Figure 2.3A, 2.3B) and there are also seeds similar to *Cyrtosperma* from the Paleocene and Eocene of the UK (M. Collinson, pers. comm.; Collinson, 1986; Figure 2.3C).

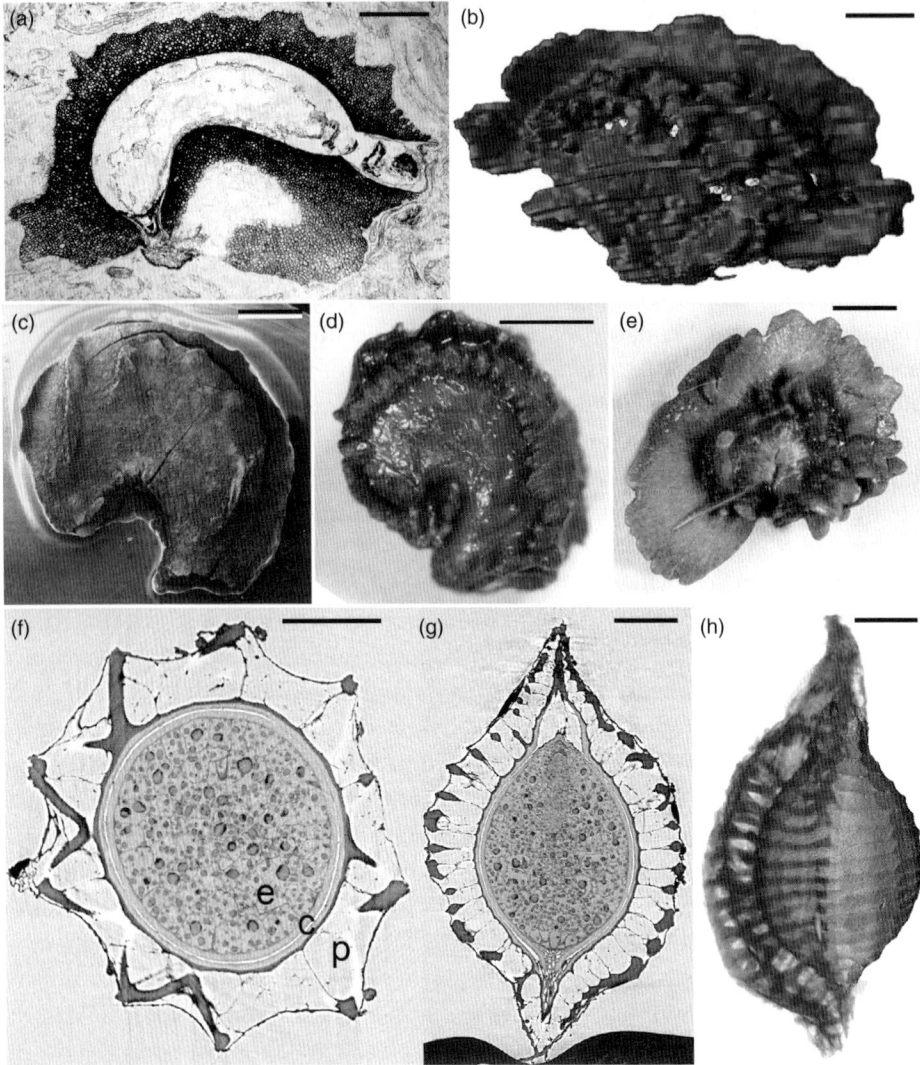

Fig 2.3 Utility of different techniques and preservation types for comparative work. A: Longitudinal section of *Keratosperma allenbyense* seed (Lasioideae, Araceae), Middle Eocene Princeton Chert, Canada. B: 3D computer reconstruction of *Keratosperma* seed. C: scanning electron micrograph of cf. *Cyrtosperma* seed from the Paleocene of the UK (photo courtesy of M. Collinson). D: Seed of extant *Cyrtosperma* (Lasioideae, Araceae). E: seed of extant *Urospatha* (Lasioideae, Araceae). F, G: digital synchrotron X-ray tomography sections of extant *Cyclanthus* seed (Cyclanthaceae) showing endosperm (e), cuticular layer (c) and palisade integument (p); F: cross section, G: longitudinal section. H: digital rendering of *Cyclanthus* seed with half of integument removed to reveal inner cuticular layer (on right). Scale bars: A, B: 0.5 mm; C: 1.0 mm; D, E: 2.0 mm; F, G, H: 0.25 mm.

Urospathites from the Oligocene and Miocene of Europe represents younger carpological evidence of lasioids (Gregor and Bogner, 1984, 1989). Subfamily Monsteroideae also has a record that extends back to the Cretaceous. Portuguese fossils include *Mayoa* pollen [(tribe Spathiphylleae; Friis et al., 2004; Hesse and Zetter, 2007; but see Hofmann and Zetter, 2010, who suggested this pollen may not be Araceae based on similarities to the Triassic taxon *Lagenella martini* (Klaus, 1960)] from the late Early Cretaceous and Late Cretaceous (Campanian-Maastrichtian) Mira locality with similarities to *Epipremnum* and *Spathiphyllum* (Monsteroideae). A stem of *Rhodospathodendron* was also described from the Late Cretaceous (Maastrichtian) of India (Bonde, 2000; Wilde et al., 2005). The only leaves are those of *Araciphyllites tertiarius* and *A. engleri* from the Middle Eocene of Germany and Pliocene of Sumatra, respectively (Kräusel, 1929; Wilde et al., 2005). Pollen of *Proxapertites* (regarded as similar to *Monstera* or *Gonatopus*) and *Spathiphyllum* date from the middle Cretaceous (Hesse and Zetter, 2007). Several types of monsteroid fruits and seeds have also been described, including infruct-escences of *Acoropsis* from Eocene Baltic Amber (Bogner, 1976; Wilde et al., 2005) and *Epipremnum spadiciflorum* from the Eocene-Oligocene of Egypt (Kräusel and Stromer, 1924; Bown et al., 1982; note that there is some doubt about whether this is indeed Araceae), and seeds of *Epipremnites* and *Scindapsites* from the Oligocene and younger of Europe (Gregor and Bogner, 1984, 1989). *Epipremnum*-like seeds were identified from the Cretaceous of Portugal (Friis et al., 2004). *Arthmiocarpus hesperus*, and infructescence from the Upper Cretaceous of South Dakota, was suggested to be Moraceae (Delevoryas, 1964), but also shows similarities to aroid spadices and should be compared with Araceae. Both early-divergent and derived clades are present in the Cretaceous, suggesting Araceae were well established by this time and some families occupied a broader geographic range than they do today.

Hydrocharitaceae is represented in the fossil record on the basis of fruits and leaves. The oldest records are *Stratiotes* fruits from the Paleocene and younger of the UK (Collinson, 1986, 1990), and fruits and seeds are found elsewhere (e.g. Miocene of the Czech Republic, Kvaček and Teodoridis, 2007). Foliage from the Oligocene of France that was described as *Vallisneria bromeliaefolia* by Saporta (1873) is now considered to be an extinct species of *Stratiotes* (Kvaček, 2003). The only vegetative remains of *Vallisneria* are those of *V. janeckii* (Bogner and Kvaček, 2009), from the early Miocene of the Czech Republic. There are also seeds, *V. ovalis* (Mai 1995), from the Eocene-Miocene of Europe. *Najas* seeds are found in the Oligocene and younger sediments of Europe (Mai, 1985; Collinson, 1988). Leaf impressions from the Eocene of France were assigned to *Ottelia* (*O. parisiensis*; Saporta, 1879). Berry (1914) compared these leaves to *Potamogeton megaphyllus* from the Eocene Claiborne flora of Georgia. Another leaf impression from Australia was also considered to be *Ottelia* (*O. praeterita*; von Müller, 1880), but von Müller reports the probable locality as being from the Wianamatta Shales, which are

Triassic in age (Herbert and Helby, 1980; Retallack et al., 2010) and therefore this fossil needs revisiting. Other freshwater genera are represented by leaves from the Eocene of Europe (Mai and Walther, 1978, 1985; Wilde, 1989; Wilde and Frankenhauser, 1998; Kvaček and Teodoridis, 2007). Seagrasses are discussed below.

Fruits and seeds of Potamogetonaceae and Ruppiaceae are well recognized from the Cretaceous onwards. Fossil seeds show a range of morphologies that are intermediate between the two families and had been interpreted as intermediates, with some taxa more similar to *Potamogeton* and others more like *Ruppia*. Revision is needed of this group of fossils in light of the recent taxonomic changes that have split these two families and placed them in different clades within Alismatales (Graham et al., 2006; APG III, 2009). For example, *Limnocarpus*, from the Upper Paleocene and onwards of Europe and Russia, is considered intermediate between *Potamogeton* and *Ruppia* (Collinson, 1982; Collinson et al., 1993). There is an extensive record, primarily of fruits, but also leaves and pollen, from the Paleocene to Pliocene of Argentina, China, Eurasia, Japan, North America and Saudi Arabia (Muller, 1981; Collinson, 1982; Ozaki, 1991; Wing et al., 1995; Manchester, 2001; Zhao et al., 2004; Gandolfo et al., 2009). Fruits are separated into two groups depending on whether they are morphologically more similar to *Potamogeton* or *Ruppia*. The *Potamogeton* group includes *Potamogeton, Eulimnocarpus, Limnocarpus, Limnocarpella, Midravalva and Palaeoruppia* (Zhao et al., 2004; Gandolfo et al., 2009). The *Ruppia* group includes *Medardus, Paleoruppia, Ruppia* and *Selseycarpus* (Zhao et al., 2004; Gandolfo et al., 2009). Manchester (2001) suggested that leaves and fruits of *Palaeopotamogeton* from the Eocene of Colorado needed further work to determine familial affinities. *Potamogeton*-like shoots with both floating and submersed leaves were described from the Maastrichtian of Mongolia (Krassilov and Makulbekov, 1995).

The only putative record of Tofieldiaceae is *Dicolpopollis* sp. pollen reported by Chmura (1973) from the Late Cretaceous of California, which she compared with *Tofieldia* and which has been used as an age constraint for dating monocot lineages (Bremer, 2000; Janssen and Bremer, 2004). The lack of ultrastructural data to support this comparison means it is not a reliable record of Tofieldiaceae (Crepet et al., 2004). Other affinities for *Dicolpopollis* have been suggested, with most species considered to be calamoid palms (Muller, 1981; Ediger et al., 1990; Harley and Morley, 1995).

Marine seagrasses are found in five different families within Alismatales: Cymodoceaceae, Hydrocharitaceae, Posidoniaceae, Ruppiaceae and Zosteraceae. These are known from fossil leaves, sometimes attached to rhizomes with roots, extending back to the late Cretaceous (Campanian). The record was most recently summarized by Van der Ham et al. (2007) and Benzecry and Brack-Hanes (2008). While some seagrass fossils have been firmly placed in one of these families, others are better considered 'seagrasses', as the characters to definitively distinguish between seagrass genera can be hard to see in fossils. Fossil seagrasses placed in Cymodoceaceae include *Thalassocharis* from the Maastrichtian of the Netherlands (Voight and Domke, 1955; Voight, 1980; van der Ham et al., 2007), as well as taxa placed in the modern genera *Cymodocea* and *Thalassodendron* from the Eocene to Pliocene of Europe, Florida and Indonesia (Laurent and Laurent, 1926; den Hartog, 1970; Lumbert et al., 1984; Ivany et al., 1990; van der Ham et al., 2007). Hydrocharitaceous seagrasses include *Thalassia* and *Thalassites* from the Eocene of Florida (Lumbert et al., 1984; Benzecry and Brack-Hanes, 2008), and *Halophila* seeds from the Miocene and Pliocene of Northern Europe (Mai, 1995). Several fossils are placed in the modern genus *Posidonia* (Posidoniaceae), from the Campanian to Eocene of Europe (Austria, Belgium, France, Germany and possibly the UK) (Chandler, 1961; den Hartog, 1970; van der Ham et al., 2007). Only one fossil is attributed to Zosteraceae: *Zostera nodosa*, from the Paleocene of Belgium (Saporta and Marion, 1878; van der Ham et al., 2007). *Archeozostera*, from the Cretaceous of Japan, is not considered to be part of Zosteraceae (Kuo et al., 1989). Other seagrasses include *Thalassocharis*, *Thalassotaenia* and *Zosterites* from the Late Cretaceous of northern Europe, several unnamed taxa from the Eocene of Florida, and *Posidocea* and *Zosterites* from the Eocene of Europe (Lumbert et al., 1984; van der Ham et al., 2007; Benzecry and Brack-Hanes, 2008). There is also indirect evidence for seagrass communities in the Cretaceous Interior Seaway in Colorado, as the formation of large limestone mounds was suggested to be a result of seagrass communities (Petta and Gerhard, 1977).

It is worth mentioning that a few taxa have been placed in Alismatales but not in any family. *Operculifructus* fruits from the Cretaceous of Mexico were described by Estrada-Ruiz and Cevallos-Ferris (2007). They resemble the infructescences of aroids and pandans, but have operculate fruits unlike modern taxa, and thus were left as *incertae cedis* within Alismatales.

2.2.5 Petrosaviales

There is no known fossil record (Figure 2.2). *Petrosavia* and *Japonolirion* are mycoheterotrophs native to high-elevation habitats in Japan, China, Malaysia and Indonesia (Cameron et al., 2003). *Petrosavia* is rare and grows in dark montane habitats while *Japonolirion* is found in alpine meadows. Both habitats have very low preservation potential and the seeds are very small (<1 mm in size; Tamura, 1998) suggesting neither *Petrosavia* nor *Japonolirion* would be likely to be fossilized.

2.2.6 Dioscoreales

There is no known fossil record of Burmanniaceae (Collinson et al., 1993; Herendeen and Crane, 1995) or Nartheciaceae. Burmanniaceae are small, often saprophytic plants that occur primarily in tropical rainforests, mountainous regions and warm-temperate forests (Maas-van de Kamer, 1998). Their small size and production of dust seeds (<1 mm in size; Maas-van de Kamer, 1998) suggests the family has a low preservation potential. Members of Nartheciaceae also have small seeds unlikely to be preserved.

Dioscoreaceae has been recognized primarily from leaves. Many of these records, especially older ones, are in need of reinvestigation (Daghlian, 1981; Conran et al., 1994) to confirm their affinities to the family. This includes Cretaceous records, like *Dioscorites cretaceous* (Berry, 1925a) from the Ripley Formation of southeastern North America. Leaves in particular can look similar to other net-veined monocot leaves, such as *Smilax* (Daghlian, 1981), and more work on venation patterns and leaf shapes is needed. Other records are more certain, such as the African *Dioscorea* leaves from the Miocene of Kenya (Jacobs and Kabuye, 1987) and from the Oligocene of Ethiopia, the latter with preserved cuticle and representing section *Lasiophyton* (A. Pan, pers. comm.). Fossil reproductive material has been identified as *Dioscoreocarpum* from the early Oligocene of Hungary (Manchester and O'Leary, 2010) and a lunate seed likely from the same plant was described from the Eocene of the Czech Republic (Kvaček, 2002). A samaroid fruit of *Dioscorea* was recognized from the Eocene Florissant flora of Colorado, USA (Manchester, 2001). Gregor (1983) reported on seeds similar to *Tacca* from the Oligocene of the Czech Republic.

2.2.7 Pandanales

While there is no known fossil record for Stemonaceae or Velloziaceae (Collinson et al., 1993; Herendeen and Crane, 1995), Pandanaceae and Triuridaceae appear in the Cretaceous and Cyclanthaceae are known from the Eocene.

Charcoalified staminate flowers with pollen have been described by Gandolfo et al. (2002) from Turonian aged clay deposits in New Jersey as *Mabelia* (two species) and *Nuhliantha* (one species) in the Triuridaceae. These flowers represent the earliest monocot macrofossils. While the flowers show a mosaic of characters, the affinity to Triuridaceae was supported by phylogenetic analysis and suggests these fossils were achlorophyllous saprophytes (Gandolfo et al., 2002).

Pandanaceae, like Triuridaceae, have a fossil record that extends back to the Cretaceous. Pandan leaves have been described from Cretaceous and Paleocene of North America, Siberia, Europe; they are most commonly placed in *Pandanites* or *Pandanus*. Kvaček and Herman (2004) and Herman and Kvaček (2010) recently reviewed the record of pandanaceous leaves and concluded that many of the records are in need of reinvestigation. Fossil Pandanaceae are typically identified based on the M-shaped cross section and spines on the margins and midribs (Kvaček and Herman 2004). However, some records lack the M-shaped cross section or spines. Also, some mapanioid sedges have very similar leaves (S. Smith, pers. obs.) and comparisons between mapanioids and pandans are needed to identify diagnostic features for each group. Epidermal features are useful for linking fossils to Pandanaceae. In addition, *Dryptopollenites semilunatus* pollen from the Late Paleocene of Australia has been suggested to have affinities with *Freycinetia* (Dowe, 1995; Greenwood and Conran, 2000). There are few reproductive structures that have been confidently placed within Pandanaceae. *Gruenbachia pandanoides* from the Lower Campanian (Cretaceous) of Austria is represented by globular infructescences with numerous basally fused phalanges that are most similar to *Pandanus* (Herman and Kvaček, 2010). Fruits of *Viracarpon* from India (Sahni, 1964; Chitaley et al., 1969; Patil, 1972; Nambudiri and Tidwell, 1978) are no longer believed to belong to Pandanaceae (Bande and Awasthi, 1986), but the affinities remain uncertain.

Fossil fruits and seeds of *Cyclanthus* (Cyclanthaceae) have been recognized from the Eocene of England and Germany (Smith et al., 2008). *Cyclanthus* fruits are discoidal and unlike any other plant. This is a good demonstration of how taphonomy can obscure identification and why it is useful to have more than one organ. Seeds were originally described from the UK as *Scirpus lakensis* and placed in Cyperaceae, but later fruits with *in situ* seeds were found from the middle Eocene Messel World Heritage Site in Germany. Smith et al. (2008, 2009a) showed that when the outer layers of *Cyclanthus* seeds were removed (Figure 2.3F-H), the resulting morphology was identical to the fossils. A few specimens with some intact outer seed coat are also known, supporting this identification.

Cyclanthodendron sahnii (Sahni and Surange, 1953) from the Eocene Deccan Intertrapean beds of India had been suggested to have affinities with Cyclanthaceae. However, these stems have since been found in organic connection with *Musocaulon indicum* pseudostems, *Heliconiaites mohgaonensis* petioles and *Tricoccites trigonum* fruits, and this *Cyclanthodendron* plant is now considered to have affinities with Zingiberales (Biradar and Bonde, 1990).

2.2.8 Liliales

Many families within Liliales lack a fossil record: Alstroemeriaceae (including Luzuriagaceae), Campynemataceae, Corsiaceae, Melanthiaceae and Philesiaceae. Colchicaceae are recognized from Pliocene pollen reported from Sahara (Muller, 1981), but there is no other fossil evidence of the family. A fossil corm or rhizome called *Gloriosites* for its resemblance to *Gloriosa* is not considered to be part of the Colchicaceae (Nordenstam, 1998). Likewise, there are few reliable records of Liliaceae, except for Miocene and Pliocene pollen from Europe (Muller, 1981).

Fossil leaves of *Petermanniopsis* were described from Victoria, Australia, on the basis of mummified material with net venation and brachyparacytic and amphi-brachyparacytic stomatal complexes (Conran et al., 1994; Conran and Christophel, 1999). Although the fossils were suggested to be most similar to *Petermannia*, they were not formally placed in Petermanniaceae (Conran et al., 1994; Conran and Christophel, 1999).

Rhipogonaceae are also recognized on the basis of southern hemisphere fossils. Net-veined leaves with cuticle were described from the Early Eocene of Tasmania, Australia, as *Rhipogonum tasmanicum* by Conran et al. (2009b). Leaves identical to extant *Rhipogonum scandens* have also been recovered from the Miocene of New Zealand (Pole, 1993, 2007).

Many leaves have been attributed to Smilacaceae, largely placed either in *Smilax* or *Smilacites*. These are from the Eocene to Miocene of North America, Europe, Japan and Australia (e.g. Sun and Dilcher 1988; Wilde, 1989; Ozaki, 1991; Scriven, 1994; Greenwood and Conran, 2000), with one potential Late Cretaceous record from Wyoming, USA (Berry, 1929). Fossil net-veined monocot leaves can be difficult to identify, and extant *Smilax* species show a large variation in leaf shape, size and venation characteristics, which can be very similar to the leaves in other families that possess ovate, reticulate-veined leaves with a cordate base such as Dioscoreaceae and even nonmonocot angiosperms like Piperaceae, Menispermaceae and Saururaceae (Daghlian, 1981; Conran et al., 1994; Greenwood and Conran, 2000). Many of these records are in need of reinvestigation and also for the recognition of venation and cuticular characters that can be used to distinguish between taxa.

2.2.9 Asparagales

There is no known fossil record for Amaryllidaceae (including Agapanthaceae and Alliaceae), Blandfordiaceae, Boryaceae, Doryanthaceae, Hypoxidaceae, Iridaceae, Ixioliriaceae, Lanariaceae, Tecophilaeaceae or Xeronemataceae. Pollen of Asteliaceae has been reported from the Upper Eocene to recent of New Zealand (Muller, 1981).

The fossil record of Asparagaceae (including Agavaceae, Aphyllanthaceae, Convallariaceae, Hesperocallidaceae, Hyacinthaceae, Laxmanniaceae, Ruscaceae and Themidaceae) is comprised of pollen or vegetative (stem and leaf) material, with no fruits or seeds to date. The oldest putative record is a corm from the Maastrichtian of the Deccan Intertrappean beds (Bonde, 2005) that was compared most closely with *Eriospermum*, but a more thorough study is still needed. Leaves (including cuticles) of *Paracordyline*, known from the Eocene of Australia, are similar to extant *Cordyline* (Conran and Christophel, 1998; Greenwood and Conran, 2000). Miocene records include leaves similar to *Dracaena* from Kenya (Jacobs and Kabuye, 1987), pollen of *Dracaena* from Europe (Muller, 1981) and stems, branches, leaves and roots of *Protoyucca* from Nevada, USA, which is most similar to *Yucca* (Tidwell and Parker, 1990; Herendeen and Crane, 1995). *Majanthemophyllum* is known from the Cretaceous and Paleocene of North America (Hollick, 1906; Bell, 1949) and Eocene–Oligocene of Europe (Knobloch and Kvaček, 1995) and although it has been suggested to have affinities with Convallariaceae (Conran and Tamura, 1998), others consider *Majanthemophyllum* to be of unknown affinities (Knobloch and Kvaček, 1995). These leaves are monocots, but further investigations are needed into their affinities and a new collection of complete leaves would help. *Smilacinites ungeri* from the Miocene of the Czech Republic of whole leaves attached to rooting structures has been compared to *Majanthemophyllum*, but no familial affinities were suggested (Kvaček et al., 2004).

For a long time, there was no fossil record of Orchidaceae except possible seeds described by Friis (1985) from the Miocene of Jutland, Denmark, which lack diagnostic characters to firmly place it in Orchidaceae (Herendeen and Crane, 1995). Conran et al. (2009a) described two new species of orchid in tribe Epidendroideae based on leaves with cuticle from the Miocene of New Zealand. There is also a bee preserved in Dominican amber with an orchid pollinium on its back that shows affinities to subtribe Goodyerinae (Ramírez et al., 2007). Ramírez et al. (2007) used this evidence to date the orchid family tree and concluded that Orchidaceae had originated *c.* 75 Ma in the Late Cretaceous. Other fossils, such as the Miocene *Eoorchis* (Schmid and Schmid, 1977), lack definitive characters linking them to Orchidaceae and are not accepted (Ramírez et al., 2007; Conran et al., 2009a).

The Xanthorrhoeaceae (including Asphodelaceae and Hemerocallidaceae) have a sparse fossil record known only from Australia and New Zealand and with affinities to the former Hemerocallidaceae. The earliest record is from the middle Eocene: leaves (with cuticle) of *Dianellophyllum* (Conran et al., 2003) that most closely resemble *Dianella*, a genus with its centre of diversity in Australia found today in South-East Asia, the Pacific, southern Africa, Madagascar and India (Clifford et al., 1998; Conran et al., 2003). Pollen of *Luminidites* is known from the Eocene to recent and shows affinities to *Phormium* (Muller, 1981; Greenwood and Conran, 2000; Raine et al., 2008).

2.3 Taxonomic summary

Monocots have been unquestionably present since the early Cretaceous. Alismatales, Pandanales and some commelinids such as palms and zingiberaleans have a robust record from early on, in contrast to groups like Liliales and Asparagales that do not definitively appear until the Eocene (Figure 2.2) and are more common in the Southern Hemisphere. For many of the reported monocot fossils, further study is needed to verify the identification. Application of techniques like 3D studies and phylogenetic analyses, in combination with new fossil finds and further studies of the morphology and anatomy of extant monocots, should help to create a reliable fossil record from which broader patterns of evolution, biogeography and response to global changes can be reconstructed.

2.3.1 Paucity of fossil Liliales and Asparagales

One of the puzzling things in looking at the fossil record of monocots is the paucity of Liliales and Asparagales. Few families of these orders have been recognized in the fossil record and fruit or seed material is virtually lacking – all the fossils are known from leaves or pollen – unlike most other families, some of which (e.g. many Alismatales) are known primarily from fruits and seeds. Without a good under-standing of monocot leaf characters, fruits and seeds provide more characters that can be used for confident identifications. These orders also have a long ghost lineage since both Pandanales and commelinids are present in the Cretaceous. A few factors might explain this: paleobiogeography, habit/habitat and the need for search images. Bremer and Janssen (2006) examined the biogeography of major monocot groups and suggested that Liliales and Asparagales were Gondwanan groups, diversifying mainly in the Southern Hemisphere. If this is true, then their fossil record should be mostly in South America, Africa, and Australia/New Zealand. Historically these are regions that have not been investigated as intensively as Europe and North America. That is starting to change, though, and for several families fossils are now known from Australia or New Zealand (e.g. Greenwood and Conran, 2000). Future paleobotanical studies should target fluvial or lacustrine sediments from the late Cretaceous or Paleocene in these areas if we hope to find older records. In addition, the ecology and habitat preferences of some groups may greatly decrease the probability of being fossilized; for example, taxa that grow primarily in forests or mountainous regions will be less likely to fossilize. Finally, without further studies of modern material, we may not recognize fossils of specific groups because of lack of comparable material.

2.4 Paleobiogeography and paleoenvironments

It is not until we have a confident and firm interpretation of the monocot fossil record, with fewer uncertainties due to lack of characters either of modern comparative material or the fossils, as well as an understanding of what key nonmolecular characters are, that we will be able to take full advantage of the paleobotanical record. Studies on the origin of groups depend on reliably identi-fied fossils to date their phylogenies; biogeographic hypotheses can be supported or rejected on the basis of when and where fossils are found.

There is increasing refinement in paleogeographic maps (e.g. Scotese, 2004) and these provide a context with which to look at long-term biogeography of monocot groups with reliable fossil records. Some studies have taken other approaches like phylogeographic methods (Bremer and Janssen, 2006), which should be used as hypotheses to be tested with the actual geological and paleontological data. Such studies may also reveal ages and areas that we should be targeting to find new fossils.

We can better understand the ancient environments in which these now-extinct plants were living by looking at the associated floral assemblage and geologic features. These fossils are rarely found in isolation and co-occurring fossils provide other lines of evidence for habitat and vegetation structure. Sedimentological features, such as proximity to a river or lake, types of paleosols, and/or geochemical or isotopic data when available, can provide environmental context for the greater landscape, precipitation, seasonality and temperature. There is an increasing amount of paleoclimatological data for the Cenozoic and paleoclimatologists have reconstructed CO_2 records back into the Precambrian. While much work has been done to look at the timing of C_4 photosynthesis in grasses and the spread of grasslands, there has been less energy devoted to understanding if and how earlier monocot evolution is linked to climate and changing geography (see Daghlian, 1981 for some discussion, but some geologic details have changed since then), such as the increased tectonic activity in the Eocene or the Cenozoic Global Climatic Optimum (Figure 2.2). Concurrent orogenies in North and South America, Europe and India dramatically changed global circulation patterns and would have created both new habitats and niches. These are questions that can only be answered by studying the fossil record.

2.5 Future directions

2.5.1 Novel techniques

Many of the problems and uncertainties are beginning to be addressed as we find new, well-preserved specimens. It is also possible that some of the families currently lack a fossil record because we have not been able to recover fossils that are there. Several families produce small seeds. Traditional collection methods (such as sieving) mean these are unlikely to be retained for study. Refining these processes, for example by using a finer sieve size when looking for seeds, may allow future recognition of these groups.

3D technology can provide new search images that will be useful for studying fossil fruits and seeds, a type of fossil that is quite common, at least for certain groups. These techniques have the capability to analyse nondestructively specimens that are otherwise rare or fragile, revealing more characters that can be used for identification. In addition, 3D technology can be used for other purposes. Studying modern material using these techniques would help provide search images, as examination of just one specimen could provide sections (comparable to histological sections; Smith et al., 2009a) for comparison. 3D morphologies can also be manipulated, mimicking taphonomic effects on an organ such as the loss of outer layers (e.g. Figure 2.3H) or the infill and casting of inner spaces.

There are several examples where this type of application has already been successful in the fossil record. Seeds of *Keratosperma allenbyense* from the Middle

Eocene Princeton Chert of British Columbia, Canada are anatomically preserved in chert (Figure 2.3A) and represent an early record of Araceae subfamily Lasioideae (Cevallos-Ferriz and Stockey, 1988; Smith and Stockey, 2003). These seeds were studied using the cellulose acetate peel technique (Galtier and Phillips, 1999). Because serial sections are made, each could be photographed and the photographs loaded into a 3D computer program (e.g. Amira®). In this way the fossil seed, buried in chert and sectioned at oblique angles, could be reconstructed in three dimensions (Figure 2.3B) allowing close comparison with the external morphology of extant lasioid seeds (Figures 2.3D, 2.3E). Features such as the number of ridges on the seed and presence of a ridge or spine were much easier to visualize from the 3D reconstruction than from peels alone. This also allows more accurate comparison between different preservation styles and modern material (Figures 2.3A–E).

Another way in which 3D studies have been applied in monocot paleobotany is to use X-ray tomography to visualize internal details of fossils that are hard to section or are rare. Fossil infructescences from the Oligocene of Egypt were assigned to the genus *Epipremnum* (Araceae) on the basis of a spadix-like structure with unilocular fruits and curved seeds (Bown et al., 1982). CT investigations have shown that fruits are multi-locular, unlike Monsteroideae (S.Y. Smith, M. Collinson and R. Abel, pers. obs.), and further studies will examine the relationships of these fossils.

Similarly, synchrotron radiation X-ray tomographic microscopy (SRXTM) has been used for studying fossil monocots (Smith et al., 2009a, 2009b). One example is *Cyclanthus*, which was described in the previous section; here seeds were initially attributed to Cyperaceae and were later reinterpreted as Cyclanthaceae on the basis of new fruiting material and taphonomic studies (Smith et al., 2008). This is a case of having the wrong search image for fossils, because taphonomic processes altered the fossils, resulting in only the cuticular envelope being preserved. SRXTM allowed digital dissection of modern *Cyclanthus* seeds (Figures 2.3F–H) that reveal the same morphology (Figure 2.3H) as the fossils (Smith et al., 2009a). For any monocot families with fruits or seeds that could be subject to loss of outer layers – e.g. a mix of hard and fleshy layers, or with layers of very different chemistries – it is important that we can search images of what these structures might look like after they are subject to transportation, abrasion, and deterioration in aquatic settings.

2.5.2 Modern comparative work

Another reason we may be missing fossils is from a lack of search images that allow us to recognize the affinities of a fossil. For that, more work is needed on modern relatives. Fruits and seeds were discussed above, where synchrotron and CT methods could prove extremely valuable. More work is needed to characterize the gross morphology and anatomy of rhizomes and other stem structures, with a clear indication of taxonomically distinctive characters that could be used to identify fossils. There is still a need for detailed SEM and TEM work on monocot pollen; Araceae (e.g. Grayum, 1992; Hesse and Zetter, 2007) and Arecaceae (Harley and Baker, 2001; Dransfield et al., 2008) have been studied in detail, but many of the other families require attention.

Monocot leaves are also often cited as problematic, because although some basic leaf architectural patterns are known (e.g. Inamdar et al., 1983; Keating, 2003; Doyle et al., 2008), most of the detailed work has focussed on nonmonocotyledonous angiosperms (e.g. Ellis et al., 2009). Cuticles may also be an important leaf feature for corroborating a monocot identity, although more work is needed to identify patterns and taxonomic significance (Tomlinson, 1974). As leaves are common fossils, it will important to determine patterns in monocot leaf architecture that can be used for identifying new fossils and reinvestigating previously reported doubtful records. Undertaking such a task would be important to determine both taxonomic usefulness and limits and any link to climate that might be useful for paleoenvironmental inferences.

In combination with detailed anatomy and morphology, development of phylogenetic techniques also provides a useful tool for assessing the monocot fossil record. It has the advantage of being more objective and permits the analysis of characters in an evolutionary context. Thus it is possible to determine the significance of characters and whether they represent apomorphies or plesiomorphies. Phylogenetic analysis also can help determine whether a fossil taxon is in the crown group of a family, or is instead a stem lineage. In addition, examining rates of evolution and combining molecular and fossil data with biogeographic knowledge may allow elucidation of broader patterns relating to diversification events and impacts of global environmental changes (e.g. Couvreur et al. 2011 used lineage-through-time analysis on Arecaceae to infer patterns and locations of palm diversification).

2.6 Acknowledgements

I would like to thank an anonymous reviewer, J. Bogner, M. Collinson, and N. Sheldon for their comments and discussion about the manuscript. Synchrotron tomography work was accomplished with the help of staff at the TOMCAT beamline, Swiss Light Source, Paul Scherrer Institute, Villigen, Switzerland. This work was supported by a Royal Society international postdoctoral fellowship, and the Michigan Society of Fellows.

2.7 References

APG II (Angiosperm Phylogeny Group II). (2003). An update of the Angiosperm Phylogeny Group classification for the orders and families of flowering plants: APG II. *Botanical Journal of the Linnean Society*, **141**, 399–436.

APG III (Angiosperm Phylogeny Group III).
(2009). An update of the Angiosperm
Phylogeny Group classification for the
orders and families of flowering plants:
APG III. *Botanical Journal of the
Linnean Society*, **161**, 105–121.

Bande, M. B. and Awasthi, N. (1986). New
thoughts on the structure and affinities
of *Viracarpon hexaspermum* Sahni from
the Deccan Intertrappean beds of India.
Studia botanica Hungarica, **19**, 13–22.

Bell, W. A. (1949). Uppermost Cretaceous
and Paleocene floras of western Alberta.
*Canada Department of Mines and
Resources, Geological Survey Bulletin*,
13, 1–231.

Benzecry, A. and Brack-Hanes, S. D. (2008).
A new hydrocharitacean seagrass from
the Eocene of Florida. *Botanical Journal
of the Linnean Society*, **157**, 19–30.

Berry, E. W. (1914). The upper Cretaceous
and Eocene floras of South Carolina and
Georgia. *US Geological Survey
Professional Paper* **84**, 1–200.

Berry, E. W. (1925a). The flora of the Ripley
Formation. *US Geological Survey
Professional Paper*, **129**, 199–226.

Berry, E. W. (1925b). Miocene Araceae
related to *Caladium* from Trinidad.
Pan-American Geologist, **44**, 38–42.

Berry, E. W. (1929). The flora of the Frontier
Formation. *United States Geological
Survey Professional Paper*, **158**, 129–135.

Biradar, N. V. and Bonde, S. D. (1990). The
genus *Cyclanthodendron* and its affinities.
In *Proceedings of the 3rd International
Organization of Palaeobotany Conference,
Melbourne, August 24th–26th 1988*, ed.
J. G. Douglas and D. C. Christophel. Ithaca:
Cornell University, pp. 51–57.

Bogner, J. (1976). Die systematische
Stellung von *Acoropsis* Conwentz, einer
fossilen Araceae aus dem Bernstein.
*Mitteilungen der Bayerischen
Staatssammlung für Palaeontologie und
Historische Geologie*, **16**, 95–98.

Bogner, J. (2001). What is *Acorus
brachystachys* Heer? *Aroideana*, **24**,
99–100.

Bogner, J. (2009). The free-floating aroids
(Araceae) – living and fossil. *Zitteliana*,
48/49, 113–128.

Bogner, J. and Kvaček, Z. (2009). A fossil
Vallisneria plant (Hydrocharitaceae)
from the Early Miocene freshwater
deposits of the Most Basin (North
Bohemia). *Aquatic Botany*, **90**, 119–123.

Bogner, J. and Mayo, S. J. (1998). Acoraceae.
In *The Families and Genera of Vascular
Plants Volume IV: Flowering Plants.
Monocotyledons: Alismatanae and
Commelinanae (except Gramineae)*, ed.
K. Kubitzki, H. Huber, P. J. Rudall, P. S.
Stevens and T. Stützel. Berlin: Springer-
Verlag, pp. 7–11.

Bogner, J., Hoffman, G. L. and Aulenback,
K. R. (2005). A fossilized aroid
infructescence, *Albertarum pueri* gen.
nov. et sp. nov., of Late Cretaceous
(Campanian) age from the Horseshoe
Canyon Formation of southern Alberta,
Canada. *Canadian Journal of Botany*,
83, 591–598.

Bogner, J., Johnson, K. R., Kvaček, Z. and
Upchurch Jr., G. R. (2007). New fossil
leaves of Araceae from the Late
Cretaceous and Paleogene of western
North America. *Zitteliana*, A**47**, 133–147.

Bonde, S. D. (2000). *Rhodospathodendron
tomlinsonii* gen. et sp. nov., an araceous
viny axis from the Nawargaon
intertrappean beds of India.
Palaeobotanist, **49**, 85–92.

Bonde, S. D. (2005). *Eriospermacormus
indicus* gen. et sp. nov. (Liliales:
Eriospermaceae): first record of a
monocotyledonous corm from the
Deccan Intertrappean beds of India.
Cretaceous Research, **26**, 197–205.

Boucot, A. J., Xu, C. and Scotese, C. R.
(2004). Phanerozoic climatic zones and
paleogeography with a consideration of

atmospheric CO_2 levels. *Paleontological Journal*, **38**, 115–122.

Bown, T. M., Kraus, M. J., Wing, S. L et al. (1982). The Fayum primate forest revisited. *Journal of Human Evolution*, **11**, 603–632.

Bremer, K. (2000). Early Cretaceous lineages of monocot flowering plants. *Proceedings of the National Academy of Sciences, USA*, **97**, 4707–4711.

Bremer, K. and Janssen, T. (2006). Gondwanan origin of major monocot groups inferred from dispersal-vicariance analysis. *Aliso*, **22**, 22–27.

Brown, R. W. (1956). Palm-like plants from the Dolores Formation (Triassic) in southwestern Colorado. *US Geological Survey Professional Paper*, **274**-H, 205–209.

Brown, R. W. (1962). Paleocene flora of the Rocky Mountains and Great Plains. *US Geological Survey Professional Paper*, **275**, 1–119.

van Bruggen, H. W. E. (1998). Aponogetonaceae. In *The Families and Genera of Vascular Plants Volume IV: Flowering Plants. Monocotyledons: Alismatanae and Commelinanae (except Gramineae)*, ed. K. Kubitzki, H. Huber, P. J. Rudall, P. S. Stevens and T. Stützel. Berlin: Springer-Verlag, pp. 260–263.

Budantsev, L. Y. and Golovneva, L. B. (2009) *Fossil Flora of the Arctic II – Paleogene Flora of Spitsbergen*. St. Petersburg: Russian Academy of Sciences, Komarov Botanical Institute. (in Russian).

Buell, M. F. (1935). Seed and seedling of *Acorus calamus. Botanical Gazette*, **96**, 758–765.

Cabrera, L. I., Salazar, G. A., Chase, M. W. et al. (2008). Phylogenetic relationships of aroids and duckweeds (Araceae) inferred from coding and noncoding plastid DNA. *American Journal of Botany*, **95**, 1153–1165.

Cameron, K. M., Chase, M. W. and Rudall, P. J. (2003). Recircumscription of the monocotyledonous family Petrosaviaceae to include *Japonolirion. Brittonia*, **55**, 214–225.

Cevallos-Ferriz, S. R. S. and Stockey, R. A. (1988). Permineralized fruits and seeds from the Princeton chert (Middle Eocene) of British Columbia: Lythraceae. *Canadian Journal of Botany*, **66**, 303–312.

Chandler, M. E. J. (1961). *The Lower Tertiary Floras of Southern England. I. Palaeocene Floras. London Clay Flora (supplement)*. Text and Atlas. London: British Museum (Natural History).

Chandler, M. E. J. (1964). *The Lower Tertiary Floras of Southern England. IV. A Summary and Survey of the Findings in the Light of Recent Botanical Observations*. London: British Museum (Natural History).

Chase, M. W., Fay, M. F., Devey, D. S. et al. (2006). Multigene analyses of monocot relationships: a summary. In *Monocots: Comparative Biology and Evolution (excluding Poales)*, ed. T. J. Columbus, E. A. Friar, C. W. Hamilton et al. Claremont, CA: Rancho Santa Ana Botanic Garden, pp. 63–75.

Chitaley, S. D., Shallom, L. J. and Mehta, N. V. (1969). *Viracarpon sahnii*, nov. spec. from the Deccan Intertrappean beds of Mahurzari. In *J. Sen Memorial Volume*, Calcutta: Botanical Society of Bengal, pp. 331–334.

Chmura, C. A. (1973). Upper Cretaceous (Campanian-Maastrichtian) angiosperm pollen from the western San Joaquin valley, California, U.S.A. *Palaeontographica Abteilung B*, **141**, 89–171, plates 21–33.

Clifford, H. T., Henderson, R. J. F. and Conran, J. G. (1998). Hemerocallidaceae. In *The Families and Genera of Vascular Plants. III. Flowering*

Plants. Monocotyledons. Lilianae (except Orchidaceae), ed. K. Kubtizki. Berlin: Springer-Verlag, pp. 245–253.

Collinson, M. E. (1982). A reassessment of fossil Potamogetonaceae fruits with description of new material from Saudi Arabia. *Tertiary Research*, **4**, 83–104.

Collinson, M. E. (1983). Palaeofioristic assemblages and palaeoecology of the Lower Oligocene Bembridge Marls, Hamstead Ledge, Isle of Wight. *Botanical Journal of the Linnean Society*, **86**, 177–225.

Collinson, M. E. (1986). The Felpham flora – a preliminary report. *Tertiary Research*, **8**, 29–32.

Collinson, M. E. (1988). Freshwater macrophytes in palaeolimnology. *Palaeogeography, Palaeoclimatology, Palaeoecology*, **62**, 317–342.

Collinson, M. E. (1990). Plant evolution and ecology during the early Cainozoic diversification. *Advances in Botanical Research*, **17**, 1–98.

Collinson, M. E., Boulter, M. C. and Holmes, P. L. (1993). 45, Magnoliophyta ('Angiospermae'). In *The Fossil Record 2*, ed. M. J. Benton. London: Chapman and Hall, pp. 809–841.

Conran, J. G. and Christophel, D. C. (1998). *Paracordyline aureonemoralis* (Lomandraceae): an Eocene monocotyledon from South Australia. *Alcheringa: An Australasian Journal of Palaeontology*, **22**, 349–357.

Conran, J. G. and Christophel, D. C. (1999). A redescription of the Australian Eocene fossil monocotyledon *Petermanniopsis* (Lilianae: aff. Petermanniaceae). *Transactions of the Royal Society of Southern Australia*, **123**, 61–67.

Conran, J. G. and Tamura, M. N. (1998). Convallariaceae. In *The Families and Genera of Vascular Plants. III. Flowering Plants. Monocotyledons. Lilianae*

(except Orchidaceae), ed. K. Kubtizki. Berlin: Springer-Verlag, pp. 186–198.

Conran, J. G., Christophel, D. C. and Scriven, L. (1994). *Petermanniopsis angleseaënsis* gen. and sp. nov.: an Australian fossil net-veined monocotyledon from Eocene Victoria. *International Journal of Plant Sciences*, **155**, 816–827.

Conran, J. G., Christophel, D. C. and Cunningham, L. (2003). An Eocene monocotyledon from Nelly Creek, Central Australia, with affinities to Hemerocallidaceae (Lilianae: Asparagales). *Alcheringa: An Australasian Journal of Palaeontology*, **27**, 107–115.

Conran, J. G., Bannister, J. M. and Lee, D. E. (2009a). Earliest orchid macrofossils: Early Miocene *Dendrobium* and *Earina* (Orchidaceae: Epidendroideae) from New Zealand. *American Journal of Botany*, **96**, 466–474.

Conran, J. G., Carpenter, R. J. and Jordan, G. J. (2009b). Early Eocene *Ripogonum* (Liliales: Ripogonaceae) leaf macrofossils from southern Australia. *Australian Systematic Botany*, **22**, 219–228.

Couvreur, T. L. P., Forest, F. and Baker, W. J. (2011). Origin and global diversification patterns of tropical rain forests: inferences from a complete genus-level phylogeny of palms. *BMC Biology*, **9**, 44.

Crepet, W. L. (1978). Investigations of angiosperms from the Eocene of North America: an aroid inflorescence. *Review of Palaeobotany and Palynology*, **25**, 241–252.

Crepet, W. L., Nixon, K. C. and Gandolfo, M. A. (2004). Fossil evidence and phylogeny: the age of major angiosperm clades based on mesofossil and macrofossil evidence from Cretaceous deposits. *American Journal of Botany*, **91**, 1666–1682.

Daghlian, C. P. (1981). A review of the fossil record of monocotyledons. *The Botanical Review*, **47**, 517–555.

Delevoryas, T. (1964). Two petrified angiosperms from the Upper Cretaceous of South Dakota. *Journal of Paleontology*, **38**, 584–586.

Dowe, J. (1995). A preliminary review of the biogeography of Australian palms. *Mooreana*, **5**, 7–22.

Doyle, J. A., Endress, P. K. and Upchurch Jr., G. R. (2008). Early Cretaceous monocots: a phylogenetic evaluation. *Acta Musei Nationalis Pragae Series B*, **64**, 59–87.

Dransfield, J., Uhl, N. W., Asmussen, C. B. et al. (2008). *Genera Palmarum: The Evolution and Classification of Palms*. London: Royal Botanic Gardens, Kew.

Ediger, V. S., Bati, Z. and Alisan, C. (1990). Paleopalynology and paleoecology of *Calamus*-like disculcate pollen grains. *Review of Palaeobotany and Palynology*, **62**, 97–105.

Ellis, B., Daly, D. C., Hickey, L. J. et al. (2009). *Manual of Leaf Architecture*. Ithaca: Cornell University Press.

Erwin, D. M. and Stockey, R. A. (1989). Permineralized monocotyledons from the Middle Eocene Princeton chert (Allenby Formation) of British Columbia: Alismataceae. *Canadian Journal of Botany*, **67**, 2636–2645.

Estrada-Ruiz, E. and Cevallos-Ferriz, S. R. S. (2007). Infructescences from the Cerro del Pueblo Formation (Late Campanian), Coahuila, and El Cien Formation (Oligocene-Miocene), Baja California Sur, Mexico. *International Journal of Plant Sciences*, **168**, 507–519.

Friis, E. M. (1985). Angiosperm fruits and seeds from the Middle Miocene of Jutland (Denmark). *Det Kongelige Danske Videnskabernes Selskab Biologiske Skrifter*, **24**, 1–165.

Friis, E. M. and Crepet, W. L. (1987). Time and appearance of floral features. In *The Origins of Angiosperms and Their Biological Consequences*, ed. E. M. Friis, W. G. Chaloner and P. R. Crane. Cambridge: Cambridge University Press, pp. 145–180.

Friis, E. M., Pedersen, K. R. and Crane, P. R. (2004). Araceae from the Early Cretaceous of Portugal: Evidence on the emergence of monocotyledons. *Proceedings of the National Academy of Sciences, USA*, **101**, 16565–16570.

Friis, E. M., Pedersen, K. R. and Crane, P. R. (2010). Diversity in obscurity: fossil flowers and the early history of angiosperms. *Philosophical Transactions of the Royal Society B*, **365**, 369–382.

Friis, E. M., Crane P. R. and Pedersen, K. R. (2011). *Early Flowers and Angiosperm Evolution*. Cambridge: Cambridge University Press.

Galtier, J. and Phillips, T. L. (1999). The acetate peel technique. In *Fossil Plants and Spores: Modern Techniques*, ed. T. P. Jones and N. P. Rowe. London: Geological Society, pp. 67–70.

Gandolfo, M. A., Nixon, K. C. and Crepet, W. L. (2000). Monocotyledons: a review of their Early Cretaceous record. In *Monocots: Systematics and Evolution*, ed. K. L. Wilson and D. A. Morrison. Melbourne: CSIRO, pp. 44–52.

Gandolfo, M. A., Nixon, K. C. and Crepet, W. L. (2002). Triuridaceae fossil flowers from the Upper Cretaceous of New Jersey. *American Journal of Botany*, **89**, 1940–1957.

Gandolfo, M. A., Zamaloa, M. C., Cúneo, N. R. and Archangelsky, A. (2009). Potamogetonaceae fossil fruits from the Tertiary of Patagonia, Argentina. *International Journal of Plant Sciences*, **170**, 419–428.

Golovneva, L. B. (1997). Morphology, systematics and distribution of the genus *Haemanthophyllum* in the Paleogene floras of the Northern Hemisphere. *Paleontologischeskii Zhurnal*, **31**, 197–207.

Gomez, B., Coiffard, C., Sender, L. M. et al. (2009). *Klitzschophyllites*, aquatic basal eudicots (Ranunculales?) from the Upper Albian (Lower Cretaceous) of Northeastern Spain. *International Journal of Plant Sciences*, **170**, 1075–1085.

Graham, S. W., Zgurski, J. M., McPherson, M. A. et al. (2006). Robust inference of monocot deep phylogeny using an expanded multigene plastid data set. In *Monocots: Comparative Biology and Evolution (excluding Poales)*, ed. T. J. Columbus, E. A. Friar, C. W. Hamilton et al. Claremont, CA: Rancho Santa Ana Botanic Garden, pp. 3–20.

Grayum, M. H. (1992). Comparative external pollen ultrastructure of the Araceae and putatively related taxa. *Monographs in Systematic Botany from the Missouri Botanical Garden*, **43**, 1–167.

Greenwood, D. R. and Conran, J. G. (2000). The Australian Cretaceous and Tertiary monocot fossil record. In *Monocots: Systematics and Evolution*, ed. K. L. Wilson and D. A. Morrison. Melbourne: CSIRO, pp. 52–59.

Gregor, H.-J. (1983). Erstnachweis der Gattung *Tacca* Forst. 1776 (Taccaceae) im Europäischen Alttertiär. *Documenta Naturae*, **6**, 27–31.

Gregor, H.-J. and Bogner, J. (1984). Fossile Araceen Mitteleuropas und ihre rezenten Vergleichsformen. *Documenta Naturae*, **19**, 1–12.

Gregor, H.-J. and Bogner, J. (1989). Neue Untersuchungen an tertiären Araceen II. *Documenta Naturae*, **49**, 12–22.

Haggard, K. K. and Tiffney, B. H. (1997). The flora of the Early Miocene Brandon Lignite, Vermont, USA. VIII. *Caldesia* (Alismataceae). *American Journal of Botany*, **84**, 239–252.

van der Ham, R. W. J. M., van Konijnenburg-van Cittert, J. H. A. and Indeherberge, L. (2007). Seagrass foliage from the Maastrichtian type area (Maastrichtian, Danian, NE Belgium, SE Netherlands). *Review of Palaeobotany and Palynology*, **144**, 301–321.

Harley, M. M. (2006). A summary of fossil records for Arecaceae. *Botanical Journal of the Linnean Society*, **151**, 39–67.

Harley, M. M. and Baker, W. J. (2001). Pollen aperture morphology in Arecaceae: application within phylogenetic analyses, and a summary of the fossil record of palm-like pollen. *Grana*, **40**, 45–77.

Harley, M. M. and Morley, R. J. (1995). Ultrastructural studies of some fossil and extant palm pollen, and the reconstruction of the biogeographical history of subtribes Iguanurinae and Calaminae. *Review of Palaeobotany and Palynology*, **85**, 153–182.

den Hartog, C. (1970). The sea-grasses of the world. *Verhandelingen der Koninklijke Nederlandse Akademie van Wetenschappen, afd. Natuurkunde, Tweede Reeks*, **2**, 1–275.

Herbert, C. and Helby, R. (1980). *A Guide to the Sydney Basin*. Geological Survey New South Wales, Bulletin 26.

Herendeen, P. S. and Crane, P. R. (1995). The fossil history of the monocotyledons. In *Monocotyledons: Systematics and Evolution*, ed. P. J. Rudall, P. J. Cribb, D. F. Cutler and C. J. Humphries. Richmond: Royal Botanic Gardens, Kew, pp. 1–21.

Herman, A. B. and Kvaček, J. (2010) *Late Cretaceous Grünbach Flora of Austria*. Wien: Naturhistorisches Museum Wien.

Herrera, F. A., Jaramillo, C. A., Dilcher, D. L., Wing, S. L. and Gómez-N., C. (2008). Fossil Araceae from a Paleocene neotropical rainforest in Colombia.

American Journal of Botany, **95**, 1569–1583.

Hesse, M. and Zetter, R. (2007). The fossil pollen record of Araceae. *Plant Systematics and Evolution*, **263**, 93–115.

Hofmann, C. C. and Zetter, R. (2010). Upper Cretaceous sulcate pollen from the Timerdyakh Formation, Vilui Basin (Siberia). *Grana*, **49**, 170–193.

Hollick, C. A. (1906). The Cretaceous flora of southern New York, and New England. *Monographs of the United States Geological Survey*, **50**, 1–219.

Inamdar, J. A., Shenoy, K. N. and Rao, N. V. (1983). Leaf architecture of some monocotyledons with reticulate venation. *Annals of Botany*, **52**, 725–735.

Ivany, L. C., Portell, R. W. and Jones, D. S. (1990). Animal-plant relationships and paleobiogeography of an Eocene seagrass community from Florida. *Palaios*, **5**, 244–258.

Jacobs, B. F. and Kabuye, C. H. S. (1987). A middle Miocene (12.2 My old) forest in the East African Rift Valley, Kenya. *Journal of Human Evolution*, **16**, 147–155.

Janssen, T. and Bremer, K. (2004). The age of major monocot groups inferred from 800+ rbcL sequences. *Botanical Journal of the Linnean Society*, **146**, 385–389.

Katz, N. Ja., Katz, S. V. and Kipiani, M. G. (1965). *Atlas and Keys of Fruits and Seeds Occurring in the Quaternary Deposits of the USSR*. Moscow: Nauka.

Keating, R. C. (2003). Leaf anatomical characters and their value in understanding morphoclines in the Araceae. *The Botanical Review*, **68**, 510–523.

Klaus, W. (1960). Sporen der Kamischen Stufe der ostalpinen Trias. *Geologisches Jahrbuch A*, **5**, 107–184.

Knobloch, E. and Kvaček, Z. (1995). A verticillate puzzle: *Nemejcia eocenica* Knobloch and Kvaček form-gen et sp.

nov. (?Angiosperms) from the Upper Eocene of Central Europe. *Review of Palaeobotany and Palynology*, **84**, 413–418.

Krassilov, V. A. and Kodrul, T. (2009). Reproductive structures associated with *Cobbania*, a floating monocot from the Late Cretaceous of the Amur Region, Russian Far East. *Acta Palaeobotanica*, **49**, 233–251

Krassilov, V. A. and Makulbekov, N. M. (1995). Maastrichtian aquatic plants from Mongolia. *Paleontological Journal*, **29**, 119–140.

Kräusel, R. (1929). Fossile Pflanzen aus dem Tertiär von Süd-Sumatra. *Verhandelingen van het Geologisch-Mijnbouwkundig Genootschap voor Nederland en Koloniën, Series Geologische*, **9**, 1–44.

Kräusel, R. and Stromer, E. (1924). Die fossilen Floren Ägyptens: Ergebnisse der Forschungsreisen Prof E. Stromers in den Wüsten Ägyptens. *Abhandlungen Bayerische Akademie der Wissenschaften, Mathematisch-Naturwissenschaftliche Abteilung*, **30**, 1–48.

Kuo, J., Seto, K., Nasu, T., Iizumi, H. and Aioi, K. (1989). Notes on Archaeozostera in relation to the Zosteraceae. *Aquatic Botany*, **34**, 317–328.

Kvaček, Z. (1995). *Limnobiophyllum* Krassilov – a fossil link between the Araceae and Lemnaceae. *Aquatic Botany*, **50**, 49–61.

Kvaček, Z. (2002). Late Eocene landscape, ecosystems and climate in northern Bohemia with particular reference to the locality of Kučlín near Bílina. *Bulletin of the Czech Geological Survey*, **77**, 217–236.

Kvaček, Z. (2003). Aquatic angiosperms from the Early Miocene Most Formation of North Bohemia (Central Europe). *Courier Forschungsinstitut Senckenberg*, **241**, 255–279.

Kvaček, Z. and Bogner, J. (2008). Twenty-million-year-old fruits and seeds of *Pistia* (Araceae) from central Europe. *Aroideana*, **31**, 90–97.

Kvaček, J. and Herman, A. B. (2004). Monocotyledons from the Early Campanian (Cretaceous) of Grünbach, Lower Austria. *Review of Palaeobotany and Palynology*, **128**, 323–353.

Kvaček, J. and Herman, A. B. (2005). Validation of *Araciphyllites austriacus* J. Kvaček et Herman (Monocotyledones: Araceae). *Journal of the National Museum, Natural History Series*, **174**, 1–5.

Kvaček, Z. and Teodoridis, V. (2007). Tertiary macrofloras of the Bohemian Massif: a review with correlations within Boreal and Central Europe. *Bulletin of Geosciences*, **82**, 383–408.

Kvaček, Z., Böhme, M., Dvorák, Z. et al. (2004). Early Miocene freshwater and swamp ecosystems of the Most Basin (northern Bohemia) with particular reference to the Bílina Mine section. *Journal of the Czech Geological Survey*, **49**, 1–40.

Laurent, L., and Laurent, J. (1926). Étude sur une plante fossile des dépôts du Tertiaire marin du sud de Célèbes. *Jaarboek Mijnwezen Nederlandsch-Indië*, **54**, 169–190.

Lumbert, S. H., den Hartog, C., Philips, R. C. and Olsen, F. S. (1984). The occurrence of fossil seagrasses in the Avon Park Formation (Late Middle Eocene), Levy Country, Florida (U.S.A.). *Aquatic Botany*, **20**, 121–129.

Maas-van der Kamer, H. (1998). Burmanniaceae. In *The Families and Genera of Vascular Plants. III. Flowering Plants. Monocotyledons. Lilianae (except Orchidaceae)*, ed. K. Kubitzki. Berlin: Springer-Verlag, pp. 154–164.

Mai, D. H. (1985). Entwicklung der Wasser- und Sumpfpflanzen-Gesellschaften Europas von der Kreide bis ins Quartär. *Flora: Morphologie, Geobotanik, Oekophysiologie*, **176**, 449–511.

Mai, D. H. (1995). *Tertiäre Vegetationsgeschichte Europas*. Jena-Stuttgart–New York: Gustav Fischer Verlag.

Mai, D. H. and Walther, H. (1978). Die Floren der Haselbacher Serie im Weisselster-Becken (Bezirk Leipzig, DDR). *Abhandlungen des Staatlichen Museums für Mineralogie und Geologie zu Dresden*, **28**, 1–200.

Mai, D. H. and Walther, H. (1985). Die obereozänen Floren des Weisselster-Beckens und seiner Randgebiete. *Abhandlungen des Staatlichen Museums für Mineralogie und Geologie zu Dresden*, **33**, 1–260.

Manchester, S. R. (2001). Update on the megafossil flora of Florissant, Colorado. *Proceedings of the Denver Museum of Nature and Science*, **4**, 137–161.

Manchester, S. R. and O'Leary, E. L. (2010). Phylogenetic distribution and identification of fin-winged fruits. *Botanical Review*, **76**, 1–82.

Mayo, S. J., Bogner, J. and Boyce, P. C. (1997). *The Genera of Araceae*. Richmond: Royal Botanic Gardens, Kew.

Mohr, B. A. R. and Rydin, C. (2002). *Trifurcatia flabellata* n. gen. n. sp., a putative monocotyledon angiosperm from the Lower Cretaceous Crato Formation (Brazil). *Mitteilungen Museum für Naturkunde Berlin, Geowissenschaftliche Reihe*, **5**, 335–344.

Mohr, B. A. R., Bernardes-de-Oliveira, M. E. C., Barale, G. and Ouaja, M. (2006). Palaeogeographic distribution and ecology of *Klitzschophyllites*, and early Cretaceous angiosperm in southern Laurasia and northern Gondwana. *Cretaceous Research*, **27**, 464–472.

Muller, J. (1981). Fossil pollen records of extant angiosperms. *Botanical Review*, **47**, 1–142.

von Müller, F. (1880). *Ottelia praeterita* F. v M. *Journal and Proceedings of the Royal Society of New South Wales*, **13**, 95–96, plate 3.

Nambudiri, E. M. V. and Tidwell, W. D. (1978). On probable affinities of *Viracarpon* Sahni from the Deccan Intertrappean flora of India. *Palaeontographica Abteilung B*, **166**, 30–43, 5 plates.

Nordenstam, B. (1998). Colchicaceae. In *The Families and Genera of Vascular Plants. III. Flowering Plants. Monocotyledons. Lilianae (except Orchidaceae)*, ed. K. Kubitzki. Berlin: Springer-Verlag, pp. 175–185.

Ozaki, K. (1991). Late Miocene and Pliocene floras in Central Honshu, Japan. *Bulletin of Kanagawa Prefectural Museum Natural Science Special Issue*, pp. 1–244.

Paleobiology Database (2010). http://www.paleodb.org

Patil, G. V. (1972). *Viracarpon chitaleyi*, sp. nov. from the Deccan Intertrappean beds of Mohgaon Kalan, India. *Botanique*, **3**, 21–26.

Petta, T. J. and Gerhard, L. C. (1977). Marine grass banks – a possible explanation for carbonate lenses, Tepee Zone, Pierre Shale (Cretaceous), Colorado. *Journal of Sedimentary Petrology*, **47**, 1018–1026.

Pneva, G. P. (1988). A new Tertiary species of *Aponogeton* (Aponogetonaceae) from Kazakhstan and Karakalpakia. *Botanicheskii Zhurnal*, **73**, 1597–1599 (in Russian).

Pole, M. (1993). Early Miocene flora of the Manuherikia Group, New Zealand. 5. Smilacaceae, Polygonaceae, Elaeocarpaceae. *Journal of the Royal Society of New Zealand*, **23**, 289–302.

Pole, M. (1999). Latest Albian-earliest Cenomanian monocotyledonous leaves from Australia. *Botanical Journal of the Linnean Society*, **129**, 177–186.

Pole, M. (2007). Monocot macrofossils from the Miocene of Southern New Zealand. *Palaeontological Electronica*, **10**, 15A, 21p.

Prasad, V., Stromberg, C. A. E., Alimohammadian, H. and Sahni, A. (2005). Dinosaur coprolites and the early evolution of grasses and grazers. *Science*, **310**, 1177–1180.

Raine, J. I., Mildenhall, D. C. and Kennedy, E. M. (2008). *New Zealand Fossil Spores and Pollen: An Illustrated Catalogue.* 3rd Edn. GNS Science miscellaneous series no. 4. http://www.gns.cri.nz/what/earthhist/fossils/spore_pollen/catalog/index.htm

Ramírez, S. R., Gravendeel, B., Singer, R. B., Marshall, C. R. and Pierce, N. E. (2007). Dating the origin of the Orchidaceae from a fossil orchid with its pollinator. *Nature*, **448**, 1042–1045.

Retallack, G. J., Sheldon, N. D., Carr, P. F. et al. (2010). Multiple Early Triassic greenhouse crises impeded recovery from Late Permian mass extinction. *Palaeogeography, Palaeoclimatology, Palaeoecology*, **308**, 233–251. doi:10.1016/j.palaeo.2010.09.022.

Riley, M. G. and Stockey, R. A. (2004). *Cardstonia tolmanii* gen. et sp. nov. (Limnocharitaceae) from the Upper Cretaceous of Alberta, Canada. *International Journal of Plant Sciences*, **165**, 897–916.

Rodriguez-de la Rosa, R. A. and Cevallos-Ferriz, S. R. S. (1994). Upper Cretaceous zingiberalean fruits with in situ seeds from southeastern Coahuila, Mexico. *International Journal of Plant Sciences*, **155**, 786–805.

Sahni, B. (1964). Revision of Indian fossil plants. Part III. Monocotyledons. *Monographs of the Birbal Sahni Institute of Palaeobotany*, **1**, 1–89.

Sahni, B. and Surange, K. R. (1953). On the structure and affinities of *Cyclanthodendron sahnii* (Rode) Sahni and Surange from the Deccan Intertrappean series. *Palaeobotanist*, **2**, 93–100.

Samylina, V. A. (1968). Early Cretaceous angiosperms of the Soviet Union based on leaf and fruit remains. *Journal of the Linnean Society of London, Botany*, **61**, 207–218.

Saporta, G. (1873). Études sur la végétation du niveau du Sud-Est de la France à l'époque tertiaire. *Annales des Sciences Naturelle, Série Botanique* **17**, 5–44.

Saporta, G. (1879). *Le monde des plantes avant l'apparition de l'homme*. Paris: G. Masson.

Saporta, G. and Marion, A.-F. (1878). Révision de la flore Heersienne de Gelinden d'après une collection appartenant au comte G. de Looz. *Mémoire Couronné et des Savants étrangers publié par l'Académie royale de Belgique*, **41**, 1–112.

Schmid, R. and Schmid, M. J. (1977). Fossil history of the Orchidaceae. In *Orchid Biology: Reviews and Perspectives*, I, ed. J. Arditti. Ithaca: Cornell University Press, pp. 25–45.

Scotese, C. R. (2004). A continental drift flip book. *Journal of Geology*, **112**, 729–741.

Scriven, L. J. (1994). *Diversity of the Mid-Eocene Maslin Bay Flora, South Australia*. PhD. Thesis, University of Adelaide.

Selling, O. H. (1974). Aponogetonaceae in the Cretaceous of South America. *Svensk Botanisk Tidskrift*, **41**, 182.

Smith, S. Y. (2006). *Morphology, Anatomy, and Phylogeny of Fossil and Extant Saururaceae: Insights from the Middle Eocene Princeton Chert*. Ph. D. dissertation, University of Alberta.

Smith, S. Y. and Stockey R. A. (2003). Aroid seeds from the Middle Eocene Princeton chert (*Keratosperma allenbyense*, Araceae): comparsions with extant Lasioideae. *International Journal of Plant Sciences*, **164**, 239–250.

Smith, S. Y. and Stockey, R. A. (2007). Establishing a fossil record for the perianthless Piperales: *Saururus tuckerae* sp. nov. (Saururaceae) from the Middle Eocene Princeton Chert. *American Journal of Botany*, **94**, 1642–1657

Smith, S. Y., Collinson, M. E. and Rudall, P. J. (2008). Fossil *Cyclanthus* (Cyclanthaceae, Pandanales) from the Eocene of Germany and England. *American Journal of Botany*, **95**, 688–699.

Smith, S. Y., Collinson, M. E., Simpson, D. A. et al. (2009a). Virtual taphonomy using synchrotron tomographic microscopy reveals cryptic features and internal structure of modern and fossil plants. *Proceedings of the National Academy of Sciences, USA*, **106**, 12013–12018.

Smith, S. Y., Collinson, M. E., Simpson, D. A. et al. (2009b). Elucidating the affinities and habitat of ancient, widespread Cyperaceae: *Volkeria messelensis* gen. et sp. nov., a fossil mapanioid sedge from the Eocene of Europe. *American Journal of Botany*, **96**, 1506–1518.

Smith, S. Y., Collinson, M. E., Rudall, P. J. and Simpson, D. A. (2010). The Cretaceous and Paleogene fossil record of Poales: review and current research. In *Diversity, Phylogeny, and Evolution in Monocotyledons: Proceedings of the Fourth International Conference on the Comparative Biology of the Monocotyledons and the Fifth International Symposium on Grass Systematics and Evolution*, ed. O. Seberg, G. Petersen, A. Barfod and J. I. Davis. Aarhus, Denmark: Aarhus University Press, pp. 333–356.

Stockey, R. A. (2006). The fossil record of basal monocots. In *Monocots: Comparative Biology and Evolution (excluding Poales)*, ed. T. J. Columbus, E. A. Friar, C. W. Hamilton et al. Claremont, California: Rancho Santa Ana Botanic Garden, pp. 91–106.

Stockey, R. A., Hoffman, G. L. and Rothwell, G. W. (1997). The fossil monocot *Limnobiophyllum scutatum*: resolving the phylogeny of Lemnaceae. *American Journal of Botany*, **84**, 355–368.

Stockey, R. A., Rothwell, G. W. and Johnson, K. R. (2007). *Cobbania corrugata* gen. et comb. nov. (Araceae): a floating aquatic monocot from the Upper Cretaceous of western North America. *American Journal of Botany*, **94**, 609–624.

Sun, Z. and Dilcher, D. (1988). Fossil *Smilax* from Eocene sediments in western Tennessee. *American Journal of Botany*, **75**, 118.

Tamura, M. N. (1998). Nartheciaceae. In *The Families and Genera of Vascular Plants. III. Flowering Plants. Monocotyledons. Lilianae (except Orchidaceae)*. Ed. K. Kubitzki. Berlin: Springer-Verlag, pp. 381–392.

Tidwell, W. D. and Parker, L. R. (1990). *Protoyucca shadishii* gen. et sp. nov., an arborescent monocotyledon with secondary growth from the Middle Miocene of northwestern Nevada, USA. *Review of Palaeobotany and Palynology*, **62**, 79–95.

Tomlinson, P. B. (1974). Development of the stomatal complex as a taxonomic character in the monocotyledons. *Taxon*, **23**, 109–128.

Veizer, J., Ala, D., Azmy, K. et al. (1999). Sr^{87}/Sr^{86}, C^{13}, and O^{18} evolution of Phanerozoic seawater. *Chemical Geology*, **161**, 59–88.

Voight, E. (1980). Upper Cretaceous bryozoan-seagrass association in the Maastrichtian of the Netherlands. In *Recent and Fossil Bryozoa*, ed. G. L. Larwood and C. Nielsen. Fredensborg: Olsen and Olsen, pp. 281–298.

Voight, E. and Domke, W. (1955). *Thalassocharis bosqueti* Debey ex Miquel, ein strukturell erhaltenes Seegras aus der holländischen Kreide. *Mitteilungen aus dem Geologischen Staatsinstitut in Hamburg*, **25**, 87–102.

Wilde, V. (1989). Untersuchungen zur Systematik der Blattreste aus dem Mitteleozän der Grube Messel bei Darmstadt (Hessen, Bundesrepublik Deutschland). *Courier Forschungsinstitut Senkenberg*, **115**, 1–213.

Wilde, V. and Frankenhauser, H. (1998). The Middle Eocene plant taphocoenosis from Eckfeld (Eifel, Germany). *Review of Palaeobotany and Palynology*, **101**, 7–28.

Wilde, V., Kvaček, Z. and Bogner, J. (2005). Fossil leaves of the Araceae from the European Eocene and notes on other aroid fossils. *International Journal of Plant Sciences*, **166**, 157–183.

Wing, S. L., Alroy, J. and Hickey, L. J. (1995). Plant and mammal diversity in the Paleocene to Early Eocene of the Bighorn Basin. *Palaeogeography, Palaeoclimatology, Palaeoecology*, **115**, 117–155.

Zachos, J., Pagani, M., Sloan, L., Thomas, E. and Billups, K. (2001). Trends, rhythms, and aberrations in global climate 65 Ma to present. *Science*, **292**, 686–693.

Zhao, L. C., Collinson, M. E. and Li, C. S. (2004). Fruits and seeds of *Ruppia* (Potamogetonaceae) from the Pliocene of Yushe Basin, Shanxi, Northern China and their ecological implications. *Botanical Journal of the Linnean Society*, **145**, 317–329.

Zhilin, S. G. (1974). *The Tertiary Floras of Ust-Urt*. Leningrad: Nauka.

3

Is syncarpy an ancestral condition in monocots and core eudicots?

Dmitry D. Sokoloff, Margarita V. Remizowa and
Paula J. Rudall

3.1 Introduction

Enclosure of the ovules within the gynoecium (angiospermy) is a key innovation of flowering plants (Endress, 1997, 2001a; Endress and Igersheim, 2000). Thus, inferring subsequent morphological transformations of the gynoecium in various lineages is important in understanding floral diversification. Traditionally, a gynoecium composed of free carpels (including the condition with a single free carpel) has been viewed as representing the primitive character state in angiosperms and this conclusion is supported by recent studies (Doyle and Endress, 2000; Endress and Igersheim, 2000; Endress, 2001a, 2001b; Armbruster et al., 2002; Endress and Doyle, 2009; Doyle and Endress, 2011). On the other hand, topologies of molecular phylogenetic trees suggest that multiple reversals have occurred in the evolution of this character, indicating multiple losses of intercarpellary fusion (e.g. Endress, 2002, 2011). Many parsimony-based hypotheses on the secondary loss of intercarpellary fusion contradict earlier views on gynoecium evolution. Examples include some well-known families where some or all members lack any sort of intercarpellary fusion (or possess a solitary free carpel) such as the monocot family Arecaceae (palms, see Rudall et al., 2011) and the eudicot families Rosaceae (roses) and Leguminosae (legumes). These families have all been difficult to place

Early Events in Monocot Evolution, eds P. Wilkin and S. J. Mayo. Published by Cambridge University Press. © The Systematics Association 2013.

in classifications based on morphology alone, mostly due to over-emphasis of the phylogenetic significance of gynoecium structure.

Inferring gynoecium evolution in monocots is highly problematic. Traditionally, the condition in which gynoecia lack any intercarpellary fusion (often termed apocarpy, but see below) has been viewed as primitive in both monocots and angiosperms in general. Intercarpellary fusion is lacking in members of three monocot lineages: some palms, some alismatids and all Triuridaceae. These three groups were therefore commonly placed close to the base of the monocot phylogeny in pre-molecular classifications (e.g. Takhtajan, 1966, 1987; Cronquist, 1981, 1988). An alternative view, also based on morphology, but using a broader range of characters, implied that the condition without any form of intercarpellary fusion is derived in monocots, probably multiple times (Dahlgren et al., 1985). There is now considerable consensus in higher-level molecular phylogenetic data in monocots (Chase, 2004, Chen et al., 2004, Davis et al., 2004, Tamura et al., 2004, Chase et al., 2006, Givnish et al., 2006, Graham et al., 2006, Iles et al., Chapter 1, this volume). Use of these molecular topologies as a basis for character optimization generally supports the latter conclusion, though some details of the monocot tree topology differ from those suggested by Dahlgren et al. (1985). Remizowa et al. (2010) used parsimonious optimizations of morphological characters onto a molecular phylogeny to infer the ancestral structure of the monocot gynoecium, using a hypothetical ancestor. They found that the results are sensitive to details of character coding and minor changes in tree topology. In the present study, we develop our earlier conclusions by adding data on a broader range of nonhypothetical outgroups.

The primary question addressed here is whether the presence of intercarpellary fusion (often termed syncarpy) represents a synapomorphy of monocots. In this respect, it is important to emphasize that at least three contrasting definitions of the term syncarpy are available. These definitions are related to the concepts of congenital and postgenital fusion. The process of congenital fusion cannot be directly observed because congenitally fused structures are *ab initio* united from inception. The presence of congenital fusion can be inferred only by placing a taxon in a phylogenetic context and inferring homologies. The fact that congenital fusion does not represent an observable process but instead an interpretative speculation has resulted in criticism of the entire concept of congenital fusion (summarized by Sattler, 1977, 1978; Timonin, 2002). In contrast, postgenital fusion represents a process that can be directly observed during development, in this case, surfaces that are initially free contact and become fused with each other (reviewed by Verbeke, 1992).

According to Leinfellner's (1950) interpretation and terminology, syncarpous gynoecia possess congenitally united carpels and the carpels of an apocarpous gynoecium can either be free or postgenitally united at floral anthesis. Thus, gynoecia in which intercarpellary fusion is exclusively postgenital are not considered syncarpous according to this definition. Alternatively, some authors use the term 'syncarpy' to describe gynoecia that possess any type of intercarpellary fusion (congenital and/or postgenital) and apocarpous for those that remain unfused (e.g. Takhtajan 2009). Finally, the term 'syncarpy' (or 'eusyncarpy') is sometimes restricted to multi-locular gynoecia with united carpels.

Leinfellner's (1950) more rigorous concept is the one that is most commonly accepted in current morphological literature (e.g. Endress, 2011); this concept is followed in the present chapter. Within this concept, only situations in which the process of intercarpellary fusion cannot be directly observed (i.e. it is congenital) are viewed as syncarpy, while examples in which (postgenital) fusion between carpels can actually be observed are considered apocarpous. Furthermore, as the occurrence of congenital fusion is a matter of evolutionary interpretation, use of the term syncarpy inevitably requires that the ancestor of angiosperms was apocarpous. Fortunately, this assumption is supported by empirical data (see above).

3.2 Material and methods

Remizowa et al. (2010) used a tree with 39 monocot terminal groups and an unnamed artificial outgroup. The overall tree topology was based on relationships presented by the Angiosperm Phylogeny Group (APG III, 2009) and Stevens (2001 onwards). Relationships within Tofieldiaceae follow Azuma and Tobe (2011). In this study, we use three contrasting tree topologies that differ in details of relationships within monocots, but are identical with respect to arrangement of the outgroups. We use 38 outgroups, including all lineages of early-divergent (ANITA- or ANA-grade) angiosperms, magnoliids and basal eudicots, and a common terminal group for all core eudicots *sensu* APG III (2009) except Gunnerales (i.e. Pentapetalae *sensu* Cantino et al., 2007). The arrangement of the outgroups follows APG III (2009). Relationships between families within orders are given according to Haston et al. (2007), except for Ranunculales where we follow Kim et al. (2004) and Endress and Doyle (2009). Within Winteraceae, *Takhtajania* and the rest of the family (Winteroideae) are accepted as two different terminal groups, based on Karol et al. (2000) and Doust and Drinnan (2004). Throughout the chapter, we use the taxonomic nomenclature accepted in APG III (2009), except for the name 'Pentapetalae' (Cantino et al., 2007), because this clade has no formal name in the classification of APG III. We agree with Endress (2010) that it would be better to treat Gunnerales as a member of the basal eudicot grade rather than include it within core eudicots.

For tree topology #1, the relationships within monocots are as shown in Remizowa et al. (2010). Tree topology #2 differs from tree topology #1 in the relative positions of Araceae and Tofieldiaceae in the monocot order Alismatales, which are as given in Iles et al. (Chapter 1, this volume). In tree topology #3, two additional changes within Alismatales are accepted according to the same source: (1) the relative positions of Scheuchzeriaceae and Aponogetonaceae are different, and (2) Araceae are split into two terminal groups, because Iles et al. (Chapter 1, this volume) found that *Gymnostachys* might be sister to the rest of Araceae. This is important because the gynoecium of *Gymnostachys* could be interpreted as monomerous, though a pseudomonomerous interpretation is also possible (Buzgo, 2001; Igersheim et al., 2001).

Maximum parsimony (MP) optimizations were constructed using WinClada (Nixon, 2002). Three gynoecium characters are considered: (1) carpels free vs. carpels united, using both congenital and postgenital fusion, (2) congenital intercarpellary fusion present vs. absent, and (3) septal (gynopleural) nectaries present vs. absent. Carpel fusion via the floral centre is not considered in characters 1 and 2. If not specified, characters for monocots are scored as in Remizowa et al. (2010). All outgroups are scored as lacking septal nectaries. The presence of intercarpellary fusion in the outgroups is scored mostly according to Endress and Igersheim (1997, 1999), Igersheim and Endress (1997, 1998), Endress et al. (2000) and Endress and Doyle (2009).

Annonaceae, Monimiaceae and Winteroideae are scored as having an apocarpous gynoecium. A detailed phylogenetic analysis showed that syncarpy is derived in Annonaceae, though in two genera the carpels are entirely congenitally united and in a third genus the gynoecium possesses a congenitally united ovary and postgenitally united distal carpel regions (Couvreur et al., 2008). Within Winteroideae, the carpels of *Zygogynum* are partially united, but members of this genus are nested within a clade of taxa with free carpels (Doust and Drinnan, 2004). The gynoecium of *Degeneria* is not scored as monomerous because of the occasional occurrence of another carpel (Takhtajan, 1980; Igersheim and Endress, 1997). Within Monimiaceae, the syncarpous genus *Tambourissa* (Endress and Igersheim, 1997) occupies a nested position in the molecular phylogeny (Renner et al., 2010). Pentapetalae *sensu* Cantino et al. (2007) are scored as ambiguous (carpels free or congenitally united). Dilleniaceae, which could be sister to all other Pentapetalae (Moore et al., 2010), are reconstructed as having an ancestrally apocarpous gynoecium without postgenital intercarpellary fusion (Horn, 2009). In contrast, the clade containing Berberidopsidales, Santalales, Caryophyllales and asterids (APG III, 2009; Moore et al., 2010) should be reconstructed as ancestrally syncarpous *sensu* Leinfellner (1950) (i.e. ancestral occurrence of congenital intercarpellary fusion). The ancestral gynoecial condition in the third clade of Pentapetalae (which includes Saxifragales, Vitales and rosids) requires further exploration.

3.3 Maximum parsimony (MP) optimizations of gynoecium evolution in monocots

There are several possible approaches to scoring intercarpellary fusion in taxa with consistently unicarpellate (monomerous) gynoecia. In the first coding (adopted by Remizowa et al., 2010), unicarpellate gynoecia are considered as a third character state, because the unicarpellate condition can occur as an extreme form in lineages with united carpels as well as in lineages with free carpels (e.g. see Rudall et al., 2005, for discussion of this aspect in Stemonaceae). In practice, coding of the unicarpellate condition as a third character state yields the same character-state reconstruction for internal nodes as scoring unicarpellate taxa as unknown (adopted by Endress and Doyle, 2009). In the second coding, unicarpellate gynoecia are considered as apocarpous (Doyle and Endress, 2000) because unicarpellate gynoecia derived from a pluricarpellate syncarpous gynoecium are apparently rare in angiosperms (though present, for example, in *Aphelia*, a genus of Centrolepidaceae in the order Poales: Sokoloff et al., 2009). We assume that the unicarpellate gynoecia of most early-divergent angiosperms and magnoliids are primarily derived from an apocarpous condition.

Regardless of tree topology and coding of unicarpellate gynoecia, our optimizations show that the occurrence of a gynoecium in which carpels are united (encompassing both postgenital and congenital intercarpellary fusion) represents an ancestral condition for monocots, with several reversals, as also suggested by Chen et al. (2004). When the unicarpellate gynoecium is scored as apocarpous without postgenital intercarpellary fusion, a gynoecium of united carpels is a synapomorphy of monocots (Fig 3.1B). When the unicarpellate gynoecium is scored as a distinct (third) character state (Fig 3.1A), it is ambiguous whether the occurrence of intercarpellary fusion is a synapomorphy of monocots or a synapomorphy of the clade comprising all angiosperms except *Amborella*, Nymphaeales and Austrobaileyales. These results are the same for all tree topologies.

In contrast, when we explore the evolution of the presence of congenital intercarpellary fusion (syncarpy *sensu* Leinfellner, 1950), the results are sensitive to tree topology. Using topology #1, the occurrence of congenital intercarpellary fusion is ancestral in monocots, with many subsequent reversals (Fig 3.2A and B, left tree). Interestingly, in the monocot order Alismatales, congenital intercarpellary fusion was first lost and then re-appeared in three independent clades according to this scenario. Using topologies #2 and #3, the evolution of congenital intercarpellary fusion is unresolved in monocots (Fig 3.2A and B, middle and right trees). Using DELTRAN optimization, apocarpy (i.e. carpels free at initiation) is interpreted as an ancestral condition for monocots. Using tree topology #2, the evolution of syncarpy is equivocal in the monocot order Alismatales

(Fig 3.2A and B, middle tree). Using tree topology #3, the absence of congenital intercarpellary fusion is ancestral for Alismatales.

The occurrence of septal (gynopleural) nectaries is an exclusively monocot character (reviewed by Endress, 1995; Smets et al., 2000; Remizowa et al., 2010) that is closely correlated with postgenital intercarpellary fusion (Hartl and Severin, 1981; van Heel, 1988; Rudall, 2002; Remizowa et al., 2006, 2010). So far, clear analogues of septal nectaries have not been found in other angiosperms, even though postgenital fusion between carpels is also found in some nonmonocot angiosperms (e.g. Endress et al., 1983; Matthews and Endress, 2005; Endress, 2011). Doyle and Endress (2000, 2011) and Endress and Doyle (2009) suggested that congenital intercarpellary fusion is an ancestral condition in monocots, indicating multiple homoplastic origins of postgenital intercarpellary fusion and septal nectaries.

Remizowa et al. (2010) highlighted an ambiguity surrounding the well-known correlation between septal nectaries and at least partially postgenital intercarpellary fusion in monocots. With the exception of relatively few taxa with triradiate septal nectaries (such as *Tofieldia* or *Borya*), there is no obvious constraint against developing gynoecia with septal nectaries in association with exclusively congenital intercarpellary fusion. A possible explanation of the fact that postgenital fusion is always present in gynoecia with septal nectaries is that septal nectaries evolved only once, and their correlation with postgenital fusion in extant monocots is inherited from this common ancestor (Remizowa et al., 2010). This hypothesis implies multiple losses of septal nectaries. Interestingly, the tree topology #3 used in the present paper (which is based on Iles et al., Chapter 1, this volume) might allow different reconstructions for the evolution of septal nectaries in monocots. When we use the same character scoring as in Remizowa et al. (2010), multiple origins of septal nectaries are inferred from MP reconstruction (Fig 3.3A). However, a change of scoring in a single terminal taxon (Aponogetonaceae) is sufficient to infer an equivocal reconstruction (Fig 3.3B); using ACCTRAN optimization, a single origin of septal nectaries is found. Scoring *Aponogeton* (Aponogetonaceae) is problematic; its carpels are almost entirely free, yet according to Igersheim et al. (2001), one of two species investigated possesses weakly differentiated septal nectaries. Other authors have also reported septal nectaries in *Aponogeton* (Daumann, 1970, Takhtajan, 2009).

Another source of ambiguity in reconstructing the evolution of septal nectaries in monocots results from using different tree topologies for the family Nartheciaceae (Dioscoreales). According to Merckx et al. (2008), Nartheciaceae are composed of two sister clades: (1) *Lophiola, Nietneria, Narthecium* and (2) *Aletris, Metanarthecium*. In clade (1), septal nectaries are lacking, while they are present in clade (2) (see also Remizowa et al., 2008). For this reason, we use both

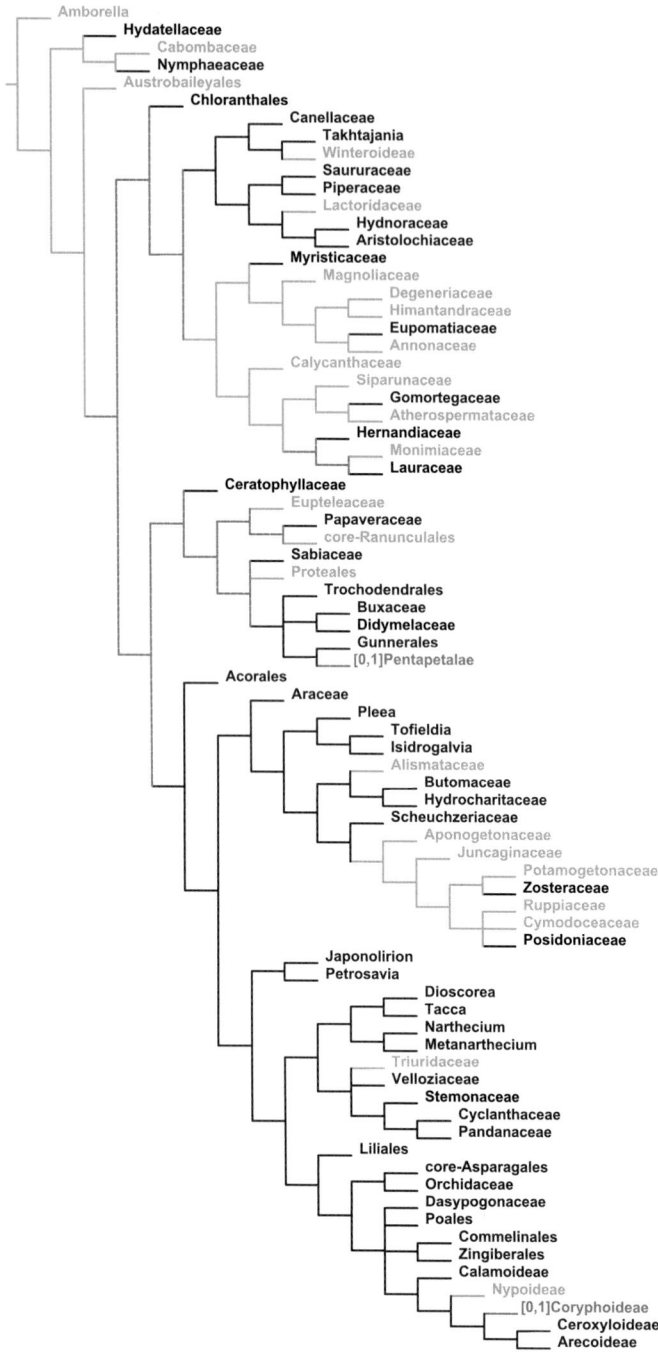

Fig 3.1A Maximum parsimony (MP) optimizations of gynoecium morphology using tree topology # 1. Black, unicarpellate gynoecia; green, gynoecia of free carpels (i.e. apocarpous without postgenital intercarpellary fusion); blue, gynoecia of united carpels irrespective of the type of intercarpellary fusion (i.e. syncarpous plus apocarpous with postgenital intercarpellary fusion); purple, ambiguity. Colour version to be found in colour plate section.

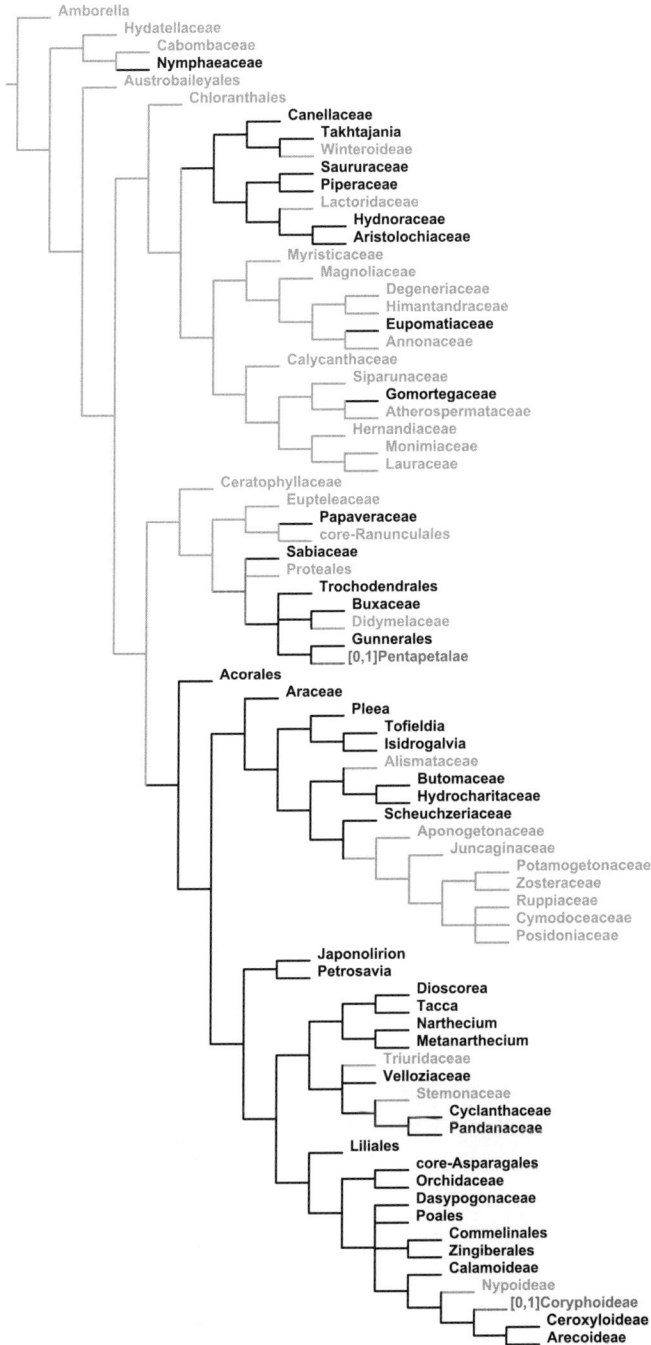

Fig 3.1B The same, but unicarpellate gynoecia are interpreted as apocarpous without postgenital intercarpellary fusion. Colour version to be found in colour plate section.

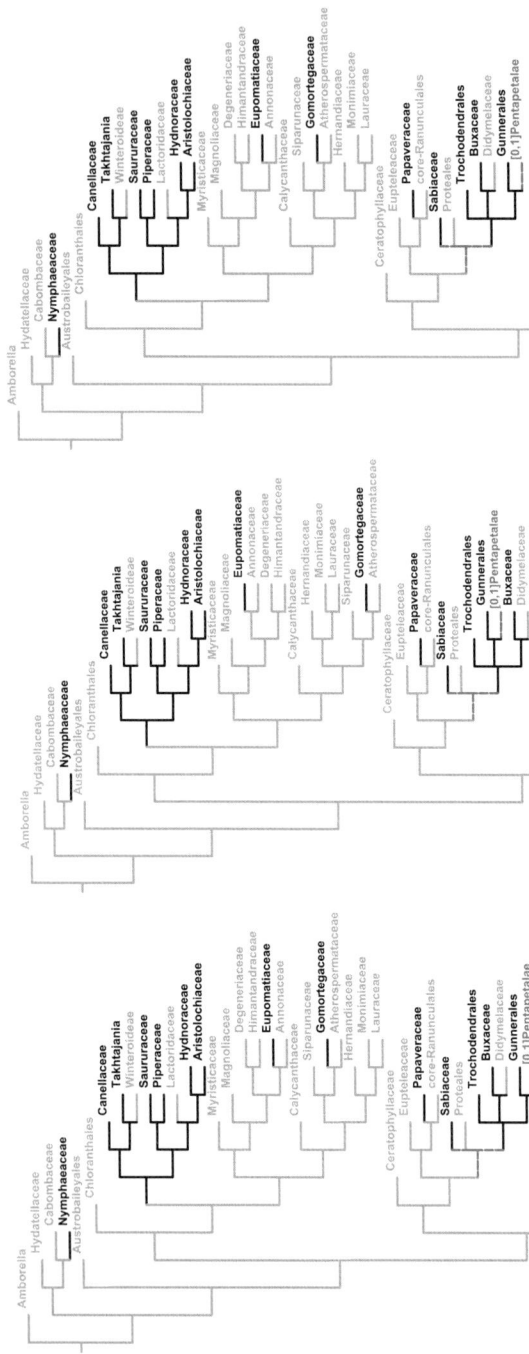

Fig 3.2A Maximum parsimony (MP) optimization of the presence vs. absence of congenital intercarpellary fusion using tree topology # 1 (left), tree topology # 2 (middle) and tree topology # 3 (right). Unicarpellate gynoecia are interpreted as apocarpous. Green, congenital intercarpellary fusion absent; blue, congenital intercarpellary fusion present; purple, ambiguity. Outgroup taxa. Colour version to be found in colour plate section.

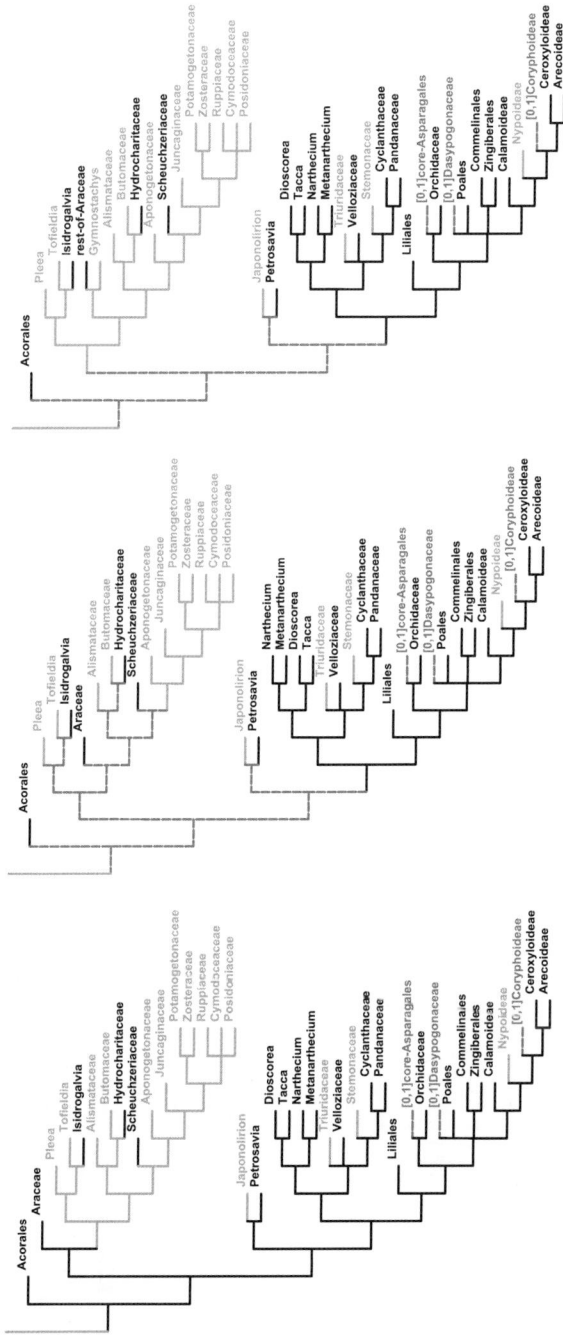

Fig 3.2B The same optimization and tree topologies. Monocot in group taxa. Colour version to be found in colour plate section.

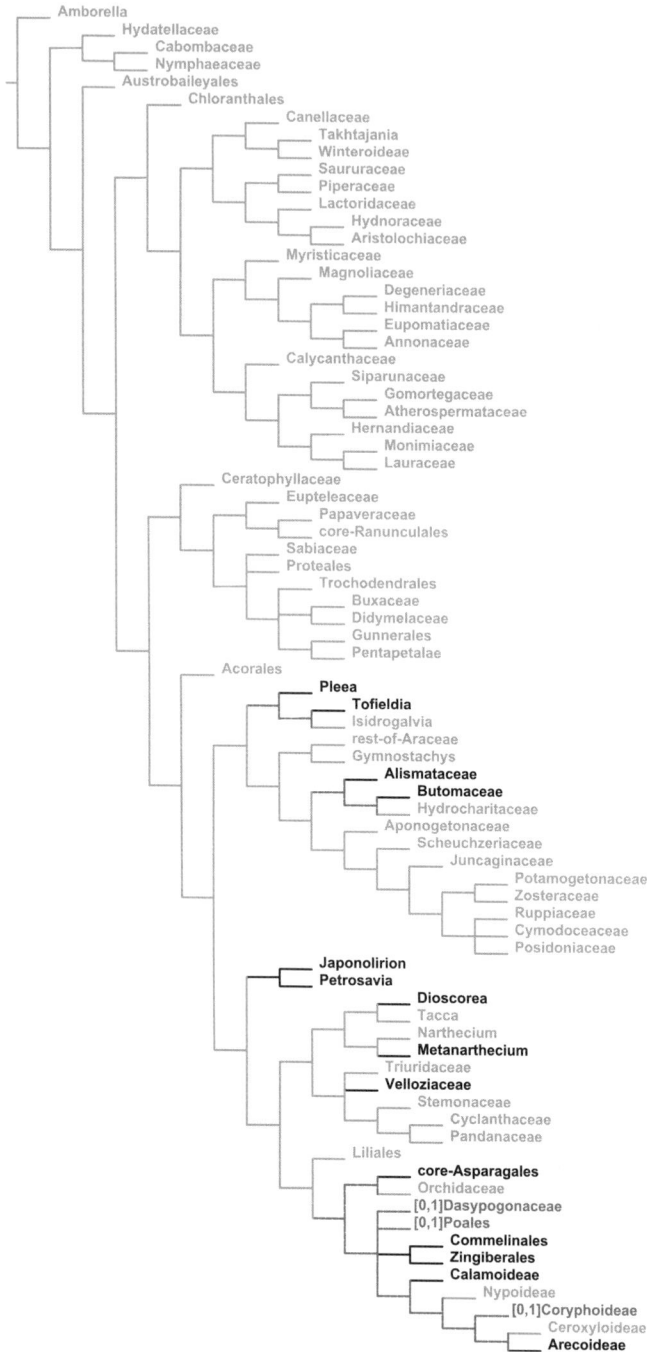

Fig 3.3A Maximum parsimony (MP) optimizations of presence vs. absence of septal (gynopleural) nectaries using tree topology # 3. Green, septal nectaries absent; blue, septal nectaries present; purple, ambiguity. Character coding following Remizowa et al. (2010). Colour version to be found in colour plate section.

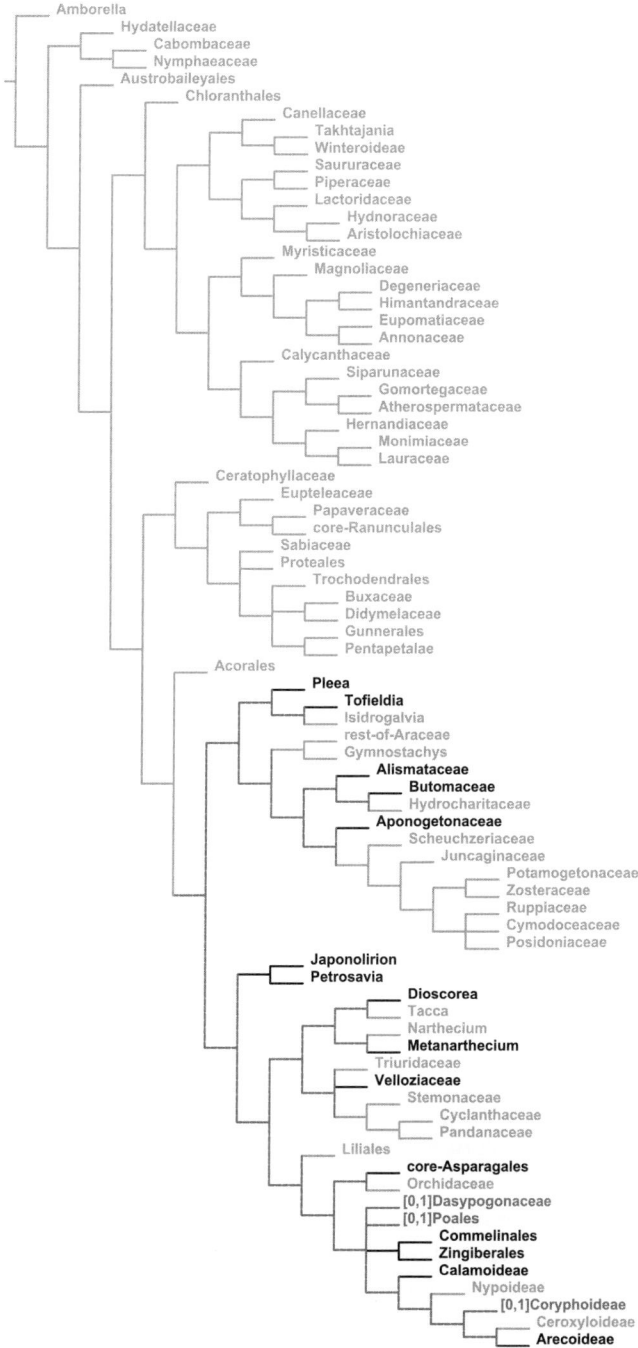

Fig 3.3B The same; Aponogetonaceae are interpreted as possessing reduced septal nectaries. Colour version to be found in colour plate section.

Narthecium and *Metanarthecium* as representatives of the family in the present paper. On the other hand, the analysis of Fuse et al. (2012; see also Zhao et al, 2012) placed *Metanarthecium* as sister to the rest of Nartheciaceae. If we accept this result and add *Aletris* to our tree, then with ACCTRAN optimization a single origin of septal nectaries in monocots is found (tree not shown).

3.4 Evolution of apocarpy in eudicots

A detailed analysis of gynoecium evolution in eudicots is beyond the scope of the present paper. However, it is useful to provide a general overview of this issue, since monocots and eudicots represent the two largest – and possibly closely related – angiosperm clades. For the early-divergent eudicot order Ranunculales, it remains unclear based on our MP character optimizations whether the apocarpous gynoecium without postgenital intercarpellary fusion represents a plesiomorphy or a reversal. However, these optimizations leave no ambiguity regarding the cases when this condition is present in some Pentapetalae, which can only be interpreted as reversals. Even if we rescore Pentapetalae as possessing an ancestrally apocarpous gynoecium without postgenital intercarpellary fusion (trees not shown), the conclusion remains the same. It is sufficient that Sabiaceae, Trochodendraceae, Gunneraceae, Myrothamnaceae and Buxaceae all possess congenitally united carpels to yield syncarpy as ancestral for Pentapetalae in the most parsimonious reconstructions, regardless of the precise placement of the apocarpous Proteales. In general, it appears that the occurrence of species-poor basal grades such as the ANITA- or ANA-grade for angiosperms and the basal eudicot grade for eudicots 'simplify' the task of ancestral character reconstruction. But are we confident that such reconstructions necessarily reflect real evolutionary scenarios? Could they be biased by nonrandom taxon extinction and/or a bottleneck effect? In our coding, we do not score the relative degree of congenital intercarpellary fusion. In the earliest-divergent lineages of eudicots with united carpels, congenital intercarpellary fusion is present only in the basal (proximal) portion of the gynoecium (Endress, 2010). Therefore, it may be better to say that MP reconstructions reveal the ancestral gynoecium morphology of Pentapetalae as proximally syncarpous.

Endress et al. (1983) listed 20 core eudicot families that contain at least one genus with apocarpous or nearly apocarpous gynoecia, which should be considered as secondarily apocarpous if the scenario outlined above is accepted. The family Leguminosae with predominantly monomerous but occasionally nonmonomerous gynoecia with free carpels can be added to this list. These families are widely scattered across the angiosperm phylogenetic tree, suggesting strong homoplasy in this character. At least for two families (Dilleniaceae and Rosaceae), detailed phylogenetic studies (Potter et al., 2007; Horn, 2009) indicate multiple origins of syncarpy derived from apocarpy (i.e. from secondary apocarpy, if the scenario outlined above is accepted). In the case of Rosaceae, subsequent reversals to (tertiary?) apocarpy in a primarily syncarpous clade are suggested. It would be difficult to propose a hypothesis on biological (adaptive) significance of such multiple changes in the direction of character transformation.

Armbruster et al. (2002) provided an MP reconstruction based on molecular phylogenetic data of gynoecium evolution in angiosperms with emphasis on magnoliids and rosids. They interpreted the results as providing evidence for numerous independent gains of syncarpy, with very few reversals to secondary apocarpy. However, their analysis suffers from methodological problems because they accepted several independent character states for the different kinds of syncarpy, based on presence or absence and type of compitum. They used up to eight character states and analysed the character as unordered. According to their scoring, members of the basal eudicot grade (Buxaceae, Sabiaceae, Trochodendaceae) possess different kinds of syncarpy and were scored in three different ways, which resulted in the conclusion that the ancestor of Pentapetalae was possibly apocarpous without postgenital intercarpellary connection. Furthermore, *Gunnera* was scored as 'functionally unicarpellate' (Armbruster et al., 2002), which is correct in a biological sense, but since the gynoecium is bicarpellate and pseudomonomerous (Endress and Igersheim, 1999; Endress, 2010), this scoring will not allow a clear homology assessment. In addition, Armbruster et al. (2002) did not include *Myrothamnus*, which has a clearly syncarpous gynoecium. Similar criticisms can be made against the analysis of gynoecium evolution in the 'Nitrogen-Fixing Clade', which also concluded that apocarpy is primitive (Fig 3 in Armbruster et al., 2002).

A potential close relationship between eudicots and monocots requires comparison between gynoecia with united carpels in basal members of both clades. As in basal eudicots, the occurrence of an internal compitum is unstable among basal monocots (Buzgo and Endress, 2000; Igersheim et al., 2001; Remizowa et al., 2006, 2008). Among basal eudicots, the gynoecium of *Sabia* (Sabiaceae) develops by a combination of congenital and postgenital intercarpellary fusion (Endress and Igersheim, 1999), as in many monocots. In Trochodendraceae and some Buxaceae, nectaries develop on the carpels, though they differ from monocot septal nectaries (Smets, 1988; Endress, 2010).

3.5 Outlook

The present study confirms our earlier conclusion (Remizowa et al., 2010) that MP optimizations of gynoecium character evolution in monocots are highly sensitive to relatively minor changes in tree topology and single-taxon changes in character scoring. This is the case with respect to the question whether congenital inter-carpellary fusion (syncarpy) represents a synapomorphy of monocots and whether the occurrence of septal nectaries is a synapomorphy of all monocots except *Acorus*. With respect to tree topology, the rooting of Alismatales has a major impact on gynoecium character reconstructions. This is not surprising because the gynoecia of Tofieldiaceae and Araceae are so different. In this respect, it is important to note that a basal position for Tofieldiaceae is only weakly supported in the analysis of Iles et al. (Chapter 1, this volume), and the branch leading to all Alismatales except Tofieldiaceae is very short. However, even if we show the basal relationships in Alismatales as a trichotomy (Araceae, Tofieldiaceae, the rest), the resulting optimizations remain similar to those with a basal position for Tofieldiaceae (not shown).

The question of stability of tree topology is significant for morphological speculation in general, not only with respect to the particular question addressed here. Some significant aspects of the tree topologies employed here (such as the positions of Ceratophyllaceae and Tofieldiaceae) are inferred from multi-gene analyses only. The positions of these taxa are different or unresolved in analyses that employ fewer molecular markers. For example, some earlier analyses placed *Ceratophyllum* as sister to the monocots (e.g. Qiu et al., 2000, Zanis et al., 2002, see however Qiu et al., 2005); according to Iwamoto and Izumidate (2010), morphology favours this topology.

The fact that some phylogenetically equivocal nodes are highly significant for reconstructions of character evolution means that further studies of angiosperm phylogeny are needed. On the other hand, it is possible that we are approaching a stage when we have learned almost all we can concerning macrophylogenetics using molecular data, at least within angiosperms (Sokoloff and Timonin, 2007). The phylogenetic placement of early-divergent and taxonomically isolated lineages such as *Ceratophyllum* could remain tentative even with greatly increased sampling of molecular markers (see also Goremykin et al., 2009). Among critical taxa, the placement of *Chloranthus* in analyses of complete plastome data depends on the method of phylogeny reconstruction (Logacheva et al., 2008). It is instructive that even with an extremely large (83 gene) data set, the relative positions of Sabiaceae and Proteales (two basal eudicot lineages with contrasting gynoecium morphology) remain ambiguous (Moore et al., 2010, Soltis et al., 2010). In the latter example, there remains little scope for usefully increasing taxon sampling, as nearly all the major basal eudicot clades are already sampled with respect to complete plastid genomes. In the light of these observations, it is likely that parsimonious reconstructions of character evolution will remain ambiguous with respect to several highly important questions, even with increased knowledge of molecular phylogenetics. Thus, alternative approaches are needed in evolutionary plant morphology. As Endress (2002) observed, character optimization using existing software provides only an initial framework for evolutionary interpretations, and apparent evolutionary directions should be discussed in a biological context.

Assessing the functional and adaptive significance of evolutionary transformations is clearly important. In general, we know much less about the functional role of morphological differences in plants than in animals, partly because of the relatively higher degree of synorganization in animals. In plants, syncarpy is a relatively rare example of a phylogenetically important character whose biological significance is well explored (Endress, 1982, 2011; Endress et al., 1983; Armbruster et al., 2002). Compitum formation is an important adaptive advantage of syncarpy. Some taxa that lack congenital intercarpellary fusion possess an extragynoecial compitum, and others form a compitum in the apical region of postgenitally united carpels. It would therefore be interesting to understand the functional bases (if any) for apocarpous gynoecia that lack an obvious extragynoecial compitum or postgenital fusion of the carpel tips in some derived monocot and eudicot lineages.

In most palms, only one ovule per flower develops into a seed, irrespective of the presence or absence of intercarpellary fusion and a compitum, so the presence of a compitum has no selective advantage. Pollen-tube growth through the receptacle has been documented in some species of the apocarpous (lacking postgenital intercarpellary connection) monocot families Alismataceae and Triuridaceae (Márquez-Guzmán et al., 1993; Wang et al., 2002, 2006; 2012; Espinosa-Matias et al., 2012), including *Lacandonia* (Triuridaceae), which possesses bisexual and cleistogamous flowers as well as in some nonmonocot angiosperms (Wang et al., 2012). It is important to clarify details of pollen-tube growth in dioecious species of Triuridaceae (see also Endress, 2011; Espinosa-Matias et al., 2012). In eudicots, the two largest apocarpous families are Rosaceae and Leguminosae. Most legumes are unicarpellate, which is biologically equivalent to syncarpy with a compitum (Endress, 1982). The case of Rosaceae is more problematic. Some Rosaceae (e.g. *Fragaria*, *Potentilla*) are characterized by a polymerous apocarpous gynoecium with uniovulate carpels that lack postgenital intercarpellary fusion; this gynoecium superficially resembles that of some Alismataceae and Triuridaceae, including the occurrence of gynobasic stylodia (though stylodium position could be due to a correlation with the uniovulate condition; see Endress and Matthews, 2006). It would be interesting to explore the possible occurrence of pollen-tube growth through the receptacle in Rosaceae. Wang et al. (2012) studied two members of Rosaceae and found no evidence for pollen-tube growth through the receptacle. Another critical factor might be the widespread occurrence of apomixis in Rosaceae. In members of the apocarpous genus *Rosa* that possess the *Rosa canina* type of meiosis, more genetic information is inherited maternally, so that the significance of pollen-tube competition is less important than in angiosperms with the normal type of meiosis.

It is tempting to explore the possibility that, in some angiosperm clades (e.g. Rosaceae, Triuridaceae), monomerous or pseudomonomerous gynoecia derived from syncarpous gynoecia became further polymerized to give rise to what super-ficially appears to be polymerous apocarpous forms. Within this highly speculative interpretation, structures that appear to be individual free carpels could therefore be viewed as highly reduced pseudomonomerous gynoecia. This unorthodox interpretation assumes that a single flower could contain more than one gynoecium. This scenario was explored with respect to the origin of Triuridaceae (Rudall et al., 2005; Rudall and Bateman, 2006). It is interesting that pseudomonomery is common in families of Rosales other than Rosaceae (see Eckardt, 1937). Interpretation of pseudomonomery in *Urtica* is not supported by convincing developmental data (Payer, 1857). More data are needed to test this hypothesis, and it is currently unlikely that it works for Rosales. From what we know about the pluriovulate carpels of Rosaceae, they appear to be perfectly normal angiosperm carpels, without any indication of a complex origin through a pseudomonomerous pathway. Shamrov and Yandovka (2008) suggested that the gynoecium of *Cerasus vulgaris* in Rosaceae is pseudomonomerous, but in our view these data could be explained by incomplete fasciation rather than pseudomo-nomery. Furthermore, *Cerasus* does not occupy a basal position in the phylogeny of Rosaceae (Potter et al., 2007).

Finally, Remizowa et al. (2010) speculated that the correlation between post-genital intercarpellary fusion and septal nectary formation could be due to similar genetic regulation of early steps of specialization of epidermal cells in both cases. This hypothesis is testable using developmental genetics. Thus, a potentially promising – but currently technically difficult – approach is to examine character suites and correlations between different characters conditioned by regulation of plant development (i.e. evolutionary developmental genetics, or evo-devo).

3.6 Acknowledgements

We are indebted to Sean Graham and Will Iles for providing us with their phylogenetic tree published in this volume. We thank Peter Endress and an anonymous reviewer for their comments on the manuscript, and Paul Wilkin for organising the symposium. The work of DDS and MVR is supported by the Russian Foundation for Basic Research, grant No. 09-04-01155.

3.7 References

APG III (Angiosperm Phylogeny Group III). (2009). An update of the Angiosperm Phylogeny Group classification for the orders and families of flowering plants: APG III. *Botanical Journal of the Linnean Society*, **161**, 105–121.

Armbruster, W. S., Debevec, E. M. and Willson, M. F. (2002). Evolution of syncarpy in angiosperms: theoretical and phylogenetic analyses of the effects of carpel fusion on offspring quantity and quality. *Journal of Evolutionary Biology*, **15**, 657–672.

Azuma, H. and Tobe, H. (2011). Molecular phylogenetic analyses of Tofieldiaceae (Alismatales): family circumscription and intergeneric relationships. *Journal of Plant Research*, **124**, 349–357.

Buzgo, M. (2001). Flower structure and development of Araceae compared with alismatids and Acoraceae. *Botanical Journal of the Linnean Society*, **136**, 393–425.

Buzgo, M. and Endress, P. K. (2000). Floral structure and development of Acoraceae and its systematic relationships with basal angiosperms. *International Journal of Plant Sciences*, **161**, 23–41.

Cantino, P. D., Doyle, J. A., Graham, S. W. et al. (2007). Towards a phylogenetic nomenclature of Tracheophyta. *Taxon*, **56**, 822–846.

Chase, M. W. (2004). Monocot relationships: an overview. *American Journal of Botany*, **91**, 1645–1655.

Chase, M. W., Fay, M. F., Devey, D. S. et al. (2006). Multigene analyses of monocot relationships: a summary. *Aliso*, **22**, 63–75.

Chen, J. M., Chen, D., Gituru, W. R., Wang, Q. F. and Guo, Y. H. (2004). Evolution of apocarpy in Alismatidae using phylogenetic evidence from chloroplast *rbc*L gene sequence data. *Botanical Bulletin of Academia Sinica*, **45**, 33–40.

Couvreur, T. L. P., Richardson, J. E., Sosef, M. S. M., Erkens, R. H. J. and Chatrou, L. W. (2008). Evolution of syncarpy and other morphological characters in African Annonaceae: A posterior mapping approach. *Molecular Phylogenetics and Evolution*, **47**, 302–318.

Cronquist, A. (1981). *An Integrated System of Classification of Flowering Plants*. New York: Columbia University Press.

Cronquist, A. (1988). *The Evolution and Classification of Flowering Plants*. New York: New York Botanical Garden.

Dahlgren, R. M. T, Clifford, H. T. and Yeo, P. F. (1985). *The Families of the Monocotyledons*. Berlin: Springer-Verlag.

Daumann, E. (1970). Das Blütennektarium der Monocotyledonen unter besonderer Berücksichtigung seiner systematischen und phylogenetischen Bedeutung. *Feddes Repertorium*, **80**, 463–590.

Davis, J. I., Stevenson, D. W., Petersen, G. et al. (2004). A phylogeny of the monocots, as inferred from *rbc*L and *atp*A sequence variation, and a comparison of methods for calculating jackknife and bootstrap values. *Systematic Botany*, **29**, 467–510.

Doust, A. N. and Drinnan, A. N. (2004). Floral development and molecular phylogeny support the generic status of *Tasmannia* (Winteraceae). *American Journal of Botany*, **91**, 321–331.

Doyle, J. A. and Endress, P. K. (2000). Morphological phylogenetic analysis of basal angiosperms: comparison and combination with molecular data. *International Journal of Plant Sciences*, **161**(Suppl.), S121–S153.

Doyle, J. A. and Endress, P. K. (2011). Tracing the early evolutionary diversification of the angiosperm flower. In *Flowers on the Tree of Life*, ed. L. Wanntorp and L. P. Ronse De Craene. Cambridge: Cambridge University Press, pp. 88–119.

Eckardt, T. (1937). Untersuchungen über Morphologie, Entwicklungsgeschichte und systematische Bedeutung des pseudomonomeren Gynoeceums. *Nova Acta Leopoldina*, **5**, 1–112.

Endress, P. K. (1982). Syncarpy and alternative modes of escaping disadvantages of apocarpy in primitive angiosperms. *Taxon*, **31**, 48–52.

Endress, P. K. (1995). Major evolutionary traits of monocot flowers. In *Monocotyledons: Systematics and Evolution*, ed. P. J. Rudall, P. J. Cribb, D. F. Cutler and C. J. Humphries. Kew: Royal Botanic Gardens, pp. 43–79.

Endress, P. K. (1997). Evolutionary biology of flowers: prospects for the next century. In *Evolution and Diversification of Land Plants*, ed. K. Iwatsuki and P. H. Raven. Tokyo: Springer, pp. 99–119.

Endress, P. K. (2001a). Origins of flower morphology. *Journal of Experimental Zoology*, **291** B, 105–115.

Endress, P. K. (2001b). The flowers in extant basal angiosperms and inferences on ancestral flowers. *International Journal of Plant Sciences*, **162**, 1111–1140.

Endress, P. K. (2002). Morphology and angiosperm systematics in the molecular era. *Botanical Review*, **68**, 545–570.

Endress, P. K. (2010). Flower structure and trends of evolution in eudicots and their major subclades. *Annals of the Missouri Botanical Garden*, **97**, 541–583.

Endress, P. K. (2011). Evolutionary diversification of the flowers in angiosperms. *American Journal of Botany*, **98**, 370–396.

Endress, P. K. and Doyle, J. A. (2009). Reconstructing the ancestral angiosperm flower and its initial specializations. *American Journal of Botany*, **96**, 22–66.

Endress, P. K. and Igersheim, A. (1997). Gynoecium diversity and systematics of the Laurales. *Botanical Journal of the Linnean Society*, **125**, 93–168.

Endress, P. K. and Igersheim, A. (1999). Gynoecium diversity and systematics of the basal eudicots. *Botanical Journal of the Linnean Society*, **130**, 305–393.

Endress, P. K. and Igersheim, A. (2000). Gynoecium structure and evolution in basal angiosperms. *International Journal of Plant Sciences*, **161** (Suppl.), S211–S223.

Endress, P. K. and Matthews, M. L. (2006). First steps towards a floral structural characterization of the major rosid subclades. *Plant Systematics and Evolution*, **260**, 223–251.

Endress, P. K., Jenny, M. and Fallen, M. E. (1983). Convergent elaboration of apocarpous gynoecia in higher advanced dicotyledons. *Nordic Journal of Botany*, **3**, 293–300.

Endress, P. K., Igersheim, A., Sampson, F. B. and Schatz, G. E. (2000). Floral structure of *Takhtajania* and its systematic position in Winteraceae. *Annals of Missouri Botanical Garden*, **87**, 347–365.

Espinosa-Matías, S., Vergara-Silva, F., Vázquez-Santana, S., Martinez-Zurita, E. and Márquez-Guzman, J. (2012). Complex patterns of morphogenesis, embryology and reproduction in *Triuris brevistylis*, a species of Triuridaceae (Pandanales) closely related to *Lacandonia schismatica*. *Botany*, **90**, 1133–1151.

Fuse, S., Lee, N.S. and Tamura, M.N. (2012). Biosystematic studies on the family Nartheciaceae (Dioscoreales) I. Phylogenetic relationships, character evolution and taxonomic re-examination. *Plant Systematics and Evolution*, **298**, 1575–1584.

Givnish, T. J., Pires, J. C., Graham, S. W. et al. (2006). Phylogeny of the monocots based on *ndh*F: evidence for widespread concerted convergence. In *Monocots: Comparative Biology and Evolution (excluding Poales)*, ed. T. J. Columbus, E. A. Friar, C. W. Hamilton et al. Claremont, CA: Rancho Santa Ana Botanic Garden, pp. 28–51.

Goremykin, V. V., Viola, R. and Hellwig, F. H. (2009). Removal of noisy characters from chloroplast genome-scale data suggests revision of phylogenetic placement of *Amborella* and *Ceratophyllum*. *Journal of Molecular Evolution*, **68**, 197–204.

Graham, S. W., Zgurski, J. M., McPherson, M. A. et al. (2006). Robust inference of monocot deep phylogeny using an expanded multigene plastid data set. In *Monocots: Comparative Biology and Evolution (excluding Poales)*, ed. T. J. Columbus, E. A. Friar, C. W. Hamilton et al. Claremont, CA: Rancho Santa Ana Botanic Garden, pp. 3–20.

Hartl, D. and Severin, I. (1981). Verwachsungen in Umfeld des Griffels bei *Allium*, *Cyanastrum* und *Heliconia* und den Monocotylen allgemein. *Beiträge zur Biologie der Pflanzen*, **55**, 235–260.

Haston, E., Richardson, J. E., Stevens, P. F., Chase, M. W. and Harris, D. J. (2007). A linear sequence of Angiosperm Phylogeny Group II families. *Taxon*, **56**, 7–12.

van Heel, W. A. (1988). On the development of some gynoecia with septal nectaries. *Blumea*, **33**, 477–504.

Horn, J. W. (2009). Phylogenetics of Dilleniaceae using sequence data from four plastid loci (*rbc*L, *inf*A, *rps*4, *rpl*16 Intron). *International Journal of Plant Sciences*, **170**, 794–813

Igersheim, A. and Endress, P. K. (1997). Gynoecium diversity and systematics of the Magnoliales and winteroids. *Botanical Journal of the Linnean Society*, **124**, 213–271.

Igersheim, A. and Endress, P. K. (1998). Gynoecium diversity and systematics of the paleoherbs. *Botanical Journal of the Linnean Society*, **127**, 289–370.

Igersheim, A., Buzgo, M. and Endress, P. K. (2001). Gynoecium diversity and systematics in basal monocots. *Botanical Journal of the Linnean Society*, **136**, 1–65.

Iwamoto, A. and Izumidate, R. (2010). Floral and vegetative development in *Ceratophyllum demersum* (Ceratophyllaceae). In *Botany 2010. Scientific Abstracts*. Providence, p. 38.

Karol, K. G., Suh, Y., Schatz, G. E. and Zimmer, E. A. (2000). Molecular evidence for the phylogenetic position of *Takhtajania* in the Winteraceae: Inference from nuclear ribosomal and chloroplast gene spacer sequences. *Annals of the Missouri Botanical Garden*, **87**, 414–432.

Kim, S., Soltis, D. E., Soltis, P. S., Zanis, M. J. and Suh, Y. (2004). Phylogenetic relationships among early-diverging eudicots based on four genes: were the eudicots ancestrally woody? *Molecular Phylogenetics and Evolution*, **31**, 16–30.

Leinfellner, W. (1950). Der Bauplan des synkarpen Gynözeums. *Österreichische Botanische Zeitschrift*, **97**, 403–436.

Logacheva, M. D., Samigullin, T. H., Dhingra, A. and Penin, A. A. (2008). Comparative chloroplast genomics and phylogenetics of *Fagopyrum esculentum* ssp. *ancestrale* – a wild ancestor of cultivated buckwheat. *BMC Plant Biology*, **8**, 59.

Márquez-Guzmán, J., Vázquez-Santana, S., Engleman, E. M., Martínez-Mena, A. and Martínez, E. (1993). Pollen development and fertilization in *Lacandonia schismatica* (Lacandoniaceae). *Annals of Missouri Botanical Garden*, **80**, 891–897.

Matthews, M. L. and Endress, P. K. (2005). Comparative floral structure and systematics in Crossosomatales (Crossosomataceae, Stachyuraceae, Staphyleaceae, Aphloiaceae, Geissolomataceae, Ixerbaceae,

Strasburgeriaceae). *Botanical Journal of the Linnean Society*, **147**, 1–46.

Merckx, V., Schols, P., Geuten, K., Huysmans, S. and Smets, E. (2008). Phylogenetic relationships in Nartheciaceae (Dioscoreales), with focus on pollen and orbicule morphology. *Belgian Journal of Botany*, **141**, 64–77.

Moore, M. J., Soltis, P. S., Bell, C. D., Burleigh, J. G. and Soltis D. E. (2010). Phylogenetic analysis of 83 plastid genes further resolves the early diversification of eudicots. *Proceedings of the National Academy of Sciences, USA*, **107**, 4623–4628.

Nixon, K. C. (2002). *Winclada vers. 1.00.08*. Ithaca, New York. Published by the author [Distributed through http://www.cladistics.com].

Payer, J. B. (1857). *Traité d'organogénie comparée de la fleur*. Paris: Librairie de Victor Masson.

Potter, D., Eriksson, T., Evans, R. C. et al. (2007). Phylogeny and classification of Rosaceae. *Plant Systematics and Evolution*, **266**, 5–43.

Qiu, Y.-L., Lee, J., Bernasconi-Quadroni, F. et al. (2000). Phylogenetic analyses of basal angiosperms based on five genes from all three genomes. *International Journal of Plant Sciences*, **161**, S3–S27.

Qiu, Y.-L., Dombrovska, O., Lee, J. et al. (2005). Phylogenetic analyses of basal angiosperms based on nine plastid, mitochondrial, and nuclear genes. *International Journal of Plant Sciences*, **166**, 815–842.

Remizowa, M. V., Sokoloff, D. D. and Rudall, P. J. (2006). Evolution of the monocot gynoecium: evidence from comparative morphology and development in *Tofieldia, Japonolirion, Petrosavia* and *Narthecium*. *Plant Systematics and Evolution*, **258**, 183–209.

Remizowa, M. V., Sokoloff, D. D. and Kondo, K. (2008). Floral evolution in the monocot family Nartheciaceae (Dioscoreales): evidence from anatomy and development in *Metanarthecium luteo-viride* Maxim. *Botanical Journal of the Linnean Society*, **158**, 1–18.

Remizowa, M. V., Sokoloff, D. D. and Rudall, P. J. (2010). Evolutionary history of the monocot flower. *Annals of the Missouri Botanical Garden*, **97**, 617–645.

Renner, S. S., Strijk, J. S., Strasberg, D. and Thébaud, C. (2010). Biogeography of the Monimiaceae (Laurales): a role for East Gondwana and long-distance dispersal, but not West Gondwana. *Journal of Biogeography*, **37**, 1227–1238.

Rudall, P. J. (2002). Homologies of inferior ovaries and septal nectaries in monocotyledons. *International Journal of Plant Sciences*, **163**, 261–276.

Rudall, P. J. and Bateman, R. M. (2006). Morphological phylogenetic analysis of Pandanales: Testing contrasting hypotheses of floral evolution. *Systematic Botany*, **31**, 223–238.

Rudall, P. J., Cunniff, J., Wilkin, P. and Caddick, L. R. (2005). Evolution of dimery, pentamery and the monocarpellary condition in Stemonaceae (Pandanales). *Taxon*, **54**, 701–711.

Rudall, P. J., Ryder, R. A. and Baker, W. J. (2011). Comparative gynoecium structure and multiple origins of apocarpy in coryphoid palms (Arecaceae). *International Journal of Plant Sciences*, **172**, 674–690.

Sattler, R. (1977). Kronröhrenentstehung bei *Solanum dulcamara* und 'kongenitale Verwachsung'. *Berichte der Deutschen Botanischen Gesellschaft*, **90**, 29–38.

Sattler, R. (1978). 'Fusion' and 'continuity' in floral morphology. *Notes from the Royal Botanic Garden Edinburgh*, **36**, 397–405.

Shamrov, I. I. and Yandovka, L. F. (2008). Development and structure of gynoecium and ovule in *Cerasus vulgaris* (Rosaceae). *Botanichesky Zhurnal*, **93**, 902–914.

Smets, E. (1988). La présence des 'nectaria persistentia' chez les Magnoliophytina (Angiospermes). *Candollea*, **43**, 709–716.

Smets, E. F., Ronse De Craene, L. P., Caris, P. and Rudall, P. J. (2000). Floral nectaries in monocotyledons: distribution and evolution. In *Monocots: Systematics and Evolution*, ed. K. L. Wilson and D. A. Morrison. Melbourne: CSIRO, pp. 230–240.

Sokoloff, D. D. and Timonin, A. C. (2007). Morphological and molecular data on the origin of angiosperms: On a way to a synthesis. *Journal of General Biology*, **68**, 83–97.

Sokoloff, D. D., Remizowa, M. V., Linder, H. P. and Rudall, P. J. (2009). Morphology and development of the gynoecium in Centrolepidaceae: the most remarkable range of variation in Poales. *American Journal of Botany*, **96**, 1925–1940.

Soltis, D. E., Moore, M. J., Burleigh, J. G., Bell, C. D. and Soltis, P. S. (2010). Assembling the angiosperm tree of life: Progress and future prospects. *Annals of the Missouri Botanical Garden*, **97**, 514–526.

Stevens, P. F. (2001 onwards). *Angiosperm Phylogeny Website. Version 9, June 2008* [and more or less continuously updated since]. http://www.mobot.org/MOBOT/research/APweb/

Takhtajan, A. L. (1966). *A System and Phylogeny of the Flowering Plants*. Moscow/Leningrad: Nauka.

Takhtajan, A. L. (1980). Degeneriaceae. In *Zhizn' rastenij [Plant Life], Volume 5, Tsvetkovye rastenia [Flowering Plants], Part 1*, ed. A. L. Takhtajan. Moscow: Prosveshchenie, pp. 121–125. [In Russian].

Takhtajan, A. L. (1987). *Systema Magnoliophytorum*. Leningrad: Nauka.

Takhtajan, A. L. (2009). *Flowering Plants*. Ed. 2. New York: Springer-Verlag.

Tamura, M. N., Yamashita, J., Fuse, S. and Haraguchi, M. (2004). Molecular phylogeny of monocotyledons inferred from combined analysis of plastid *matK* and *rbcL* gene sequences. *Journal of Plant Research*, **117**, 109–120.

Timonin, A. C. (2002). Sattler's dynamic morphology: an acme or a reverie? *Wulfenia*, **9**, 9–18.

Verbeke, J. A. (1992). Fusion events during floral morphogenesis. *Annual Reviews of Plant Physiology and Plant Molecular Biology*, **43**, 583–598.

Wang, X. F., Tao, Y. B. and Lu, Y. T. (2002). Pollen tubes enter neighbouring ovules by way of receptacle tissue, resulting in increased fruit-set in *Sagittaria potamogetifolia* Merr. *Annals of Botany*, **89**, 791–796.

Wang, X.-F., Tan, Y.-Y., Chen, J.-H. and Lu, Y.-T. (2006). Pollen tube reallocation in two preanthesis cleistogamous species, *Ranalisma rostratum* and *Sagittaria guyanensis* ssp. *lappula* (Alismataceae). *Aquatic Botany*, **85**, 233–240.

Wang, X.-F., Armbruster, W. S. and Huang, S.-Q. (2012). Extragynoecial pollen tube growth in apocarpous angiosperms is phylogenetically widespread and probably adaptive. *New Phytologist*, **193**, 253–260.

Zanis, M. J., Soltis, D. E., Soltis, P. S., Mathews, S. and Donoghue, M. J. (2002). The root of the angiosperms revisited. *Proceedings of the National Academy of Sciences, USA*, **99**, 6848–6853.

Zhao, Y.-M., Wang, W. and Zhang, S.-R. (2012). Delimitation and phylogeny of *Aletris* (Nartheciaceae) with implications for perianth evolution. *Journal of Systematics and Evolution* **50**, 135–145.

4

Diversification of pollen and tapetum in early-divergent monocots

CAROL A. FURNESS

4.1 Introduction

Monocot pollen has traditionally been viewed as monosulcate, dull and boring when compared with pollen of other seed plants. This apparent lack of diversity can at least partially be explained by the fact that monocot pollen does not acetolyse well and has therefore been neglected in studies of comparative pollen morphology, though some notable studies were carried out, for example, Radulescu (1970, 1973), Argue (1974, 1976) and Zavada (1983). However, advances in monocot phylogenetics during the late twentieth and early twenty-first centuries (e.g. Chase et al., 1995, 2000, 2006; Stevenson and Laconte, 1995; Davis et al., 2004) have led to a renaissance in the study of monocot pollen and monocots are now probably the palynologically best-known major angiosperm clade. Detailed studies of exine ontogeny, fossil pollen and the pollen morphology of some large, taxonomically difficult monocot genera have provided useful systematic data (e.g. Le Thomas et al., 2001; Schols et al., 2001, 2003, 2005a, 2005b; Hesse and Zetter, 2007; Smith et al., 2009). Pollen and anther characters including microsporogenesis, pollen apertures and tapetum type have been shown to be of value in monocot systematics, particularly in the lilioid orders Asparagales, Liliales and Dioscoreales, and also in some commelinids such as Arecaceae and Cyperaceae (Rudall et al., 1997, 2000; Caddick et al., 1998; Harley and Baker, 2001; Simpson et al., 2003). Thus it is reasonable to assume that pollen could help to elucidate the problematic

Early Events in Monocot Evolution, eds P. Wilkin and S. J. Mayo. Published by Cambridge University Press. © The Systematics Association 2013.

systematics of Alismatales. Reciprocally, improved phylogenies of early-divergent monocots (Chen et al., 2004; Chase et al., 2006) could help to shed light on pollen evolution within this group.

The aim of this paper is to investigate pollen evolution in Alismatales by comparing their pollen and tapetum characters with those of *Acorus*, the putative sister to all other monocots (Chase et al., 2006), and the lilioid orders Dioscoreales, Pandanales and Petrosaviales. Alismatales represent a key clade in homology assessments because they are sister to all other monocots except Acorales (Chase et al., 2006). They formerly contained 14 families of mainly aquatic or emergent monocots, recently reduced to 13 by the inclusion of Limnocharitaceae in Alismataceae (APG II, 2003; APG III, 2009). Recent molecular analyses indicate either two or three main clades within Alismatales; the large pantropical family Araceae (113 genera), the alismatids or aquatic clade (57 genera) and finally Tofieldiaceae (four genera), which are sister to the alismatids, at least in some analyses (Chen et al., 2004; Chase et al., 2006). The order Petrosaviales is sister to all other monocots except Alismatales and Acorales. Petrosaviales contain two familes: Japonoliriaceae and Petrosaviaceae, each with only a single genus (Chase et al., 2006). The association of Cyclanthaceae, Pandanaceae, Stemonaceae and Velloziaceae in the order Pandanales was first suggested by *rbc*L analysis (Chase et al., 1995) and confirmed by subsequent analyses, with the addition of Triuridaceae (Chase et al., 2000; Davis et al., 2004). Most of these families had formerly been associated with different monocot groups, except for Cyclanthaceae and Pandanaceae (Dahlgren and Clifford, 1982). The current circumscription of the lilioid order Dioscoreales is also relatively recent, comprising three families (Nartheciaceae, Burmanniaceae and Dioscoreaceae), following morphological and molecular analyses by Caddick et al. (2000, 2002).

Pollen and tapetal characters for each of the above orders are summarized below. These are followed by more detailed discussion of the evolution of individual characters based on optimisations (Figs 4.1–4.6) using Winclada (Nixon, 2002). Pollen and tapetum characters (Furness and Banks, 2010) were optimized onto a phylogeny of early-divergent monocots using a tree topology based on molecular phylogenetic studies (Chen et al., 2004; Chase et al., 2006; see also Remizowa et al., 2006).

4.2 Pollen and tapetal characters of early-divergent monocots

4.2.1 Alismatales

Pollen of most Alismatales is spheroidal to ellipsoidal in shape, with a wide range in diameter from *c*. 12–200 μm (Furness and Banks, 2010). Highly unusual filamentous pollen (that can be over 1000 μm long and *c*. 10–20 μm wide) characterizes the

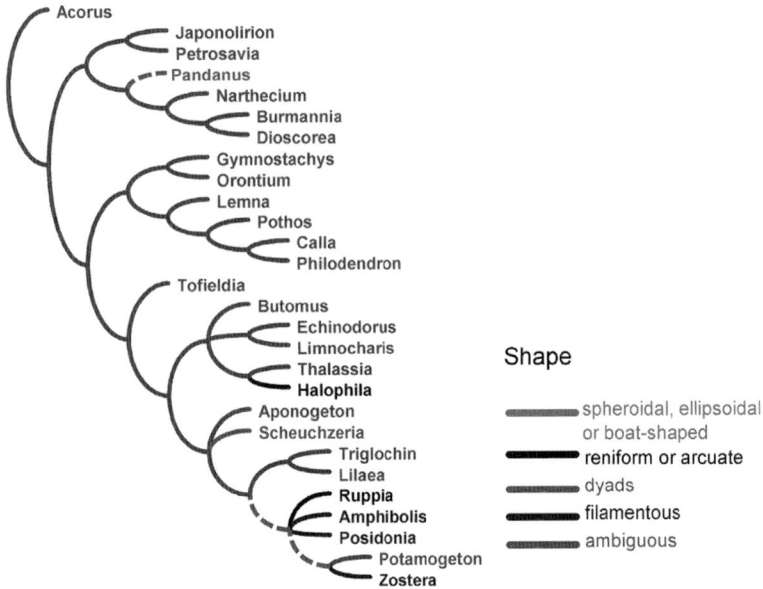

Fig 4.1 Pollen shape.

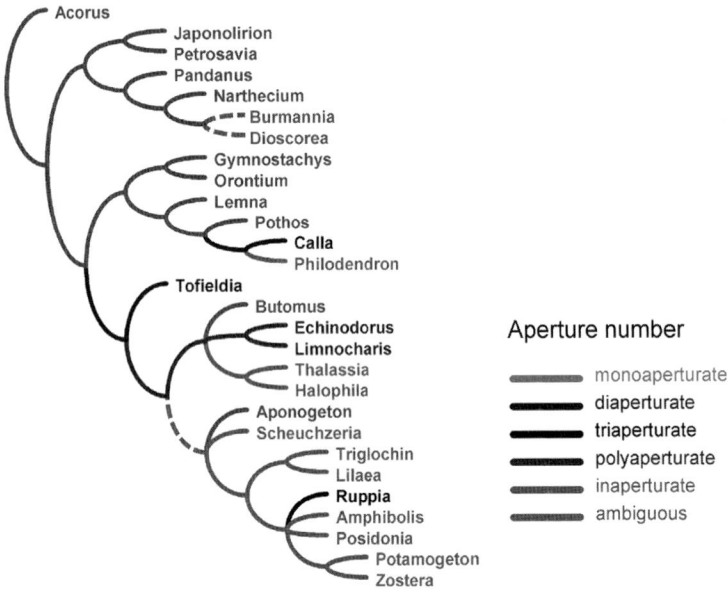

Fig 4.2 Pollen aperture number.

Figs 4.1–4.6 Pollen and tapetum characters optimized onto a phylogenetic tree of early-divergent monocots using Winclada (Nixon 2002) showing unambiguous changes. Colour versions to be found in colour plate section.

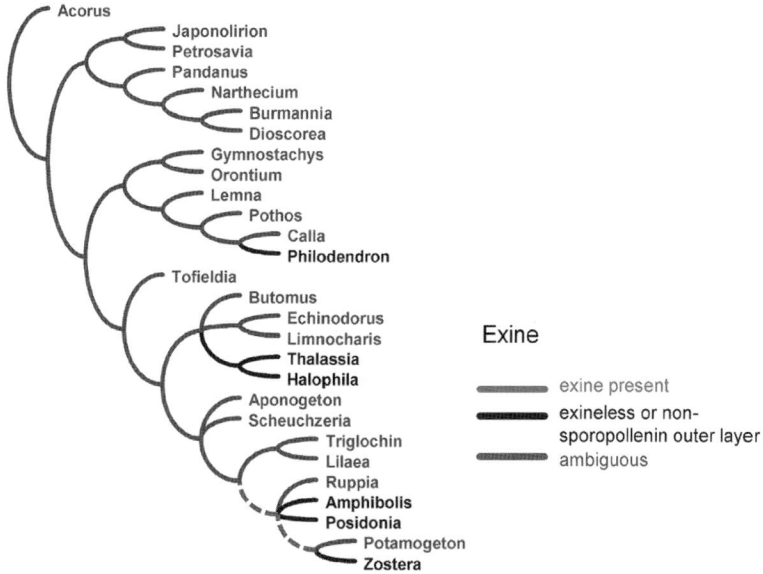

Fig 4.3 Pollen exine.

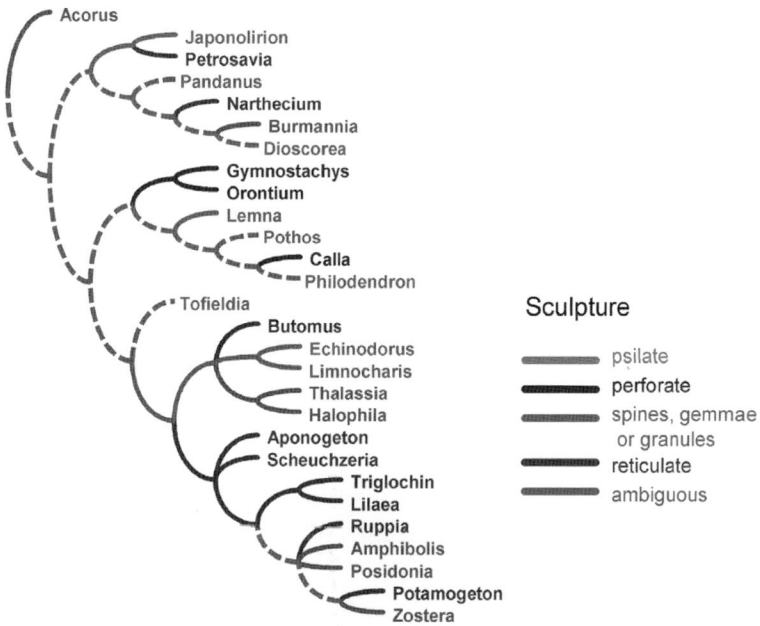

Fig 4.4 Pollen sculpture.

Figs 4.1–4.6 (*cont.*) Colour versions to be found in colour plate section.

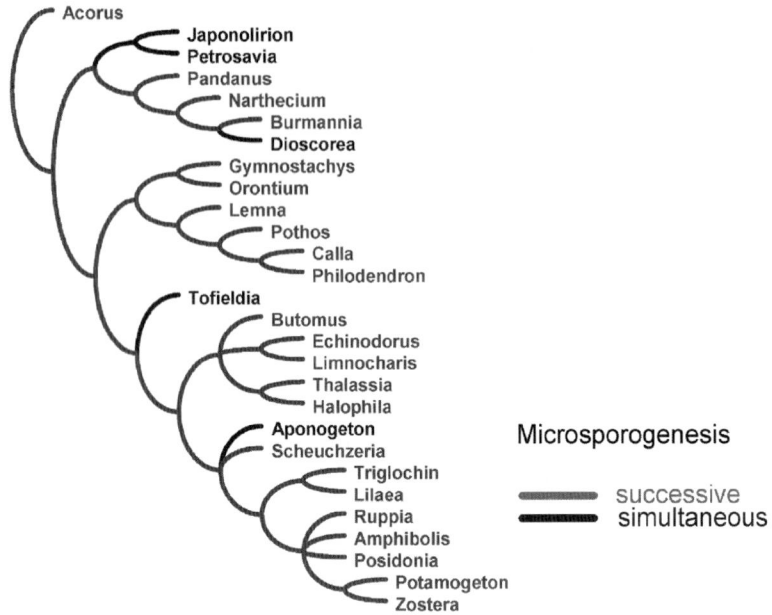

Fig 4.5 Microsporogenesis type.

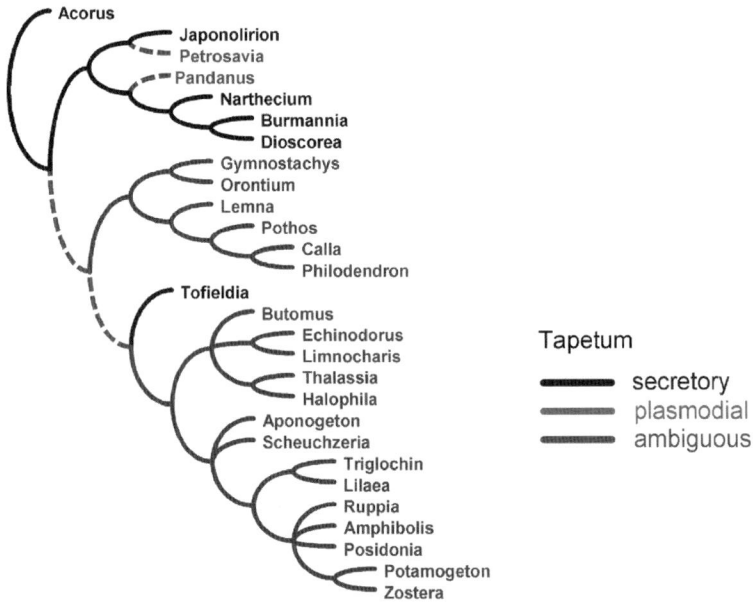

Fig 4.6 Tapetum type. (Reproduced with permission from Furness and Banks, 2010.)

Figs 4.1–4.6 (*cont.*) Colour versions to be found in colour plate section.

seagrasses (Cymodoceaceae, Posidoniaceae and Zosteraceae: Ducker and Knox, 1976; Ducker et al., 1978; Pettitt et al., 1984; Fig 4.7). The extreme length of this pollen is hypothesized to be an adaptation to hydrophilous pollination, enabling it to function as an efficient 'search vehicle' and be captured effectively by the stigma (Cox, 1983). Adaptation to hydrophilous pollination is also shown by the linear tetrads of ellipsoidal or reniform pollen grains enclosed in mucilaginous tubes produced by *Halophila* (Hydrocharitaceae), and the curved (arcuate) pollen grains of *Ruppia* (Ruppiaceae) that are held together in chains by trapped air bubbles (Verhoeven, 1979; Pettitt, 1980, 1981). The most common pollen aperture type in Alismatales is inaperturate (Furness and Banks, 2010). This has been linked to the aquatic or semi-aquatic habit of many alismatids, in that a thick protective exine is not required to prevent dessication, and thus the pollen tube can emerge from any point through the thin outer wall (omniaperturate pollen *sensu* Thanikai-moni, 1978). Other alismatids have weakly defined apertures termed 'tenuitate', for example, *Althenia* (Potamogetonaceae) and *Ruppia* (Díez et al., 1988; Harley, 2004). Pollen of the expanded Alismataceae is pantoporate, with multiple potential germination sites, though the pores may be indistinct due to their granular or spiny membranes (Chanda et al., 1988; Furness and Banks, 2010; Fig 4.8). Monosulcate pollen is relatively rare in Alismatales, particularly in alismatids, where it occurs only in *Aponogeton* (Aponogetonaceae) and *Butomus* (Butomaceae) (Furness and Banks, 2010; Fig 4.9). Unusual diaperturate pollen occurs in Tofieldiaceae (Takahashi and Kawano, 1989; Handa et al., 2001). Apertures in Araceae are diverse: mono-sulcate, ulcerate, diaperturate, zona-aperturate (ring-shaped) and inaperturate (Grayum, 1992). Exine reduction is common in Alismatales, often associated with the aquatic or semi-aquatic habit. Extreme reduction occurs in the seagrasses and some Hydrocharitaceae where sporopollenin is absent and the pollen wall is composed of polysaccharide with a smooth surface (Pettitt and Jermy, 1975; Ducker et al., 1978; Pettitt, 1981; Pettitt et al., 1984). The outer wall layer in inaperturate (omniaperturate) Aroideae (Araceae) pollen is also composed of polysaccharide and may facilitate emergence of the pollen tube in the warm, moist environment of the spadix (Hesse, 2006). Other Alismatales, for example *Potamogeton* (Potamoge-tonaceae), have a reticulate surface and the exine is reduced to muri supported by short columellae (Nunes et al., 2009). Spiny pollen characterizes the enlarged Alismataceae (Furness and Banks, 2010; Fig 4.8). As in many monocots, endexine is thin or difficult to distinguish in many Alismatales (Furness and Banks, 2010).

Pollen in most Alismatales develops from successive microsporogenesis, where a callose wall is deposited after the first meiotic division forming a distinct dyad stage, and the resulting tetrads are usually tetragonal (Fig 4.10). Exceptions are *Aponoge-ton* and *Tofieldia* (Tofieldiaceae), where microsporogenesis is simultaneous with callose deposition delayed until after the completion of the second meiotic division and the tetrads are usually tetrahedral (Huynh, 1976; Van Bruggen, 1985; Furness

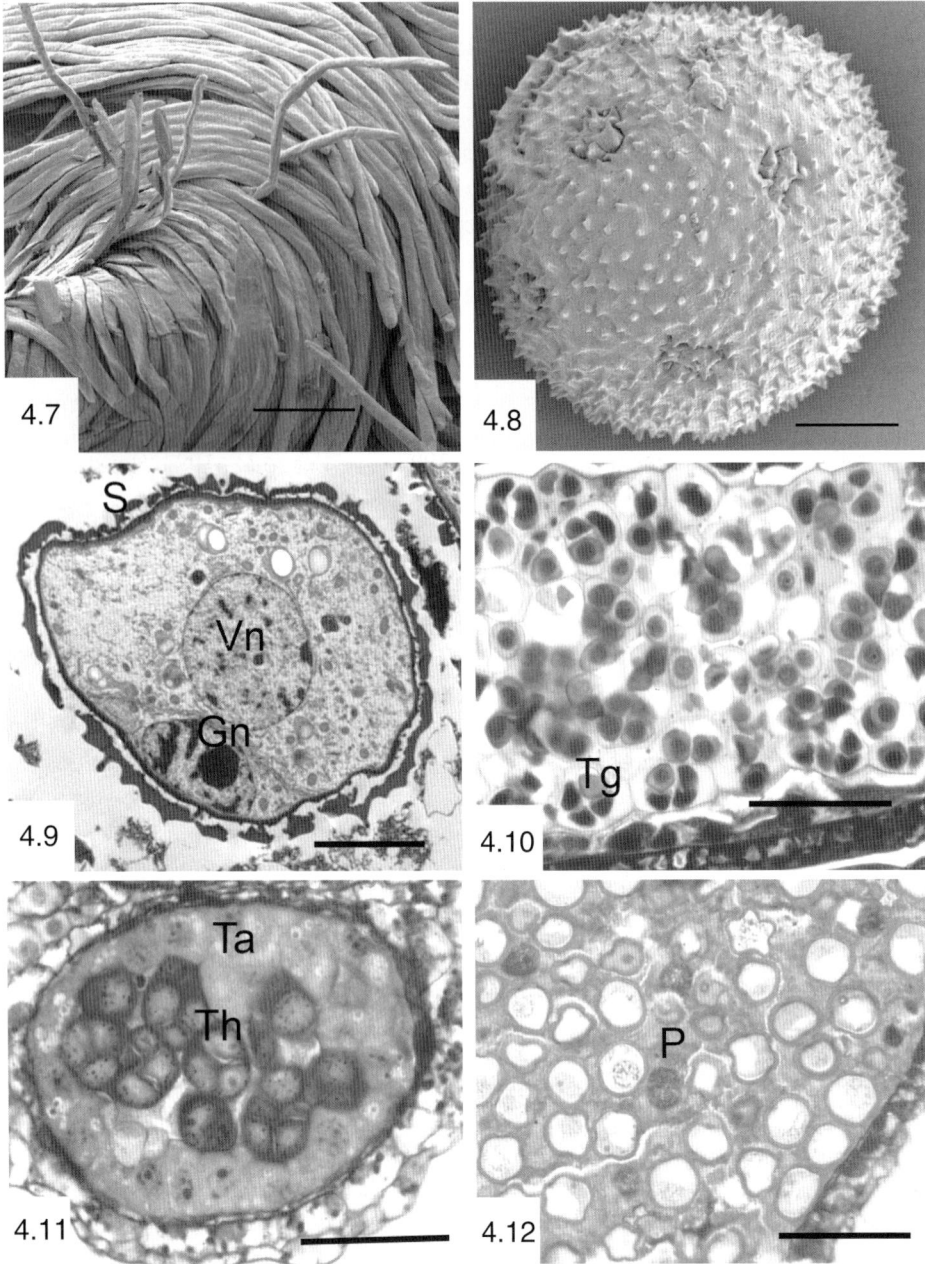

Figs 4.7–4.12 Pollen and tapetum of Alismatales.

Fig 4.7 Filamentous inaperturate pollen of *Zostera novazelandica* (Zosteraceae) in an anther locule (SEM).

Fig 4.8 Spheroidal pantoporate pollen of *Limnophyton fluitans* (Alismataceae) with a microechinate surface and pore membranes (SEM).

and Rudall, 1999; Furness and Banks, 2010; Fig 4.11). The tapetum in Alismatales is almost uniformly plasmodial with the tapetal cells intruding into the anther locule and fusing to form a multi-nucleate plasmodium surrounding the developing microspores (Fig 4.12). A secretory tapetum is however recorded in *Tofieldia* (Furness and Banks, 2010). In the secretory type the tapetal cells remain at the edge of the locule and secrete into the locule so that nutrients and sporopollenin precursors pass to the developing microspores via the locular fluid.

4.2.2 Petrosaviales

Pollen of Petrosaviales is spheroidal to ellipsoidal in shape and *c.* 25 × 22 μm in size with monosulcate apertures (Furness and Banks, 2010; Fig 4.13). *Japonolirion* (Japonoliriaceae) has gemmate pollen that lacks a tectum and the gemmae rest on the foot layer (Takahashi and Kawano, 1989; Caddick et al., 1998; Handa et al., 2001). In contrast, pollen of *Petrosavia* (Petrosaviales) is finely reticulate and semitectate with tall columellae supporting the muri (Caddick et al., 1998; Handa et al., 2001). Simultaneous microsporogenesis is recorded in both genera and a secretory tapetum in *Japonolirion* though tapetum type is unknown in *Petrosavia* (Caddick et al., 1998; Furness and Rudall, 1999; Tobe, 2008).

4.2.3 Pandanales

Pollen and tapetal characters in Pandanales were reviewed by Furness and Rudall (2006) and the description below is largely based on this work and references therein. The order contains five families and includes a range of pollen morphological diversity. Most pollen is ovoid or ellipsoidal although unusual reniform pollen also occurs, together with ovoid pollen, in all genera of Pandanaceae except *Sararanga*. Pollen is dispersed as permanent tetrads in *Vellozia* (Velloziaceae) and in all other taxa as monads with size *c.* 11–30 × 12–40 μm. Apertures are monosulcate (with an operculum in *Barbacenia*, *Talbotia*: Velloziaceae, Fig 4.14), ulcerate (*Carludovica*, *Cyclanthus*: Cyclanthaceae, Fig 4.15; Pandanaceae), or

---

Caption for Figs 4.7–12 (*cont.*)

Fig 4.9 Section through a monosulcate pollen grain of *Aponogeton distachyus* (Aponogetonaceae) in an anther locule, showing the vegetative and generative cells (TEM).

Fig 4.10 Anther locule of *Echinodorus grandiflorus* (Alismataceae) containing tetrads produced by successive microsporogenesis (LM).

Fig 4.11 Anther locule containing tetrads of *Aponogeton distachyus* produced by simultaneous microsporogenesis, surrounded by the tapetum (LM).

Fig 4.12 Vacuolate microspores of *Echinodorus grandiflorus* in an anther locule surrounded by a plasmodial tapetum (LM). Gn = generative nucleus; P = plasmodium; S = sulcus; Ta = tapetum; Tg = tetragonal tetrad; Th = tetrahedral tetrad; Vn = vegetative nucleus. Scale bars 50 μm in Figs 4.7, 4.10, 4.11; 5 μm in 4.8, 4.9 and 20 μm in 4.12.

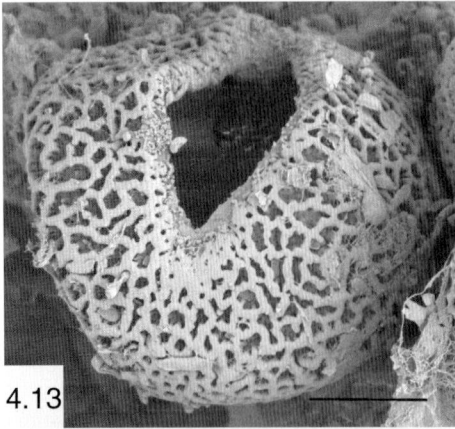

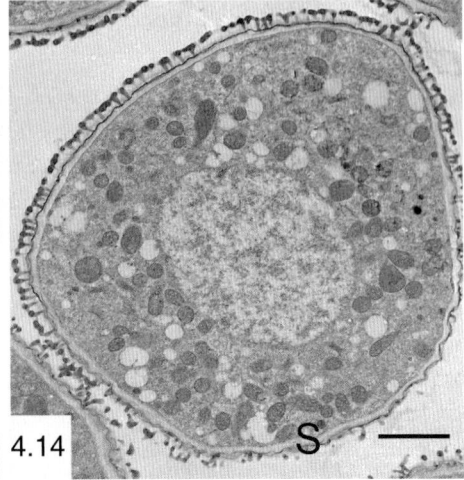

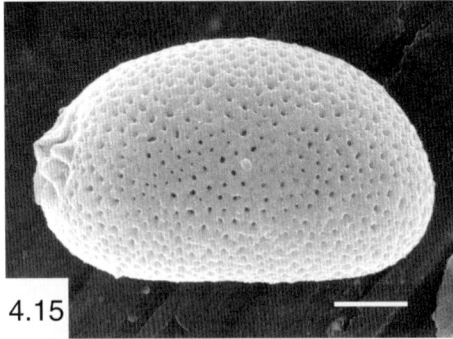

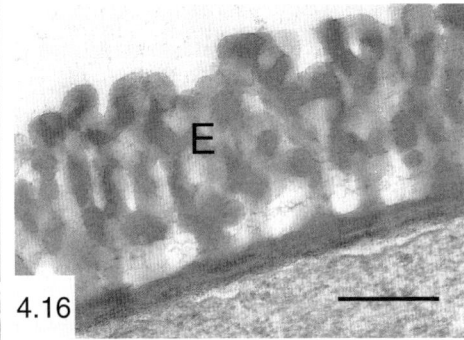

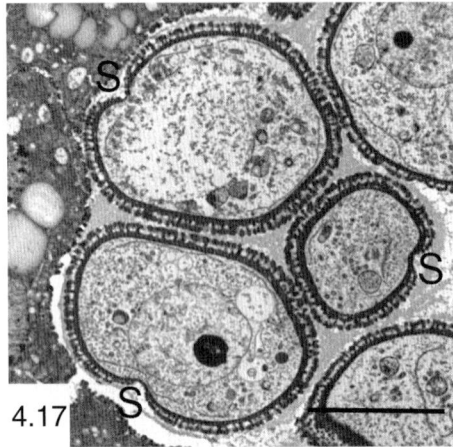

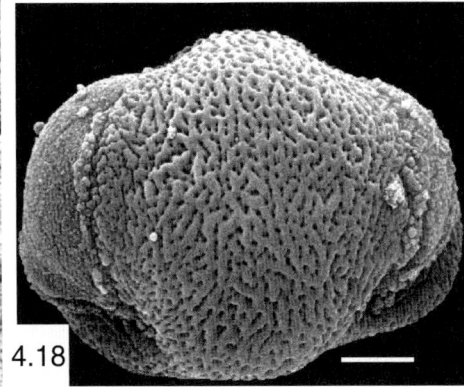

Figs 4.13–4.18 Pollen and microspores of Petrosaviales, Pandanales and Dioscoreales.

Fig 4.13 Monosulcate pollen of *Petrosavia stellaris* (Petrosaviaceae) (SEM).

Fig 4.14 Section of a monosulcate operculate microspore of *Talbotia elegans* (Velloziaceae) (TEM).

Fig 4.15 Ulcerate pollen of *Carludovica palmata* (Cyclanthaceae) (SEM).

Fig 4.16 Columellate/granular wall structure in pollen of *Stemona* sp. aff. *tuberosa* (Stemonaceae).

Fig 4.17 Section through a tetrahedral tetrad of microspores of *Dioscorea deltoides* (Dioscoreaceae) showing the distal sulci developing (TEM).

Fig 4.18 Disulculate pollen of *Dioscorea sylvatica* (Dioscoreaceae) (SEM). E = exine; S = sulcus. Scale bars 5 μm in Figs 4.13, 4.15, 4.18; 2 μm in 4.14; 0.5 μm in 4.16; and 10 μm in 4.17.

inaperturate (*Pentastemona, Stichoneuron*: Stemonaceae; Triuridaceae; *Vellozia*: Velloziaceae). Most Pandanales have columellate exines though exines are columellate/granular in some *Stemona* species (Stemonaceae, Fig 4.16) and a single layered homogeneous exine occurs in *Pandanus* (Pandanaceae). Reduction of the exine to a layer with gemmae on top occurs in *Pentastemona* and in Triuridaceae where the exine between the gemmae is extremely thin. A thin endexine occurs in pollen of most Pandanales, except for Cyclanthaceae. Perforate and reticulate are the most common surface sculpturing types though there is a great deal of variability, especially in the large genus *Pandanus*, including psilate, echinate, scabrate and verrucate. Triuridaceae and some Stemonaceae have unusual spiny-gemmate pollen.

Microsporogenesis in Pandanales is exclusively of the successive type. Records of tapetum type are somewhat confused. Most records are for the secretory type, but the plasmodial type is also recorded in some Pandanaceae and an invasive nonsyncytial type (where the tapetal cells invade the locule but do not fuse to form a multi-nucleate plasmodium) possibly occurs in some Pandanacae and Triuridaceae, and in *Talbotia* (Velloziaceae).

4.2.4 Dioscoreales

Data on pollen morphology and pollen development in Dioscoreales described below are taken from studies by Caddick et al. (1998), Schols et al. (2005a, 2005b), Merckx et al. (2008) and references therein. Pollen is spheroidal to ellipsoidal with a maximum diameter of *c*. 15–50 μm. Apertures are monosulcate (Nartheciaceae, Taccaceae, the early divergent *Stenophora* clade of *Dioscorea*: Dioscoreaceae, Fig 4.17), disulculate (*Dioscorea* excluding the *Stenophora* clade, Fig 4.18), or 1-2-porate or inaperturate (Burmanniaceae). Unusual 4–5-porate pollen occurs in *Trichopus* (Dioscoreaceae). Sculpturing is quite diverse, including striate pollen (unusual in monocots though found in some *Dioscorea* and some Araceae: Grayum, 1992), together with psilate, perforate, microreticulate, reticulate, rugulate and gemmate surfaces. Exines are usually semitectate and columellate though homogeneous exines occur in Burmanniaceae. Microsporogenesis is successive apart from Dioscoreaceae which have the simultaneous type (Fig 4.17). The tapetum is uniformly secretory.

4.2.5 Acorales

Pollen of *Acorus* (Acoraceae) is ellipsoidal, *c*. 15–20 μm in diameter, with a monosulcate aperture (Grayum, 1992; Rudall and Furness, 1997). The surface is smooth (psilate) with small scattered perforations, and the exine is composed of a thick tectum supported by short, sparse columellae, with a thin foot layer and thin endexine (Grayum, 1992; Rudall and Furness, 1997). Microsporogenesis is successive and the tapetum is secretory (Grayum, 1991; Rudall and Furness, 1997).

4.3 Evolution of pollen and tapetal characters

4.3.1 Pollen shape and size

Spheroidal or ellipsoidal pollen (pollen may also appear boat-shaped due to dehy-dration) is common in early-divergent monocots (Fig 4.1). Outside Alismatales reniform pollen is an apomorphy for some Pandanaceae and pollen is shed in tetrads in Velloziaceae. Reniform or arcuate pollen has evolved twice independently in Alismatales, in Hydrocharitaceae and Ruppiaceae (Fig 4.1). Pollen shed as dyads is an apomorphy for *Scheuchzeria* (Scheuchzeriaceae: Fig 4.1). Extremely long fila-mentous pollen may have a single origin and is a possible synapomorphy for the seagrasses (Cymodoceaceae, Posidoniaceae and Zosteraceae), possibly with a rever-sal to spheroidal pollen in *Potamogeton*, though this requires further testing (Fig 4.1).

4.3.2 Apertures

Aperture number has been optimized onto the phylogeny as this groups single apertures (monosulcate and ulcerate) together as one character state (Fig 4.2). Monoaperturate pollen (Fig 4.2) occurs in Acorales, Petrosaviales, many Pandanales, some Dioscoreales and many Araceae (Alismatales). Within Dioscor-eales pollen is 1–2-porate or inaperturate (Burmanniaceae) and monosulcate or disulculate (Dioscoreaceae). Diaperturate pollen also occurs in *Calla* (Araceae) and *Tofieldia* (Tofieldiaceae). There is a possible trend from a single aperture to two apertures, perhaps related to increased germination efficiency (Fig 4.2). A further increase in aperture number to polyaperturate is a synapomorphy for Alismataceae s.l. (including their sister lineage, the former Limnocharitaceae, Fig 4.2). Monoaperturate pollen is relatively rare in alismatids being restricted to *Butomus* and *Aponogeton*. On the other hand, inaperturate pollen is common and is a possible synapomorphy for many later-branching alismatids (Fig 4.2): Scheuchzeriaceae, Juncaginaceae, Potamogetonaceae, and the seagrasses (Cymodoceaceae, Posidoniaceae, Zosteraceae). Inaperturate pollen could have evolved at least three times independently in Alismatales: in the Araceae subfamily Aroideae (e.g. *Philodendron*), Hydrocharitaceae and the later-branching alismatids (Fig 4.2). In these groups inaperturate pollen is thin-walled or exineless, or the exine is restricted to muri with large lumina, permitting many potential germination sites (omniaperturate pollen). The three apertures found in *Ruppia* pollen are weakly delimited and the exine is a loose reticulum. Omniaperturate pollen fits the general trend towards increasing potential pollen germination sites and hence germination efficiency described above. Indeed, it has been hypothe-sized that increasing pollen aperture number is a general trend throughout angiosperms (Ressayre et al., 2002) and an increase from one to three apertures is a possible key innovation in the eudicots (Furness and Rudall, 2004).

4.3.3 Exine

Exineless pollen, or pollen with a nonsporopollenin outer layer, has evolved in tandem with inaperturate (omniaperturate) pollen, though not all inaperturate pollen is exineless. The exine may be reduced to a coarse reticulum with short columellae and large lumina, as in *Potamogeton*, for example. Reduction in the sporopollenin exine is characteristic of monocots occupying moist or wet habitats (the most extreme example being the seagrasses), or at least carrying out pollination in such an environment (as in the spadix of Aroideae). An interesting example of reduction in the amount of exinous sporopollenin occurs in Alismataceae, where the protuberances covering the pores in *Echinodorus* are hollow, developmental studies indicating that these protuberances are homologous with the columellae (Furness and Banks, 2010). The optimization indicates that exineless pollen originated at least three times independently in early-divergent monocots (Fig 4.3). The presence of solid exines in early-branching taxa (for example, *Acorus*) and the restriction of sporopollenin exine loss to the late-branching Araceae (Aroideae) and the later-branching alismatids, indicate a general trend from pollen possessing an exine to loss of exine (Fig 4.3).

4.3.4 Sculpture

Pollen sculpture is often polymorphic within a genus, hence the high degree of ambiguity in the optimization of this character, for example, in *Pandanus*, *Dioscorea*, some Araceae and *Tofieldia* (Fig 4.4). Infrageneric diversity is, however, reduced in the alismatid clade. Here there are three main sculpturing types: spines (an apomorphy for the expanded Alismataceae), psilate or smooth, and reticulate. Smooth pollen has arisen at least twice independently within the alismatids and is correlated with inaperturate (omniaperturate) exineless pollen (Fig 4.4). Since the wall of such pollen is composed of polysaccharide, sculptured features are absent, and the surface is smooth. This is therefore nonhomologous with the smooth pollen surface in *Acorus* and *Burmannia*, where the sporopollenin exine has a smooth outer surface. This illustrates the possible misinterpretation that may arise from the examination of surface features using scanning electron microscopy alone without due consideration as to how they are formed. Pollen with a smooth polysaccharidic outer layer has also evolved independently in Aroideae (e.g. *Philodendron*). Reticulate pollen has arisen at least twice independently in the alismatids, in *Butomus* and in a number of later-branching families: Aponogetonaceae, Scheuchzeriaceae, Juncaginaceae, Ruppiaceae and Potamogetonaceae (Fig 4.4). These are inaperturate (omniaperturate) except for *Butomus* and *Aponogeton*, which are monosulcate, and *Ruppia* which has three weakly delimited tenuitate apertures. Reduction of the sporopollenin exine to a reticulum, often with large lumina, is correlated with inaperturate (omniaperturate) pollen, but not completely so, and from the evidence in the optimization, the change to a reticulate surface may have potentially occurred before aperture loss (Fig 4.4). It is also striking that Potamogetonaceae, Juncaginaceae and Ruppiaceae share some characters (a reticulate surface and the presence of a sporopollenin exine) that are absent in the seagrasses, though *Potamogeton* is sister to *Zostera* (Zosteraceae) in the phylogenetic tree. Spheroidal or ellipsoidal pollen occurs in both Potamoge- tonaceae and Juncaginaceae, and *Althenia* (Potamogetonaceae) has a single tenui- tate aperture (Díez et al., 1988), similar to the three tenuitate apertures in *Ruppia*. Exineless filamentous pollen is a potential synapomorphy for the seagrass families and may indicate a close relationship between them, whereas *Potamogeton* is potentially wrongly placed and possibly should be closer to Juncaginaceae and Ruppiaceae.

4.3.5 Microsporogenesis

Successive microsporogenesis is common in early-divergent monocots, though not in angiosperms in general (Fig 4.5). The predominance of the simultaneous type in early-divergent angiosperms and in land plants in general (including gymnosperms) indicates that simultaneous microsporogenesis is plesiomorphic in angiosperms (Furness et al., 2002). Simultaneous microsporogenesis occurs in Petrosaviales, Dioscoreaceae, Tofieldiaceae and Aponogetonaceae (Fig 4.5). The presence of other tetrad shapes in *Aponogeton* as well as the more characteristic tetrahedral ones (Furness and Banks, 2010; Fig 4.11) indicates the variability that can occur in tetrads produced via simultaneous microsporogenesis in monocots (Nadot et al., 2008).

4.3.6 Tapetum

A secretory tapetum is plesiomorphic in monocots, as in angiosperms in general (Furness and Rudall, 2001). However, a plasmodial tapetum is a potential synapo- morphy for Alismatales, with a reversal to the secretory type in *Tofieldia* (Fig 4.6). *Tofieldia* thus has two potential apomorphies (simultaneous microsporogenesis and a secretory tapetum) shared with Petrosaviales and some Dioscoreales. This could represent experimentation in an early-divergent lineage. A reversal from the secretory tapetum type to the plasmodial and invasive nonsyncytial types also occurs in some Pandanales.

4.4 Conclusions

Among early-divergent monocots there is a trend for increasing variation in pollen shape, an increase in aperture number (or potential germination sites), reduction of the sporopollenin exine and increasing diversity in surface sculpturing. Microsporogenesis is predominantly successive with occasional rever-sals to simultaneous, and a predominantly plasmodial tapetum has evolved. Highly specialized pollen has evolved in alismatids, probably as an adaptation to an aquatic or semi-aquatic habit, culminating in the filamentous inaperturate exineless pollen of the late-branching seagrass families. This pollen is so unusual, indeed unique, among pollen of seed plants, that it alone is sufficient to refute the idea that monocot pollen is monosulcate, dull and boring. Considering the early-divergent monocots as a whole, their pollen is clearly not all monosulcate, and is diverse and interesting.

4.5 Acknowledgements

Thanks to Carlos Magdalena and other staff of the Royal Botanic Gardens, Kew, and Dmitry Sokoloff (Moscow State University) for plant material; Hannah Banks (Kew) for SEM images of pollen of *Limnophyton*, *Petrosavia* and *Zostera*; Peter Schols (ex K. U. Leuven) for the SEM image of *Dioscorea sylvatica* and to Paula Rudall (Kew) for helpful comments. Figures 4.1–4.6 are taken from Furness and Banks (2010) with kind permission from the University of Chicago Press.

4.6 References

APG II (Angiosperm Phylogeny Group II). (2003). An update of the Angiosperm Phylogeny Group classification for the orders and families of flowering plants: APG II. *Botanical Journal of the Linnean Society*, **141**, 399–436.

APG III (Angiosperm Phylogeny Group). (2009). An update of the Angiosperm Phylogeny Group classification for the orders and families of flowering plants: APG III. *Botanical Journal of the Linnean Society*, **161**, 105–121.

Argue, C. L. (1974). Pollen studies in the Alismataceae (Alismaceae). *Botanical Gazette*, **135**, 338–344.

Argue, C. L. (1976). Pollen studies in the Alismataceae with special reference to taxonomy. *Pollen et Spores*, **18**, 161–201.

Caddick, L., Furness, C. A., Stobart, K. L. and Rudall, P. J. (1998). Microsporogenesis and pollen morphology in Dioscoreales and allied taxa. *Grana*, **37**, 321–336.

Caddick, L. R., Rudall, P. J., Wilkin, P. and Chase M. W. (2000). Yams and their allies: systematics of Dioscoreales. In *Systematics and Evolution of Monocots. Proceedings of the 2nd International Monocot Conference*, ed. K. L. Wilson and D. A. Morrison. Melbourne: CSIRO, pp. 475–487.

Caddick, L. R., Rudall, P. J., Wilkin, P., Hedderson, T. A. J. and Chase, M. W. (2002). Phylogenetics of Dioscoreales based on a combined analysis of morphological and molecular data. *Botanical Journal of the Linnean Society*, **138**, 123–144.

Chanda, S., Nilsson, S. and Blackmore, S. (1988). Phylogenetic trends in the Alismatales with reference to pollen grains. *Grana*, **27**, 257–272.

Chase, M. W., Duvall, M. R., Hills, H. G. et al. (1995). Molecular phylogenetics of Lilianae. In *Monocotyledons: Systematics and Evolution*, ed. P. J. Rudall, P. J. Cribb, D. F. Cutler and C. J. Humphries. Kew: Royal Botanic Gardens, pp. 109–137.

Chase, M. W., Soltis, D. E., Soltis, P. S. et al. (2000). Higher-level systematics of the monocotyledons: an assessment of current knowledge and a new classification. In *Monocots: Systematics and Evolution*, ed. K. L. Wilson and D. A. Morrison. Melbourne: CSIRO, pp. 3–16.

Chase, M. W., Fay, M. F., Devey, D. S. et al. (2006). Multigene analysis of monocot relationships: a summary. *Aliso*, **22**, 63–75.

Chen, J-M., Chen, D., Gituru, W. R., Wang, Q-F. and Guo, Y-H. (2004). Evolution of apocarpy in Alismatidae using phylogenetic evidence from chloroplast *rbcL* gene sequence data. *Botanical Bulletin of the Academica Sinica*, **45**, 33–40.

Cox, P. A. (1983). Search theory, random motion, and the convergent evolution of pollen and spore morphology in aquatic plants. *American Naturalist*, **121**, 9–31.

Dahlgren, R. M. T. and Clifford, T. (1982). *The Monocotyledons: A Comparative Study*. London: Academic Press.

Davis, J. I., Stevenson, D. W., Petersen, G. P. et al. (2004). A phylogeny of the monocots, as inferred from *rbcL* and *atpA* sequence variation, and a comparison of methods for calculating jackknife and bootstrap values. *Systematic Botany*, **29**, 467–510.

Díez, M. J., Talavera, S. and García-Murillo, P. (1988). Contributions to the palynology of hydrophytic, non-entomophilous angiosperms. 1. Studies with LM and SEM. *Candollea*, **43**, 147–158.

Ducker, S. C. and Knox, R. B. (1976). Submarine pollination in seagrasses. *Nature*, **263**, 705–706.

Ducker, S. C., Pettitt, J. M. and Knox, R. B. (1978). Biology of Australian seagrasses: pollen development and submarine pollination in *Amphibolis antarctica* and *Thalassodendron ciliatum* (Cymodeaceae). *Australian Journal of Botany*, **26**, 265–285.

Furness, C. A. and Banks, H. (2010). Pollen evolution in the early-divergent monocot order Alismatales. *International Journal of Plant Sciences*, **171**, 713–739.

Furness, C. A. and Rudall, P. J. (1999). Microsporogenesis in monocotyledons. *Annals of Botany*, **84**, 475–499.

Furness, C. A. and Rudall, P. J. (2001). The tapetum in basal angiosperms: early diversity. *International Journal of Plant Sciences*, **162**, 375–392.

Furness, C. A. and Rudall, P. J. (2004). Pollen aperture evolution – a crucial factor for eudicot success? *Trends in Plant Science*, **9**, 1360–1385.

Furness, C. A. and Rudall, P. J. (2006). Comparative structure and development of pollen and tapetum in Pandanales. *International Journal of Plant Sciences*, **167**, 331–348.

Furness, C. A., Rudall, P. J. and Sampson, F. B. (2002). Evolution of microsporogenesis in angiosperms. *International Journal of Plant Sciences* **163**, 235–260.

Grayum, M. H. (1991). Systematic embryology of the Araceae. *Botanical Review*, **57**, 167–203.

Grayum, M. H. (1992). Comparative external ultrastructure of the Araceae and putatively related taxa. *Monographs in Systematic Botany from the Missouri Botanical Garden*, **43**, 1–167.

Handa, K., Tsuji, S. and Tamura, M. N. (2001). Pollen morphology of Japanese Asparagales and Liliales (Lilianae).

Japanese Journal of Historical Botany, **9**, 85–125.

Harley, M. M. (2004). Triaperturate pollen in the monocotyledons: configurations and conjectures. *Plant Systematics and Evolution*, **247**, 75–122.

Harley, M. M. and Baker, W. J. (2001). Pollen aperture morphology in Arecaceae: application within phylogenetic analyses, and a summary of the fossil record of palm-like pollen. *Grana*, **40**, 45–77.

Hesse, M. (2006). Reasons and consequences of the lack of a sporopollenin ektexine in Aroideae (Araceae). *Flora*, **201**, 421–428.

Hesse, M. and Zetter, R. (2007). The fossil pollen record of Araceae. *Plant Systematics and Evolution*, **263**, 93–115.

Huynh, K-L. (1976). Arrangement of some monosulcate, disulcate, trisulcate, dicolpate, and tricolpate pollen types in the tetrads, and some aspects of evolution in the angiosperms. In *The Evolutionary Significance of the Exine*, ed. I. K. Ferguson and J. Muller. New York and London: Academic Press, pp. 101–124.

Le Thomas, A., Suárez-Cervera, M. and P. Goldblatt (2001). Ontogeny of the exine in pollen of *Aristea* (Iridaceae). *Grana*, **40**, 35–44.

Merckx, V., Schols, P., Geuten, K., Huysmans, S. and Smets, E. (2008). Phylogenetic relationships in Nartheciaceae (Dioscoreales), with focus on pollen and orbicule morphology. *Belgian Journal of Botany*, **141**, 64–77.

Nadot, S., Furness, C. A., Sannier, J. et al. (2008). Phylogenetic comparative analysis of microsporogenesis in angiosperms with a focus on monocots. *American Journal of Botany*, **95**, 1426–1436.

Nixon, K. C. (2002). *Winclada, Ver.1.00.08. Published by the author*, Ithaca, NY.

Distributed through http://www.cladistics.com.

Nunes, E. L. P., Bona, C. and Coan, A. I. (2009). Release of developmental constraints on tetrad shape is confirmed in inaperturate pollen of *Potamogeton*. *Annals of Botany*, **104**, 1011–1015.

Pettitt, J. M. (1980). Reproduction in seagrasses: nature of the pollen and receptive surface of the stigma in the Hydrocharitaceae. *Annals of Botany*, **45**, 257–271.

Pettitt, J. M. (1981). Reproduction in seagrasses: pollen development in *Thalassia hemprichii, Halophila stipulacea* and *Thalassodendron ciliatum. Annals of Botany*, **48**, 609–622.

Pettitt, J. M. and Jermy, A. C. (1975). Pollen in hydrophilous angiosperms. *Micron*, **5**, 377–405.

Pettitt, J. M., McConchie, C. A., Ducker, S. C. and Knox, R. B. (1984). Reproduction in seagrasses: pollen wall morphogenesis in *Amphibolis antarctica* and wall structure in filiform grains. *Nordic Journal of Botany*, **4**, 199–216.

Radulescu, D. (1970). Recherches morpho-palynologiques sur les espèces d'Iridaceae. *Acta horti botanici Bucurestiensis*, 1968, 311–350.

Radulescu, D. (1973). Recherches morpho-palynologiques sur la famille Liliaceae. *Acta horti botanici Bucurestiensis*, 1973, 133–248.

Remizowa, M., Sokoloff, D. and Rudall, P. J. (2006). Evolution of the monocot gynoecium: evidence from comparative morphology and development in *Tofieldia, Japonolirion, Petrosavia* and *Narthecium. Plant Systematics and Evolution*, **258**, 183–209.

Ressayre, A., Raquin, C., Mignot, A., Godelle, B. and Gouyon, P-H. (2002). Correlated variation in microtubule distribution, callose deposition during male post-meiotic cytokinesis, and pollen aperture number across *Nicotiana* species (Solanaceae). *American Journal of Botany*, **89**, 393–400.

Rudall, P. J. and Furness, C. A. (1997). Systematics of *Acorus*: ovule and anther. *International Journal of Plant Sciences*, **158**, 640–651.

Rudall, P. J., Furness, C. A., Chase, M. W. and Fay, M. F. (1997). Microsporogenesis and pollen sulcus type in Asparagales. *Canadian Journal of Botany*, **75**, 408–430.

Rudall, P. J., Stobart, K. L., Hong, W-P. et al. (2000). Consider the lilies: systematics of Liliales. In *Monocots: Systematics and Evolution*, ed. K. L. Wilson and D. A. Morrison. Melbourne: CSIRO, pp. 347–359.

Schols, P., Furness, C. A., Wilkin, P., Huysmans, S. and Smets, E. (2001). Morphology of pollen and orbicules in some *Dioscorea* species and its systematic implications. *Botanical Journal of the Linnean Society*, **136**, 295–311.

Schols, P., Furness, C. A., Wilkin, P. et al. (2003). Pollen morphology of *Dioscorea* (Dioscoreaceae) and its relation to systematics. *Botanical Journal of the Linnean Society*, **143**, 375–390.

Schols, P., Furness, C. A., Merckx, V., Wilkin, P. and Smets, E. (2005a). Comparative pollen development in Dioscoreales. *International Journal of Plant Sciences*, **166**, 909–924.

Schols, P., Wilkin, P., Furness, C. A., Huysmans, S. and Smets, E. (2005b). Pollen evolution in yams (*Dioscorea*: Dioscoreaceae). *Systematic Botany*, **30**, 750–758.

Simpson, D. A., Furness, C. A., Hodkinson, T. R., Muthama Muasya, A. and Chase, M. W. (2003). Phylogenetic relationships in Cyperaceae subfamily Mapanioideae inferred from pollen and plastid DNA sequence data. *American Journal of Botany*, **90**, 1071–1086.

Smith, S. Y., Collinson, M. E., Simpson, D. A. et al. (2009). Elucidating the affinities and habitat of ancient, widespread Cyperaceae: *Volkeria messelensis* gen. et sp. nov., a fossil mapanioid sedge from the Eocene of Europe. *American Journal of Botany*, **96**, 1506–1518.

Stevenson, D. W. and Laconte, H. (1995). Cladistic analysis of monocot families. In *Monocotyledons: Systematics and Evolution*, ed. P. J. Rudall, P. J. Cribb, D. F. Cutler and C. J. Humphries. Kew: Royal Botanic Gardens, pp. 543–578.

Takahashi, M. and Kawano, S. (1989). Pollen morphology of the Melanthiaceae and its systematic implications. *Annals of the Missouri Botanical Garden*, **76**, 863–876.

Thanikaimoni, G. (1978). Palynological terms: proposed definitions – 1. *Proceedings of the 4th International Palynological Conference, Lucknow (1976–1977)*, **1**, 228–239.

Tobe, H. (2008). Embryology of *Japonolirion* (Petrosaviaceae, Petrosaviales): a comparison with other monocots. *Journal of Plant Research*, **121**, 407–416.

Van Bruggen, H. W. E. (1985). Monograph of the genus *Aponogeton* (Aponogetonaceae). *Bibliotheca Botanica*, **33**, 1–76.

Verhoeven, J. T. A. (1979). The ecology of *Ruppia*-dominated communities in Western Europe. I. Distribution of *Ruppia* representatives in relation to their autecology. *Aquatic Botany*, **6**, 197–268.

Zavada, M. S. (1983). Comparative morphology of monocot pollen and evolutionary trends of apertures and wall structures. *Botanical Review*, **49**, 331–379.

5

Macroecological correlates of global monocot species richness

F. Andrew Jones, Benjamin Sobkowiak, C. David L. Orme, Rafaël Govaerts and Vincent Savolainen

5.1 Introduction

Biological diversity is not evenly distributed across the globe. The factors that determine contemporary species distribution can be attributed to a combination of biotic (e.g. ecological) and abiotic (e.g. climatic) constraints on species distributions, macroevolutionary patterns of speciation and extinction, and the interaction of these ecological and evolutionary processes with geological processes (such as continental drift, mountain building and island formation) that occur across a range of spatial and temporal scales. Knowledge of the climatic factors that are correlated with species richness and diversity can generate insights into the mechanisms maintaining diversity within biotic assemblages, as well as mechanisms that increase or inhibit diversification of lineages of living organisms across space and time. A wide range of taxa, including vertebrates, invertebrates and plants exhibit macroecological correlations between species richness and climatic variables, with higher richness associated with areas of higher temperature, precipitation and energy (Qian, 1998; Hawkins et al., 2003; Davies et al., 2007; Qian and Ricklefs, 2008). That strong correlations are found between diverse groups of organisms and comparatively few climatic variables suggests that general macroevolutionary and macroecological mechanisms underlie the origin and maintenance of biological diversity within and among different taxonomic groups (Ricklefs, 2006).

Early Events in Monocot Evolution, eds P. Wilkin and S. J. Mayo. Published by Cambridge University Press. © The Systematics Association 2013.

Knowledge of those processes responsible for the origin and the maintenance of diversity has implications beyond simply understanding ecological systems. Predicting future responses of biological diversity to climate change and anthropogenic disturbance poses significant challenges, particularly in species-rich areas of the tropics (Thomas et al., 2004; Hannah and Lovejoy, 2007). Predicting future responses in the tropics is challenging because of the idiosyncratic and individualistic nature of species response to variation in the biotic and abiotic environment (Bush and Colinvaux, 1990; Stewart et al., 2010). In tropical ecosystems, where hundreds to thousands of species of plants can co-exist on spatial scales of $<1 \text{ km}^2$ (Leigh et al., 2004), developing species-specific understanding of climatic tolerances is a daunting task. One way forward is to explore the extent to which species within the same taxon (order, family, and genus) respond in similar or different ways to climatic gradients. Such an approach would allow for generalizations to be made within related groups and potentially identify taxonomic groups that are more vulnerable to extinction as climatic conditions change.

That species richness is broadly correlated with climatic variables has been recognized for a long period of time (Pianka, 1966; Hawkins et al., 2003; Currie et al., 2004). Much research has been aimed at describing global ecological and climatic correlates of diversity in angiosperms and vertebrates (Currie, 1991; Francis and Currie, 2003; Kreft and Jetz, 2007). Here, we undertake a similar approach to understand the relationship between climatic variables and global species richness in a large group of flowering plants, the monocots. Monocots are an excellent model system for addressing the ecological and evolutionary determinants of species diversity, distribution and diversification. The monocots are a large monophyletic plant group that includes approximately a quarter of all flowering plants (70 000+ species, 2839 genera, 77 families, 10 orders) *sensu* APG III (2009). Monocot species exhibit a wide range of environmental tolerances and occupy almost all major terrestrial biomes, from rainforests to savannas to tundra.

Knowledge of the climatic, evolutionary and geographical factors that determine species richness can shed light on the role that past environmental change may have played in determining current patterns of species distributions, as well as provide insight into anticipated changes in local and regional biota as a function of current and future global climate change. Insight into how these factors determine patterns and levels of diversity within particular taxonomic groups and how the latter interact with biogeographical gradients can be used in a comparative context to determine the relative importance of different macroecological and macroevolutionary factors in determining species diversity and diversification (Mittelbach et al., 2007).

In this chapter, we ask: What are the climatic correlates of monocot species richness across the globe? We limit our study to an examination of the ecological and geographical correlates of global patterns of contemporary monocot species

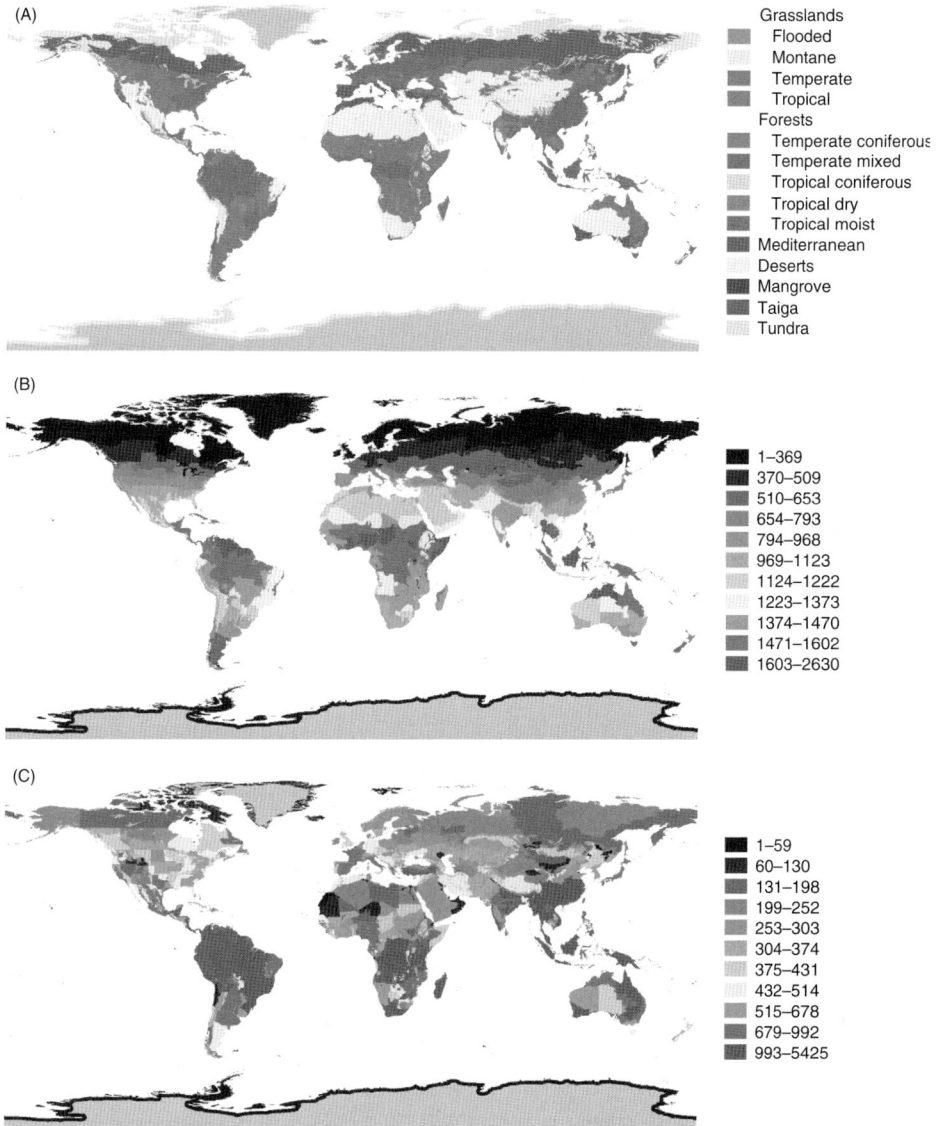

(A)

Grasslands
- Flooded
- Montane
- Temperate
- Tropical

Forests
- Temperate coniferous
- Temperate mixed
- Tropical coniferous
- Tropical dry
- Tropical moist

Mediterranean
Deserts
Mangrove
Taiga
Tundra

(B)

- 1–369
- 370–509
- 510–653
- 654–793
- 794–968
- 969–1123
- 1124–1222
- 1223–1373
- 1374–1470
- 1471–1602
- 1603–2630

(C)

- 1–59
- 60–130
- 131–198
- 199–252
- 253–303
- 304–374
- 375–431
- 432–514
- 515–678
- 679–992
- 993–5425

Fig 5.1 Global maps showing the distribution of A: biomes. B: actual evapotranspiration (AET values). C: monocot species richness (numbers of species per spatial region). Colour version to be found in colour plate section.

richness as measured by the number of species of monocots co-occurring within a particular area. Although species richness varies by orders of magnitude globally, a common pattern present in almost all broadly defined taxonomic groups is a latitudinal gradient in species diversity. Monocots also show a strong latitudinal gradient, with some exceptions discussed below (Fig 5.1). Species diversity is highest at the equator and generally decreases with increasing latitude. The latitudinal gradient in species richness has been attributed to many different abiotic, biotic, historical and geographical processes that vary with latitude (Pianka, 1966; Stevens, 1989; Hillebrand, 2004; Qian and Ricklefs, 2007; Schemske et al., 2009), but in general, higher species diversity is found in areas with higher rainfall, higher temperatures, higher solar radiation and fewer days of hard freezes, all of which also tend to correlate with latitude.

We generated a new global data set on the spatial distribution of monocots by refining species mapped to TDWG level 3 spatial units into appropriate areas of appropriate biomes within these units. In the absence of species-level data, the biome associations of species were inferred from genus-level data. The placement of genera within biomes and biomes within TDWG level 3 units then allows us to assign climatic variables to the refined areas in which the genera occur. To examine the climatic variables associated with monocot richness we take a correlative approach using a combination of linear models and climatic, environmental and topographic variables recorded within global regions defined as World Wildlife Fund ecoregions (Olson et al. 2001). With these data in hand, we then ask the following questions:

(1) What are the broad patterns of global species richness in the monocots?

(2) What environmental factors (temperature, water and elevation) are correlated with species richness for all monocots and within particular monocot orders?

(3) How do patterns differ within and among monocot orders and across latitudes, continents and biomes?

5.2 Materials and methods

5.2.1 Construction of the global monocot generic distribution database based on biome associations

The development of the World Monocot Checklist by Kew Botanic Gardens in 2006 ('World Checklist of Monocotyledons', http://www.kew.org/wcsp/monocots/ and http://www.kew.org/data/grasses-syn/) provides a definitive database of all accepted monocot species and genera along with their geographic distributions and the number of valid species in each genus (Govaerts, 2004). The geographical locations that we used in this analysis place each genus to at least the third level of the Taxonomic Database Working Group (TDWG) geographic code system for recording plant distributions (Brummitt, 2001). Generally speaking, TDWG level 3 assigns one or more geographic unit to taxa where the unit is usually equivalent to a country or, in the case of large countries, subdivisions, such as states within the USA and Brazil, and subdivided regions of Russia. The median area of TDWG level 3 units is around 130 000 km^2, roughly the area of Java, but areas range from small islands ($\sim$ 2 km^2) to large regions ($\sim$ 4 million km^2, Northern Brazil).

While a complete database of the broad geographic locations of members of a particular taxonomic group is a significant step in developing a better understanding of the biogeography of a large taxonomically diverse group, the large area found in certain regions is problematic for modelling the climatic variables associated with species richness. For example, the TDWG level 3 unit of California in the USA is large (407 748 km^2) and treating the entire state as a whole misses much of the environmental and biotic heterogeneity found within it. There are many different biomes in California (deserts, grasslands, woodlands, conifer forests, alpine tundra, chaparral scrub) that have different environmental conditions and therefore harbour many different plant species. Understanding the environmental correlates of species richness within and among large biogeographic regions would be improved if species or genera could be more narrowly placed into vegetation types that reflect their actual distribution across the landscape. Ideally, one would want species-specific information on range sizes and presence–absence across a landscape, but this is problematic when examining a large taxonomic group with many species. As yet, such comprehensive range maps exist for only certain monocot species and genera such as orchids and palms (Pridgeon et al., 1999; Dransfield et al., 2008).

Here we refine the distribution of monocot species and genera by assigning them to biomes within TDWG level 3 entities. The assignment of a species or genus to a particular biome then allows us to define more narrowly the climatic and environmental correlates associated with the region in which a species or genus is present. The narrower distribution of genera then allows for analyses of species richness in a more biologically meaningful and accurate way and allows for more precise determination of those factors that may promote or limit species richness.

We undertook an extensive literature survey to document the biome associations of monocot genera, assigning all accepted monocot genera into one or more of 14 major terrestrial biomes defined by Olson et al. (2001) in the World Wildlife Fund (WWF) map of terrestrial ecoregions (Olson et al., 2001) (Fig 5.1A). Data were drawn from ecological and taxonomic descriptions accessed from more than 200 published sources (references available from the authors). The spatial regions used in the analyses below have therefore been created from the intersection of the TDWG level 3 entities and the biome definitions underlying the WWF terrestrial ecosystems (Olson et al., 2001) giving a total of 1003 regions available for analyses. Species richness within each region is calculated as the sum of the species of all genera that are found within a specific TDWG level 3 province and assigned to a biome within the region. Species richness may be over-estimated within biomes nested within TDWG units when a species in a genus inhabits more than one biome present within that spatial unit.

5.2.2 Ecological variables and linear models

Based on exploratory analyses and knowledge of environmental variables that have been shown previously to be important in determining plant-species richness patterns (Qian, 1998; Kreft and Jetz, 2007), we calculated the means and ranges of nine climatic and environmental variables within each spatial unit using the zonal statistics tool in ArcGIS 9.2 (ESRI, 2009). These variables included four that are associated with temperature (mean annual temperature, mean temperature of the coldest month, mean temperature of the warmest month, temperature range), and three describing annual levels of precipitation (mean annual precipitation, precipitation in the wettest month, precipitation in the driest month), all of which were taken from the BIOCLIM data set (10 arc minute resolution; Hijmans et al., 2005). We further included a measure of actual evapotranspiration (AET, 30 arc minute resolution; Ahn and Tateishi, 1994), which is a combination of measured evaporation and transpiration and therefore integrates information on solar radiation, temperature, primary production and moisture. Mean elevation of the region was also included in our analyses (GTOPO30, 30 arc second resolution; US Geological Survey, 2003). Finally, three variables were included that account for the effect of area, vegetation type and biogeographic history between units: region area, biome type and continent. These are used as categorical variables in the analyses.

We generated linear models to determine the most important individual climatic, environmental and biogeographic variables that predict monocot species richness among different regions. Log transformation of species richness (S = log (richness + 1)), area, moisture variables and AET improved model fit and were therefore included in all subsequent analyses. The fit of individual models was determined by examining the proportion of the model deviance explained. The percentage of deviance explained by the model is calculated by (1 – residual model deviance/null model deviance) * 100. Deviance here is simply –2 times the log likelihood of the model. We consider the proportion of null model deviance explained because it allows us to compare likelihoods and model fit among different models where the independent variables change (for example, all monocot richness vs. richness in a single order). For each individual variable, we then report model parameter estimates, standard errors of estimates, t-values, and P values as a test of statistical significance, Akaike's Information Criterion (AIC, Burnham and Anderson, 2004) and percentage of model deviance explained (Table 5.1).

In order to test the hypothesis that heterogeneity exists in the response of richness within particular orders to climatic factors, we also undertook a series of multiple regressions (generalized linear models, GLMs) comparing global patterns to

Table 5.1 Linear model estimates of the effect of nine independent variables on log species richness in the monocots.

Variable	Estimate	Std. Error	t	P	% Deviance	AIC
NULL	5.60	0.04	141.8	***	0.0	3045
AREA						
Log(km^2)	0.25	0.01	20.75	***	31.38	2691.8
ENERGY						
Mean AET	0.62	0.03	19.14	***	39.25	1428.17
Mean annual temperature	0.19	0.04	5.29	***	31.83	2062.42
Min temperature coldest month	0.0005	0.0002	2.46	**	0.007	2558.1
Max temperature warmest month	0.002	0.0005	4.57	***	2.33	2543.5
WATER						
Annual precipitation	0.47	0.03	14.67	0.00	39.00	2137.12
Precipitation wettest month	0.48	0.03	13.87	0.00	37.71	2155.61
Precipitation driest month	0.10	0.02	5.32	0.00	26.54	2084.78
ELEVATION						
Elevation	0.15	0.02	7.23	0.00	32.75	2608.25

richness at the ordinal level. We used a simplified model containing area of the spatial unit, elevation, mean AET, mean precipitation and mean annual temperature, and compared the magnitude and effect of these variables for the species richness of all monocots and for species richness within the nine most diverse orders of monocots (Table 5.2: Alismatales, Arecales, Asparagales, Commelinales, Dioscoreales, Liliales, Pandanales, Poales and Zingiberales). All GLMs were performed using the GLM function in the R statistical software package (R Development Core Team, 2010) and assumed normally distributed errors. Finally, second-order polynomial regression (Fig 5.2) was used to describe how richness changes in these orders as a function of absolute latitude (L): $S = a + bL + cL^2$).

Table 5.2 Linear model parameter estimates and standard error (SE) of the effect of area, climatic variables and elevation on global patterns of species richness in major monocot orders. Parameter estimates in bold are statistically significant effects ($P < 0.01$)

Order	Intercept (SE)	Area (SE)	AET (SE)	Precipitation (SE)	Temperature (SE)	Elevation (SE)	Model Deviance
All monocots	−0.4835 (0.2901)	**0.2284** (**0.01485**)	**0.294** (**0.06233**)	**0.2684** (**0.04952**)	**0.00143** (**0.0002799**)	**0.00017** (**0.00003**)	43.19
Alismatales	**−5.074** (**0.4317**)	**0.2427** (**0.02209**)	**0.5766** (**0.09274**)	**0.2513** (**0.07369**)	0.0004409 (0.0004164)	−0.00007 (0.00004)	35.7
Arecales	**−6.127** (**0.4328**)	**0.2455** (**0.02215**)	−0.281 (0.09298)	**0.7493** (**0.07388**)	**0.00779** (**0.0004175**)	0.00005 (0.00004)	53.55
Asparagales	**−4.802** (**0.4972**)	**0.3137** (**0.02544**)	**0.4224** (**0.1068**)	**0.3669** (**0.08487**)	**0.00335** (**0.0004796**)	**0.00040** (**0.00005**)	38.66
Commelinales	**−6.598** (**0.4806**)	**0.2382** (**0.02459**)	**0.9082** (**0.1033**)	−0.1834 (0.08204)	**0.005118** (**0.0004636**)	**0.00020** (**0.00005**)	44.79
Dioscoreales	**−7.577** (**0.4291**)	**0.263** (**0.02195**)	**0.5094** (**0.09217**)	**0.2643** (**0.07324**)	**0.005901** (**0.0004139**)	**0.00033** (**0.00004**)	52.48
Liliales	0.8661 (0.5415)	**0.105** (**0.02771**)	**0.3597** (**0.1163**)	**−0.2527** (**0.09244**)	**−0.005574** (**0.0005224**)	**0.00015** (**0.00005**)	22.70
Pandanales	**−4.531** (**0.3656**)	**0.1977** (**0.01871**)	−0.1363 (0.07854)	**0.4629** (**0.0624**)	**0.003811** (**0.0003526**)	0.00008 (0.00003)	33.50
Poales	**0.726** (**0.3055**)	**0.1904** (**0.01563**)	**0.3434** (**0.06563**)	0.06221 (0.05215)	−0.0004974 (0.0002947)	0.00002 (0.00003)	24.74
Zingiberales	**−8.171** (**0.4698**)	**0.245** (**0.02404**)	−0.04854 (0.1009)	**0.8501** (**0.08019**)	**0.005697** (**0.0004531**)	**0.00030** (**0.00005**)	48.17

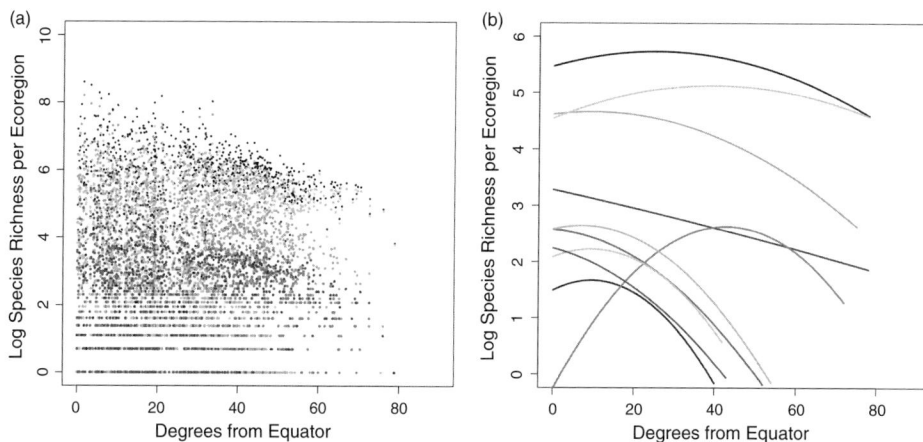

Fig 5.2 Latitudinal gradients in species richness within different monocot orders A: individual data points for orders (see below for colour scheme). B: lines representing the predicted relationship between species number and latitude (degrees from the equator) for orders in A determined using polynomial regression. Black = all monocots (Multiple $R^2 = 0.02$), red = Dioscoreales ($R^2 = 0.31$), blue = Pandanales ($R^2 = 0.08$), green = Liliales ($R^2 = 0.32$), orange = Asparagales ($R^2 = 0.07$), grey = Commelinales ($R^2 = 0.30$), light blue = Zingiberales ($R^2 = 0.04$), pink = Poales ($R^2 = 0.02$), purple = Alismatales ($R^2 = 0.07$), brown = Arecaceae ($R^2 = 0.11$). Colour version to be found in colour plate section.

5.3 Results and discussion

5.3.1 Patterns of species richness across all monocots

Of the 77 accepted monocot plant families, Pontederiaceae, Potamogetonaceae and Ruppiaceae were not included in these analyses because Olson et al. (2001) does not include freshwater spatial units, making it difficult to place these taxa in biomes within TDWG regions. Therefore, only terrestrial genera are examined here. Another limitation of our data is that at the time of writing, biome associations for Orchidaceae remain marginally incomplete due to unavailability of suitable references for some newer genera. Overall, however, our data set contains biome associations, and spatial and climatic data for 97% of all terrestrial monocot genera.

The number of species found in each region was summed and projected onto the WWF region shapefile in ArcMap (Fig 5.1C) to display monocot species richness across regions and also to display patterns of different environmental variables across the globe. Monocot species richness within regions ranges from 1 to 5425 species, with 63 regions containing no species largely because these areas are defined as ice or rock. The mean number of species found in occupied regions was 456, standard deviation was 514 and median richness was 337. Richness generally declines as one moves away from the equator, although at mid-latitudes

(a)

(b)

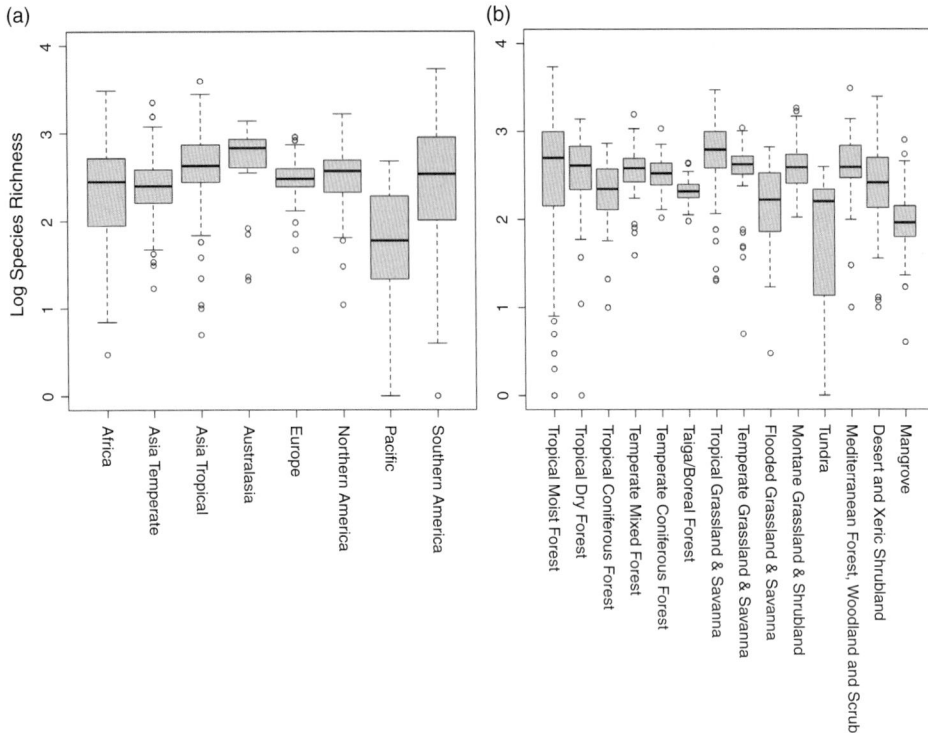

Fig 5.3 Box plots of log species richness of monocots by A: continent and B: biome.

some areas occur with distinctly higher richness than surrounding regions, namely the Mediterranean ecosystems of Europe, Southwestern Australia and South Africa. Not surprisingly, the lowest richness was found around the Arctic and Antarctic tundra (Figs 5.1, 5.3B). Of the single variables examined that correlated with species richness, area was a strong predictor of species richness in monocots (Table 5.1) and it explained 31% of model deviance in species richness. It has long been recognized that area is one of the strongest predictors of diversity (Arrhenius, 1921; Gleason, 1922; Preston, 1960; MacArthur and Wilson, 1967).

There is significant variation among orders of monocots in patterns of richness across broad latitudinal gradients (Fig 5.2A). Monocots as a whole show a general decline in species numbers with increasing latitude (Fig 5.2B, black line). Many orders that are largely restricted to tropical and subtropical areas (Arecales, Zingiberales, Dioscoreales) have the strongest latitudinal gradients, with declines observed in richness at higher latitudes. Several factors could combine to produce the strong latitudinal gradients observed here. The gradient in these orders could be due in part to the failure of species within these orders to colonize and diversify within high-latitude areas. This is due to the fact that species in these orders have strong niche conservatism linked with adaptation to climatic variables associated with low latitudes (Wiens and Donoghue, 2004). For example, species of tropical origin may have an inability to tolerate drier or cold climates, which prevents the successful colonization of high latitudes. For example, when we examine the multiple regression models for the climatic variables associated with richness for Arecaceae and Zingiberales (Table 5.2), we find that the slopes of the relationship between richness and temperature and precipitation for these species are stronger, and in some cases much stronger, than they are for most other orders of monocots and for monocots as a whole. This implies an overall greater response of species within these orders to global temperature and precipitation gradients, which could imply strong niche conservatism and explain why their distribution is limited to tropical and subtropical ecosystems.

Despite the strong latitudinal gradients in some orders, as mentioned above and shown in Fig 5.2, our analysis also shows that significant variation exists in the strength and the direction of the latitudinal gradient for different orders. For example, within Liliales, which includes families such as Liliaceae, Melanthiaceae and Colchicaceae, there is an opposite trend of increasing richness with increasing latitude. A weak latitudinal trend is also seen in the order Poales, which includes the families Poaceae, Cyperaceae, Bromeliaceae and Eriocaulaceae (Fig 5.2B, pink line). In the Poales and Liliales, richness peaks at middle latitudes. For Liliales, peaks in richness are seen in temperate coniferous forests, Mediterranean, temperate grassland and savannah biomes. For Poales, peaks are also seen in temperate grasslands and Mediterranean biomes. The effect of climatic variables on richness in these groups can shed some light on these patterns. For example, within the Liliales, a negative relationship exists between species richness, and temperature and precipitation (Table 5.2), implying that diversity decreases with increasing temperature and latitude, which sheds light on the pattern observed in Fig 5.2B. For the Poales, temperature and precipitation were found to have no significant effect on species richness observed within regions (Table 5.2), which could partially explain the weak latitudinal gradient in these groups. However, it must also be emphasized that while this pattern exists across the order, different patterns exist at the familial level as the grasses and sedges are more diverse than the bromeliads, and have larger ranges. Therefore, further examination of these patterns could benefit from a hierarchical examination of the effect of climatic variables at the family level.

This result also points to the likelihood that niche conservatism and dispersal limitation may limit the distribution and diversification of lineages within particular taxa, but that temperate lineages may also be limited in colonizing and diversifying in the tropics. These results point to the complex nature of latitude and climatic correlates of richness within the monocots and the importance of considering geographical and historical correlates when examining global patterns in richness. Our results also suggest the need to examine differences among particular functional traits within groups, and the effect that these have had on the distribution of taxa and their ability to colonize and diversify within new habitats (Edwards et al., 2010). For example, to what extent did the evolution of rhizomes or C4 photosynthesis enable the colonization of drier and colder environments?

5.3.2 Geographical variation in richness

Our data set also reveals that significant variation exists in monocot species richness as a function of larger biogeographic realm or continent (Fig 5.3). South America and Tropical Asia have the highest monocot richness, while the largely island biome of the Pacific is the least speciose. Within particular biomes, tropical moist forests and tropical grasslands have the greatest richness, while lowest richness is found in mangrove habitats and in tundra biomes (Fig 5.3B). Broadly speaking, as evidenced from the maps produced in Fig 5.1 and from an analysis of latitudinal patterns, the regions with the greatest richness of monocots are concentrated in the tropics. The highest richness of monocots can be found in northern tropical South America and in the Indo-Malay peninsula. Specifically, the most species-rich area for monocots identified in this analysis is located in tropical moist rainforest in Ecuador. This area is characterized by having an annual mean temperature of 21.3 °C, annual precipitation of 2315 mm and a mean elevation of 1037 m. This area contains numerous species from all monocot orders and richness is particularly high in palm species and genera (Bjorholm et al., 2005). Our analysis found that this region is also highest in members of the Asparagales, probably due to high orchid diversity. The region has long been known to be an area of extremely high plant diversity (Humboldt, 1845–1858; Kreft and Jetz, 2007). Previous modelling studies of plant diversity within the Neotropics have revealed that much of the diversity is correlated with the topographical complexity of the region. Complex topology in the Andes is hypothesized to have created great environmental heterogeneity in climatic variables and presumably a greater number of niches for species with different environmental tolerances (Janzen, 1967; Distler et al., 2009). However, our data set does not include information on plant species distributions within regions (for example, we have no information on elevational limits of species) to be able to make this generalization. Instead, we rely on variation among different regions in making all predictions. Despite this, we still see elevational trends in predicting species richness which are consistent with larger trends in latitudinal diversity.

High species richness varies geographically across latitudinal and longitudinal gradients within orders. For example, the highest richness of Dioscoreales and Poales is found in Southeastern Brazil; highest richness of palms is located in Borneo; highest richness of Liliales is found in temperate forests in Turkey, Pandanales in tropical moist forests of New Guinea, Zingiberales in tropical moist forests of Thailand, and Alismatales in Colombian tropical moist forest. High richness also occurs in the savannah and tropical forests of the African subcontinent and although the number of genera found there is not as high as in South America and Asia, these are still significant areas of monocot richness, probably due to the importance of grasses in these systems.

5.3.3 Richness and environmental heterogeneity

Previous studies have demonstrated the strong correlation that exists between measures of energy in a particular environment (e.g. solar radiation and temperature) and species richness (Currie, 1991; Hawkins et al., 2003), although the reasons behind this relationship are unclear. One strictly ecological explanation of the relationship between diversity and energy is the energy–biomass hypothesis (Wright, 1983; Davies et al., 2004; Clarke and Gaston, 2006). This hypothesis states that high primary production in the tropics simply supports more populations of plants, which enables greater species diversity. Our analysis of species–biome relationships within the monocots largely supports these findings, with a strong positive relationship observed between mean AET for a region and mean annual temperatures (Table 5.1).

Measures of actual evapotranspiration provide an integrated measure that can reveal ecologically important climatic variables that are linked to environmental energy, as well as providing information on water availability and primary production. AET was the strongest predictor of species richness both in terms of explaining >39% species-richness model deviance followed by mean annual temperature, which explained 32% of model deviance when examined individually. Minimum temperature of the coldest month and maximum temperature of the warmest month, although statistically significant, do not explain more than 2% of model deviance in species richness. Precipitation also plays a strong role in determining patterns of species richness in the monocots. Mean annual precipitation was the strongest such predictor of species richness (model deviance 39%), followed by mean precipitation in the wettest month (31%), mean precipitation in the driest month (26%), with a strong positive relationship observed between each of these measures (Table 5.1).

Finally, the mean elevation of a particular area was also positively correlated with richness, and this explained approximately 32% of the model deviation when examined alone. However, our model implies a simple linear relationship between elevation and species richness, when in fact, most documented studies of richness across a variety of spatial scales, from local to global, have found that there is no simple monotonic relationship between elevation and diversity (Hillebrand, 2004). Instead, it is likely that a humped distribution better predicts this relationship, with diversity peaking at some intermediate altitude and declining at low and high altitudes, although this relationship can take different forms, depending upon the scale of the analysis (Nogues-Bravo et al., 2008). Furthermore, a strong relationship also exists between species richness and the standard deviation of elevation within a region (data not shown). As mentioned above, much of this effect may be driven by the complex climatic nature of mountain environments which often offer a great diversity of habitats and heterogeneity which may further allow for high levels of species to co-exist in a given area. Species diversity is positively correlated with the heterogeneity, range and types of habitats found in a particular area. A given region containing a high range of elevation may therefore encompass a greater range of available habitats within it. Furthermore, altitude affects a greater number of ecological factors, including the temperature and precipitation of a region, which will create a greater variety of available niches. In our analysis, the ranges of bioclimatic variables (standard deviation) were not shown to be strong predictors of richness. This is likely due to the fact that much of the actual variation is averaged away when looking at such a broad area as used in our study.

The linear models of species richness within monocot orders and environmental variables revealed significant heterogeneity and variation in: (1) the strength of response of particular orders to climatic and environmental variables, (2) the direction of response of each order to environmental variables and (3) the overall explanatory power of the simple model to explain species-richness patterns within particular orders. Below, we discuss the overall patterns within and among particular orders and the importance of this finding for understanding global patterns of species richness in an evolutionary context.

5.3.4 Predicting the effect of climate change on species richness

One goal of macroecological studies of diversity has been to identify those climatic variables that determine levels of species richness within broad taxonomic groups (birds, mammals, amphibians, angiosperms) to arrive at an understanding of general principles of the ecological factors underlying biological diversity. For example, several previous studies in plants have used a correlative approach similar to the one presented here (Qian, 1998; Qian and Ricklefs, 2000; Kreft and Jetz, 2007; Sommer et al., 2010). These studies also found that many of the same variables that contribute to richness in our analysis (moisture, temperature, evapotranspiration) are generally also important when one looks at patterns across all angiosperms. The correlative approach used here to identify the important climatic drivers of biodiversity is attractive because by understanding the effects of climate on diversity, it may also be possible to forecast levels of species extinctions or changes in biodiversity patterns expected in the future, given a variety of different climate change scenarios (Pereira et al., 2010; Sommer et al., 2010).

Even within related groups of organisms, such as the monocots, there exists wide natural variation in the observed correlations of richness levels among different

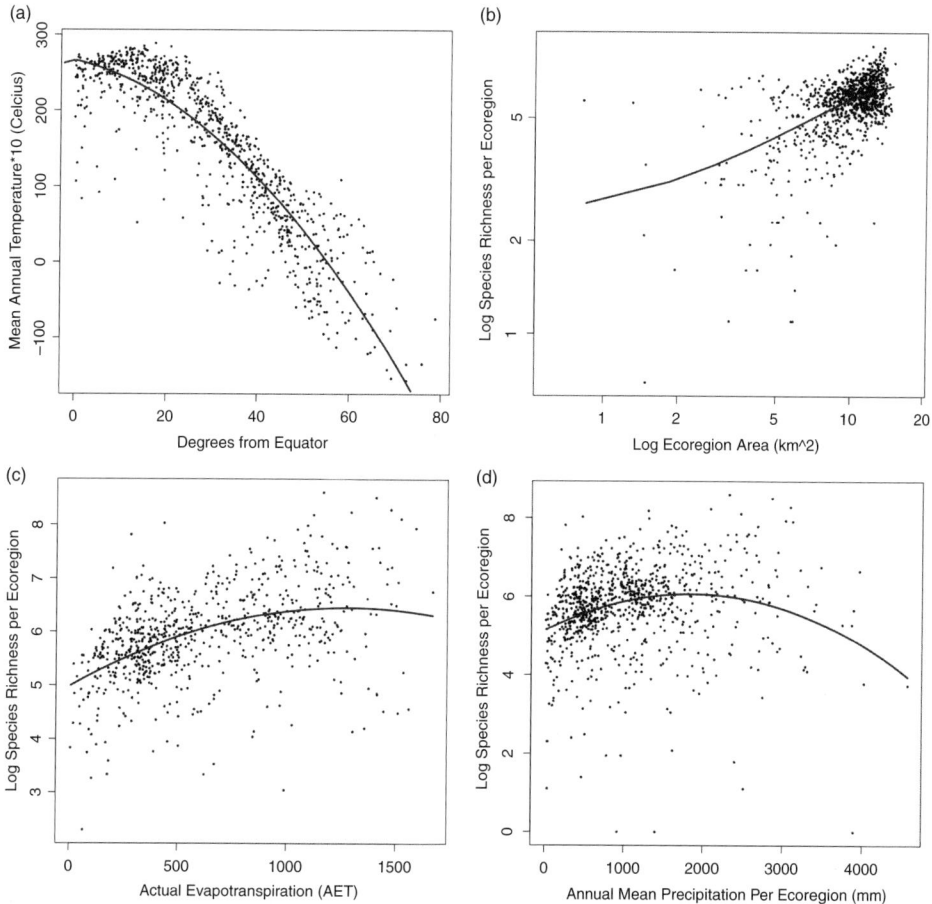

Fig 5.4 Regression models between A: latitude and mean annual temperature (Multiple $R^2 = 0.79$), B: log ecoregion area and log monocot species richness (Multiple $R^2 = 0.31$), C: mean actual evapotranspiration and log species richness (Multiple $R^2 = 0.22$), D: mean annual precipitation and log species richness (Multiple $R^2 = 0.07$).

orders to climatic variability. As is shown in Table 5.2, we found that a relatively simple linear model of species richness for all monocots made different predictions regarding the strength and direction of the effects of climate on richness patterns. When all the monocots are considered as a whole and richness patterns are regressed on the environmental variables, species richness is expected to be highest in areas of high temperature, high AET, high precipitation (Fig 5.4) and elevation.

The climatic relationships are not always consistent when one examines richness patterns within different monocot orders. The strength of the relationship between richness and a climatic variable (the effect size), the sign of the relationship, and the overall explanatory power of the different variables may differ among orders. When we analysed the effect of the same climate variables on richness within different orders of monocots and compared this to the model considering all monocots, a different picture emerged. For example, the model for all mono-cots predicts an increase in richness as an effect of increased temperature. How-ever, the order-level models predict that richness within Liliales is more sensitive to increases in temperature. Species richness in Liliales is negatively correlated with annual mean temperature, a result that is in contrast to the pattern observed in other orders, but may also be a consequence of using a linear predictor to predict a nonlinear relationship between richness and temperature (for example, see Fig 5.4C). Moreover, in some groups, such as Alismatales and Poales, temperature is not a significant predictor of richness at all (Table 5.2).

Our results clearly show that there can be wide variation in the ability of climatic variables to explain species richness. Model deviance explained for the GLMs for the order analysis ranged from 24.7% for Poales to 53.5% for Arecales. That we see variation in both the strength of the effects of climate variables on richness, as well as the direction of the effect and large variation in the overall explanatory power of a simple model within particular groups, suggests a number of things. First of all, future attempts at modelling species richness of monocots should take into account variation that exists within taxonomic groups. This variation has import-ant biological implications as different families and orders vary in importance within ecological communities. This variation is likely driven by different traits present within groups that enable survival in different environments, so biological traits should also be considered in such an analysis.

The results presented here have implications for predicting the response of species richness to changing climate. Most modelling studies that simply consider species richness within particular areas completely ignore the individualistic nature of species response to climatic variables. Because species differ from one another in their range sizes, in their level of adaptation to local climatic conditions, in their ability to migrate or evolve *in situ* to changing climate and in their past responses to environmental change (Stewart et al., 2010), it is simplistic to expect that one can use purely correlative models to predict changes in biological diver-sity across global scales by using input from global circulation models. Clearly, species-specific differences in responses to climatic variables would be expected to create differences among species in their vulnerability to and probability of extinction during climate change (Thomas et al., 2004).

We have shown that variation exists at the order level and it would not be surprising to find that significant variation exists at all taxonomic levels. However, as in the above analysis, species are treated as a single component of a local community without regard to their relative abundance. Species numbers in an area are comparatively easy to quantify. Less tractable is the determination of relative abundances of those species, and finally, even less tractable is an understanding of how interactions among species both within and among different trophic levels ultimately determine presence–absence within a particular area or their relative abundances.

One approach to overcoming the limitations of treating all species as equal would be to model individual species responses to environmental conditions using niche-modelling approaches. However, particularly for plants, lack of information on species occurrences can limit the applicability of these methods across large groups such as the monocots. Determining the extent to which species response to climatic and environmental conditions is phylogenetically conserved within and among genera and families might lead to better predictive models, not only of the response of species in the past to changing climate, but also of the effects of climatic change on biodiversity in the future.

5.4 Acknowledgements

This research was supported by generous funding from the Leverhulme Trust awarded to VS.

5.5 References

Ahn, C. H. and Tateishi, R. (1994). Development of a global 30-minute grid potential evapotranspiration dataset. *Journal of the Japanese Society of Photogrammetry and Remote Sensing*, **33**, 12–21.

APG III (Angiosperm Phylogeny Group III). (2009). An update of the angiosperm phylogeny group classification for the orders and families of flowering plants: APG III. *Botanical Journal of the Linnean Society*, **161**, 105–121.

Arrhenius, O. (1921). Species and area. *Journal of Ecology*, **9**, 95–99.

Bjorholm, S., Svenning, J.-C., Skov, F. and Balslev, H. (2005). Environmental and spatial controls of palm (Arecaceae) species richness across the Americas. *Global Ecology and Biogeography*, **14**, 423–429.

Brummitt, R. K. (2001). *World Geographical Scheme for Recording Plant Distributions*. Pittsburgh, PA: Hunt Institute for Botanical Documentation, Carnegie Mellon University.

Burnham, K. P. and Anderson, D. R. (2004). *Model Selection and Multimodel Inference: A Practical Information-theoretic Approach*, New York: Springer.

Bush, M. B. and Colinvaux, P. A. (1990). A pollen record of a complete glacial cycle from lowland Panama. *Journal of Vegetation Science*, **1**, 105–118.

Clarke, A. and Gaston, K. J. (2006). Climate, energy and diversity. *Proceedings of the Royal Society B: Biological Sciences*, **273**, 2257–2266.

Currie, D. J. (1991). Energy and large-scale patterns of animal- and plant-species richness. *The American Naturalist*, **137**, 27–49.

Currie, D. J., Mittelbach, G. G., Cornell, H. V. et al. (2004). Predictions and tests of climate-based hypotheses of broad-scale variation in taxonomic richness. *Ecology Letters*, **7**, 1121–1134.

Davies, R. G., Orme, C. D. L., Storch, D. et al. (2007). Topography, energy and the global distribution of bird species richness. *Proceedings of the Royal Society B: Biological Sciences*, **274**, 1189–1197.

Davies, T. J., Barraclough, T. G., Savolainen, V. and Chase, M. W. (2004). Environmental causes for plant biodiversity gradients. *Philosophical Transactions of the Royal Society of London. Series B: Biological Sciences*, **359**, 1645–1656.

Distler, T., Jørgensen, P. M., Graham, A., Davidse, G. and Jiménez, I. (2009). Determinants and prediction of broad-scale plant richness across the western neotropics. *Annals of the Missouri Botanical Garden*, **96**, 470–491.

Dransfield, J., Uhl, N., Asmussen, C. et al. (2008). *Genera Palmarum, Evolution and Classification of the Palms*. Kew: Royal Botanic Gardens.

Edwards, E. J., Osborne, C. P., Strömberg, C. a. E., Smith, S. A. and Consortium, C. G. (2010). The origins of C4 grasslands: Integrating evolutionary and ecosystem science. *Science*, **328**, 587–591.

ESRI (2009). *ArcGIS 9.2* [Geographical Information System] http://www.esri.com/

Francis, A. P. and Currie, D. J. (2003). A globally consistent richness-climate relationship for angiosperms. *The American Naturalist*, **161**, 523–536.

Gleason, H. A. (1922). On the relation between species and area. *Ecology*, **3**, 158–162.

Govaerts, R. (2004). The monocot checklist project. *Taxon*, **53**, 144–146.

Hannah, L. and Lovejoy, T. E. (2007). Conservation, climate change, and tropical forests. In *Tropical Forest Responses to Climate Change*, ed. Bush,

M. B. and Flenley, J. R. Chichester, UK: Praxis, pp. 367–378.

Hawkins, B. A., Field, R., Cornell, H. V. et al. (2003). Energy, water, and broad-scale geographic patterns of species richness. *Ecology*, **84**, 3105–3117.

Hijmans, R. J., Cameron, S. E., Parra, J. L., Jones, P. G. and Jarvis, A. (2005). Very high resolution interpolated climate surfaces for global land areas. *International Journal of Climatology*, **25**, 1965–1978.

Hillebrand, H. (2004). On the generality of the latitudinal diversity gradient. *The American Naturalist*, **163**, 192–211.

Humboldt, A. (1845–1858). *Kosmos: Entwurf einer physischen Weltbeschreibung*, Stuttgart/Tubingen, Germany, Cotta.

Janzen, D. H. (1967). Why mountain passes are higher in the tropics. *The American Naturalist*, **101**, 233–249.

Kreft, H. and Jetz, W. (2007). Global patterns and determinants of vascular plant diversity. *Proceedings of the National Academy of Sciences*, **104**, 5925–5930.

Leigh, E. G., Davidar, P., Dick, C. W. et al. (2004). Why do some tropical forests have so many species of trees? *Biotropica*, **36**, 447–473.

MacArthur, R. H. and Wilson, E. O. (1967). *The Theory of Island Biogeography*, Princeton, NJ: Princeton University Press.

Mittelbach, G. G., Schemske, D. W., Cornell, H. V. et al. (2007). Evolution and the latitudinal diversity gradient: Speciation, extinction and biogeography. *Ecology Letters*, **10**, 315–331.

Nogues-Bravo, D., Araujo, M. B., Romdal, T. and Rahbek, C. (2008). Scale effects and human impact on the elevational species richness gradients. *Nature*, **453**, 216–219.

Olson, D.M., Dinerstein, E., Wikramanayake, E.D. et al. (2001). Terrestrial ecoregions of the world: A new map of life on earth. *Bioscience*, **51**, 933–938.

Pereira, H. M., Leadley, P. W., Proença, V. et al. (2010). Scenarios for global biodiversity in the 21st century. *Science*, **330**, 1496–1501.

Pianka, E. R. (1966). Latitudinal gradients in species diversity: A review of concepts. *The American Naturalist*, **100**, 33–46.

Preston, F. W. (1960). Time and space and the variation of species. *Ecology*, **41**, 612–627.

Pridgeon, A. M., Cribb, P. J., Chase, M. W. and Rasmussen, F. N. (1999). *Genera Orchidacearum. Volume 1, General Introduction, Apostasioideae, Cypripedioideae*. Oxford: Oxford University Press.

Qian, H. (1998). Large-scale biogeographic patterns of vascular plant richness in North America: An analysis at the generic level. *Journal of Biogeography*, **25**, 829–836.

Qian, H. and Ricklefs, R. E. (2000). Large-scale processes and the Asian bias in species diversity of temperate plants. *Nature*, **407**, 180–182.

Qian, H. and Ricklefs, R. E. (2007). A latitudinal gradient in large-scale beta diversity for vascular plants in North America. *Ecology Letters*, **10**, 737–744.

Qian, H. and Ricklefs, R. E. (2008). Global concordance in diversity patterns of vascular plants and terrestrial vertebrates. *Ecology Letters*, **11**, 547–553.

R Development Core Team. 2010. *R: A Language and Environment for Statistical Computing*. Vienna: R Foundation for Statistical Computing.

Ricklefs, R. E. (2006). Evolutionary diversification and the origin of the diversity-environment relationship. *Ecology*, **87**, 3–13.

Schemske, D. W., Mittelbach, G. G., Cornell, H. V., Sobel, J. M. and Roy, K. (2009). Is there a latitudinal gradient in the importance of biotic interactions? *Annual Review of Ecology, Evolution, and Systematics*, **40**, 245–269.

Sommer, J., Kreft, H., Kier, G. et al. (2010). Projected impacts of climate change on regional capacities for global plant species richness. *Proceedings of the Royal Society B: Biological Sciences*, **277**, 2271–2280.

Stevens, G. C. (1989). The latitudinal gradient in geographical range: How so many species coexist in the tropics. *The American Naturalist*, **133**, 240–256.

Stewart, J. R., Lister, A. M., Barnes, I. and Dalén, L. (2010). Refugia revisited: Individualistic responses of species in space and time. *Proceedings of the Royal Society B: Biological Sciences*, **277**, 661–671.

Thomas, C. D., Cameron, A., Green, R. E. et al. (2004). Extinction risk from climate change. *Nature*, **427**, 145–148.

US Geological Survey (2003). *GTOPO30 [Global Digital Elevation Model]. Earth Resources Observation and Science (EROS) Center*. Available at: http://eros. usgs.gov/#/Find_Data/Products_ and_Data_Available/gtopo30_info (accessed December 2012).

Wiens, J. J. and Donoghue, M. J. (2004). Historical biogeography, ecology and species richness. *Trends in Ecology and Evolution*, **19**, 639–644.

Wright, D. H. (1983). Species-energy theory: An extension of species-area theory. *Oikos*, **41**, 496–506.

6

In time and with water … the systematics
of alismatid monocotyledons

DONALD H. LES AND NICHOLAS P. TIPPERY

6.1 Introduction

> *In time and with water, everything changes.*
>
> (Leonardo da Vinci)

In a way, Leonardo da Vinci's succinct characterization of water also applies appositely to the aquatic subclass Alismatidae Takht. ('alismatids'). Indeed, the long evolutionary history of this remarkably diverse group of monocotyledons has resulted in numerous adaptations to facilitate an existence in water, and these represent a substantial departure from their terrestrial counterparts.

Arguably, alismatids represent the greatest adaptive radiation of freshwater plants on Earth. Here recognized to include 11 families, 57 genera and approximately 477 species (Table 6.1), the group contains a rich diversity of floral and vegetative habits, pollination systems (including the largest concentration of water-pollinated species) and also the only occurrences of marine plants in the angiosperms.

Formal studies of alismatid relationships began 40 years ago, initially by phenetic and cladistic analyses of nonmolecular data. The first macromolecular-based phylogeny of the subclass did not appear until 1993, when roughly half of the families and fewer than 20% of genera were evaluated using *rbc*L sequence data (Les et al., 1993). The most comprehensive phylogenetic survey of alismatids

Early Events in Monocot Evolution, eds P. Wilkin and S. J. Mayo. Published by Cambridge University Press. © The Systematics Association 2013.

Table 6.1 Taxonomic synopsis of subclass Alismatidae showing families (bold font; number of genera indicated in brackets) and subordinate genera (approximate number of species in parentheses).

Alismataceae Vent. [17]

Albidella Pichon (1), *Alisma* L. (11), *Astonia* S.W.L. Jacobs (1), *Baldellia* Parl. (2), *Burnatia* Micheli (1), *Butomopsis* Kunth (1), *Caldesia* Parl. (4), *Damasonium* Mill. (4), *Echinodorus* Rich. and Engelm. ex A. Gray (28), *Helanthium* Engelm. ex Benth. and Hook.f. (3), *Hydrocleys* Rich. (5), *Limnocharis* Humb. and Bonpl. (2), *Limnophyton* Miq. (4), *Luronium* Raf. (1), *Ranalisma* Stapf (2), *Sagittaria* L. (40), *Wiesneria* Micheli (3)

Aponogetonaceae Planch. [1]

Aponogeton L.f. (50)

Butomaceae Mirb. [1]

Butomus L. (1)

Cymodoceaceae Vines [6]

Amphibolis C.Agardh (2), *Cymodocea* K.D.Koenig (4), *Halodule* Endl. (6), *Ruppia* L. (4), *Syringodium* Kütz. (2), *Thalassodendron* Hartog (2)

Hydrocharitaceae Juss. [17]

Apalanthe Planch. (1), *Appertiella* C.D.K. Cook and L. Triest (1), *Blyxa* Noranha ex Thouars (9), *Egeria* Planch. (3), *Elodea* Michx. (5), *Enhalus* Rich. (1), *Halophila* Thouars (10), *Hydrilla* Rich. (1), *Hydrocharis* L. (3), *Lagarosiphon* Harv. (9), *Limnobium* Rich. (2), *Najas* L. (39), *Nechamandra* Planch. (1), *Ottelia* Pers. (21), *Stratiotes* L. (1), *Thalassia* Banks ex K.D. Koenig (2), *Vallisneria* L. (18)

Juncaginaceae Rich. [3]

Cycnogeton Endl. (1), *Tetroncium* Willd. (1), *Triglochin* L. (15)

Maundiaceae Nakai [1]

Maundia F.Muell. (1)

Posidoniaceae Vines [1]

Posidonia K.D. Koenig (5)

Potamogetonaceae Bercht. and J. Presl [7]

Althenia F.Petit (1), *Groenlandia* Fourr. (1), *Lepilaena* Harv. (5), *Potamogeton* L. (100), *Pseudalthenia* (Graebn.) Nakai (1), *Stuckenia* Börner (3), *Zannichellia* L. (6)

Scheuchzeriaceae F. Rudolphi [1]

Scheuchzeria L. (1)

Zosteraceae Dumort. [2]

Phyllospadix Hook. (5), *Zostera* L. (9)

was published in 1997 and included all recognized families, 83% of genera and 15% of the species. Phylogenetic studies consistently have supported the monophyly of alismatids, as we will discuss.

In this chapter we provide an overview of phylogenetic research directed at the elucidation of relationships with and within the alismatids since these initial efforts, and summarize how various genetic loci have been used to evaluate high, intermediate and low taxonomic levels in the group. A main objective is to provide a comprehensive, comparative evaluation of existing systematic information and to identify those groups that would benefit most from additional study.

6.2 New data

Many of the phylogenetic trees presented in the discussion have been taken directly from pertinent published literature, but have been adapted and redrawn to summarize the results as succinctly as possible. In these instances a brief summary of the original methodology used to derive the trees is provided in the text and/or figure captions. Because these trees have been redrawn, we have not provided values of internal nodal support and refer the reader to the original publications for more specific details of these analyses.

In addition, we also include several phylogenetic trees from our own ongoing work, which are based on new, unpublished data and analyses. In these instances, a consistent analytical methodology was used, which is summarized briefly as follows. Nucleotide sequence data were obtained from specimens preserved in CTAB (Rogstad, 1992) or dried in silica. In addition to our own collections, supplementary materials were provided by J. Bogner (*Aponogeton*), C. Bove (*Najas*), C. B. Hellquist (*Elodea*), W. Iles (*Maundia*), S. W. L. Jacobs (various species), C. Kasselmann (*Aponogeton*), D. Padgett (*Vallisneria*), D. Perleberg (*Najas*), C. Phiri (*Vallisneria*) and H. W. E. van Bruggen (*Aponogeton*). Genomic DNA was extracted using a standard method (Doyle and Doyle, 1987) then amplified and sequenced for select gene regions following Les et al. (2008), using previously published primers (Baldwin 1992; Johnson and Soltis 1995; Bremer et al., 2002; Les et al., 2008, 2009; Tippery et al., 2008). Molecular data were analysed using both equally weighted maximum parsimony (MP) and maximum likelihood (ML) methods. Heuristic MP tree searches and bootstrap (BS) analyses were performed in PAUP* ver. 4.0b10 (Swofford, 2002) using the parameters given in Les et al. (2009). Single-tree and bootstrap (1000 replicates) ML analyses were conducted in GARLI ver. 0.97.r737 (Zwickl, 2006) on data partitioned among different gene regions using models recommended by jModelTest ver. 0.1.1 under the AIC criterion (Posada, 2008). The resulting trees are depicted as MP strict consensus trees with the captions indicating the specific ML evolutionary model applied, nodal support values and the following tree statistics: CI (consistency index), CI$_{exc}$ (consistency index excluding uninformative sites) and RI (retention index). These newly generated trees appear in Figs 6.2, 6.4, 6.11, 6.15, and 6.16. GenBank accession numbers for all sequences used in our analyses (including those newly generated) are provided in the Appendix.

6.3 Circumscription of Alismatidae

The spate of phylogenetic studies conducted over the past several decades has helped to clarify the circumscription of Alismatidae substantially. All pertinent molecular phylogenetic studies to date (e.g. Fig 6.1) consistently have resolved at least the core alismatid families (see Les et al., 1997) as a clade, thus confirming their monophyly. To this end, necessary taxonomic alterations have included the removal of two enigmatic families (Triuridaceae Gardner, Petrosaviaceae Hutch.) once placed traditionally with alismatids (Tomlinson, 1982), but since demonstrated to resolve outside the group phylogenetically (Cameron et al., 2003; Li and Zhou, 2007).

In recent taxonomic syntheses (e.g. APG I, 1998; APG II, 2003; APG III, 2009), alismatids were treated as a single order (Alismatales Dumort.), which, along with the more traditional families, also included Araceae Juss. and Tofieldiaceae Takht. In that interpretation, the Alismatales were monophyletic with Tofieldiaceae regarded as the sister group to the more traditional alismatid taxa and Araceae as the sister to that clade (Judd et al., 2008).

That circumscription has become accepted by many, due in large part to the results of several published studies which have rendered this same congruent topology (Fig 6.1A) from analyses of a variety of molecular data sets, including *rbc*L, *ndh*F, combined *mat*K + *rbc*L sequences and combined cpDNA (*atp*B + *rbc*L) + nuclear 18S rDNA sequences (Chase et al., 1993; Tamura et al., 2004; Givnish et al., 2006; Soltis et al., 2007).

However, that proposed circumscription of Alismatales is not without inconsistencies and there also exists a fair amount of conflicting data. A large amount of cpDNA sequence data (Fig 6.1B) resolves Araceae, not Tofieldiaceae, as the sister group of core alismatids, and some combined data (Fig 6.1F) resolve Araceae and Tofieldiaceae as their sister clade (Li and Zhou, 2009). More problematic are nuclear *PHYC* sequences (Duvall and Ervin, 2004) which place Tofieldiaceae completely outside of the group (Fig 6.1C) and some combined DNA data, which place Acoraceae Martinov as the sister group of core alismatids (Fig 6.1D, E). In fact, the most diverse data set (which includes chloroplast, mitochondrial and nuclear gene sequences) also is one that resolves Acoraceae as the sister group of alismatids (Fig 6.1D), an interesting result given that Acoraceae have become accepted widely as the sister group to all other monocotyledons.

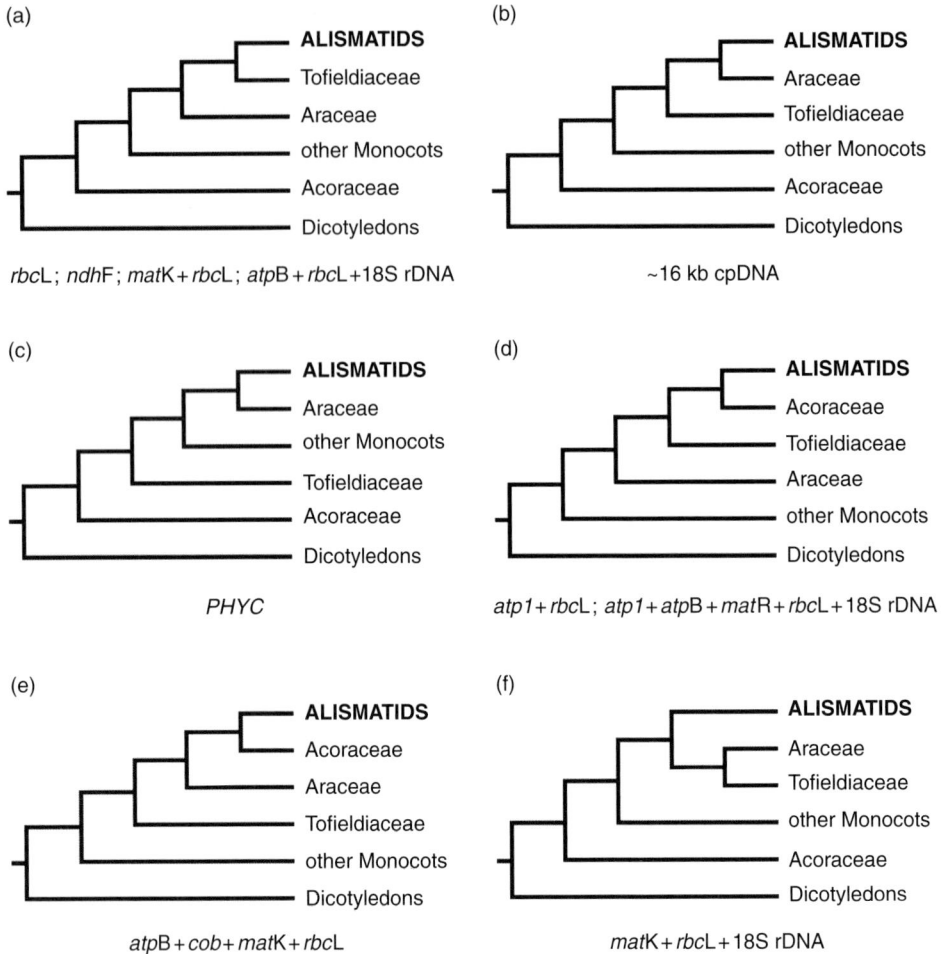

Fig 6.1 Generalized cladograms based on several studies incorporating various sources of DNA sequence data, showing different placements of alismatids among other monocotyledons. The trees shown are simplified substantially from the original studies cited; in most cases, the dicotyledons comprise a paraphyletic sister grade. A: Tofieldiaceae resolve as the sister group of alismatids in analyses of cpDNA loci including *rbc*L (MP combinable component consensus, Chase et al., 1993), *ndh*F (MP; Givnish et al., 2006), combined *mat*K and *rbc*L sequences (MP strict consensus; Tamura et al., 2004) and combined analyses (BI majority rule consensus) of cpDNA (*atp*B + *rbc*L) and nuclear 18S rDNA sequences (Soltis et al., 2007). B: Araceae resolve as the sister group of alismatid monocotyledons in analyses based on 16 kb of cpDNA sequence data; Tofieldiaceae represent the next sister group (ML; Graham et al. 2006). C: Araceae resolve as the sister group of alismatid monocotyledons in analysis of nuclear *PHYC* sequences (BI consensus); the majority of remaining monocots represents the next sister group (Duvall and Ervin, 2004). D: Acoraceae resolve as the sister group of alismatid monocotyledons (Tofieldiaceae represent the next sister group) in combined analyses (MP strict consensus) of one mtDNA

The focus of this chapter is not to pursue an extensive discussion on the problems associated with phylogenetic reconstruction at deep taxonomic levels. Yet, given these discrepancies, we advocate that a more conservative circumscription of alismatids be adopted, specifically one that is limited to those core families that consistently resolve as a clade. Accordingly, we follow here the circumscription of the group *sensu stricto* as advocated by Haynes and Les (2005), where Alismatidae are recognized at the level of subclass, and comprise two distinct clades that represent the orders Alismatales and Potamogetonales Dumort. (= Zosterales Nakai). This group essentially represents the same one as indicated by the analysis of Davis et al. (2004). The circumscription is similar to that indicated by the classification system of Thorne and Reveal (2007), but substitutes the rank of subclass (Alismatidae) for superorder (Alismatanae Takht.), in exclusion of Araceae and Tofieldiaceae.

The realization that extant alismatids comprise two distinct clades was first indicated by a phylogenetic analysis of single-gene (*rbc*L) sequence data for 11 taxa representing 9 of the accepted families (Les et al., 1993). Subsequently, that result was maintained in more comprehensive *rbc*L surveys, including representatives of 25 genera from 15 families (Les and Haynes, 1995), and then 69 species from 47 genera (Les et al., 1997). We now have evaluated an even larger *rbc*L data set consisting of 167 *rbc*L sequences representing 158 alismatid taxa and 9 outgroup (Arales Dumort.) taxa (Fig 6.2). This most comprehensive analysis of alismatid taxa to date is consistent with previous studies in resolving the same two major clades (designated as clade I and clade II), which are recognized here as the orders Alismatales and Potamogetonales. Nine subclades (A–I) also are evident and this notation is used in the remainder of the text to facilitate discussion.

In addition to the single-gene *rbc*L studies, other data sets also confirm the fundamental subdivision of the Alismatidae into two clades. Analyses of nearly 14 kb of cpDNA sequence data (Iles et al., 2009), as well as combined mtDNA (*atp*1 + *cob*) and cpDNA (*rbc*L) sequence data (Petersen et al., 2006) all provide highly congruent phylogenetic topologies (Fig 6.3) that agree with the results of the various *rbc*L analyses. Furthermore, these two major clades are associated with fundamentally different derivations of floral structure, which have been termed

---

Caption for Fig 6.1 (*cont.*) locus (*atp*1) and one cpDNA locus (*rbc*L) (Davis et al. 2004) and in combined analyses of mitochondrial (*atp*1 + *atp*B + *mat*R), chloroplast (*rbc*L), and nuclear (18S rDNA) sequences (MP; Qiu et al., 2000). E: Acoraceae resolve as the sister group of alismatid monocotyledons in combined analyses (MP) of two mtDNA loci (*atp*1 + *cob*) and two cpDNA loci (*mat*K + *rbc*L); Araceae represent the next sister group (Petersen et al., 2006). F: An Araceae/Tolfieldiaceae clade resolves as the sister group of alismatid monocotyledons in combined analyses (MP strict consensus) of cpDNA (*mat*K + *rbc*L) and 18S rDNA sequences (Li and Zhou, 2007).

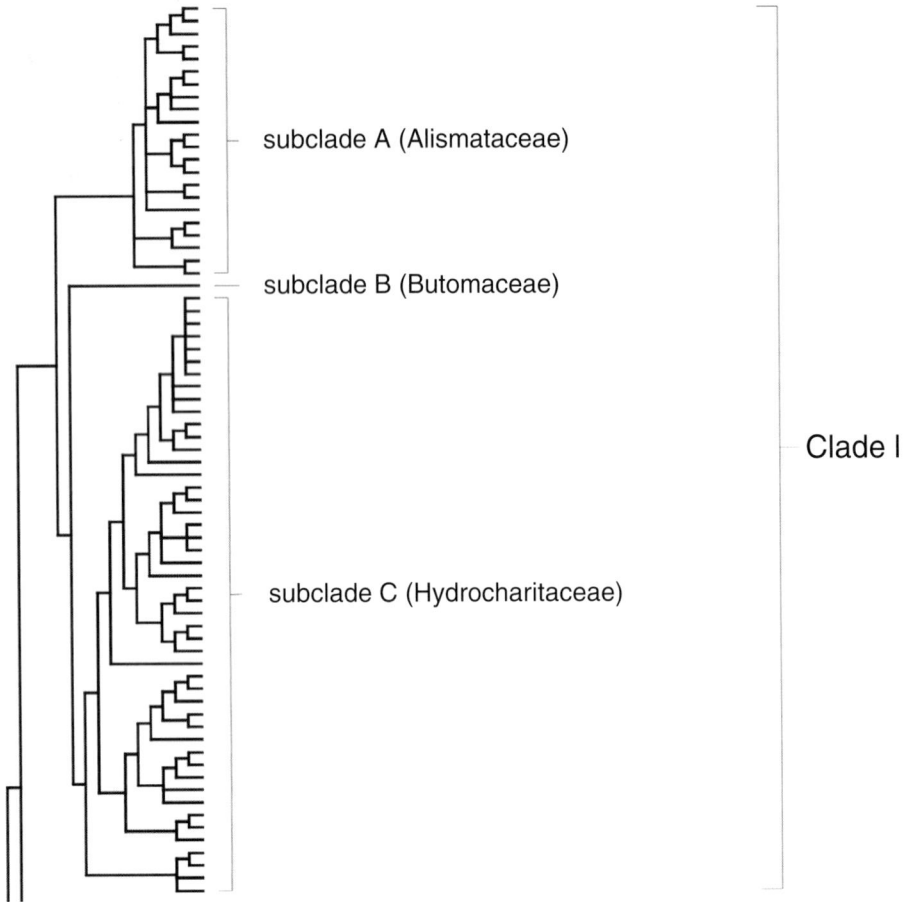

Fig 6.2 Phylogenetic analysis of Alismatidae using *rbc*L sequence data for 167 taxa (current study). The tree identifies two major clades (I, II), which represent the orders Alismatales and Potamogetonales, respectively. Nine subclades (A–I) have been identified in order to facilitate discussion throughout the text. The MP strict consensus tree is shown. Because of space constraints, nodal support values were excluded from this tree. However, expanded details of several subclades are shown in Figs 6.10, 6.17, 6.18 and 6.23, in which the values of MP and ML (GTR+I+G model) nodal BS support are indicated above and below branches respectively ('-' indicates <50% support). In addition, several text discussions provide specific support values for relevant portions of this tree that have not been shown in any of the additional figures. Tree statistics: CI = 0.36; CI_{exc} = 0.32; RI = 0.86; ln$L$ = –12283.

'petaloid' (Alismatales) and 'tepaloid' (Potamogetonales) forms (Posluszny et al., 2000). Taken together, these data indicate that this circumscription of Alismatales and Potamogetonales does appear to designate cohesive and phylogenetically meaningful clades.

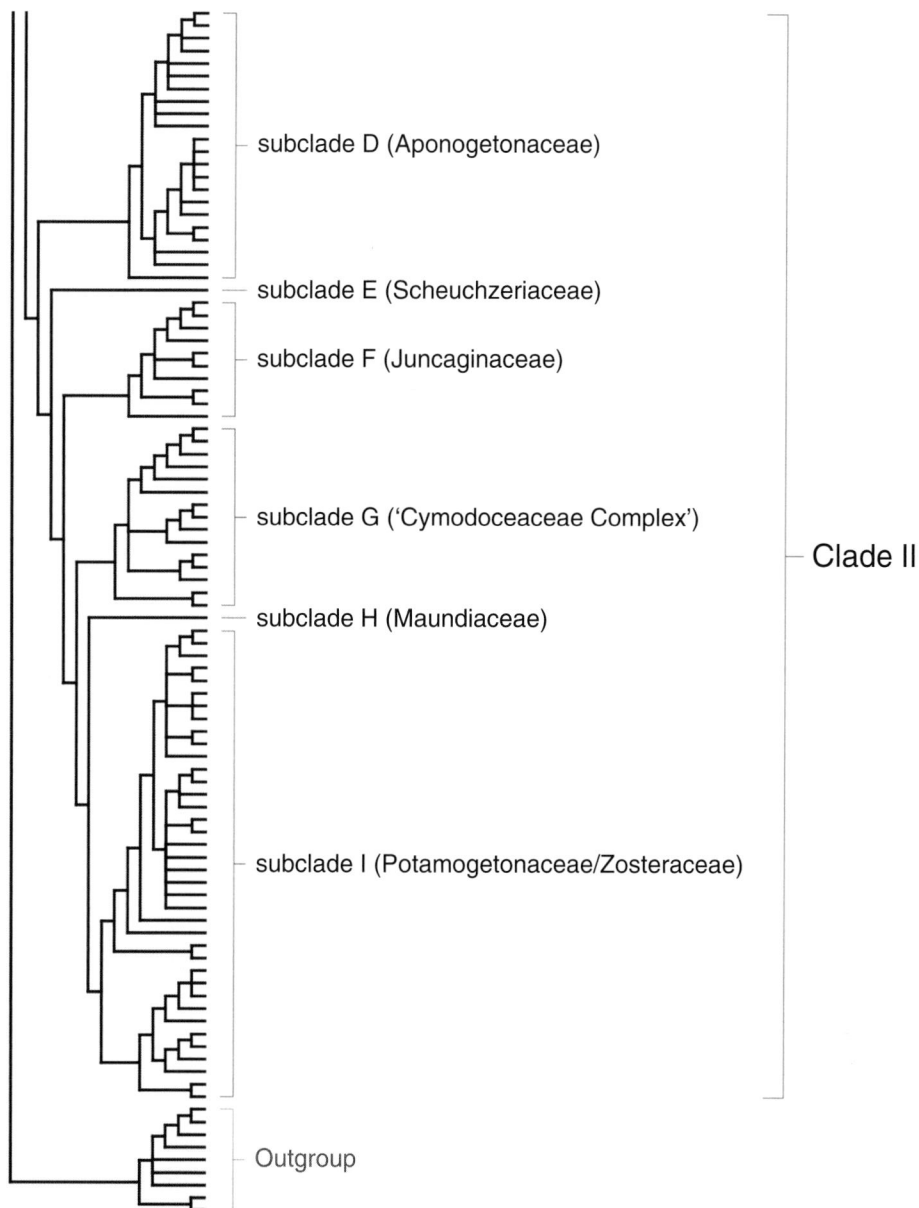

Fig 6.2 (*cont.*)

6.4 Alismatales (clade I)

The circumscription of Alismatales followed here includes three families (Alismataceae, Butomaceae and Hydrocharitaceae) and differs from the concept presented by Haynes and Les (2005) by transferring various genera included

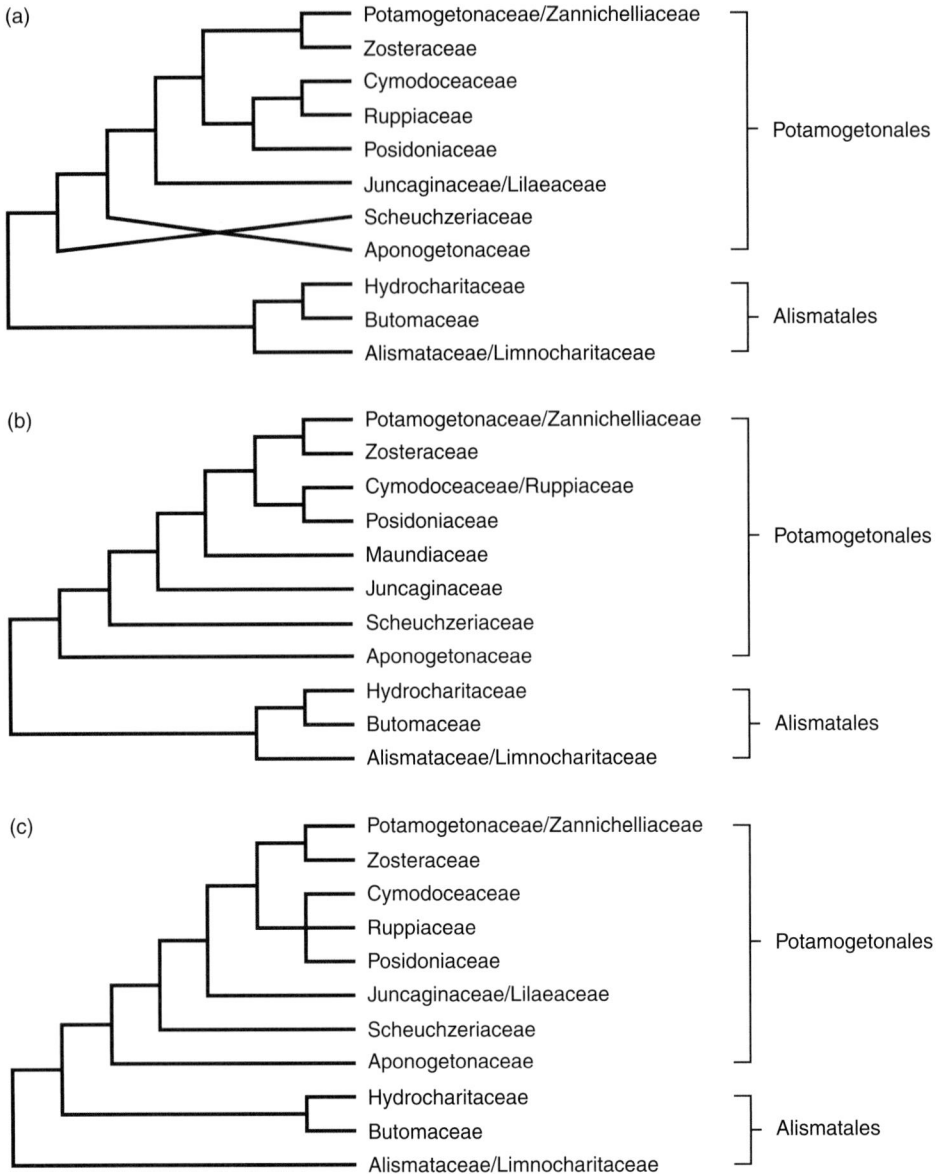

Fig 6.3 Phylogenetic relationships among alismatid families. A: Resolution of two distinct clades corresponding to Potamogetonales and Alismatales *sensu stricto* (s.s.) resulting from analysis of *rbc*L sequence data (weighted MP strict consensus; Les et al., 1997). B: Topology highly congruent with previous tree (with addition of Maundiaceae) resulting from parsimony/likelihood analyses of roughly 14 kb of DNA sequence data (ML; Iles et al., 2009). C: Relationships derived from combined analysis (MP) of two mitochondrial loci (*atp*1, *cob*) and chloroplast (*rbc*L) data differ slightly from previous analyses primarily by the resolution of Alismatales s.s. as a grade rather than a clade (Petersen et al., 2006). None of the conflicts observed among these trees is strongly supported.

formerly in Limnocharitaceae and Najadaceae to Alismataceae and Hydrocharitaceae, respectively.

The small family Limnocharitaceae Takht. was established less than 60 years ago. The three component genera (*Butomopsis*, *Hydrocleys* and *Limnocharis*) traditionally had been placed in Butomaceae, but were viewed also as having affinities with Alismataceae (Cronquist, 1981). In this sense, Limnocharitaceae were regarded as intermediate phylogenetically between Alismataceae and Butomaceae. However, in recent treatments, Limnocharitaceae were merged with Alismataceae essentially because 'Convincing evidence for the monophyly of Alismataceae s.s. is lacking . . .' (APG III, 2009).

The 47-genera *rbc*L sequence analysis by Les et al. (1997) was consistent with the latter conclusion by placing two of the three genera of Limnocharitaceae (*Hydrocleys*, *Limnocharis*) within a clade containing eight genera of Alismataceae; in that study, however, the two families were maintained as separate, awaiting more conclusive evidence necessitated by the low level of internal support obtained for critical branches. That study at least clarified that neither *Hydrocleys* nor *Limnocharis* was closely allied with Butomaceae, as some earlier authors had believed.

We since have been able to procure material of the third genus assigned to Limnocharitaceae (*Butomopsis*) and have revisited the issue of relationships by analysing a combined data set of *mat*K + *rbc*L + nrITS sequences (Fig 6.4). The results of this analysis yielded a strongly supported clade comprising the three former Limnocharitaceae genera, which was embedded within Alismataceae by both ML (not shown) and MP analyses (Fig 6.4). Although the precise placement of this clade amongst other Alismataceae was not strongly supported (BS = 62–66%), the result paralleled those obtained by the earlier *rbc*L analysis (Les et al., 1997) and even by the large cpDNA data set analysis (Iles et al., 2009), which also provided only weak support. Nevertheless, because none of these analyses resolved Limnocharitaceae as a sister clade to Alismataceae, the consistent placement of the former as a clade within the latter leads us to recommend that the groups be merged.

A similar hesitancy has characterized taxonomic dispositions of the genus *Najas* in recent years. Once typifying the order Najadales Dumort., which included such families as Cymodoceaceae, Potamogetonaceae, Ruppiaceae Horan. and Zosteraceae (Cronquist, 1981), *Najas* (Najadaceae Juss.) actually was found to possess peculiar seed-coat structures present elsewhere only among some members of Hydrocharitaceae (Shaffer-Fehre, 1991a, 1991b). Shortly afterwards, analyses of *rbc*L sequence data pointed similarly to the relationship of *Najas* and Hydrocharitaceae and also indicated its substantial taxonomic distance from other families with which it long had been associated (Les et al., 1993, 1997). Yet, the level of internal support for critical branches in the resulting phylogenies remained

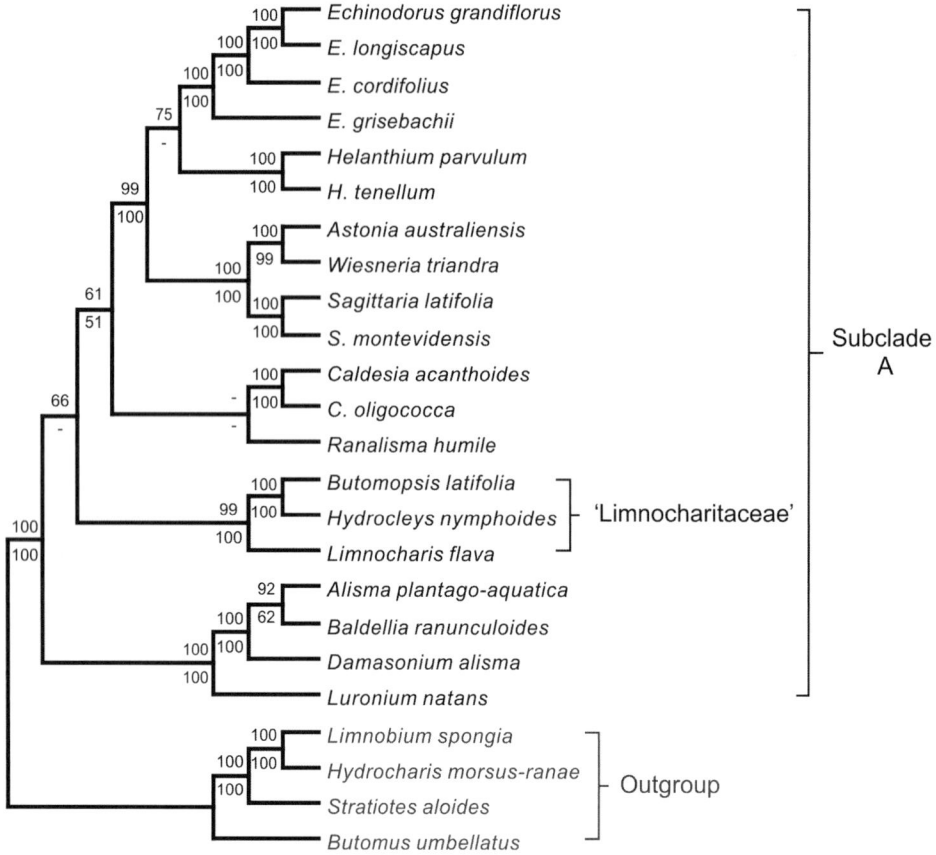

Fig 6.4 Phylogenetic relationships of Alismataceae and genera assigned formerly to Limnocharitaceae (*Butomopsis*, *Hydrocleys* and *Limnocharis*) as indicated by MP/ML analysis of combined *mat*K/*trn*K (ML: TVM+G model) and nrITS/*rbc*L (ML: GTR+I+G model) data (current study). MP strict consensus tree is shown. Values of MP and ML nodal BS support are indicated above and below branches respectively ('–' indicates <50% support). Tree statistics: CI = 0.66; CI_{exc} = 0.60; RI = 0.70; $\ln L$ = –26078.

low, and the merger of Najadaceae and Hydrocharitaceae was deferred pending the acquisition of more persuasive evidence. Although not definitive, those studies did clarify that Najadaceae were not allied with Potamogetonaceae as formerly believed, but instead were placed phylogenetically either within Hydrocharitaceae or at least as its sister group.

More convincing evidence finally was obtained by a combined analysis of morphological and molecular (nrDNA, cpDNA) data (Les et al., 2006), which provided strong internal support for the inclusion of *Najas* within Hydrocharitaceae in a position with *Hydrilla* and among other members of subfamily Hydrilloideae Luerss. Confirmation of this general result came from analysis of *atp1* + *cob* +

*rbc*L sequences (Petersen et al. 2006), which also placed *Najas* in Hydrocharita-ceae (Hydrilloideae) in proximity to *Hydrilla*. A similar position is indicated also in our 167-taxon *rbc*L analysis (Fig 6.2). The inclusion of *Najas* in Hydrocharitaceae is also strongly indicated in analyses of large amounts of cpDNA data (Iles et al., 2009).

Li and Zhou (2009) challenged the association of *Najas* in Hydrocharitaceae by the results of a combined *rbc*L-morphology analysis, despite admitting that the morphological characters used in their analysis were highly homoplasious. That analysis placed *Najas* as the sister group of Zosteraceae, but with low internal support. This association of *Najas* is not tenable. The inclusion of *Najas* in Hydrocharitaceae is supported by *mat*K (Tanaka et al., 1997), combined cpDNA, nrITS, morphology data (Les et al., 2006) and by the results of a 17-gene phylogeny (Iles et al., 2009). Furthermore, *rbc*L data alone (without morphological data) consistently have indicated the same association (Les et al., 1993, 1997).

None of these studies supports the placement of *Najas* with Zosteraceae, which is reminiscent of some results obtained by Les and Haynes (1995), who clearly demonstrated that homoplasious morphological data associated with Alismatidae can produce a misleading phylogenetic signal, which can significantly alter the placement of genera. It is evident that the association of Najadaceae and Zosteraceae obtained by Li and Zhou (2009) similarly is an artefact of homo-plasious morphological data. Conversely, all of the molecular results, as well as some morphological analyses (Les and Haynes, 1995), resolve *Najas* within or as the sister to Hydrocharitaceae. Given the preponderance of data in support of this association, we have unhesitatingly accepted the transfer of *Najas* to Hydrocharitaceae.

As a consequence of the preceding discussion, we have delimited the order Alismatales as comprising only the families Alismataceae, Butomaceae and Hydrocharitaceae, along with the dissolution of Limnocharitaceae and Najadaceae for the reasons given.

6.4.1 Alismataceae (subclade A)

We delimit Alismataceae as comprising 17 genera and 113 species (Table 6.1). A number of intergeneric phylogenetic studies have been conducted on the family, which have included analyses of one gene/two genera (Les et al., 1993), 4 genes/2 genera (Petersen et al., 2006), 1 gene/10 genera (Les et al., 1997), 17 genes/3 genera (Iles et al., 2009) and 3 genes/14 genera (Fig 6.4). The last, which is newly reported here, represents the most comprehensive study to date with respect to taxon sampling because it surveys 14/17 (82%) of the recognized genera, excluding only *Albidella*, *Burnatia* and *Limnophyton*. The inclusion of the last in Alismata-ceae was verified independently by Keener (2005), who demonstrated a close association of *Limnophyton* and *Wiesneria* using 5S-NTS sequence data.

Lehtonen and Myllys (2008) independently evaluated the position of *Albidella* using combined morphological and DNA sequence data. Although they advocated the removal of the genus from *Echinodorus*, their results showed major inconsistencies with morphology resolving *Albidella* as the sister to *Echinodorus*, *mat*K placing *Albidella* within *Echinodorus* and nuclear DNA placing it far removed from *Echinodorus* as the sister to all other Alismataceae. We also have experienced some difficulties evaluating the sequences reported for *Albidella* and recommend that the phylogenetic placement of this genus be re-evaluated with additional data.

It also is noteworthy that our combined data analysis of 14 Alismataceae genera (Fig 6.4) resolves *Helanthium* as the sister group to *Echinodorus* with weak support. Even if this result is accepted, we still advocate that these groups be retained as separate genera (rather than subgenera), because they are distinct morphologically (Lehtonen and Myllys, 2008) and are characterized by a substantial amount of genetic divergence in all molecular analyses.

Taken together, the results from all of these analyses indicate consistently that Alismataceae are monophyletic, at least in exclusion of *Burnatia*, which remains unsurveyed phylogenetically for any locus.

6.4.1.1 Infrageneric relationships

Five of seventeen genera in Alismataceae are monotypic, thus obviating the need for infrageneric phylogenetic evaluation (Table 6.1). Of the three bitypic genera (Table 6.1), where interspecific relationships similarly are uncomplicated, both species of *Ranalisma* (Fig 6.4) and *Baldellia* (Fig 6.6) have been analysed phylogenetically to confirm their monophyly; *Limnocharis laforestii* Duchass. ex Griseb. remains unsurveyed.

All three species of *Helanthium* have been evaluated thoroughly (Lehtonen, 2006, 2008; Lehtonen and Myllys, 2008), using a combination of *mat*K + nrITS + LEAFY + 5S-NTS sequences and morphological data (Fig 6.5). Additional population sampling (and perhaps also the use of higher-resolution genetic markers) will be required to satisfactorily resolve the taxonomic status of *H. bolivianum* (Rusby) Lehtonen & Myllys, which DNA sequences indicate to be paraphyletic. Some of the

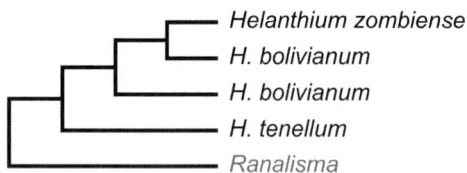

Helanthium zombiense
H. bolivianum
H. bolivianum
H. tenellum
Ranalisma

Fig 6.5 Relationships in *Helanthium* based on combined analysis (MP) of *mat*K + nrITS + LEAFY + 5S-NTS sequences (adapted from Lehtonen and Myllys, 2008).

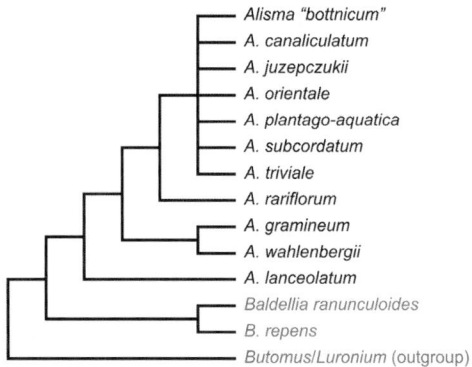

Fig 6.6 Interspecific relationships in *Alisma*/*Baldellia* as indicated by MP analysis of nrITS sequence data (redrawn from Jacobson and Hedrén, 2007).

smaller genera still in need of denser taxonomic coverage include *Caldesia, Damasonium, Hydrocleys, Limnophyton* and *Wiesneria*.

The three largest genera (*Alisma, Echinodorus* and *Sagittaria*) have been fairly well studied phylogenetically, at least enough to propose circumscriptions that arguably reflect monophyletic taxa. Jacobson and Hedrén (2007) evaluated all 11 species of *Alisma* (and both species of *Baldellia*) using nrITS/*trn*L sequences and RAPD data (Fig 6.6). In the case of *Alisma*, nrITS sequences provided low resolution for seven of the species; however, further resolution was achieved by evaluation of RAPD data (Jacobson and Hedrén, 2007). The *trn*L locus also was evaluated, but provided minimal information (due to the low number of parsimony-informative sites) and the phylogenetic signal reportedly conflicted with that obtained for the two nuclear data sets (Jacobson and Hedrén, 2007). It would be informative to obtain additional cpDNA data for *Alisma* to determine whether other chloroplast loci indicate similar discrepancies due possibly to factors such as hybridization.

Echinodorus has been studied quite thoroughly (Lehtonen, 2006, 2008; Lehtonen and Myllys, 2008) by evaluating morphological data for all 28 species and multiple DNA sequences (*mat*K + nrITS + LEAFY + 5S-NTS) for 23 of the 28 recognized species (Fig 6.7). These studies provide a reasonable overview of phylogenetic relationships in *Echinodorus* and support its distinction from *Helanthium*. Yet, despite this fair amount of systematic evaluation, several questions remain unsettled, including the appropriate taxonomic disposition of several putative new species, as well as a reconsideration of the taxonomy for several existing species (e.g. *E. grandiflorus, E. cordifolius, E. macrophyllus* (Kunth) Micheli), which resolve as poly- or paraphyletic (Fig 6.7).

Relationships among most (35 of 40) *Sagittaria* species have been evaluated using 5S-NTS data (Fig 6.8; Keener, 2005), which provides a reasonable hypothesis

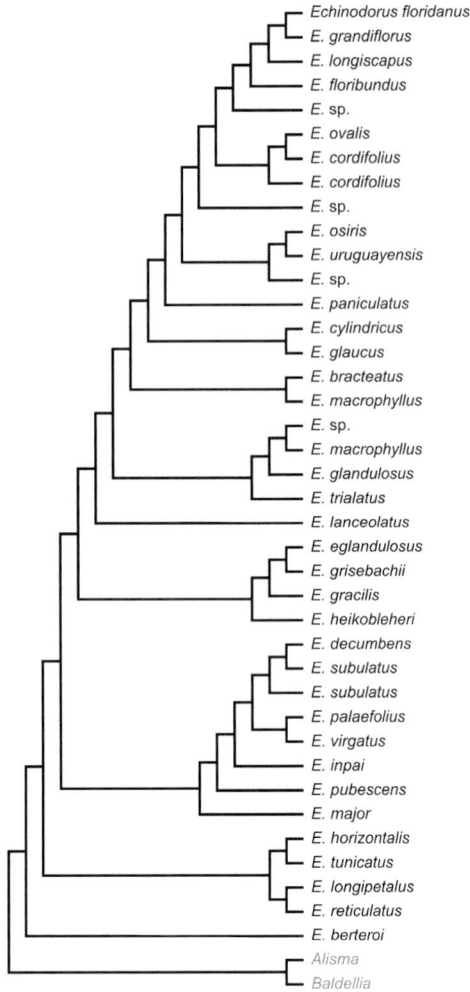

Fig 6.7 Phylogenetic relationships in *Echinodorus* indicated by MP analysis of combined morphological data and DNA (5S-NTS + LEAFY + *mat*K + nrITS) sequences (adapted from Lehtonen and Myllys, 2008).

of interspecific relationships in the genus. Here it would be desirable to analyse at least some cpDNA data in addition, to verify the species groups rendered by the nuclear sequence data analysis. Also, the taxonomic status of *Lophotocarpus* T.Durand has not been settled fully by the 5S-NTS data, which resolve several taxa assigned to the former *Lophotocarpus* (*S. montevidensis*) as a sister clade to the remainder of *Sagittaria* (Fig 6.8). Whether members of this clade differ materially from *Sagittaria* to warrant the re-establishment of *Lophotocarpus* should be given further consideration.

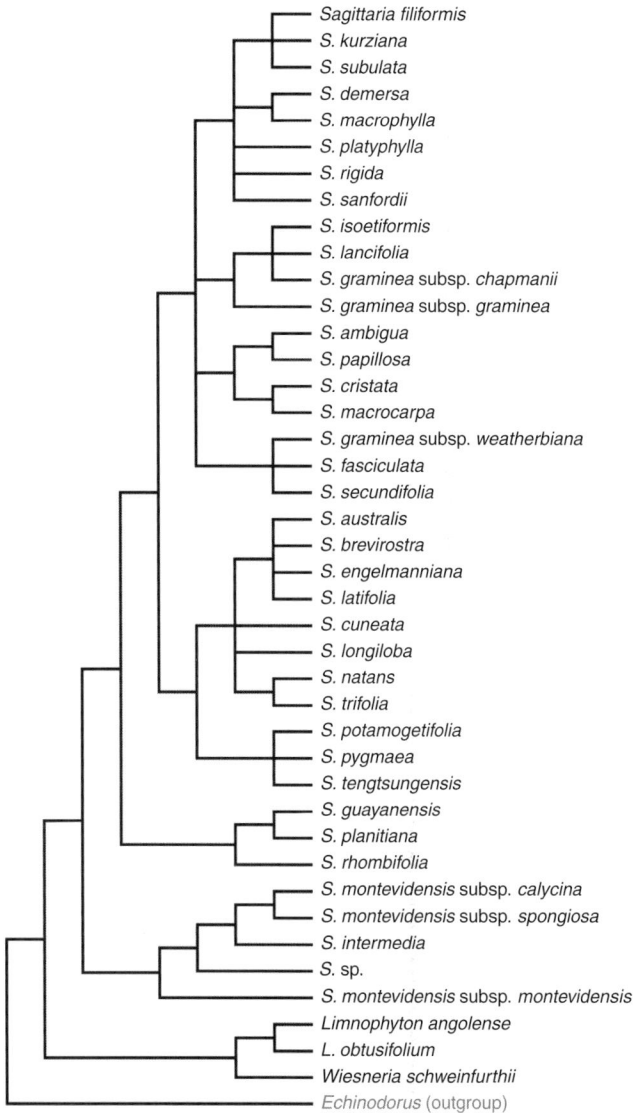

Fig 6.8 Interspecific relationships in *Sagittaria* based on analysis (MP strict consensus) of 5S-NTS data (adapted from Keener, 2005).

6.4.2 Butomaceae (subclade B)

Butomaceae solely comprise the monotypic *Butomus umbellatus*, which resolves in an isolated position between the families Alismataceae and Hydrocharitaceae, specifically as the sister group to the latter (Figs 6.2–6.4). This placement of the genus is supported in analyses incorporating up to 14–16 kb of cpDNA data

(Graham et al., 2006; Iles et al., 2009) and also by results obtained from combined cpDNA/mtDNA (Petersen et al., 2006) and *atp*1 + *rbc*L (Davis et al., 2004) data analyses. *Butomus* itself consists of two distinct diploid and triploid cytotypes (Krahulcová and Jarolímová, 1993); however, most authors recognize only a single species in the genus (Cook, 1996).

6.4.3 Hydrocharitaceae (subclade C)

The 'hydrocharits' contain 17 genera and roughly 127 species (Table 6.1). Sixteen of the genera have been included in phylogenetic studies (Figs 6.2, 6.9, 6.10) with only *Appertiella* remaining unsurveyed by any type of molecular data analysis. A consistent result of these phylogenetic analyses is the inclusion of the three marine genera (*Enhalus, Halophila* and *Thalassia*) within a single clade. These studies also place the genera within four natural groups, which have been designated as distinct subfamilies (*Anacharidoideae* Thomé, *Hydrilloideae*, *Hydrocharitoideae* Eaton, and *Stratiotoideae* Luerss.). These groups have been recovered consistently by different phylogenetic analyses incorporating not only single and multiple gene sequences (cpDNA, mtDNA and nrDNA), but morphological data as well (Fig 6.9). The most comprehensive taxon coverage has been achieved by analysis of *rbc*L data (Figs 6.2, 6.10), which includes 48 species, thereby providing an overview of interspecific relationships for nearly 40% of the family.

6.4.3.1 Infrageneric relationships

Six genera are monotypic (Table 6.1); however, at least one of these (*Hydrilla*) probably consists of at least one additional species (L. Benoit, pers. comm.), and one genus (*Appertiella*) has remained elusive in attempts to procure material for molecular studies. Both species in each of the bitypic genera (*Limnobium, Thalassia*) have been surveyed and phylogenetic analyses confirm that these genera are monophyletic (Figs 6.2, 6.10). Monophyly also is indicated by analyses of 2/3 of the species of *Hydrocharis*, and 3/5 of the species of *Elodea* (Figs 6.10, 6.11). *Elodea* requires more intensive study, especially given that our preliminary analysis of the genus (Fig 6.11) has verified the presence of natural interspecific hybrids between *E. canadensis* and *E. nuttallii*, thus corroborating earlier crossing studies that demonstrated the interfertility of these species (Les and Philbrick, 1993). Although *Elodea* itself is monophyletic, *Egeria* introduces additional problems, with the two surveyed species failing to resolve as a clade in analyses incorporating either cpDNA (Figs 6.2, 6.10) or nrITS data (Fig 6.11). These results raise the question of whether *Elodea* and *Egeria* might better be combined and treated taxonomically as a single genus. The elucidation of this question and clarification of generic limits for these genera eventually will require the inclusion of the remaining unsampled species of *Elodea* (*E. callitrichoides* Casp., *E. potamogeton* Espinosa) and *Egeria* (*E. heterostemon* S. Koehler and C.P. Bove) in phylogenetic analyses.

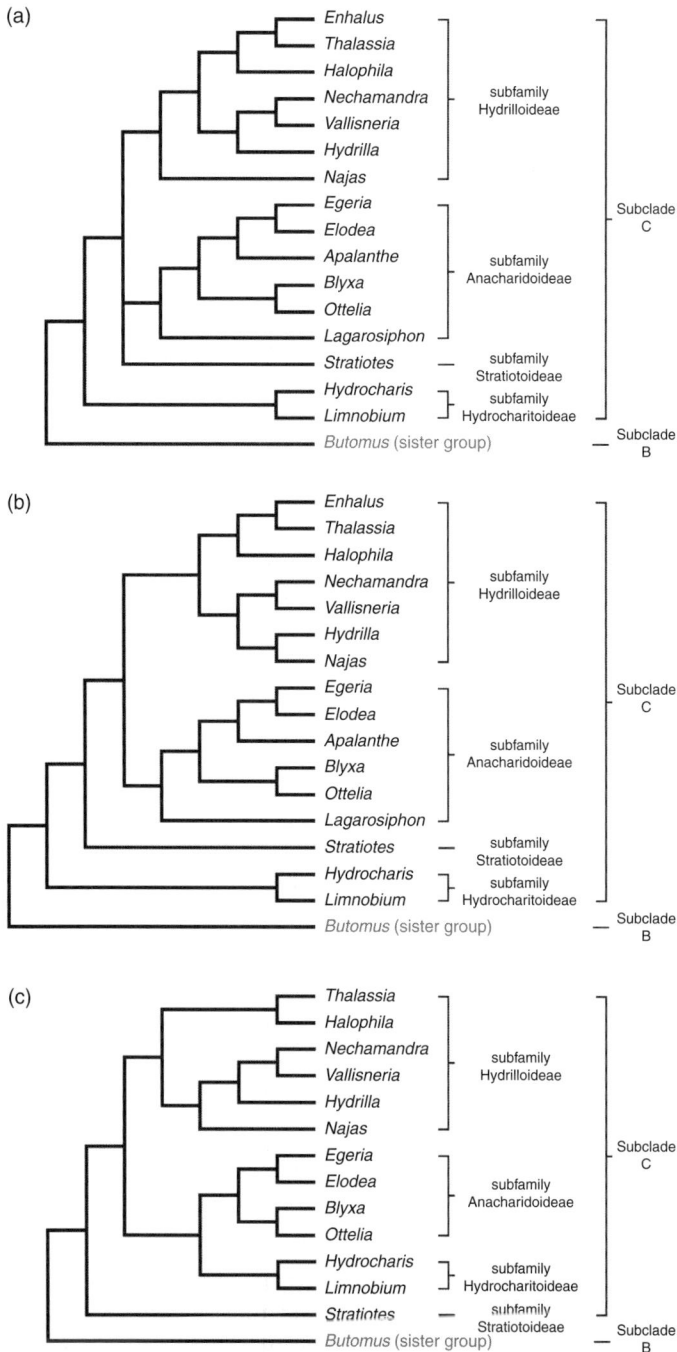

Fig 6.9 Similar intergeneric relationships in Hydrocharitaceae showing the consistent resolution of four groups, which have been designated as subfamilies. A: *rbc*L data (weighted MP strict consensus; Les et al. 1997). B: *mat*K + *rbc*L + *trn*K introns + nrDNA + morphology showing slightly different placement of *Najas* and resolution of *Stratiotes* (MP; Les et al., 2006). C: cladogram derived from *atp*1 + *cob* + *rbc*L data (MP; Petersen et al., 2006) closely resembles previous results, with the exception of slightly different positions of *Stratiotes* and subfamily Hydrocharitoideae.

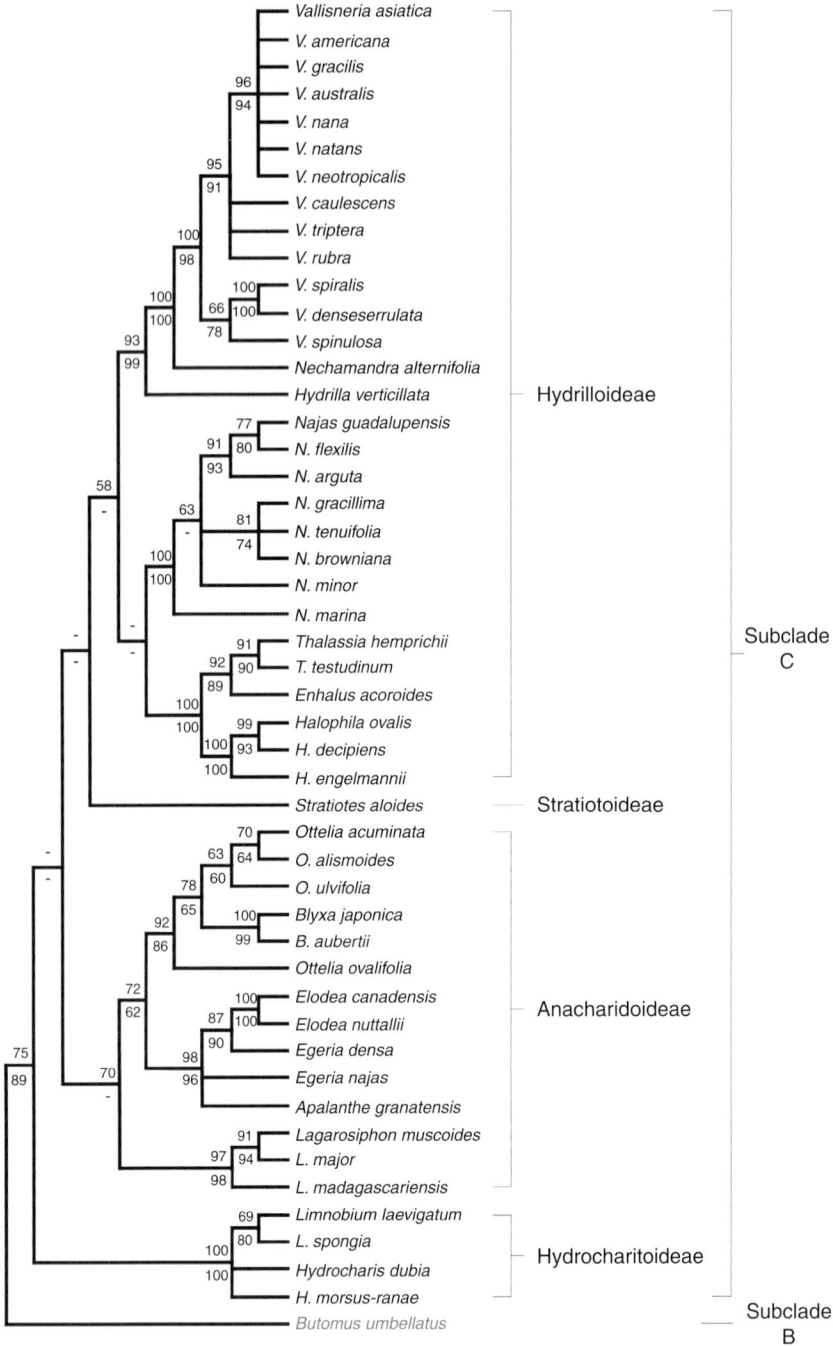

Fig 6.10 Expanded version of 'subclade C' (see Fig 6.2) showing details of relationships in Hydrocharitaceae as indicated by the results of a 167-taxon *rbc*L analysis.

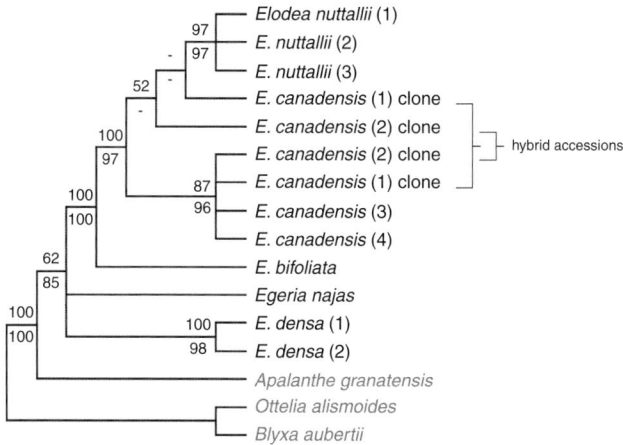

Fig 6.11 Phylogenetic relationships of *Apalanthe*, *Egeria* and *Elodea* as indicated by MP/ML analysis (ML: TrN+G model) of nrITS data (current study). MP strict consensus tree is shown. The association of several cloned nrITS alleles indicates the existence of interspecific hybrids involving *E. canadensis* and *E. nuttallii*. Nodal BS support values are indicated above (MP) and below (ML) branches respectively ('–' indicates <50% support). Tree statistics: CI = 0.85; CI_{exc} = 0.73; RI = 0.77; lnL = –3617.

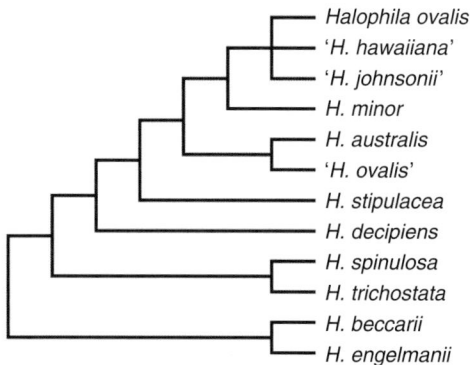

Fig 6.12 Phylogenetic relationships in *Halophila* as indicated by MP analysis of nrITS data (adapted from Waycott et al., 2002).

Several of the larger hydrocharit genera have been studied fairly comprehensively. Up to 14 species have been recognized in *Halophila* by some authors (e.g. Green and Short, 2003), but the actual number remains uncertain. Waycott et al. (2002) analysed nrITS sequences for 11 *Halophila* species (Fig 6.12), and concluded that two species (*H. hawaiiana* Doty and B.C.Stone, *H. johnsonii* N.J.Eiseman) could not be distinguished from *H. ovalis*. They also found that specimens identified as *H. ovalis* fell into two clades, with one set of accessions resolving with *H. australis* Doty and B.C.Stone (Fig 6.12). The misplaced '*H. ovalis*' material could indicate a cryptic species (Waycott et al. 2002); thus the true number of species in *Halophila* probably is 13. This comprehensive study of *Halophila* confirmed that the genus is monophyletic and also indicated that a phylogenetic reduction series exists, which proceeds from larger, more complex-leaved species to more diminutive, simpler forms. However, it would be useful to obtain additional data for *Halophila* (such as cpDNA sequences) to provide a means of comparison with the nrITS data (Waycott et al., 2006).

Najas is a genus of 39 species worldwide (Cook, 1996). As already discussed, *Najas* recently has been transferred to Hydrocharitaceae as a result of the placement of the genus indicated by numerous phylogenetic studies. A comprehensive, formal phylogenetic analysis of the genus itself has not been conducted to date; however, *Najas* presently is the focus of systematic investigations by the senior author. One investigation based on analysis of combined nrITS and cpDNA sequence data (*mat*K + *rbc*L + *trn*K) for eight *Najas* species emphasized North American taxa (Fig 6.13), but provided support for the maintenance of two subgenera (*Caulinia* (Willd.) A.Braun, *Najas* L.) and also demonstrated the distinctness of sections *Americanae* Magnus and *Euvaginatae* Magnus (Les et al., 2010). Additional phylogenetic studies currently are underway by the senior author and now include roughly half of the species; eventually the completion of this work should yield a much-improved phylogenetic understanding of the genus. These molecular studies also have indicated for the first time the occurrence of interspecific hybridization in *Najas*, which had long gone undetected due to the high degree of morphological similarity between the hybrids and their maternal parents (Les et al., 2010).

Vallisneria (18 species) also is well-studied phylogenetically as a result of analyses incorporating various molecular loci (nrITS, *rbc*L, *trn*K 5′ intron sequences) and morphological data (Les et al. 2008). The study by Les et al.

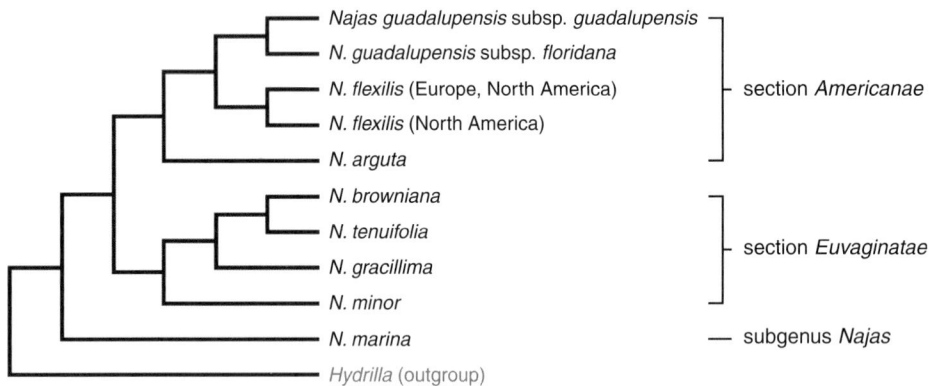

Fig 6.13 Systematic relationships in the genus *Najas*, inferred from phylogenetic analysis (MP strict consensus) of combined DNA sequence data (adapted from Les et al., 2010).

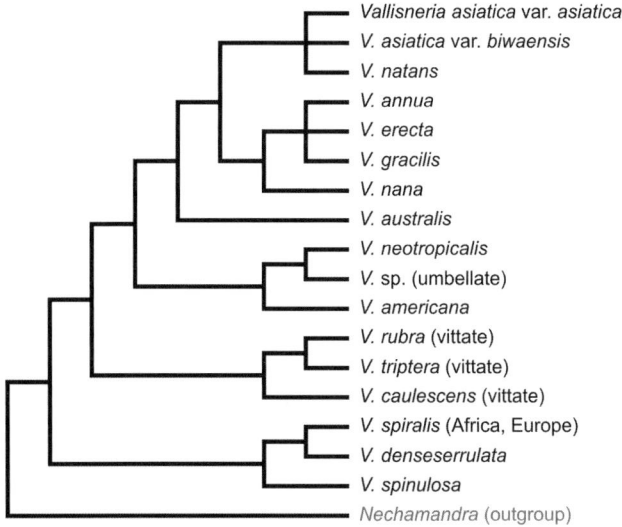

Fig 6.14 Interspecific relationships in *Vallisneria* as supported by phylogenetic analysis (MP strict consensus) of combined molecular data (redrawn from Les et al., 2008). This analysis excluded only *V. aethiopica* (Africa), which might be closely related to *V. spiralis* (see text). The vittate (caulescent) ingroup species are indicated; the remainder comprises rosulate taxa.

(2008) summarized phylogenetic relationships for 17 of the 18 species recognized worldwide (Fig 6.14) and excluded only the African *V. aethiopica* Fenzl. We since have acquired and analysed material attributed to *V. aethiopica*, but found it to be indistinguishable genetically from *V. spiralis* (Les and Tippery unpublished). However, the status of the morphologically distinctive *V. aethiopica* requires further evaluation.

Phylogenetic analyses of both molecular and morphological data (Les et al. 2008) placed the former genus *Maidenia* Domin (now *Vallisneria rubra*) within a clade of several leafy-stemmed (i.e. 'vittate') *Vallisneria* taxa, which was embedded among the rosulate species in all analyses using molecular data (Fig 6.14). That study also proposed that several characters formerly thought to be fairly informative taxonomically were highly homoplasious, probably as the result of selection to maintain the efficiency of the pollination system (Les et al. 2008).

6.4.3.2 Poorly studied taxa

Blyxa, *Lagarosiphon* and *Ottelia* represent the three hydrocharit genera most in need of more comprehensive study. Three of the nine *Lagarosiphon* species available for phylogenetic estimation resolve as a clade (Figs 6.2, 6.10); however, additional species sampling is necessary to further corroborate the monophyly of the genus. Only 7 of the 30 or so species in *Blyxa* and *Ottelia* have been analysed simultaneously, and the preliminary results are insufficient to demonstrate that these genera resolve as distinct clades (Figs 6.2, 6.10). An extensive systematic study of both genera is encouraged.

6.5 Potamogetonales (clade II)

Results of recent phylogenetic analyses have warranted a number of alterations to the taxonomy of this order, such as the removal of Najadaceae and its transfer to Alismatales as already discussed. The taxonomic concept of Potamogetonales presented here includes six families (Aponogetonaceae, Juncaginaceae, Maundiaceae, Potamogetonaceae, Scheuchzeriaceae, Zosteraceae) in addition to the 'Cymodoceaceae complex', which also includes Posidoniaceae and Ruppiaceae (tentatively). Although Zannichelliaceae Chevall. also have been recognized as distinct in many past treatments, we have decided to merge this family with Potamogetonaceae for reasons given below.

The disposition of the monotypic *Maundia triglochinoides* (Maundiaceae), which typically has been included in Juncaginaceae (Cook, 1996), warrants additional explanation. This genus was not included in the *rbc*L analysis of Alismatidae by Les et al. (1997), who consequently could not evaluate its phylogenetic position. Iles et al. (2009) procured material of *Maundia* and included it in an analysis of key alismatid families for which approximately 14 kb of cpDNA data were evaluated. Their analysis distinguished *Maundia* from Juncaginaceae as an isolated sister group to the clade containing the Cymodoceaceae complex, Potamogetonaceae, Zannichelliaceae and Zosteraceae; however, Juncaginaceae (four genera) was represented only by *Triglochin* in that analysis. These results inspired some authors (APG III, 2009) to consider the transfer of *Maundia* (monotypic) to a distinct family (Maundiaceae); however, a further evaluation of the question was urged. A recent analysis of *rbc*L, *mat*K and *atp*1 data for various monocots including *Cycnogeton*, *Lilaea* Bonpl. (monotypic), *Maundia*, *Tetroncium* (monotypic) and *Triglochin*, also failed to resolve *Maundia* with other Juncaginaceae, thereby strengthening arguments to remove it from that family (von Mering and Kadereit, 2010). The same study also advocated the merger of *Lilaea* with *Triglochin*, leaving Juncaginaceae with only three genera (*Cycnogeton*, *Tetroncium* and *Triglochin*).

Our 167-taxon *rbc*L analysis (Fig 6.2) also resolved *Maundia* as distinct from Juncaginaceae, which was represented in our analysis by *Tetroncium*, *Triglochin/Lilaea* (7 species), and *Cycnogeton* (2 species). In this analysis (Fig 6.2), *Maundia* resolved as the sister group to the Potamogetonaceae/Zosteraceae clade, but only with weak internal support (<53% ML or MP BS support for critical nodes; values not shown on tree). To provide additional insight, we conducted an analysis of

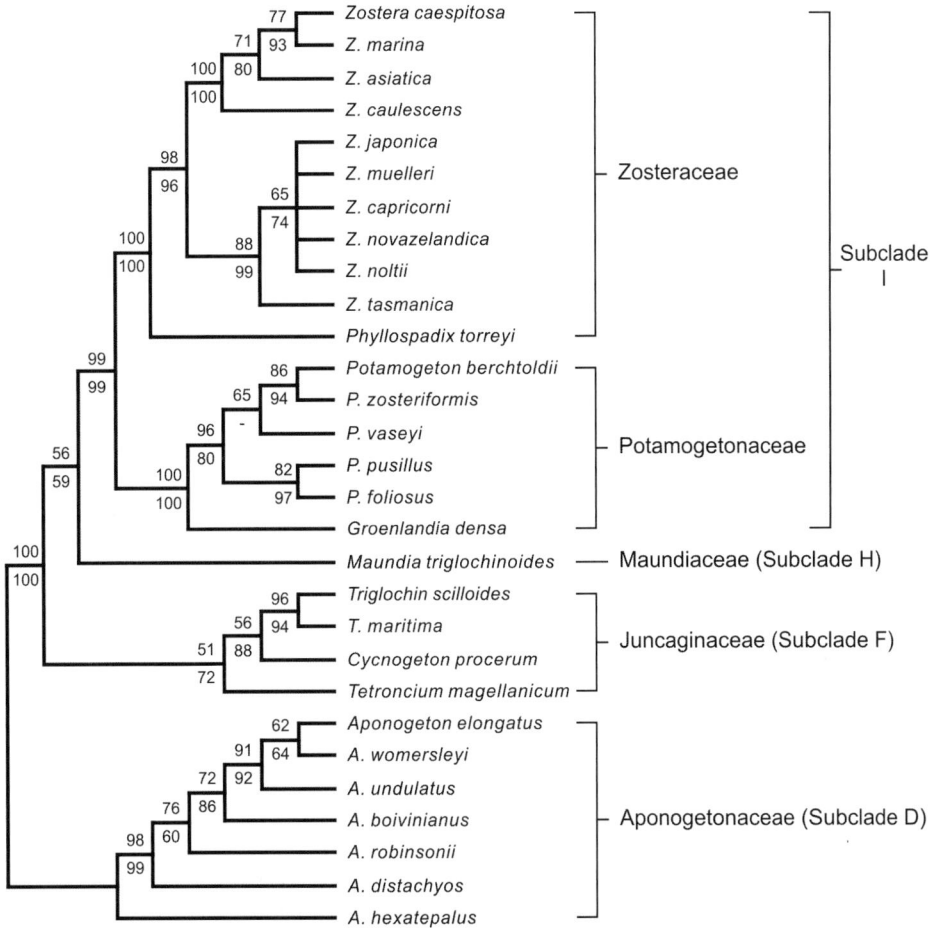

Fig 6.15 Phylogenetic placement of *Maundia* (Maundiaceae) as determined by analysis of *trn*K 5′ intron sequence data (current study). The MP strict consensus tree situates *Maundia* as the sister group of the Potamogetonaceae/Zosteraceae clade, but only with weak internal support. BS support values are indicated above (MP) and below (ML; TVM+G model) nodes respectively ('–' indicates <50% support). Tree statistics: CI = 0.78; CI$_{exc}$ = 0.72; RI = 0.91; ln*L* = –3523.

*trn*K 5′ intron sequence data, which included *Maundia* and representatives of all three genera of Juncaginaceae (*Cycnogeton*, *Tetroncium* and *Triglochin* [including *Lilaea*]). That analysis (Fig 6.15) further supported the results obtained by Iles et al. (2009), von Mering and Kadereit (2010), and our own 167-taxon analysis by resolving *Maundia* outside of Juncaginaceae as the sister group to the Potamogetonaceae/Zosteraceae clade (but again with fairly weak support). The lack of strong support for the position of *Maundia* in all of these analyses is somewhat disconcerting; however, it does seem apparent that the genus at least does not associate closely with members of Juncaginaceae. For this reason it seems appropriate to accept the taxonomic recognition of *Maundia* as a distinct, monotypic family, as advocated by von Mering and Kadereit (2010).

6.5.1 Aponogetonaceae (subclade D)

Aponogetonaceae consistently resolve as the sister group to all other members of Potamogetonales in various relevant analyses (Figs. 6.2, 6.3). Only one notable exception occurs, wherein the phylogenetic position of the family is exchanged with Scheuchzeriaceae (Fig 6.3A). In that anomalous analysis, *rbc*L data were analysed by MP using step-matrix weighting (Les et al., 1995). However, in other analyses (including our 167-taxon MP/ML *rbc*L analysis), the former position is retained (Figs. 6.2, 6.3) and is accepted here to represent the most reasonable phylogenetic placement of Aponogetonaceae within the subclass.

6.5.1.1 Infrageneric relationships

Les et al. (2005) evaluated phylogenetic relationships of 21 Aponogetonaceae taxa, focusing on the Australian species. In that study, which incorporated morphological and molecular (*mat*K, nrITS, *trn*K 5′ intron) sequence data, the genus was structured phylogenetically as four major clades, each recognized taxonomically as a section. That analysis resolved the Australian *Aponogeton hexatepalus* (section *Viridis* Les, M.Moody and S.W.L.Jacobs) as sister to the remainder of the genus, which mainly comprised clades with strong geographical affinities, such as section *Flavida* Les, M.Moody and S.W.L.Jacobs (Australia), section *Aponogeton* L.f. subsection *Aponogeton* (Asia) and section *Aponogeton* subsection *Polystachys* A.Camus (Madagascar). Numerous instances of hybridization and polyploidy present technical difficulties that must be considered when evaluating phylogenetic relationships in *Aponogeton* (Les et al., 2005).

Since the study by Les et al. (2005), we have acquired a total of 29 species (roughly 60% of the estimated 50 species worldwide), which has allowed us to expand our original survey of *Aponogeton*. In this analysis, relationships were re-evaluated using combined nrITS and *trn*K 5′ intron sequence data. The resulting phylogenetic relationships (Fig 6.16) were similar to those reported by Les et al. (2005) with several noteworthy differences. The Australian taxa remained in two groups, with *A. hexatepalus* as the sister to the rest of the genus. The newly included *A. womersleyi* (New Guinea) resolved within the major Australian clade as considered by Les et al. (2005); however, it was found to be nearly identical to *A. cuneatus*, which was described recently by Jacobs et al. (2006). Because *A. womersleyi* has never been reported in Australia, the taxonomic status of *A. cuneatus* will require further evaluation to determine whether these species are distinct. Our expanded taxon analysis is consistent with the original study by

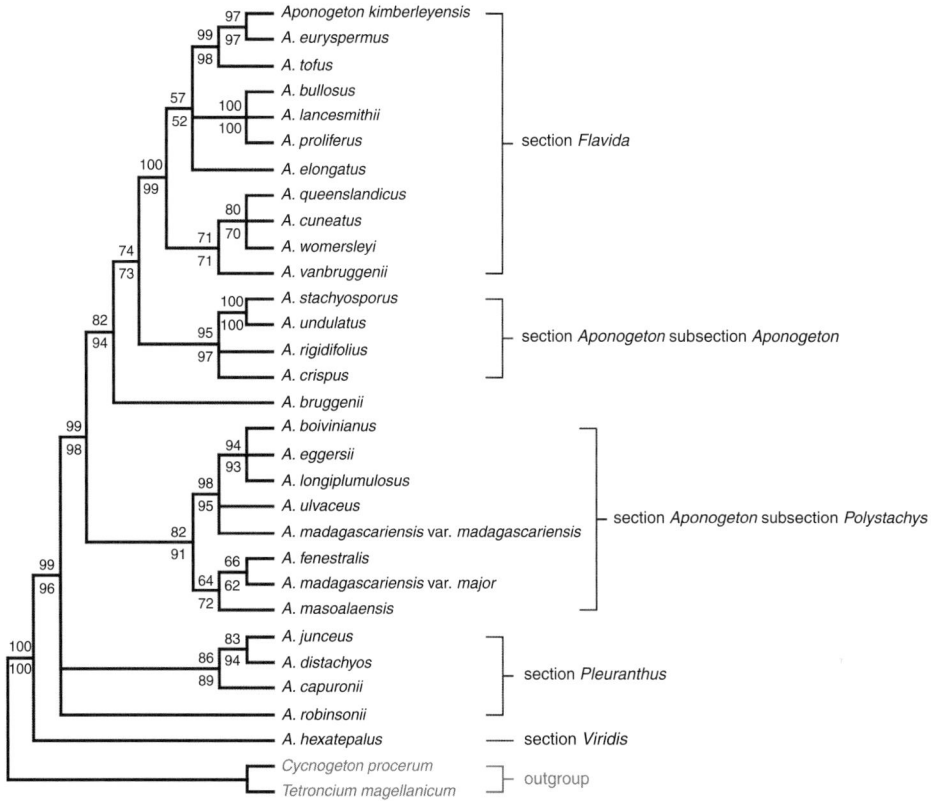

Fig 6.16 Interspecific relationships in *Aponogeton* (Aponogetonaceae) reconstructed by MP/ML analysis of combined nrITS (ML: TIM3+G model) + *trn*K 5′ intron (ML: TPM1uf+I model) sequence data (current study). MP strict consensus tree is shown. Values indicating the degree of internal (BS) support are indicated above (MP) and below (ML) nodes respectively ('–' indicates <50% support). Tree statistics: CI = 0.79; CI_{exc} = 0.67; RI = 0.79; $\ln L$ = –7025.

indicating an Asian clade (i.e. *Aponogeton* subsection *Aponogeton*) as sister to the major Australian clade (section *Flavida*); however, an additional Asian (Indian) species (*A. bruggenii*) resolved as the sister group to the Australian/Asian clade (Fig 6.16). A major Malagasy clade (*Aponogeton* subsection *Polystachys*; now represented by at least eight taxa) similarly is resolved; however, one Madagascar species (*A. capuronii*) was situated as the sister group to a clade of species from continental Africa. Although the continued inclusion of taxa undoubtedly will further clarify phytogeographical affinities of *Aponogeton* species, these preliminary studies indicate that most Australian and Malagasy species were derived from Asian progenitors, and that continental African species probably were derived from Madagascar.

Aponogeton eventually will require further taxonomic refinements. The monophyly of section *Pleuranthus* could not be confirmed due to our inability to resolve the position of *A. robinsonii* with certainty. Although section *Aponogeton* is paraphyletic as circumscribed, its two major subsections do resolve as distinct subclades, and most other recognized sections and subsections of the genus resolve as clades in phylogenetic analyses (Fig 6.16).

6.5.2 Scheuchzeriaceae (subclade E)

This monotypic family of bog-plants (*Scheuchzeria palustris*) resolves consistently as the sister group to Potamogetonales in exclusion of Aponogetonaceae (Figs 6.2–6.3; however, see discussion under Aponogetonaceae). Old and New World populations differ somewhat by fruit and stigma/style characters (Fernald, 1923); however, these morphological differences generally are treated taxonomically at the subspecific level. A genetic survey of this species on a worldwide basis could shed additional light on patterns of divergence in this widespread taxon.

6.5.3 Juncaginaceae (subclade F)

Following the circumscription proposed by von Mering and Kadereit (2010), Juncaginaceae contain three genera: *Cycnogeton* (8 species), *Tetroncium* (monotypic) and *Triglochin* (~20 species). To summarize intergeneric relationships we have conducted an updated evaluation of *rbc*L data for a third of the taxa, which includes *Tetroncium*, *Cycnogeton* (2 species) and *Triglochin* (7 species). The resulting topology (Figs 6.2, 6.17) resolves each genus as a clade and is consistent with results of several previously published analyses (Les et al., 1997; Petersen et al., 2006; von Mering and Kadereit, 2010), as well as our analysis of *trn*K 5' intron sequence data (Fig 6.15). In all instances *Triglochin* and *Cycnogeton* are sister clades, with *Tetroncium* sister to the remainder of the family.

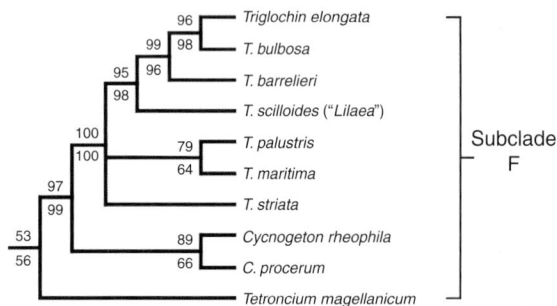

Fig 6.17 Expanded version of 'subclade F' (see Fig 6.2) showing details of relationships in Juncaginaceae as indicated by the results of a 167-taxon *rbc*L analysis.

6.5.3.1 Infrageneric relationships

Analysis of *rbc*L data (Figs 6.2, 6.17) provides the most inclusive molecular phylo-genetic survey of Juncaginaceae available to date. It is apparent from this analysis that the taxon once segregated as the genus *Lilaea* (*T. scilloides*) clearly nests within *Triglochin*, as reported previously by von Mering and Kadereit (2010), and there is no compelling reason to accept this taxon as anything other than a highly specialized member of *Triglochin*. The inclusion of additional *Cycnogeton* and *Triglochin* species will be necessary before any more definitive evaluation of interspecific relationships can be made in these genera. Additional work in this area currently is underway by von Mering, who now has acquired molecular data for at least 15 *Triglochin* species (personal communication).

6.5.4 'Cymodoceaceae complex' (subclade G)

Les et al. (1997) designated as the 'Cymodoceaceae complex' a group comprising the families Cymodoceaceae, Posidoniaceae and Ruppiaceae. These three families had not been associated together in treatments prior to that study, where they resolved as a weakly supported clade (40% MP bootstrap) by analysis of *rbc*L sequence data. The same clade was recovered in our 167-taxon analysis, but with higher internal support (Fig 6.18). This clade also received a moderate level of support (77% MP BS) in the *rbc*L analysis by von Mering and Kadereit (2010). In addition to these single-gene analyses, further corroborative evidence supporting the phylogenetic integrity of the Cymodoceaceae complex has been provided by Iles et al. (2009), whose analysis of large amounts of cpDNA sequence data resolved the same group as a clade with high internal support (ML BS >95%). A combined analysis of *cob*, *atp*1 and *rbc*L sequences (Petersen et al., 2006) also yielded this same clade. The high degree of consistency obtained among these various molecular analyses, and the high level of internal support obtained in at least some of the analyses provides substantial evidence to indicate that these three families associate as a natural clade.

Li and Zhou (2009) challenged the inclusion of *Ruppia* within the Cymodoceaceae complex based upon their combined analysis of *rbc*L sequences and morphological data. In that analysis, *Ruppia* associated with Zannichelliaceae and Potamogetonaceae rather than with Cymodoceaceae and Posidoniaceae. However, several aspects of their study render that result untenable. First, the internal support for the proposed association remained weak (64% MP BS; <50% for all the other critical nodes). The addition of morphological data also influenced the degree of internal support for some of their 'strongly supported' groups (such as Potamogetonaceae and Zosteraceae), which was lower than values obtained for those same groups in analyses using *rbc*L data only (Les et al., 1997). Moreover, all pertinent analyses of *rbc*L data alone have resolved the Cymodoceaceae complex

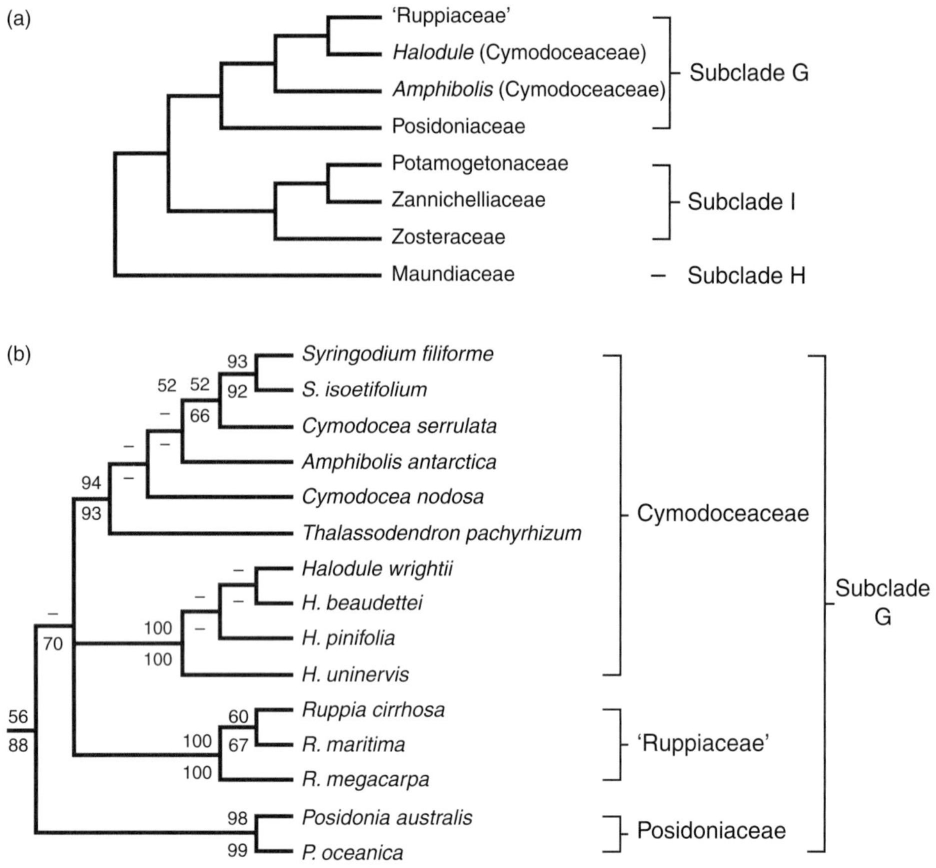

(a)

(b)

Fig 6.18 Intergeneric relationships in the 'Cymodoceaceae complex'. A: Results from ML analysis of approximately 14 kb of cpDNA sequence data support the Cymodoceaceae 'complex' as a clade with >95% BS support (adapted from Iles et al., 2009). Posidoniaceae are sister to the remainder of the complex with the same high level of support. B: Expanded version of 'subclade G' (see Fig 6.2) shows similar relationships as indicated by the results of a 167-taxon *rbc*L analysis (current study). The failure of Ruppiaceae to resolve as a clade distinct from Cymodoceaceae in such studies provides ample justification to merge these families.

as a clade, indicating that the morphological data indeed were responsible for influencing the different placement of Ruppiaceae in the one instance reported by Li and Zhou (2009). This result is not surprising given that several studies (Les and Haynes, 1995; Les et al., 1997) have demonstrated the misleading effect of homoplasious morphological data on phylogenetic reconstruction in alismatids.

Li and Zhou (2009) acknowledged that morphological data for the alismatids were highly homoplasious and discussed in some detail the probable convergence of a number of characters included in their analysis. They also called for the need to acquire additional DNA sequence data besides *rbc*L, which are now available. In this case a large amount of cpDNA sequence data (Iles et al., 2009) and combined cpDNA/mtDNA sequences (Petersen et al., 2006) have corroborated the results indicated by the initial *rbc*L analysis of Les et al. (1997) and never have returned results that would support those obtained by Li and Zhou (2009). As a consequence, we continue to recognize the Cymodoceaceae complex as a meaningful clade based on the best available current evidence.

What is less certain is whether the Cymodoceaceae complex should be recognized taxonomically as a single family rather than in its present disposition as comprising three families. In evaluating *rbc*L data (Fig 6.18; Les et al., 1997; von Mering and Kadereit, 2010) or multiple cpDNA sequences (Iles et al. 2009), *Posidonia* consistently resolves as a sister clade to the remaining genera; however, the distinction of Ruppiaceae and Cymodoceaceae is not evident either in those analyses or by analysis of combined cpDNA/mtDNA sequence data (Petersen et al., 2006), which all are equivocal in distinguishing Ruppiaceae and Cymodoceaceae as separate clades. Given these circumstances it seems reasonable and appropriate to merge Ruppiaceae with Cymodoceaceae (which has nomenclatural priority) as we have done here.

6.5.4.1 Cymodoceaceae

We regard Cymodoceaceae as a seagrass clade comprising six genera: *Amphibolis*, *Cymodocea*, *Halodule*, *Ruppia*, *Syringodium* and *Thalassodendron*. The genera *Amphibolis*, *Syringodium* and *Thalassodendron* each contain only two species, but both species have been analysed phylogenetically only in *Syringodium* (Fig 6.18) as a test of monophyly. At least for *rbc*L data (which represents the densest taxonomic coverage of any systematic evaluation for this group to date), it is difficult to distinguish *Amphibolis*, *Cymodocea*, *Syringodium* and *Thalassodendron* as distinct. Rather, the species of these genera analysed thus far appear to be quite closely related (by exhibiting relatively low pairwise levels of genetic divergence) and associate together in one clade (Fig 6.18). These observations could justify the merger of these species into a single genus, of which *Cymodocea* would have nomenclatural priority. However, before such a nomenclatural modification is adopted, it would be advisable to include the four species (*Amphibolis griffithii* (J.M.Black) Hartog, *Cymodocea angustata* Ostenf., *C. rotundata* Asch. and Schweinf., *Thalassodendron ciliatum* (Forssk.) Hartog), which have not yet been evaluated in phylogenetic analyses of this group, as well as to reconsider the morphological characters that have been used to define these genera in the past. It is noteworthy in this respect, that *Syringodium isoetifolium*, *Amphibolis antarctica*, *A. griffithii* and *Thalassodendron ciliatum* all were named originally as species of *Cymodocea*.

In contrast, *Halodule* and *Ruppia* appear to be monophyletic, with each genus resolving as a distinct, strongly supported clade within the family (Fig 6.18).

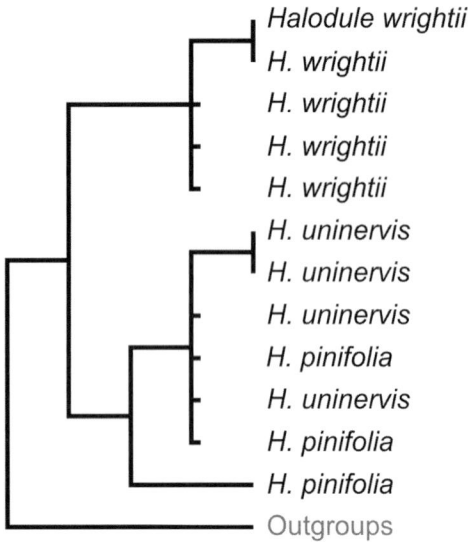

Fig 6.19 Interspecific relationships in *Halodule* as indicated by MP analysis of nrITS sequence data (adapted from Waycott et al., 2006).

Relationships within *Halodule* have been investigated in some detail by Waycott et al. (2006), who analysed nrITS and *trn*L sequence data for multiple populations representing three of the six currently recognized species. The results of those analyses were inconclusive, but indicated that there may only be as few as two defensible species in the genus (Fig 6.19). At the least, that study found no justification to support the continued taxonomic segregation of *H. pinifolia* from *H. uninervis*. Additional populations of *Halodule* need to be investigated and additional genetic loci included in its phylogenetic evaluation so that a comprehensive assessment of relationships can be made for the genus.

Cook (1996) estimated that *Ruppia* contained from 2–10 species. A recent phylogenetic study of *Ruppia* by Ito et al. (2010) indicated that there are at least four distinct species, with several potentially additional taxa indicated within the *R. maritima* 'complex' (Fig 6.20). Polyploidy has complicated efforts to evaluate relationships in the *Ruppia maritima* complex, which includes diploid, triploid and tetraploid cytotypes (Ito et al., 2010). Consequently, molecular analyses of this group should take into account the possibility of potentially paralogous loci, which could present interpretational difficulties.

6.5.4.2 Posidoniaceae

Phylogenetic relationships in *Posidonia* (the sole genus of Posidoniaceae) have been investigated using combined nrITS and *trn*L sequence data together with morphological analyses (Waycott et al., 2006). Those analyses (Fig 6.21) have

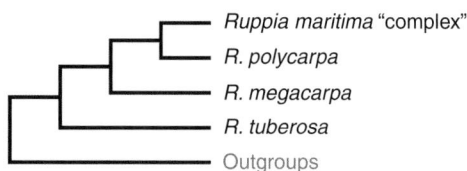

Fig 6.20 Interspecific relationships in *Ruppia* as indicated by MP analysis of *PHYB* data (simplified from Ito et al., 2010). In that same study, cladograms constructed from combined cpDNA data (*mat*K + *rbc*L + *rpo*B + *rpo*C1) showed a similar topology but with the positions of *R. megacarpa* and *R. tuberosa* reversed.

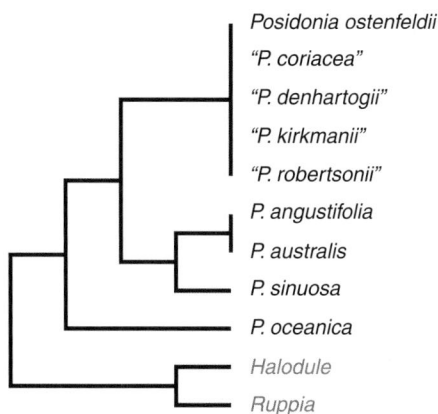

Fig 6.21 Phylogenetic relationships in *Posidonia* as indicated by MP analysis of combined nrITS and *trn*L sequence data (adapted from Waycott et al., 2006).

confirmed the monophyly of the genus and support the recognition of five distinct species within it. They also warranted the merger of four taxa (*P. coriacea* M.L.Cambridge and J.Kuo, *P. denhartogii* J.Kuo and M.L.Cambridge, *P. kirkmanii* J.Kuo and M.L.Cambridge, and *P. robertsonii* J.Kuo and M.L.Cambridge) with *P. ostenfeldii* Hartog, from which they appear to be indistinguishable both genetically and morphologically (Waycott et al., 2006). The morphological and genetic data consistently indicate that the Mediterranean *Posidonia oceanica* is the sister species to all of the remaining (Australian) taxa (Waycott et al., 2006).

6.5.5 Maundiaceae (subclade H)

As described above, the phylogenetic placement of Maundiaceae is somewhat problematic due to the lack of pertinent internal support generated in some molecular data analyses. In the *rbc*L analysis performed by von Mering and Kadereit (2010), *Maundia* associated as the sister group to the

Potamogetonaceae/Zosteraceae clade, but with weak (<70%) internal (MP BS) support. In fact, no critical branch (those delimiting the family from Juncaginaceae) received more than 71% support. Our 167-taxon *rbc*L analysis (Fig 6.2) resolved *Maundia* in the same position, but with even less support (ML/MP BS values no higher than 57% at critical nodes). A similar placement (and similarly low support vales of <59%) was obtained in our *trn*K 5′ intron analysis (Fig 6.15); however, that analysis unfortunately did not include any representatives of the Cymodoceaceae complex.

The only analysis providing reasonable internal support for the position of *Maundia* was by Iles et al. (2009), who analysed roughly 14 kb of cpDNA sequence data. In that study *Maundia* resolved in a similar position (Fig 6.3B), specifically as the sister group to a clade that included the Cymodoceaceae complex, Potamogetonaceae and Zosteraceae, and with a high level of internal support (ML bootstrap >95%). Consequently, we have accepted the well-supported topology from the study by Iles et al. (2009) as representing the most reasonable hypothesis among the two alternative placements of the family indicated by existing molecular phylogenetic analyses. In any event the alternative placements of Maundiaceae do not differ substantially with respect to other alismatid taxa.

6.5.6 Potamogetonaceae/Zosteraceae (subclade I)

This group was among the best-supported clades (100% MP BS) recovered in the original *rbc*L analysis by Les et al. (1997). Understandably, the same group (subclade I) also was resolved with high support (99–100% ML/MP BS) in our 167-taxon *rbc*L analysis (Fig 6.2) and by the large cpDNA data analysis (Iles et al., 2009) and combined *cob*, *atp*1, *rbc*L sequence analysis (Petersen et al., 2006). This consistent, well-supported phylogenetic topology provides strong evidence that Potamogetonaceae (including Zannichelliaceae) and Zosteraceae represent a cohesive, sister group relationship.

6.5.6.1 Potamogetonaceae

Although the latter two studies (Iles et al., 2009; Petersen et al., 2006) resolved Zannichelliaceae as the sister clade to Potamogetonaceae, the minimal taxon sampling in both cases was insufficient to determine with certainty whether either family was monophyletic. Evidence to the contrary appeared in the *rbc*L study by Les et al. (1997), where two genera of Zannichelliaceae (*Lepilaena*, *Zannichellia*) resolved as a clade (with 100% MP BS support) among other Potamogetonaceae. Comparable support (99–100% ML/MP BS) for the same result was maintained in our 167-taxon *rbc*L analysis (Fig 6.2), which included nearly a quarter of the estimated 100 *Potamogeton* species. Furthermore, cladograms based on combined *trn*L, *psb*A-*trn*H and 5S-NTS sequence data (Fig 6.22) also grouped Zannichellia-ceae with Potamogetonaceae (Lindqvist et al. 2006).

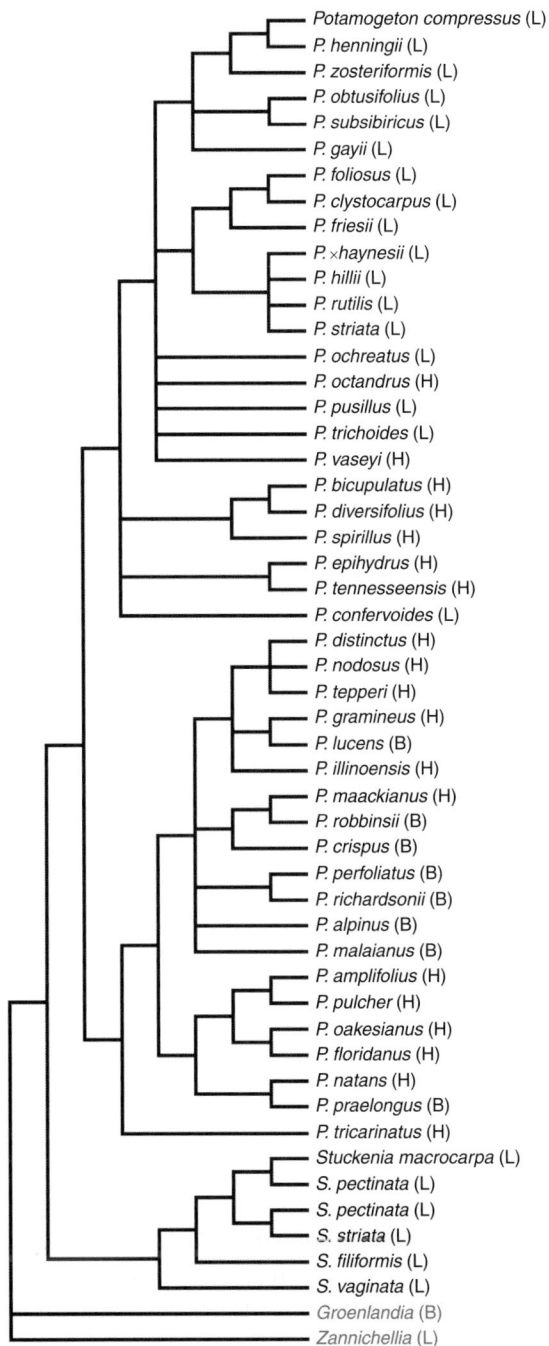

Fig 6.22 Phylogenetic relationships in Potamogetonaceae as indicated by analysis (jackknife majority rule consensus) of combined nuclear (5S-NTS) and cpDNA (*trn*L + *psb*A-*trn*H) sequence data (redrawn from Lindqvist et al., 2006). Basic morphological types are indicated as H (heterophyllous), B (broad-leaved homophyllous) or L (linear-leaved homophyllous).

Superficially, Zannichelliaceae are distinctive morphologically, and non-molecular data never have indicated the phyletic proximity of the family with Potamogetonaceae (Les and Haynes, 1995). However, Potamogetonaceae are extremely diverse morphologically and possess some characters (e.g. fruit morphology) that do not differ fundamentally from Zannichelliaceae. Despite the high degree of reduction that characterizes this taxon, there also are certain features (such as the presence of unusual, coiled cotyledons in *Groenlandia* and *Zannichellia*) that potentially are synapomorphic (Haynes et al., 1998).

Philosophically, there is no compelling reason why aquatic plants should be classified into numerous, depauperate families, especially when there is strong evidence indicating very close relationships among them. In such instances, the recognition of larger, more diverse families would be more in line with taxonomic concepts applied to the majority of terrestrial families, which often contain diverse assemblages of genera. For these reasons, we have elected to merge Zannichelliaceae with Potamogetonaceae. In this circumscription, Potamogetonaceae are recognized as including seven genera: *Althenia, Groenlandia, Lepilaena, Potamogeton, Pseudalthenia, Stuckenia* and *Zannichellia*. Phylogenetic relationships have been evaluated among five of these genera (excluding *Althenia, Pseudalthenia*) using *rbc*L sequence data (Fig 6.2; Les et al., 1997) and among four of the genera (further excluding *Lepilaena*) using combined *trn*L, *psb*A-*trn*H and 5S-NTS sequence data (Lindqvist et al., 2006).

In all cases, *Potamogeton* (in exclusion of *Stuckenia*) resolves as monophyletic. The distinctness of these genera is evidenced by the substantial level of genetic divergence indicated in every pertinent molecular phylogenetic analysis conducted to date (e.g. Fig 6.2; Les et al., 1997; Lindqvist et al., 2006). Some opposition to segregating *Potamogeton* and *Stuckenia* has been expressed by Zhang et al. (2008), who conducted a phylogenetic analysis of *Potamogeton* using *trn*T–*trn*F spacer sequence data; yet, their own results also clearly resolve these groups as distinct, strongly supported clades. Zhang et al. (2008) based their scepticism on one anomalous accession of *S. filiformis* (Pers.) Börner, which resolved within *Potamogeton* (with *P. gramineus*). However, two other *S. filiformis* accessions resolved appropriately within *Stuckenia*, as also did the accession of this species analysed by Lindqvist et al. (2006). Zhang et al. (2008) suggested that the anomalous placement of the single *S. filiformis* accession with *P. gramineus* could indicate hybridization. However, given the extensive level of genetic divergence between these genera and the lack of any reported morphological indications of hybridity in that accession, this situation requires further evaluation, especially to exclude contamination as a possible explanation.

Thus far, each surveyed genus (*Groenlandia, Lepilaena, Potamogeton, Stuckenia* and *Zannichellia*) is distinct genetically; however, their inter-relationships have not been resolved adequately. A complete picture of intergeneric relationships in

Potamogetonaceae will not be possible until species from the unsurveyed genera *Althenia* and *Pseudalthenia* are included in phylogenetic analyses, along with additional data from other genetic loci.

6.5.6.2 Infrageneric relationships

Infrageneric relationships have been poorly studied in *Lepilaena* and *Zannichellia*, and these genera require additional systematic evaluation. *Althenia, Groenlandia* and *Pseudalthenia* arguably are monotypic (Cook, 1996).

Lindqvist et al. (2006) evaluated roughly half of the estimated 100 species of *Potamogeton* using a combination of nuclear and cpDNA molecular markers. Their resulting phylogenetic tree divided the genus into two clades, which roughly represent morphological groups corresponding to narrow-leaved (upper clade) and broader-leaved (lower clade) taxa (Fig 6.22). This result parallels those reported earlier by Iida et al. (2004), who also recovered two major clades of linear-leaved and broad-leaved taxa (based on noncoding cpDNA sequences) in their study of Japanese *Potamogeton*.

The occurrence of heterophyllous species in each of the major clades recovered by Lindqvist et al. (2006) also is consistent with earlier hypotheses (based on patterns of secondary metabolites), suggesting that *Potamogeton* species are heterophyllous ancestrally and that linear-leaved and broad-leaved homophyllous species are multiply derived from heterophyllous ancestors (Les and Sheridan, 1990).

The phylogenetic overviews of *Potamogeton* by Iida et al. (2004) and Lindqvist et al. (2006) have presented useful starting points for evaluating relationships in this large and complex genus. However, a comprehensive overview of inter-relationships in the genus will not be achieved until additional taxa are evaluated and complications related to polyploidy and hybridization have been dealt with effectively. Now that preliminary estimates of relationships in *Potamogeton* are available, we recommend that intensive studies proceed following a clade-by-clade approach, which incorporates increased taxon sampling (including multiple populations of species), additional genetic loci, and evaluation of hybridization and paralogy. A number of striking anomalies, such as the association of the morphologically disparate *P. praelongus* and *P. natans* (Iida et al., 2004; Lindqvist et al., 2006) also need to be better reconciled. Even analyses of relatively small portions of the genus have disclosed substantial taxonomic discrepancies (Les et al., 2009) and it is apparent that a large amount of work still is necessary before a satisfactory phylogenetic picture of this genus emerges.

6.5.6.3 Zosteraceae

Only two sister genera (*Phyllospadix, Zostera*) currently are recognized in this family due to the merger of the former genus *Heterozostera* (Setch.) Hartog which was found to nest phylogenetically within *Zostera* (Les et al., 2002; Tanaka et al.,

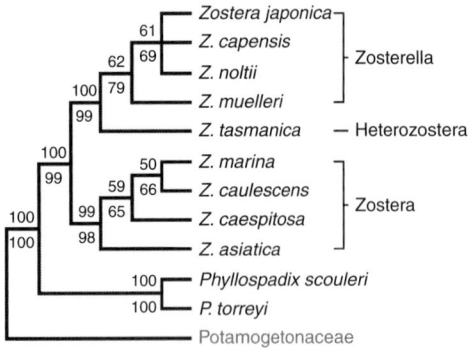

Fig 6.23 Relationships in Zosteraceae as resolved by analysis of *rbc*L data (see 'subclade I', Fig 6.2). Subgenera of *Zostera* are delimited by brackets.

2003). The monophyly of both genera has been confirmed by phylogenetic analysis of morphological data (Les et al., 2002), as well as by phylogenies generated using molecular data sets (Les et al., 2002; Tanaka et al., 2003).

Zostera is subdivided into three subgenera: subgenus *Zosterella* (Asch.) Ostenf., which consistently resolves as the sister clade to the former genus *Heterozostera* (now recognized as the monotypic subgenus *Heterozostera* Setch.); in turn, subgenus *Zostera* L. represents the sister group to the subgeneric *Heterozostera/Zosterella* clade (Fig 6.23). All currently recognized *Zostera* species now have been analysed phylogenetically if one excludes several new taxa named by Kuo (2005), and *Z. pacifica* S.Watson (Coyer et al., 2008), which are weakly defined morphologically and perhaps warrant recognition as infraspecific taxa instead of distinct species.

Although the phylogenetic relationships of *Zostera* now are fairly well understood and can be evaluated quite comprehensively using available *mat*K, nrITS and *rbc*L data (Les et al., 2002; Tanaka et al., 2003), *Z. capensis* remains somewhat problematic. Morphological data place *Z. capensis* within subgenus *Zosterella* (Hartog, 1970; Les et al., 2002). However, the *mat*K sequences reported by Tanaka et al. (2003) for *Z. capensis* and *Z. tasmanica* are identical, placing the former species anomalously within subgenus *Heterozostera*. Furthermore, the result from *mat*K conflicts with that we obtained using data from a different chloroplast locus (*rbc*L), which resolved *Z. capensis* in agreement with morphology (Figs 6.2, 6.23). Moreover, although the *rbc*L sequence was obtained directly from material collected in the native (African Cape) range of *Z. capensis* (Forest et al., 2007), it closely matches or is identical to *rbc*L sequences reported for *Z. japonica* (Appendix), another member of subgenus *Zosterella*.

Both *Z. capensis* and *Z. japonica* are extremely similar morphologically (Hartog, 1970), and differ primarily by whether the seed striae are distinct (*Z. capensis*) or indistinct (*Z. japonica*). These observations raise the possibility of whether *Z. capensis* might represent an early introduction of *Z. japonica* to Africa, rather than a novel species, but this question deserves further evaluation. In any event, *mat*K and *rbc*L are both chloroplast genes, which share the same evolutionary history. Consequently, the different phylogenetic relationships indicated must be incorrect for either one of the sequences or for both sequences. Here, it is more likely that the *rbc*L sequence yields the correct phylogeny, as the result agrees with the morphological distinctions that characterize subgenus *Zosterella*. The summary phylogeny of *Zostera* provided by *rbc*L data (Fig 6.23) reflects this interpretation.

Despite its small size (five species), phylogenetic relationships within *Phyllospadix* have not yet been evaluated using either morphological or molecular data. Comparable molecular data (*rbc*L sequences in this case) exist for only two species (*P. scouleri*, *P. torreyi*), which have been included in the *rbc*L analysis summarized in Fig 6.23. Only one sequence (*mat*K) is available for a third species (*P. iwatensis* Makino), which precluded its inclusion in that assessment. No molecular data have yet been reported for either *P. japonicus* Makino or *P. serrulatus* Rupr. ex Asch.

6.6 Summary

Ongoing phylogenetic studies of alismatids consistently demonstrate that the group is monophyletic; however, it remains uncertain whether their immediate sister group is Acoraceae, Araceae or Tofieldiaceae. Phylogenetic analyses also corroborate that alismatids comprise two major clades, which we recognize taxonomically as: Alismatales (petaloid taxa) and Potamogetonales (tepaloid taxa).

Alismatids comprise 11 distinct and phylogenetically defensible families. There has been no compelling evidence from any recent study that justifies the continued segregation of several families, namely Limnocharitaceae, Ruppiaceae and Zannichelliaceae. We recommend that these families be merged with Alismataceae, Cymodoceaceae and Potamogetonaceae, respectively. A consistent pattern of interfamilial relationships in Alismatidae is recovered in phylogenetic analyses. The associations of Potamogetonaceae with Zosteraceae, Posidoniaceae with Cymodoceaceae (including Ruppiaceae) and Butomaceae with Hydrocharitaceae are indicated repeatedly. The placement of *Najas* (Najadaceae) within Hydrocharitaceae now has been firmly established by analyses of a large amount of molecular data. Four subclades are resolved consistently within Hydrocharitaceae, which correspond to previously proposed subfamilies. The monotypic Maundiaceae clearly are distinct from Juncaginaceae; however, the precise relationship of the family to Cymodoceaceae remains uncertain. The dissolution of *Lilaea* and its merger with *Triglochin* is well supported by various phylogenetic studies.

Intergeneric relationships have been elucidated quite well for most families. However, the precise relationship of some genera (e.g. *Albidella*) remains unsettled or has not been evaluated sufficiently (e.g. *Althenia, Lepilaena, Pseudalthenia* and *Zannichellia*).

Interspecific relationships have been clarified considerably in many major alismatid genera, such as *Alisma, Echinodorus, Halophila, Vallisneria* and *Zostera*. Although good progress also has been made in elucidating relationships in some of the larger genera, including *Aponogeton, Najas, Potamogeton* and *Sagittaria*, additional taxon sampling and data are needed for more comprehensive systematic coverage. Several large genera (*Blyxa, Lagarosiphon, Ottelia*) remain very poorly understood systematically.

6.7 Acknowledgements

This material is based upon work supported by the National Science Foundation under Grant No. DEB-0841658. Any opinions, findings, and conclusions or recommendations expressed in this material are those of the authors and do not necessarily reflect the views of the National Science Foundation.

6.8 References

APG I (Angiosperm Phylogeny Group I). (1998). An ordinal classification for the families of flowering plants. *Annals of the Missouri Botanical Gardens*, **85**, 531–553.

APG II (Angiosperm Phylogeny Group II). (2003). An update of the Angiosperm Phylogeny Group classification for the orders of flowering plants: APG II. *Botanical Journal of the Linnean Society*, **141**, 399–436.

APG III (Angiosperm Phylogeny Group III). (2009). An update of the Angiosperm Phylogeny Group classification for the orders and families of flowering plants: APG III. *Botanical Journal of the Linnean Society*, **161**, 105–121.

Baldwin, B. G. (1992). Phylogenetic utility of the internal transcribed spacers of ribosomal DNA in plants: an example from the Compositae. *Molecular Phylogenetics and Evolution*, **1**, 3–16.

Bremer, B., Bremer, K., Heidari, N. et al. (2002). Phylogenetics of asterids based on 3 coding and 3 non-coding chloroplast DNA markers and the utility of non-coding DNA at higher taxonomic levels. *Molecular Phylogenetics and Evolution*, **24**, 273–300.

Cameron, K. M., Chase, M. W. and Rudall, P. J. (2003). Recircumscription of the monocotyledonous family Petrosaviaceae to include *Japonolirion*. *Brittonia*, **55**, 214–225.

Chase, M. W., Soltis, D. E., Olmstead, R. G. et al. (1993). Phylogenetics of seed plants: an analysis of nucleotide sequences from the plastid gene *rbc*L. *Annals of the Missouri Botanical Garden*, **80**, 528–580.

Cook, C. D. K. (1996). *Aquatic Plant Book*, 2nd edn. Amsterdam: SPB Academic Publishing bv.

Coyer J. A., Miller, K. A., Engle, J. M. et al. (2008). Eelgrass meadows in the California Channel Islands and adjacent coast reveal a mosaic of two species, evidence for introgression and variable clonality. *Annals of Botany*, **101**, 73–87.

Cronquist, A. (1981).*An Integrated System of Classification of Flowering Plants*. New York: Columbia University Press.

Davis, J. I., Stevenson, D. W., Petersen, G. et al. (2004). A phylogeny of the monocots, as inferred from *rbc*L and *atp*A sequence variation, and a comparison of methods for calculating jackknife and bootstrap values. *Systematic Botany*, **29**, 467–510.

Doyle, J. J. and Doyle, J. L. (1987). A rapid DNA isolation procedure for small quantities of fresh leaf tissue. *Phytochemical Bulletin*, **19**, 11–15.

Duvall, M. R. and Ervin, A. B. (2004). 18S gene trees are positively misleading for monocot/dicot phylogenetics. *Molecular Phylogenetics and Evolution*, **30**, 97–106.

Fernald, M. L. (1923). The American variety of *Scheuchzeria palustris*. *Rhodora*, **25**, 177–179.

Forest, F., Grenyer, R., Rouget, M. et al. (2007). Preserving the evolutionary potential of floras in biodiversity hotspots. *Nature*, **445**, 757–760.

Givnish, T. J., Pires, J. C., Graham, S. W. et al. (2006). Phylogeny of the monocotyledons based on the highly informative plastid gene *ndhF*: evidence for widespread concerted convergence. In *Monocots: Comparative Biology And Evolution (Excluding Poales)*, ed. J. T. Columbus, E. A. Friar, C. W. Hamilton, et al. Claremont, CA: Rancho Santa Ana Botanic Garden, pp. 28–51.

Graham, S. W., Zgurski, J. M., McPherson, M. A. **et al.** (2006). Robust inference of monocot deep phylogeny using an expanded multigene plastid data set. *Aliso*, **22**, 3–20.

Green, E. P. and Short, F. T. (2003). *World Atlas Of Seagrasses*. Berkeley, CA: University of California Press.

Hartog, C. D en. (1970). *The Sea-grasses Of The World*. Amsterdam: North-Holland Publishing Co.

Haynes, R. R. and Les, D. H. (2005). Alismatales. In *Nature Encyclopedia of Life Sciences*. London: Nature Publishing Group.

Haynes, R. R., Les, D. H. and Holm-Nielsen, L. B. (1998). Zannichelliaceae. In *The Families And Genera Of Vascular Plants, Volume IV, Flowering Plants: Monocotyledons, Alismatanae and Commelinanae (Except Gramineae)*, ed. K. Kubitzki. Berlin: Springer-Verlag, pp. 470–474.

Iida, S., Kosugeb, K. and Kadono, Y. (2004). Molecular phylogeny of Japanese *Potamogeton* species in light of noncoding chloroplast sequences. *Aquatic Botany*, **80**, 115–127.

Iles, W., Smith, S. Y. and Graham, S. W. (2009). Robust resolution of the backbone of Alismatales phylogeny [abstract]. Vancouver: Botany 2009.

Ito, Y., Ohi-Toma, T., Murata, J. and Tanaka, N. (2010). Hybridization and polyploidy of an aquatic plant, *Ruppia* (Ruppiaceae), inferred from plastid and nuclear DNA phylogenies. *American Journal of Botany*, **97**, 1156–1167.

Jacobs, S. W. L., Les, D. H., Moody, M. L. and Hellquist, C. B. (2006). Two new species of *Aponogeton* (Aponogetonaceae), and a key to species from Australia. *Telopea*, **11**, 129–134.

Jacobson, A. and Hedrén, M. (2007). Phylogenetic relationships in *Alisma*

(Alismataceae) based on RAPDs, and sequence data from ITS and *trnL*. *Plant Systematics and Evolution*, **265**, 27–44.

Johnson, L. A. and Soltis, D. E. (1995). Phylogenetic inference in Saxifragaceae sensu stricto and *Gilia* (Polemoniaceae) using *matK* sequences. *Annals of the Missouri Botanical Garden*, **82**, 149–175.

Judd, W. S., Campbell, C. S., Kellogg, E. A., Stevens, P. F. and Donoghue, M. J. (2008). *Plant Systematics. A Phylogenetic Approach*, 3rd ed. Sunderland, MA: Sinauer.

Keener, B. R. (2005). *Molecular Systematics and Revision of the Aquatic Monocot Genus* Sagittaria *(Alismataceae)*. Ph.D. dissertation, Tuscaloosa, AL: The University of Alabama.

Krahulcová, A. and Jarolímová, V. (1993). Ecology of two cytotypes of *Butomus umbellatus* I. Karyology and breeding. *Folia Geobotanica et Phytotaxonomica*, **28**, 385–411.

Kuo, J. (2005). A revision of the genus *Heterozostera* (Zosteraceae). *Aquatic Botany*, **81**, 97–140.

Lehtonen, S. (2006). Phylogenetics of *Echinodorus* (Alismataceae) based on morphological data. *Botanical Journal of the Linnean Society*, **150**, 291–305.

Lehtonen, S. (2008). An integrative approach to species delimitation in *Echinodorus* (Alismataceae) and the description of two new species. *Kew Bulletin*, **63**, 525–563.

Lehtonen, S. and Myllys, L. (2008). Cladistic analysis of *Echinodorus* (Alismataceae): simultaneous analysis of molecular and morphological data. *Cladistics*, **24**, 218–239.

Les, D. H. and Haynes, R. R. (1995). Systematics of subclass Alismatidae: a synthesis of approaches. In *Monocotyledons: Systematics And Evolution*, ed. P. J. Rudall, P. J. Cribb,

D. F. Cutler, and C. J. Humphries. Kew: Royal Botanic Gardens, pp. 353–377.

Les, D. H. and Philbrick, C. T. (1993). Studies of hybridization and chromosome number variation in aquatic angiosperms: evolutionary implications. *Aquatic Botany*, **44**, 181–228.

Les, D. H. and Sheridan, D. J. (1990). Biochemical heterophylly and flavonoid evolution in North American *Potamogeton* (Potamogetonaceae). *American Journal of Botany*, **77**, 453–465.

Les, D. H., Garvin, D. K. and Wimpee, C. F. (1993). Phylogenetic studies in the monocot subclass Alismatidae: evidence for a reappraisal of the aquatic order Najadales. *Molecular Phylogenetics and Evolution*, **2**, 304–314.

Les, D. H., Cleland, M. A. and Waycott M. (1997). Phylogenetic studies in Alismatidae, II: evolution of marine angiosperms ('seagrasses') and hydrophily. *Systematic Botany*, **22**, 443–463.

Les, D. H., Moody, M. L., Jacobs, S. W. L. and Bayer, R. J. (2002). Systematics of seagrasses (Zosteraceae) in Australia and New Zealand. *Systematic Botany*, **27**, 468–484.

Les, D. H., Moody, M. L. and Jacobs, S. W. L. (2005). Phylogeny and systematics of *Aponogeton* (Aponogetonaceae): the Australian species. *Systematic Botany*, **30**, 503–519.

Les, D. H., Moody, M. L. and Soros, C. (2006). A reappraisal of phylogenetic relationships in the monocotyledon family Hydrocharitaceae. In *Monocots: Comparative Biology and Evolution*, ed. J. T. Columbus, E. A. Friar, C. W. Hamilton et al. Claremont, California: Rancho Santa Ana Botanic Garden, pp. 211–230.

Les, D. H., Jacobs, S. W. L., Tippery, N. P. et al. (2008). Systematics of *Vallisneria* (Hydrocharitaceae). *Systematic Botany*, **33**, 49–65.

Les, D. H., Murray, N. M. and Tippery, N. P. (2009). Systematics of two imperiled pondweeds (*Potamogeton vaseyi, P. gemmiparus*) and taxonomic ramifications for subsection *Pusilli* (Potamogetonaceae). *Systematic Botany*, **34**, 643–651.

Les, D. H., Sheldon, S. P. and Tippery, N. P. (2010). Hybridization in hydrophiles: natural interspecific hybrids in *Najas* L. (Hydrocharitaceae). *Systematic Botany*, **35**, 736–744.

Li, X.-X. and Zhou, Z.-K. (2007). The higher-level phylogeny of monocots based on *matK*, *rbcL* and 18S rDNA sequences. *Acta Phytotaxonomica Sinica*, **45**, 113–133.

Li, X. and Zhou, Z. (2009). Phylogenetic studies of the core Alismatales inferred from morphology and *rbcL* sequences. *Progress in Natural Science*, **19**, 931–945.

Lindqvist, C., De Laet, J., Haynes, R. R. et al. (2006). Molecular phylogenetics of an aquatic plant lineage, Potamogetonaceae. *Cladistics*, **22**, 568–588.

Petersen, G., Seberg, O., Davis, J. I. and Stevenson, D. W. (2006). RNA editing and phylogenetic reconstruction in two monocot mitochondrial genes. *Taxon*, **55**, 871–886.

Posada, D. (2008). jModelTest: Phylogenetic model averaging. *Molecular Biology and Evolution*, **25**, 1253–1256.

Posluszny, U., Charlton, W. A. and Les, D. H. (2000). Modularity in Helobial flowers. In *Systematics and Evolution of Monocots*, ed. K. L. Wilson and D. Morrison. Victoria: CSIRO Publishing, pp. 63–74.

Qiu, Y.-L., Lee, J., Bernasconi-Quadroni, F. et al. (2000). Phylogeny of basal angiosperms: analyses of five genes from three genomes. *International Journal of Plant Science*, **161**(S6), S3–S27.

Rogstad, S. H. (1992). Saturated NaCl-CTAB solution as a means of field preservation of leaves for DNA analyses. *Taxon*, **41**, 701–708.

Shaffer-Fehre, M. (1991a). The endotegmen tuberculae: an account of little-known structures from the seed coat of the Hydrocharitoideae (Hydrocharitaceae) and *Najas* (Najadaceae). *Botanical Journal of the Linnean Society*, **107**, 169–188.

Shaffer-Fehre, M. (1991b). The position of *Najas* within the subclass Alismatidae (Monocotyledones) in the light of new evidence from seed coat structures in the Hydrocharitoideae (Hydrocharitales). *Botanical Journal of the Linnean Society*, **107**, 189–209.

Soltis, D. E., Gitzendanner, M. A. and Soltis, P. S. (2007). A 567-taxon data set for angiosperms: the challenges posed by Bayesian analyses of large data sets. *International Journal of Plant Sciences*, **168**, 137–157.

Swofford, D. L. (2002). *PAUP*. Phylogenetic analysis using parsimony (*and other methods)*, ver. 4. Sunderland, MA: Sinauer Associates.

Tamura, M. N., Yamashita, J., Fuse, S. and Haraguchi, M. (2004). Molecular phylogeny of monocotyledons inferred from combined analysis of plastid *matK* and *rbcL* gene sequences. *Journal of Plant Research*, **117**, 109–120.

Tanaka, N., Setoguchi, H. and Murata, J. (1997). Phylogeny of the family Hydrocharitaceae inferred from *rbcL* and *matK* gene sequence data. *Journal of Plant Research*, **110**, 329–337.

Tanaka, N., Kuo, J., Omori, Y., Nakaoka, M. and Aioi, K. (2003). Phylogenetic relationships in the genera *Zostera* and *Heterozostera* (Zosteraceae) based on *matK* sequence data. *Journal of Plant Research*, **116**, 273–279.

Thorne, R. F. and Reveal, J. (2007). An updated classification of the class Magnoliopsida ('Angiospermae'). *Botanical Review*, **73**, 67–182.

Tippery, N. P., Les, D. H., Padgett, D. J. and Jacobs, S. W. L. (2008). Generic circumscription in Menyanthaceae: A phylogenetic evaluation. *Systematic Botany*, **33**, 598–612.

Tomlinson, P. B. (1982). *Anatomy Of The Monocotyledons VII. Helobiae (Alismatidae) (Including The Seagrasses)*. Oxford: Clarendon Press.

von Mering, S. and Kadereit, J. W. (2010). Phylogeny, systematics, and recircumscription of Juncaginaceae – a cosmopolitan wetland family. In *Diversity, Phylogeny, and Evolution in the Monocotyledons*, ed. O. Seberg, G. Petersen, A. S. Barfod, and J. I. Davis.

Aarhus: Aarhus University Press, pp. 55–79.

Waycott, M., Freshwater, D. W., York, R. A., Calladine, A. and Kenworthy, W. J. (2002). Evolutionary trends in the seagrass genus *Halophila* (Thouars): Insights from molecular phylogeny. *Bulletin of Marine Science*, **71**, 1299–1308.

Waycott, M., Procaccini, G., Les, D. H. and Reusch, T. B. H. (2006). Seagrass evolution, ecology and conservation: a genetic perspective. In *Seagrasses: Biology, Ecology And Conservation*, ed. A. W. D. Larkum, R. J. Orth and C. M. Duarte. Dordrecht: Springer-Verlag, pp. 25–50.

Zhang, T., Wang, Q., Li, W., Cheng, Y. and Wang, J. (2008). Analysis of phylogenetic relationships of *Potamogeton* species in China based on chloroplast *trnT-trnF* sequences. *Aquatic Botany*, **89**, 34–42.

Zwickl, D. J. (2006). *Genetic Algorithm Approaches for the Phylogenetic Analysis of Large Biological Sequence Datasets Under the Maximum Likelihood Criterion*. Ph.D. thesis. Austin, TX: University of Texas.

6.9 Appendix

Voucher and GenBank accession number information for sequences newly analysed in this study. Paragraphs separate groups discussed in the text (see Fig 6.2). Order of information is as follows: taxon, authority, voucher information (only for sequences newly reported here), GenBank accession numbers (ITS, *matK/trnK*, *rbcL*). Asterisks (*) indicate sequences newly reported; dashes (–) indicate sequences not applicable to our study; slashes (/) indicate discontinuous sequences that were joined in our analyses; numbers in brackets ([]) indicate additional published sequences for the same or related taxa that were not used in our study.

A. Alismataceae. *Alisma plantago-aquatica* L. (DQ339085, AF542573, L08759); *Astonia australiensis* (Aston) S.W.L.Jacobs, *Jacobs 9670* (NSW), (AY335952, HQ456456*/HQ456457*, HQ456499*); *Baldellia ranunculoides* (L.) Parlatore,

Charlton s.n. (MANCH), (HQ456394*, HQ456458*, U80677 [DQ859163]); *Butomopsis latifolia* Kunth, *Jacobs 9257* (NSW), (HQ456395*, HQ456459*, HQ456500*); *Caldesia acanthocarpa* (F.Muell.) Buchenau, *Jacobs 8209* (NSW), (HQ456396*, HQ456460*, HQ456501*); *C. oligococca* (F.Muell.) Buchenau, *Jacobs 8203* (NSW), (HQ456397*, HQ456461*, HQ456502* [AY277799]); *Damasonium alisma* Mill., *Charlton s.n.* (MANCH), (HQ456398*, HQ456462*, U80678); *Echinodorus amazonicus* Rataj, *Charlton s.n.* (MANCH), (-, -, HQ456503*); *E. cordifolius* (L.) Griseb. (EF088079, EF088127, DQ859164); *E. grandiflorus* (Cham. and Schltdl.) Micheli (EF088070, EF088118, U80679); *E. grisebachii* Small (EF088046, EF088095, -); *E. longiscapus* Arechav. (EF088068, EF088116, -); *E. osiris* Rataj (-, -, DQ859165); *Helanthium parvulum* (Engelm.) Small, *Moody 381* (CONN), (HQ456399*, HQ456463*, HQ456504*); *H. tenellum* (Mart.) Britton, *Charlton s.n.* (MANCH), (EF088056, EF088105, HQ456505*); *Hydrocleys nymphoides* (Willd.) Buchenau, *Les s.n.* (CONN), (HQ456423*, HQ456470*, U80716 [AB004900]); *Limnocharis flava* (L.) Buchenau, *Les s.n.* (CONN), (HQ456400*, HQ456464*, U80717 [AB088807]); *Luronium natans* (L.) Raf., *Charlton s.n.* (MANCH), (HQ456401*, HQ456465*, U80680); *Ranalisma humile* (Kuntze) Hutch., *Charlton s.n.* (MANCH), (HQ456402*, HQ456466*, U80681); *R. rostratum* Stapf (AY395986, AY952415, AY952438); *Sagittaria lancifolia* L., *Les s.n.* (CONN), (-, -, HQ456506*); *S. latifolia* Willd., *Les s.n.* (MANCH), (HQ456403*, HQ456467*, L08767); *S. montevidensis* Cham. and Schltdl., *Les s.n.* (CONN), (HQ456404*, HQ456468*, HQ456507*); *Wiesneria triandra* Micheli, *Cook s.n.* (Z), (AY335953, HQ535983*, U80682).

B. Butomaceae. *Butomus umbellatus* L., *Les 499* (CONN), (AY870346, HQ456469*, U80685 [AY149345]).

C. Hydrocharitaceae. *Apalanthe granatensis* (Humb. and Bonpl.) C.D.K.Cook and Urmi-König (AY870362, -, U80693); *Blyxa aubertii* L.C.Richard (AY870359, -, U80694 [AB088810 – *B. echinosperma* (C.B.Clarke) Hook.f.]); *B. japonica* Maxim. ex Asch. and Gürke (-, -, AB004886); *Egeria densa* Planch. (AY330707/AY870360, -, U80695 [AB004887]); *E. najas* Planch. (AY330708, -, DQ859166); *Elodea bifoliata* St.John, *Les 801* (CONN), (HQ456405*, -, -); *E. canadensis* Michx., (1) *Les 803* (CONN), (HQ456406-HQ456413*, -, -); (2) *Les 813* (CONN), (HQ456414-HQ456421*, -, -); (3) *Les 814* (CONN), (HQ456422*, -, -); (4) (AY330704, -, DQ859167); *E. nuttallii* (Planch.) H.St.John (AY330706/AY870361/EF526382, -, U80696 [AB004888]); *Enhalus acoroides* (L.f.) Rich. ex Steud. (-, -, U80697 [AB004889]); *Halophila decipiens* Ostenf. (-, -, U80698); *H. engelmannii* Asch. (-, -, U80699); *H. ovalis* (R.Br.) Hook.f. (-, -, AB004890 [DQ859168]); *Hydrilla verticillata* (L.f.) Casp. (-, -, U80700 [AB004891/GU135149/GU135242]); *Hydrocharis dubia* (Blume) Backer (-, -, AB004892); *H. morsus-ranae* L. (AY335962, AY870375/AY874445, U80701); *Lagarosiphon madagascariensis* Casp. (-, -, AB004893); *L. major* (Ridley) Moss (-, -, U80703); *L. muscoides* Harv.

(–, –, U80702); *Limnobium laevigatum* (Humb. and Bonpl. ex Willd.) Heine (–, –, AB004894); *L. spongia* (Bosc.) Steud., *Cook s.n.* (Z), (AY335963, HQ456471*, U80704); *Najas arguta* Kunth (–, –, HM240485); *N. browniana* Rendle (–, –, HM240486); *N. flexilis* (Willd.) Rostk. and W.L.E.Schmidt (–, –, HM240489); *N. gracillima* (A.Braun ex Engelm.) Magnus (–, –, HM240490); *N. guadalupensis* (Spreng.) Magnus (–, –, DQ859169 [DQ859170]); *N. marina* L. (–, –, U80705); *N. minor* All. (–, –, HM240506); *N. tenuifolia* R.Br. (–, –, HM240507); *Nechamandra alternifolia* (Roxburgh ex Wight) Thwaites (–, –, U80706 [AB506768]); *Ottelia acuminata* (Gagnep.) Dandy (–, –, AY952435); *O. alismoides* (L.) Pers. (AY870358, –, U80707 [AB004895]); *O. ovalifolia* Rich. (–, –, DQ859171); *O. ulvifolia* Walp. (–, –, U80708); *Stratiotes aloides* L., *Les s.n.* (CONN), (AY870357, HQ456472*, U80709 [AB004896]); *Thalassia hemprichii* (Ehrenb.) Asch. (–, –, U80710 [AB004897]); *T. testudinum* Banks ex K.Koenig (–, –, U80711); *Vallisneria americana* Michx. (–, –, EF143005); *V. asiatica* Miki (–, –, AB004898 [EF143007/ EF155532]); *V. australis* S.W.L.Jacobs and Les (–, –, EF143008); *V. caulescens* F.M.Bailey and F.Muell. (–, –, EF143009); *V. denseserrulata* Makino (–, –, EF143010); *V. gracilis* F.M.Bailey (–, –, EF143012 [EF143006 – *V. annua* S.W.L. Jacobs and K.A.Frank]); *V. nana* R.Br. (–, –, EF143013); *V. natans* (Lour.) Hara (–, –, EF143014); *V. neotropicalis* Marie-Vict. (–, –, EF143015); *V. rubra* (Rendle) Les and S.W.L.Jacobs (–, –, EF143004); *V. spinulosa* S.Z.Yan (–, –, EF143017); *V. spiralis* L. (–, –, U80712 [DQ859177/EF143018]); *V. triptera* S.W.L.Jacobs and K.A.Frank (–, –, EF143019).

D. Aponogetonaceae. *Aponogeton boivinianus* Baill. ex Jum., *van Bruggen s.n.* (CONN), (HQ456425*, HQ456475*, HQ456508*); *A. bruggenii* S.R.Yadav and R.S. Govekar, *Herr s.n.* (CONN), (HQ456426*, HQ456476*, –); *A. bullosus* H.Bruggen, *Jacobs 8572 and Les 595* (CONN, NSW), (AY926318, AY926279/AY926344, HQ456509*); *A. capuronii* H.Bruggen, *van Bruggen s.n.* (CONN), (HQ456427*, HQ456477*, HQ456510*); *A. crispus* Thunb. (AY926288, AY926263/AY926328, DQ859162); *A. cuneatus* S.W.L.Jacobs (AY926291, AY926264/AY926329, –); *A. decaryi* Jum., *Eggers s.n.* (CONN), (–, –, HQ456511*); *A. distachyos* L.f. (AY926320, AY926281/AY926346, U80684); *A. eggersii* Bogner and H.Bruggen, *Schöpfel s.n.* (CONN), (HQ456437*, HQ456483*, –); *A. elongatus* F.Muell. ex Benth. (AY926296, AY926266/AY926331, U68091 [U80683]); *A. euryspermus* Hellq. and S.W.L.Jacobs, *Jacobs 8839* (NSW), (AY926310, AY926275/AY926340, HQ456512*); *A. fenestralis* (Pers.) Hook.f., *Kasselmann s.n.* (CONN), (–, HQ456487*, AB088808); *A. hexatepalus* H.Bruggen, *Sainty 434337* (NSW), (AY926321, AY926282/AY926347, HQ456513*); *A. junceus* Lehm. ex Schltr., *Viljoen s.n.* (CONN), (HQ456441*, HQ456486*, HQ456514*); *A. kimberleyensis* Hellq. and S.W.L.Jacobs (AY926309, AY926274/AY926339, –); *A. lancesmithii* Hellq. and S.W.L.Jacobs, *Jacobs 8567 and Les 590* (CONN, NSW), (AY926316, AY926277/AY926342, HQ456515*); *A. longiplumulosus* H.Bruggen, *Jacobs 8534 and Les 560* (CONN, NSW), (AY926284,

AY926260/AY926325, HQ456516*); *A. madagascariensis* (Mirb.) H.Bruggen, *Jacobs 8536 and Les 562* (CONN, NSW), (AY926285, AY926261/AY926326, HQ456517*); *A. madagascariensis* var. *major* (Baum) H.Bruggen, *Kasselmann s.n.* (CONN), (HQ456444*, HQ456488*, –); *A. masoalaensis* Bogner, *Bogner 2087* (M), (HQ456445*, HQ456489*, –); *A. proliferus* Hellq. and S.W.L.Jacobs, *Jacobs 8523 and Les 549* (CONN, NSW), (AY926315, AY926276/AY926341, HQ456518*); *A. queenslandicus* H.Bruggen, *Jacobs 8524 and Les 550* (CONN, NSW), (AY926293, AY926265/AY926330, HQ456519*); *A. rigidifolius* H.Bruggen, *Jacobs 8529 and Les 555* (CONN, NSW), (AY926287, AY926262/AY926327, HQ456520*); *A. robinsonii* A.Camus, *Bogner 2905* (M), (AY926319, AY926280/AY926345, HQ456521*); *A. stachyosporus* de Wit, *Jacobs 8538 and Les 564* (CONN, NSW), (AY926304, AY926272/AY926337, HQ456522*); *A. tofus* S.W.L.Jacobs (AY926301, AY926270/AY926335, –); *A. ulvaceus* Baker, *Eggers s.n.* (M), (AY926283, AY926259/AY926324, HQ456523*); *A. undulatus* Roxb., *van Bruggen s.n.* (CONN), (HQ456451*, HQ456492*, HQ456524*); *A. vanbruggenii* Hellq. and S.W.L.Jacobs, *Jacobs 8542 and Les 568* (CONN, NSW), (AY926300, AY926269/AY926334, HQ456525*); *A. womersleyi* H.Bruggen, *Bleher s.n.* (M), (HQ456452*, HQ456493*, –).

E. Scheuchzeriaceae. *Scheuchzeria palustris* L. (–, –, U03728).

F. Juncaginaceae. *Cycnogeton procerum* Buchenau (AY926323, AY926349, U80713); *Tetroncium magellanicum* L. (AY926322, AY926348, GQ452337); *Triglochin barrelieri* Loisel. (–, –, GQ452331); *T. bulbosa* L. (–, –, AM234996); *T. elongata* Buchenau (–, –, GQ452332); *T. maritima* L., *Les s.n.* (CONN), (HQ456455*, HQ456495*, U80714 [AB088811/GQ452333]); *T. palustris* L. (–, –, DQ859176 [GQ452334]); *T. rheophila* Aston (–, –, GQ452335); *T. scilloides* (Poir.) Mering and Kadereit, *Philbrick 3031* (WCSU), (HQ456453*, HQ456494*, U80715); *T. striata* Ruiz and Pav. (–, –, GQ452336).

G. Cymodoceaceae 'complex'. Cymodoceaceae. *Amphibolis antarctica* (Labill.) Asch. (–, –, U80686); *Cymodocea nodosa* (Ucria) Asch. (–, –, U80688); *C. serrulata* (R.Br.) Asch. and Magnus (–, –, U80687); *Halodule beaudettei* (Hartog) Hartog (–, –, U80689); *H. pinifolia* (Miki) Hartog (–, –, U80690); *H. uninervis* (Forssk.) Asch. (–, AY952424, AY952436); *H. wrightii* Asch. (–, –, AY787476/AY787477); *Syringodium filiforme* Kütz (–, –, U03727); *S. isoetifolium* (Asch.) Dandy (–, –, U80691); *Thalassodendron pachyrhizum* Hartog (–, –, U80692). Posidoniaceae. *Posidonia australis* Hook.f. (–, –, U80718); *P. oceanica* (L.) Delile (–, –, U80719). Ruppiaceae. *Ruppia cirrhosa* (Pentagna) Grande (–, –, DQ859175); *R. maritima* L. (–, , U03729); *R. megacarpa* R.Mason (–, –, U80728).

H. Maundiaceae. *Maundia triglochinoides* F.Muell., *Stanberg and Sainty LS 80* (NSW), (HQ456454*, HQ456496*, GQ452330).

I. Potamogetonaceae/Zosteraceae. Potamogetonaceae. *Groenlandia densa* (L.) Fourr., *Philbrick 4585* (WCSU), (–, HQ456497*, U80720 [AB196954]); *Lepilaena australis* J.Drum. ex Harv. (–, –, U80729); *Potamogeton alpinus* Balb.

(–, –, AB196845); *P. amplifolius* Tuck. (–, –, L08765); *P. berchtoldii* Fieber (–, GQ247475, –); *P. compressus* L. (–, –, AB196846); *P. confervoides* Rchb. (–, –, U80721); *P. crispus* L. (–, –, AB196847 [U80722]); *P. cristatus* Regel and Maack (–, –, AB196939); *P. dentatus* Hagström (–, –, AB196940 [AB332412]); *P. distinctus* A.Benn. (–, –, AB196941 [AB004901/AB088809]); *P. foliosus* Raf. (–, GQ247493, –); *P. fryeri* A.Benn. (–, –, AB196942); *P. gramineus* L. (–, –, AB196943 [U80723/ EF174581]); *P. lucens* L. (–, –, DQ859173); *P. maackianus* A.Benn. (–, –, AB196944 [AB506769]); *P. malaianus* Miq. (–, –, AB196945 [EU741053/ EU741054]); *P. natans* L. (–, –, AB196946 [DQ859174]); *P. obtusifolius* Mert. and W.D.J.Koch (–, –, AB196947); *P. octandrus* Poir. (–, –, AB196948); *P. oxyphyllus* Miq. (–, –, AB196949); *P. perfoliatus* L. (–, –, AB196951 [U80724/EU741051]); *P. prae-longus* Wulfen (–, –, AB196952); *P. pusillus* L. (–, GQ247502, AB196950 [AB250148]); *P. richardsonii* (A.Benn.) Rydb. (–, –, U03730); *P. robbinsii* Oakes (–, –, U80725); *P. vaseyi* J.W.Robbins ex A.Gray (–, GQ247506, –); *P. zosteriformis* Fernald, *Les 494* (CONN), (GQ247438, HQ456498*, U80726); *Stuckenia pectinata* (L.) Börner (–, –, U80727 [AB196953]); *Zannichellia palustris* L. (–, –, U03725 [AB196955]). Zosteraceae. *Phyllospadix scouleri* Hook. (–, –, DQ859172); *P. torreyi* S.Watson (–, AY077975, U80731); *Zostera asiatica* Miki (–, AB125360, AB125352); *Z. caespitosa* Miki (–, AB125359, AB125351); *Z. capensis* Setch. (–, –, AM235166); *Z. capricorni* Asch. (–, AY077983, –); *Z. caulescens* Miki (–, AB125358, AB125350); *Z. japonica* Asch. and Graebn. (–, AB125361, AB125353 [AY077964]); *Z. marina* L. (–, AB125357, AB125348 [U80734/AB125349]); *Z. muelleri* Irmisch ex Asch. (–, AY077984, AY077962); *Z. noltii* Hornemann (–, AY077981, U80733); *Z. novaze-landica* Setch. (–, AY077982, –); *Z. tasmanica* Martens ex Asch. (–, AY077979, U80730).

OUTGROUP. Araceae. *Anchomanes difformis* Engl. (–, –, L10254); *Ariopsis peltata* J.Graham (–, –, L10255); *Gymnostachys anceps* R.Br. (–, –, M91629); *Lasia spinosa* Thwaites (–, –, L10250); *Montrichardia arborescens* Schott (–, –, L10248); *Pistia stratiotes* L. (–, –, M96963); *Symplocarpus foetidus* (L.) Salisb. (–, –, L10247); *Xanthosoma sagittifolium* (L.) Schott (–, –, L10246). Lemnaceae. *Lemna minuta* Kunth (–, –, M91630).

7

Evolution of floral traits in relation to pollination mechanisms in Hydrocharitaceae

Norio Tanaka, Koichi Uehara and Jin Murata

7.1 The relationship between pollen and stigma morphology and pollination mechanisms

7.1.1 Pollen morphology

Pollen morphology has been considered to reflect phylogenetic relationships and is also used in systematic studies (Nilsson, 1990; Harley et al., 1991; Crane et al., 1995). However, there are some cases in which the exine sculpture or structure appears to be correlated with the pollination mechanisms, as discussed below.

Greatly reduced exines have been reported in many hypohydrophilous angiosperms, for example in the monocotyledonous species *Halodule wrightii* and *Thalassodendron ciliatum* (Cymodoceaceae) and in the dicotyledonous species *Ceratophyllum demersum* (Ceratophyllaceae) (Pettit and Jermy, 1975; Pettitt, 1981). Wodehouse (1935) suggested that the thickness of the exine has been reduced during the course of evolution from a terrestrial ancestor to an aquatic descendant, since a thick exine is not necessary in hypohydrophilous pollination. *Ruppia maritima* (Ruppiaceae) and *Potamogeton pectinatus* (Potamogetonaceae), which must be pollinated by bubbles under water (Verhoeven, 1979), share a reticulate exine and a thin continuous foot layer. Schwanitz (1967)

Early Events in Monocot Evolution, eds P. Wilkin and S. J. Mayo. Published by Cambridge University Press. © The Systematics Association 2013.

suggested that the reticulate exine in *Ruppia maritima* is instrumental in keeping the pollen afloat. On the other hand, both *Potamogeton natans* and *P. pectinatus* have a reticulate exine despite being anemophilous (Pettitt and Jermy, 1975).

Walker (1976) stated that elaborate pollen sculpture seems to correlate with entomophily, while psilate pollen grains are largely characteristic of anemophilous plants. However, there is no hint of anemophily in Araceae, a family that in subfamily Aroideae has largely psilate pollen (Grayum, 1992; Mayo et al., 1997; Hesse, 2000). Apart from the reduced pollen exine of hypohydrophilous species that have evolved in parallel, it is unclear whether the apparent correlation between pollen characters and pollination mechanisms is a result of adaptive evolution. Many descriptions of pollen morphology were reported also in Hydrocharitaceae (reviewed in Tanaka et al., 2004), but it is not clear whether pollen morphology really correlates with pollination mechanisms.

7.1.2 Stigma morphology

Morphological and physiological stigmatic features affect reproductive biology and breeding systems (Heslop-Harrison, 2000; Yang et al., 2002; Sigrist and Sazima, 2004). However, there are only a few reports on adaptive evolution between stigma surface morphology and pollination mechanisms (Guo and Huang, 1999). Based on an analysis of floral characters of about 100 species of angiosperms, Lee (1978) pointed out the possibility of elaborate stigmas and oblate pollen grains being adapted for wind pollination. Yang et al. (2002) reported an adaptation in *Pedicularis* species (Orobanchaceae) involving stigmatic surface morphology and other aspects of floral biology, and further referred to the importance of studying stigma characteristics and their significance in pollination adaptation. In some hypohydrophilous species, it has been reported that pollen germination requires stigmatic exudate (*Zostera marina*: De Cock, 1980; *Posidonia australis*: McConchie and Knox, 1989; *Zannichellia palustris*: Guo et al., 1990), and filiform stigmas are commonly found in *Najas* and *Halophila* (Hydrocharitaceae), Zosteraceae, Posidoniaceae and Cymodoceaceae (Ducker et al., 1978, Pettitt, 1980, McConchie and Knox, 1989; Guo and Huang, 1999).

7.2 The diverse pollination mechanisms of the family Hydrocharitaceae

Hydrocharitaceae is an aquatic family consisting of 18 genera and *c.* 80 species. The family is very diverse in its characteristic pollination mechanisms; these have been divided into five types by Cook (1982) and Tanaka (2000):

(1) Entomophily (Fig 7.1, a–e): flowers bloom above the water surface and are insect-pollinated, e.g. *Apalanthe, Blyxa, Egeria, Hydrocharis, Ottelia* and *Stratiotes*;

(2) Anemophily (Fig 7.1: f–g): stamens of the male flowers of *Limnobium laevigatum* are exserted and the pollen grains are powdery, with the female flowers having large papillose stigmas.

(3) Hypohydrophily (Fig 7.1: h–j): male and female flowers bloom under water and the pollen grains drift and pollinate the stigmas of the female flowers under water, e.g. *Halophila*, *Najas* and *Thalassia*;

(4) Pollen-epihydrophily (Fig 7.1: k–n): male flowers become detached from the parent plant and release pollen grains that float on the water surface and pollinate the stigmas of the female flowers, e.g. *Elodea* and *Hydrilla* (Tanaka, 2000, 2003); this mechanism is known only in Hydrocharitaceae;

(5) Male flower-epihydrophily (Fig 7.1: o–q): male flowers become detached from the parent plant, float on the water surface and directly pollinate the stigmas of the female flowers, e.g. *Appertiella*, *Enhalus*, *Lagarosiphon*, *Nechamandra* and *Vallisneria*; this is a mechanism restricted to Hydrocharitaceae;

7.3 Floral traits of Hydrocharitaceae

7.3.1 Pollen morphology

Pollen grains of Hydrocharitaceae are all inaperturate and show morphological diversity in shape and size (spherical to elliptical, 28 to 205 μm) and varied exine sculpture and structure (Tanaka et al., 2004).

7.3.1.1 Entomophilous species

The entomophilous pollen grains have projections such as spines or baculae and obvious and well-developed exine structures. It is considered that these exines are variations of tectate-columellate pollen exine.

Egeria densa (Fig 7.2a, b): Pollen grains are spherical, *c.* 68.0 μm in diameter. The exine of the pollen grains consists of spines and two distinct layers: an outer low electron-dense layer that consists of fine granules under the spine and rough granules elsewhere, and an inner basal highly electron-dense layer, which consists of rough granules.

Blyxa japonica (Fig 7.2c, d): Pollen grains are spherical, *c.* 35.0 μm in diameter and are not acetolysis-resistant. The exine consists of spines and a thick aggregation of clavae connected to each other at the base.

Ottelia alismoides (Fig 7.2e, f): Pollen grains are *c.* 65.0 μm in diameter. The exine consists of spines and two distinct layers: an outer layer that consists of fine granules under the spines and an inner layer that consists of rough granules.

Fig 7.1 Flowers of Hydrocharitaceae. a–e: Entomophilous species. a: Male flower of *Egeria densa*. b: Hermaphrodite flower of *Blyxa echinosperma*. c: Hermaphrodite flower of *Ottelia alismoides*. d: Female flower of *Stratiotes aloides*. e: Male flower of *Hydrocharis dubia*. f–g: Anemophilous species: f: Female flower of *Limnobium laevigatum*. g: Male flower of *L. laevigatum*. h–j: Hypohydrophilous species. h: Female flower of *Thalassia hemprichii*. i: Male flower of *T. hemprichii*. j: Female flower of *Halophila ovalis*. k–n: Pollen-epihydrophilous species. k: Spathe of *Hydrilla verticillata* with male flower, which is wrapped in a bubble, probably oxygen produced by photosynthesis, and surfaces due to its buoyancy. l: Male flower of *H. verticillata* on the surface, scattering pollen grains on the water. m: Pollen grains of *H. verticillata* drifting towards female flower on the water surface, being sucked into stigmas between the perianth segments. n: Male flower and pollen grains of *Elodea nuttallii*. o–q: Male flower-epihydrophilous species. o: Male flowers of *Vallisneria asiatica* drifting around a female flower. p: Male flowers of *V. asiatica*. q: Male flowers of *Enhalus acoroides* attached around stigmas of female flower. Colour version to be found in colour plate section.

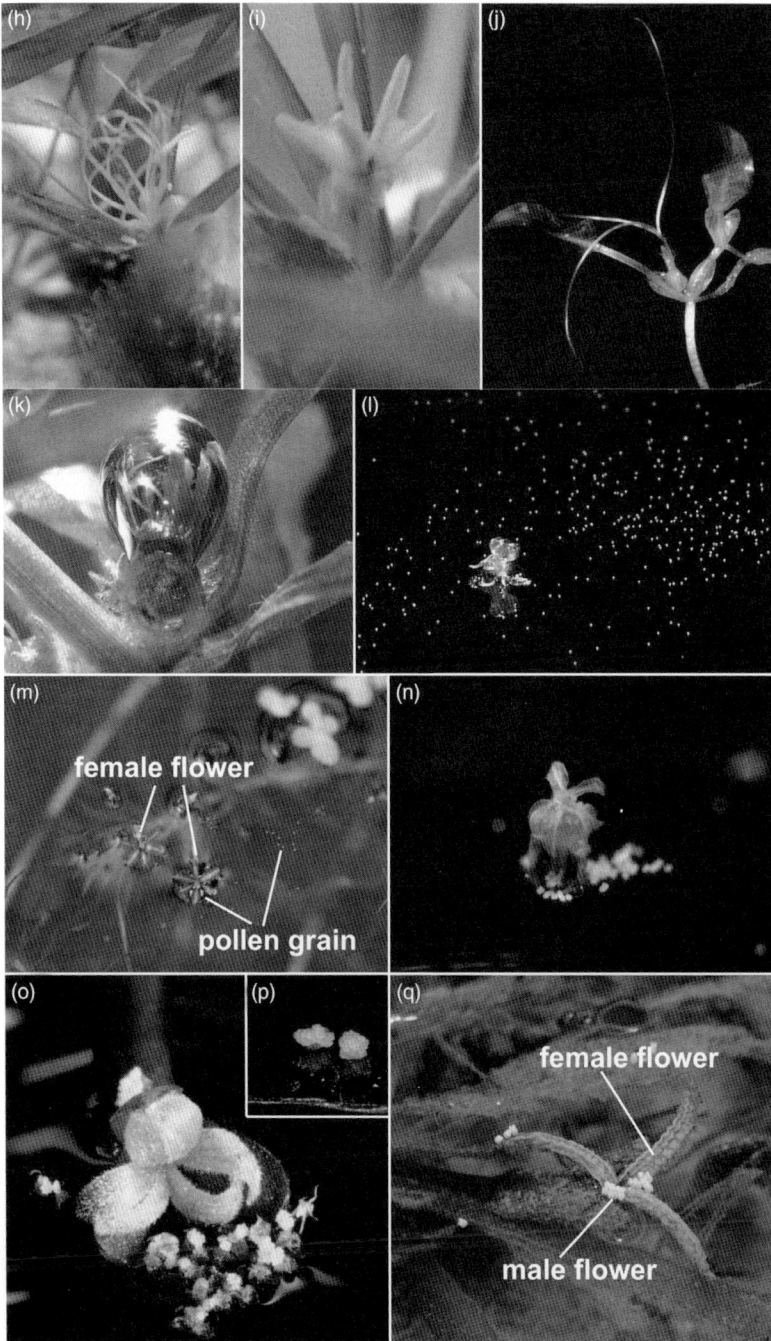

Fig 7.1 (*cont.*) Colour version to be found in colour plate section.

Fig 7.2 Pollen grains of Hydrocharitaceae. a–j: Entomophilous species. a: Pollen grain of *Egeria densa*. SEM. Bar = 10 μm. b: Exine structure of *E. densa*. TEM. Bar = 1 μm. c: Pollen grain of *Blyxa japonica*. SEM. Bar = 10 μm. d: Exine structure of *B. japonica*. TEM. Bar = 1 μm. e: Pollen grain of *Ottelia alismoides*. SEM. Bar = 10 μm. f: Exine structure of *O. alismoides*. TEM. Bar = 1 μm. g: Pollen grain of *Stratiotes aloides*. SEM. Bar = 10 μm. h: Exine structure of *S. aloides*. TEM. Bar = 1 μm. i: Pollen grain of *Hydrocharis dubia*. SEM. Bar = 10 μm. j: Exine structure of *H. dubia*. TEM. Bar = 1 μm. k-l: Anemophilous species. k: Pollen grain of *Limnobium laevigatum*. SEM. Bar = 5 μm. l: Exine structure of *L. laevigatum*. TEM. Bar = 1 μm.

Stratiotes aloides (Fig 7.2g, h): Pollen grains are spherical, *c.* 50.0 μm in diameter. The exine consists of baculae: rough granules, a partial tectum and an inner continuous basal layer.

Hydrocharis dubia (Fig 7.2i, j): Pollen grains are spherical, *c.* 30.0 μm in diameter. The exine consists of spines, columellae and a tectum on the intine.

7.3.1.2 Anemophilous species

Limnobium laevigatum (Fig 7.2k, l): Pollen grains are spherical, reticulate and *c.* 28.0 μm in diameter. The exine consists of columellae and tecta. Minute spines are observed on the surface of the reticulate tectum. This is a kind of tectate-columellate pollen exine.

7.3.1.3 Hypohydrophilous species

The pollen grains of the hypohydrophilous species commonly have a reduced exine structure, irrespective of either the presence or absence of globules.

Thalassia hemprichii (Fig 7.3a): Pollen grains are spherical, *c.* 145.0 μm in diameter and are not acetolysis-resistant. No continuous exinous layer was observed. Fibrous material is attached to the intine. These results are similar to the results reported by Pettitt (1980).

Najas oguraensis (Fig 7.3b): Pollen grains are elliptical, major axis *c.* 75.0 μm in length and minor axis *c.* 31.0 μm in length, and are not acetolysis-resistant. No continuous exinous layer is observed. Fibrous material is observed on the intine as in *Thalassia hemprichii*. The exine character of *N. oguraensis* is the same as *Najas flexilis* (Pettitt and Jermy, 1975).

Halophila: Pettitt (1980) described the mature pollen grains of *H. stipulacea* as ellipsoid or somewhat reniform. The pollen is emitted in strings with multiple grains contained within a transparent moniliform tube. The pollen grains of *H. ovalis*, *H. stipulacea* and *H. decipiens* have reduced exine (Pettitt and Jermy, 1975; Pettitt, 1980).

7.3.1.4 Pollen-epihydrophilous species

Elodea nuttallii (Fig 7.3c–e): Pollen grains are shed as tetragonal tetrads, *c.* 193.0 μm in length at the polar axis of the tetrad and are not acetolysis-resistant. Tightly arranged small spines like arrowheads on columns are observed on an irregular reticulate structure. A continuous basal layer is observed on the intine.

Hydrilla verticillata (Fig 7.3f–h): Pollen grains are spherical, *c.* 205.0 μm in diameter and are not acetolysis-resistant. The exine consists of small spines like arrowheads on columns that are tightly arranged on a discontinuous granulate exinous layer.

7.3.1.5 Male flower-epihydrophilous species

Lagarosiphon muscoides (Fig 7.3i): Pollen grains are spherical, echinate, *c.* 15.0 μm in diameter, not acetolysis-resistant. Spines, *c.* 0.4 μm in height, are observed.

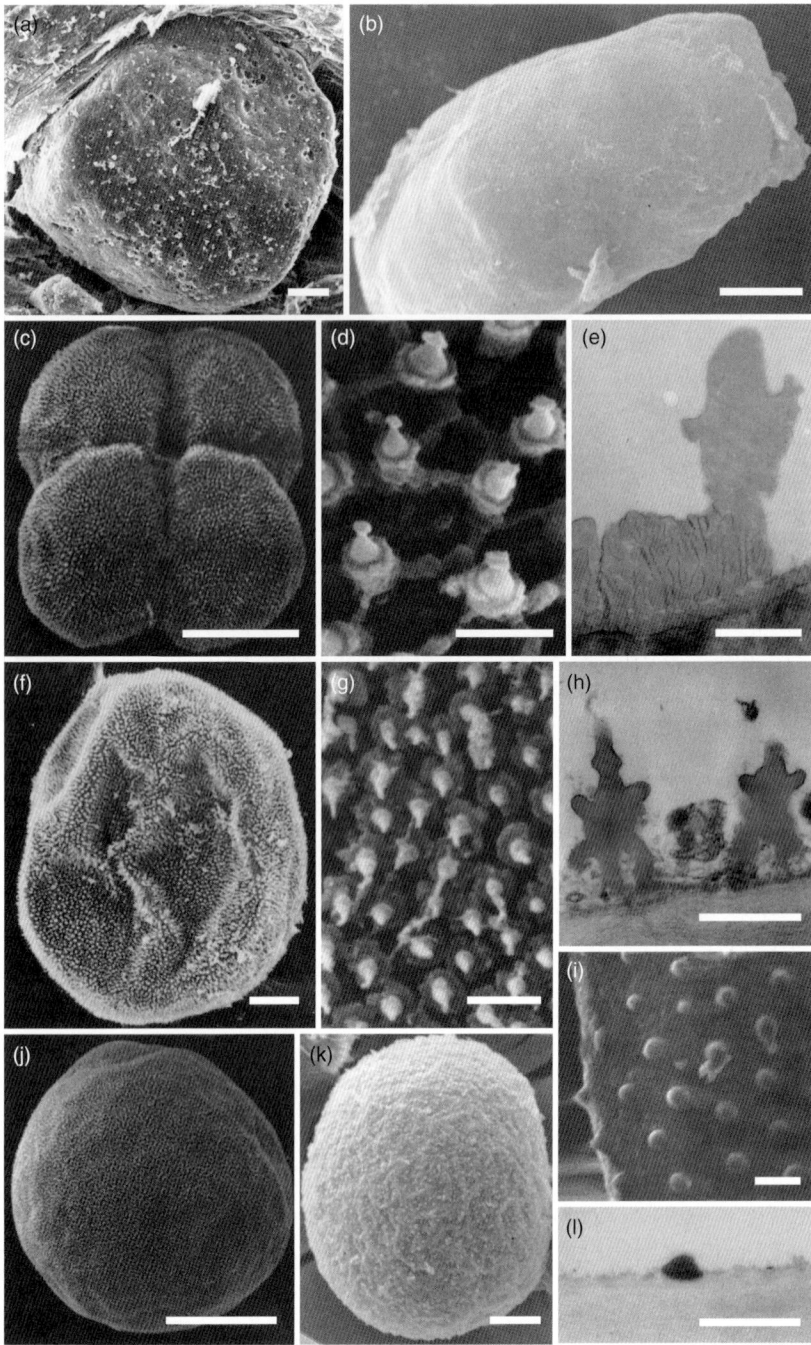

Fig 7.3 Pollen grains of Hydrocharitaceae. a–b: Hypohydrophilous species. a: Pollen grain of *Thalassia hemprichii*. SEM. Bar = 10 μm. b: Pollen grain of *Najas oguraensis*. SEM. Bar = 5 μm. c–h: Pollen-epihydrophilous species. c: Pollen tetrad of *Elodea nuttallii*. SEM. Bar = 50 μm. d: Exine sculpture of *E. nuttallii*. SEM. Bar = 2 μm. e: Exine structure of *E. nuttallii*. TEM. Bar = 2 μm. f: Pollen grain of *Hydrilla verticillata*. SEM. Bar = 10 μm. g: Exine sculpture of *H. verticillata*. SEM. Bar = 1 μm. h: Exine structure of *H. verticillata*. TEM. Bar = 1 μm. i–l: Male flower-epihydrophilous species. i: Exine sculpture of *Lagarosiphon muscoides*. SEM. Bar = 1 μm. j: Pollen grain of *Enhalus acoroides*. SEM. Bar = 50 μm. k: Pollen grain of *Vallisneria asiatica*. SEM. Bar = 10 μm. l: Exine structure of *V. asiatica*. TEM. Bar = 1 μm.

Enhalus acoroides (Fig 7.3j): Pollen grains are spherical, reticulate, *c.* 150.0 μm in diameter, not acetolysis-resistant. There is a reticulate and not discontinuous exine structure on the intine. These results are consistent with the report of Pettitt (1980).

Vallisneria asiatica (Fig 7.3k, l): Pollen grains are spherical, *c.* 144.0 μm in diameter, and not acetolysis-resistant. Gemmate exinous structures on the intine are irregularly found.

7.3.2 Stigma morphology

In Hydrocharitaceae, stigma surface morphology has been described for various taxa. Most genera have stigmas with adaxial papillae (*Thalassia*: Pettitt, 1976; *Enhalus*: Svedelius, 1904; *Egeria*: Cook and Urmi-König, 1984a; *Elodea*: Cook and Urmi-König, 1985; *Apalanthe*: Cook, 1985; *Stratiotes*: Cook and Urmi-König, 1983b; *Nechamandra*: Cook and Lüönd, 1982c; *Appertiella*: Cook and Triest, 1982; *Lagarosiphon*: Symoens and Triest, 1983; *Najas*: Haynes, 1977; *Limnobium*: Cook and Urmi-König, 1983a; *Ottelia*: Cook and Urmi-König, 1984b; *Hydrocharis*: Cook and Lüönd, 1982b; *Halophila*: Ravikumar and Ganesan, 1990; *Hydrilla*: Cook and Lüönd, 1982a, except for *Vallisneria* and *Maidenia* that have stigmas with trichomes (Lowden, 1982; Tomlinson, 1982; McConchie, 1983), *Halophila* with a partially papillose stigma (Ravikumar and Ganesan, 1990) and *Najas* with stigmas sometimes with additional short teeth interpreted as 'barren stigmas' (Tomlinson, 1982). The stigma surface morphology of *Blyxa* has not previously been reported (Cook et al., 1981; Tomlinson, 1982), nor have the evolutionary processes and the relationships between pollination mechanisms and stigma morphology been discussed on the basis of phylogenetic relationships.

In this study, all genera of Hydrocharitaceae were observed to have dry stigmas, as defined by Heslop-Harrison and Shivanna (1977). In most genera of Hydrocharitaceae, papillae covering the adaxial surface of stigmas were observed using SEM (Fig 7.4a–d). The stigma of *Blyxa japonica* is also papillose – this is the first time the stigma has been described in this genus – the papillae ranging from 50 to 160 μm in height. *Vallisneria asiatica* has stigma trichomes measuring 180 μm in height (Fig 7.4e). *Halophila ovalis* has trichotomous filiform stigmas and small domed projections, 10 μm in both diameter and height, only on the folded adaxial side of each stigma branch except the distal part (Fig 7.4f). *Najas minor* has dichotomous filiform stigmas and small domed projections, 10 μm in both diameter and height, on the adaxial side of each stigma branch. Based on these results, the stigmatic surface morphology of Hydrocharitaceae can be classified into three types: (1) stigmas with papillae only on the adaxial side (*Blyxa, Elodea, Enhalus, Limnobium, Nechamandra, Ottelia, Thalassia, Egeria, Hydrilla, Hydrocharis* and *Lagarosiphon*); (2) stigmas with trichomes only on the adaxial side (*Vallisneria*) and (3) filiform stigmas partially covered with small domed projections (*Halophila* and *Najas*).

Fig 7.4 Stigmas of Hydrocharitaceae. SEM. Bar = 100 μm. a–d: Stigmas with papillae.
a: *Hydrocharis dubia*. b: *Limnobium laevigatum*. c: *Thalassia hemprichii*. d: *Hydrilla verticillata*.
e: Stigma with trichomes. *Vallisneria asiatica*. f: Filiform stigma. *Halophila ovalis*. Small domed
projections occur only on the folded adaxial side of each part of the branched stigma.

7.4 The evolution of pollination mechanisms in Hydrocharitaceae

Two molecular phylogenetic studies have been used to assess the evolutionary development of pollination mechanisms in Hydrocharitaceae. Tanaka et al. (1997) analysed the evolution of pollination mechanisms in Hydrocharitaceae using the chloroplast genes *mat*K and *rbc*L, and Les et al. (2006) added a nuclear ribosomal ITS region. In this study, we conducted the character-state reconstruction of pollination mechanisms on the basis of the topology connecting the highly supported four clades in both trees as a polytomy. The character-state reconstruction of pollination mechanisms is shown in Fig 7.5. This shows that entomophily arose from a common ancestor and is a plesiomorphic character in Hydrocharitaceae. In addition, hypohydrophily originated once, while pollen-epihydrophily and male flower-epihydrophily evolved in parallel twice and three times respectively, from both entomophilous and hypohydrophilous lineages.

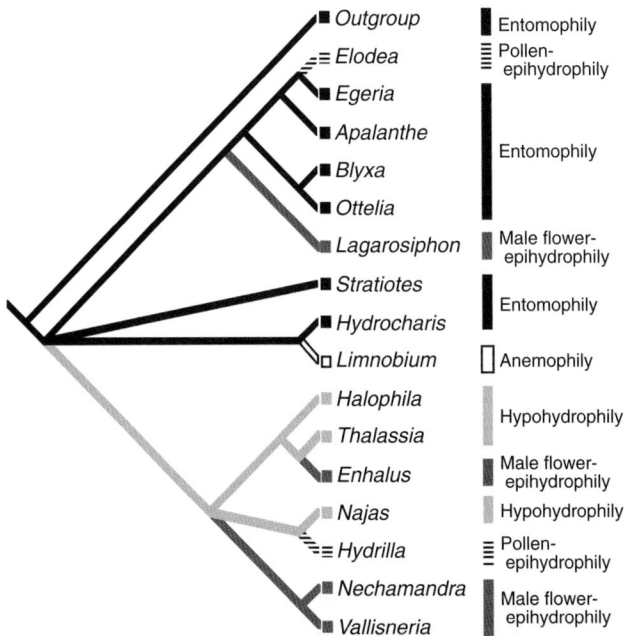

Fig 7.5 Character evolution of pollination mechanisms in Hydrocharitaceae. The character-state reconstruction is based on the topology connecting the highly supported four clades in Tanaka et al. (1997) and Les et al. (2006). *Maidenia* is treated as an ingroup of *Vallisneria* on the basis of Les et al. (2006). Polytomies of the tree were interpreted as 'hard polytomies' in MacClade software (Maddison and Maddison, 2003).

7.5 The correlation between floral traits and pollination mechanisms

7.5.1 Pollen morphology and pollination mechanisms

7.5.1.1 Entomophilous species

The exine in the entomophilous pollen grains of Hydrocharitaceae, although varied, is characterized by projections like spines or baculae and obvious, well-developed structures (Fig 7.2a–j). Proctor et al. (1996) stated that exine ornamentation, such as that of echinate pollen often observed in entomophilous species, plays a role in aggregating the grains into large clumps and allows more efficient pollen transfer. Walker (1976) also suggested that the sculpturing in entomophilous taxa aids in attaching pollen to the pollinator and in combination with oil droplets produces functional pollen polyads which makes possible a number of fertilizations from a single act of pollination. In addition, the molecular phylogeny shows that entomophily is a plesiomorphic character in Hydrocharitaceae (Fig 7.5). The other pollination mechanisms, derived from entomophily in some genera, such as the pollen-epihydrophilous genus *Elodea* and the anemophilous genus *Limnobium*, have pollen grains with a smoother surface than that of entomophilous pollen (Fig 7.2a, c, e, g, i). We therefore speculate that during the course of the evolution of entomophily from a common ancestor, the projecting exine sculpture was retained as an adaptive character in entomophily, despite the varied exine structures.

7.5.1.2 Anemophilous species

In Hydrocharitaceae, an exine structure with typical tecta and columellae was observed only in *Hydrocharis* and *Limnobium* (Fig 7.2j, l). This reflects their close relationship in the molecular phylogeny (Fig 7.5). However, the spines on the reticulate pollen grains of *L. laevigatum* are much smaller than those of *H. dubia* pollen, and the entire surface of *L. laevigatum* pollen is smoother than that of *H. dubia* pollen (Fig 7.2i, k). Whitehead (1968) stated that a smooth exine sculpture was necessary for effective wind pollination. Wodehouse (1935) reported that most pollen of northern European anemophilous species is either smooth or close to it, while strong sculpturing is characteristic of zoophilous species. Thus, although the pollen grains of *Hydrocharis* and *Limnobium* are similar in having exines with typical tecta and columellae, according to the character evolution (Fig 7.5) the smaller spines of *Limnobium* may have evolved to adapt to anemophily.

7.5.1.3 Hypohydrophilous species

With respect to exine structure, a 'discontinuous' exine structure is derived at the root of Group 2 and is constant in its constituent taxa (Fig 7.6a). In contrast, exine sculpture is consistently characterized as having 'little or no sculpture' in hypohydrophily, but differs in *Enhalus* and *Hydrilla*, which have different pollination mechanisms

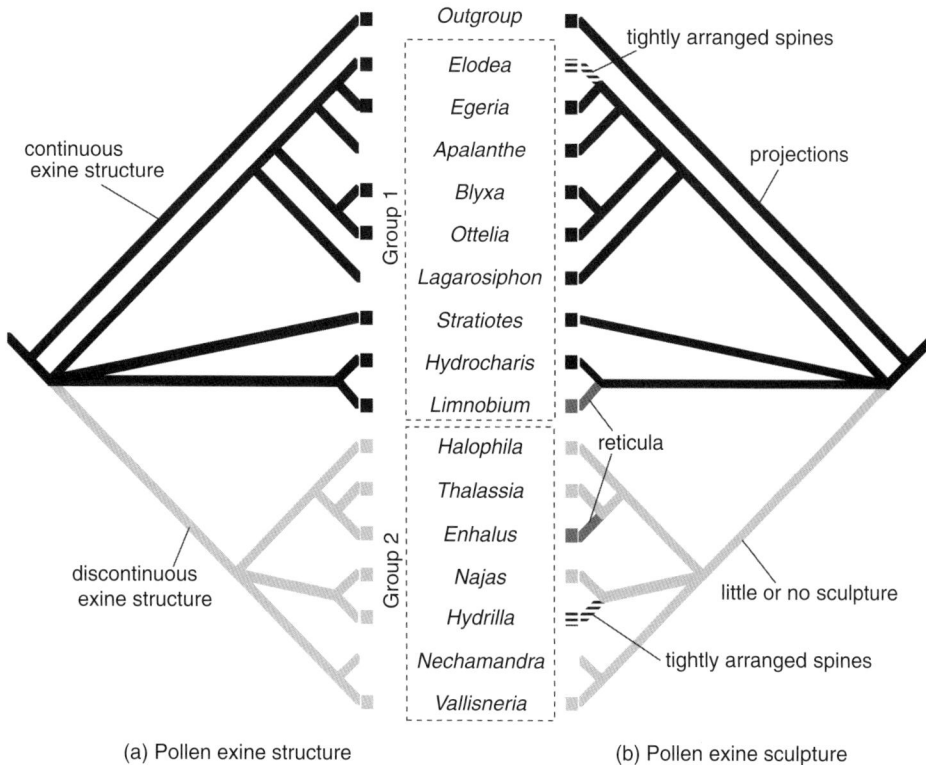

Fig 7.6 Character evolution of pollen exine structure (a) and sculpture (b) in Hydrocharitaceae. Shading of branches indicates character states in pollen exine structure (a) and pollen exine sculpture (b). Each state is shown in each branch. The genera are divided into two groups: Group 1 mainly consists of entomophily cluster; Group 2 hypohydrophily cluster. Tree reconstruction as in Fig 7.5.

(Figs 7.5, 7.6b). Therefore, we consider a 'discontinuous' exine structure and 'little or no sculpture' to be an adaptive characteristic of hypohydrophilous species. Similarly, reduced exine structures have also been reported in other lineages, including the hypohydrophilous species *Halodule wrightii* (Cymodoceaceae) and *Ceratophyllum demersum* (Ceratophyllaceae) (Pettitt and Jermy, 1975; Takahashi, 1995). Wodehouse (1935) suggested that the exine of the pollen of most terrestrial angiosperms is simply unnecessary in the pollen of aquatic plants. The exine is believed to protect the male spore and subsequent gametophyte from desiccation and other hazards of subaerial dispersal (Heslop-Harrison, 1976).

7.5.1.4 Pollen-epihydrophilous species

Two pollen-epihydrophilous species, *Elodea nuttallii* and *Hydrilla verticillata*, have similar exine sculptures with tightly arranged spines like arrowheads on columns, but their exine structures differ. *Elodea nuttallii* has a continuous exine basal layer (Fig 7.3e). *Hydrilla verticillata* has only a discontinuous exine basal layer (Fig 7.3h). In the character-state reconstruction for exine structure, *Elodea* and *Hydrilla* retain the plesiomorphic characters of their respective groups (Fig 7.6a). The exines of *Elodea* and *Hydrilla* are derived from 'projections' and 'little or no sculpture', respectively, but have evolved to a similar sculpture (Fig 7.6b). The modification of their exine sculpture character to 'tightly arranged spines' is synchronized with the development of the pollination mechanism. Consequently, we propose that the very similar exine sculptures of *Elodea* and *Hydrilla* have evolved as an adaptation to a common pollination mechanism. Conversely, it is probable that the difference in the exine structure of *Elodea* and *Hydrilla* is the result of their different phylogenies.

   Although the pollen grains of species with other pollination mechanisms, such as *Vallisneria asiatica* and *Enhalus acoroides*, sink as soon as they come in contact with water (unpublished data), the pollen grains of *Hydrilla verticillata* and *Elodea nuttalii* float on the surface for 20 h or longer (Tanaka, 2000). The pollen grains of *Elodea* and *Hydrilla* share the ability to drift on the water surface during pollination. Therefore, the similarity of exine sculpture is very likely an adaptation to floating on water.

7.5.1.5 Male flower-epihydrophilous species

Male flower-epihydrophilous species, *Lagarosiphon muscoides*, *Enhalus acoroides* and *Vallisneria asiatica*, differ in exine sculpture. However, both *E. acoroides* and *V. asiatica* commonly have discontinuous exine structure. There are many examples of hypohydrophilous species with reduced exines, for example, *Najas oguraensis* (Fig 7.3b) and *Halophila stipulacea* (Pettitt, 1980). Because the exine protects the male spore and gametophyte from desiccation and other hazards of subaerial dispersal (Heslop-Harrison, 1976), it is thought that airborne pollen grains must necessarily have a stable exine structure. Therefore, we believe that the reduced exines are correlated with underwater pollination, as stated in the previous section. However, a reduced exine structure has also been reported in terrestrial plants, for example, *Canna* (Skvarla and Rowley 1970) and *Heliconia* (Kress et al., 1978), which grow in rainforests where the humidity exceeds 90% (Kress, 1986). Kress (1986) examined the viability of *Heliconia* pollen grains under high humidity and reported that all grains remained viable for 24 h in an extremely humid atmosphere. Furness and Rudall (1999) postulated that an omniaperturate condition with reduced exine, which is found not only in many monocots, for example *Trillium* of the family Trilliaceae (Takahashi, 1982, 1983, 1987), but also in basal dicot families like the Lauraceae (Walker, 1976), may be an adaptation to increase germination efficiency: the pollen tube is formed more quickly and can emerge from any point on the pollen surface. An elaborate sporopollenin layer may obstruct the production of a pollen tube quickly in the moist and warm environment of the spathe/spadix environment typical for Aroideae in Araceae (Hesse, 2006a, 2006b). *Vallisneria* and *Enhalus* are thought to be pollinated in the very humid atmosphere just above the water surface, and a very humid atmosphere would permit a reduced exine despite an aerial pollination mechanism. Observation of the exine structure of *Lagarosiphon* may reveal the adaptive significance of the reduced exine in the male flower-epihydrophilous species.

7.5.2 Stigma surface morphology and pollination mechanisms

In contrast to pollen morphology, the stigmatic surface morphology is not closely related to pollination mechanisms (Fig 7.7). In entomophilous, anemophilous, pollen-epihydrophilous and male flower-epihydrophilous genera, no clear

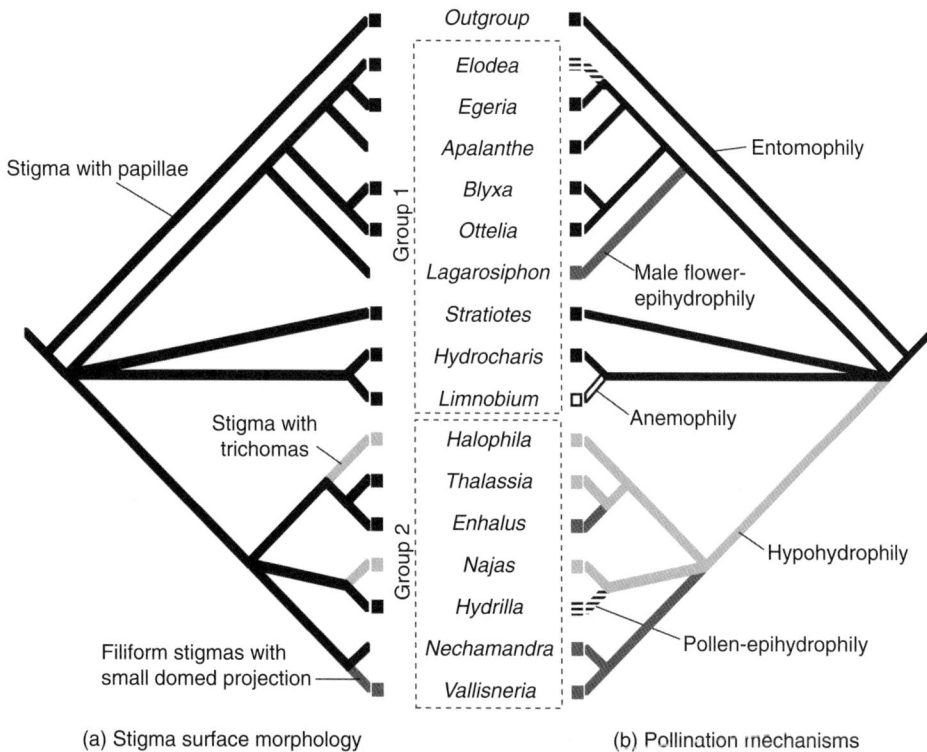

(a) Stigma surface morphology   (b) Pollination mechanisms

Fig 7.7 Character evolution of stigma surface morphology (a) and pollination mechanisms (b) in Hydrocharitaceae. Shading of branches indicates character states. Each state is shown in each branch. The genera are divided into two groups: Group 1 mainly consists of entomophily cluster; Group 2 hypohydrophily cluster. Tree reconstruction as in Fig 7.5.

relationships are observed between the stigma morphology and pollination mechanisms. The hypohydrophilous genera, *Halophila* and *Najas* apparently share a common stigmatic character and are grouped into type 3 as stated previously, but the other hypohydrophilous genus *Thalassia* has the papillose stigma of type 1. Based on these data, we cannot confirm a correlation between stigma surface morphology and pollination mechanisms. However, we observed a difference in structural strength of the stigmas, particularly their papillae, in the process of preparing plant material for SEM observation and focussed only on the surface morphology of stigma. It would be necessary to observe the membrane structure and other aspects, for example, the whole shape of the stigma, density and geometry of papillae and so on for further clarification.

7.6 Conclusions

The aquatic plant family Hydrocharitaceae exhibits various pollination mechanisms. Stigmatic morphology shows no definite correlation with pollination mechanisms, while, in contrast, pollen morphology is generally so correlated. Selection pressures acting on pollination mechanisms have reduced the exine structure in hypohydrophilous species, resulting in various exine sculptures that are adapted to different pollination mechanisms in entomophilous, anemophilous and pollen-epihydrophilous species. In the evolution of pollination mechanisms in the family, strong selection pressure acted on the transfer of pollen to stigmas under diverse environmental conditions. Meanwhile, only a weak selection pressure acted on the attachment of pollen to stigmas (Fig 7.8).

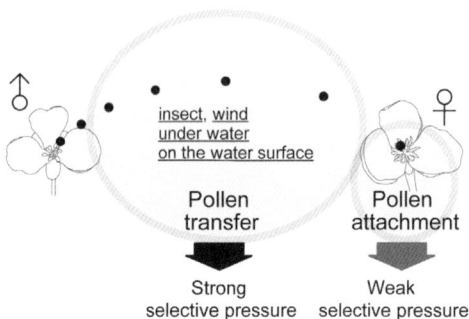

Fig 7.8 Conceptual diagram of evolution of pollination mechanisms in Hydrocharitaceae. In the evolution of pollination mechanisms in the family, strong selection pressure acted on the transfer of pollen to stigmas under diverse environmental conditions. However, only a weak selection pressure acted on the attachment of pollen to stigmas (illustration by Shoh Nagata).

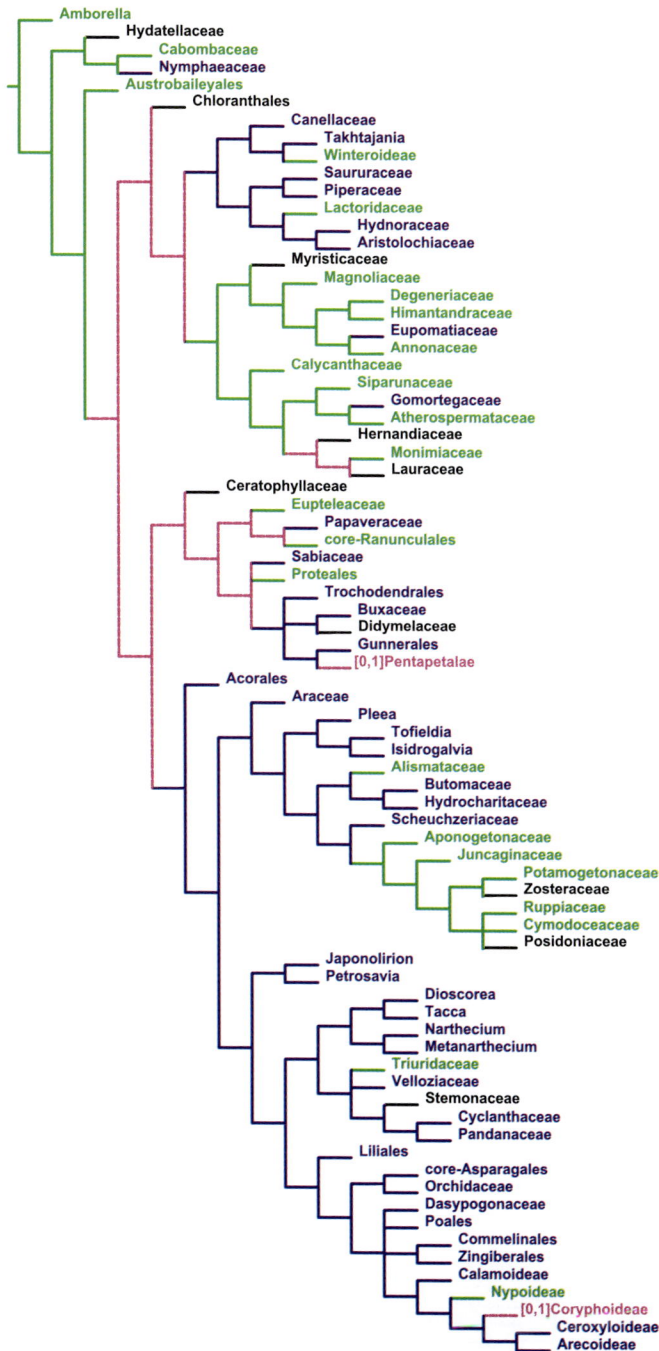

Fig 3.1A Maximum parsimony (MP) optimizations of gynoecium morphology using tree topology # 1. Black, unicarpellate gynoecia; green, gynoecia of free carpels (i.e. apocarpous without postgenital intercarpellary fusion); blue, gynoecia of united carpels irrespective of the type of intercarpellary fusion (i.e. syncarpous plus apocarpous with postgenital intercarpellary fusion); purple, ambiguity.

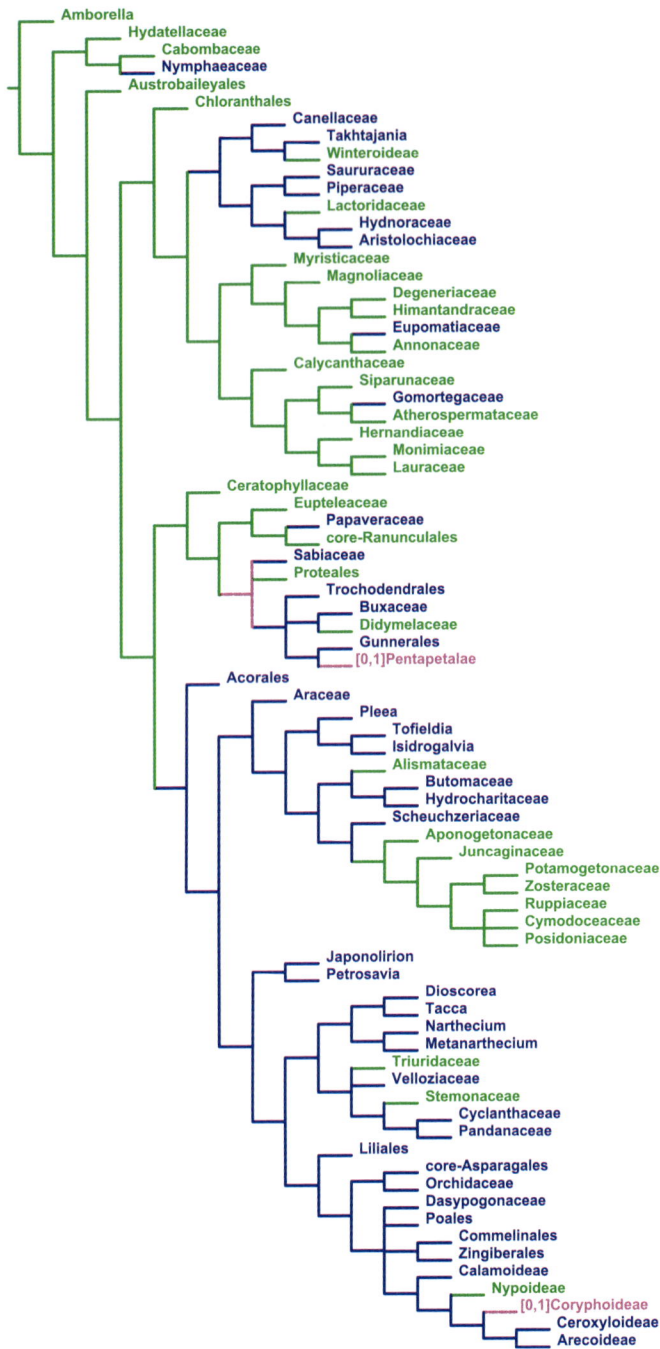

Fig 3.1B The same, but unicarpellate gynoecia are interpreted as apocarpous without postgenital intercarpellary fusion.

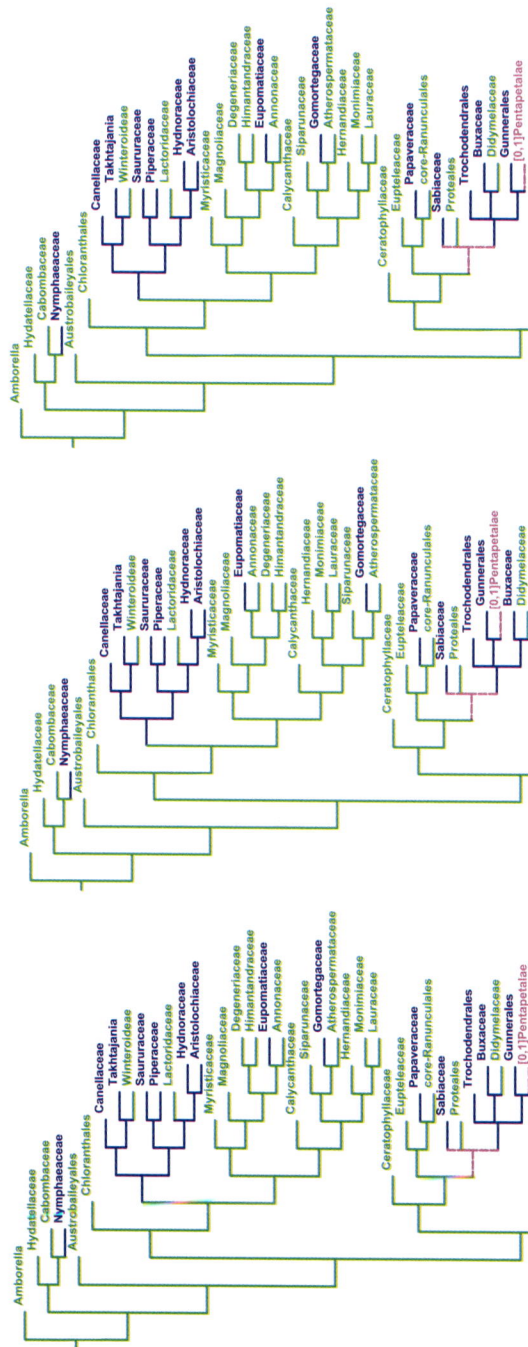

Fig 3.2A Maximum parsimony (MP) optimization of the presence vs. absence of congenital intercarpellary fusion using tree topology # 1 (left), tree topology # 2 (middle) and tree topology # 3 (right). Unicarpellate gynoecia are interpreted as apocarpous. Green, congenital intercarpellary fusion absent; blue, congenital intercarpellary fusion present; purple, ambiguity. Outgroup taxa.

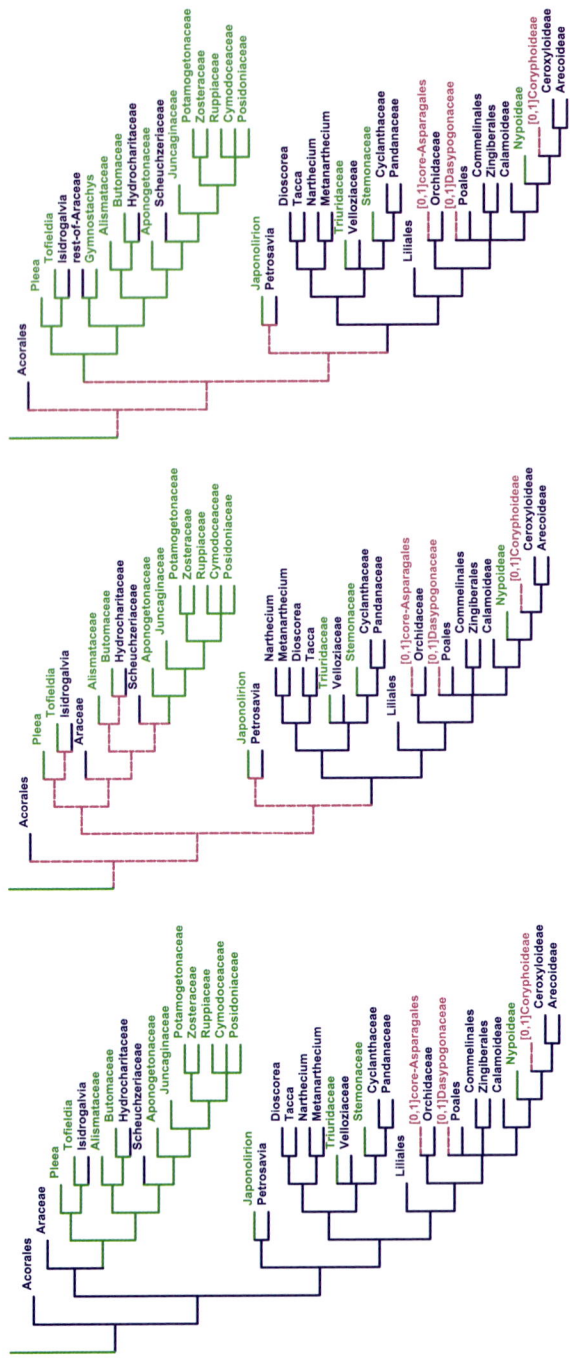

Fig 3.2B The same optimization and tree topologies. Monocot in group taxa.

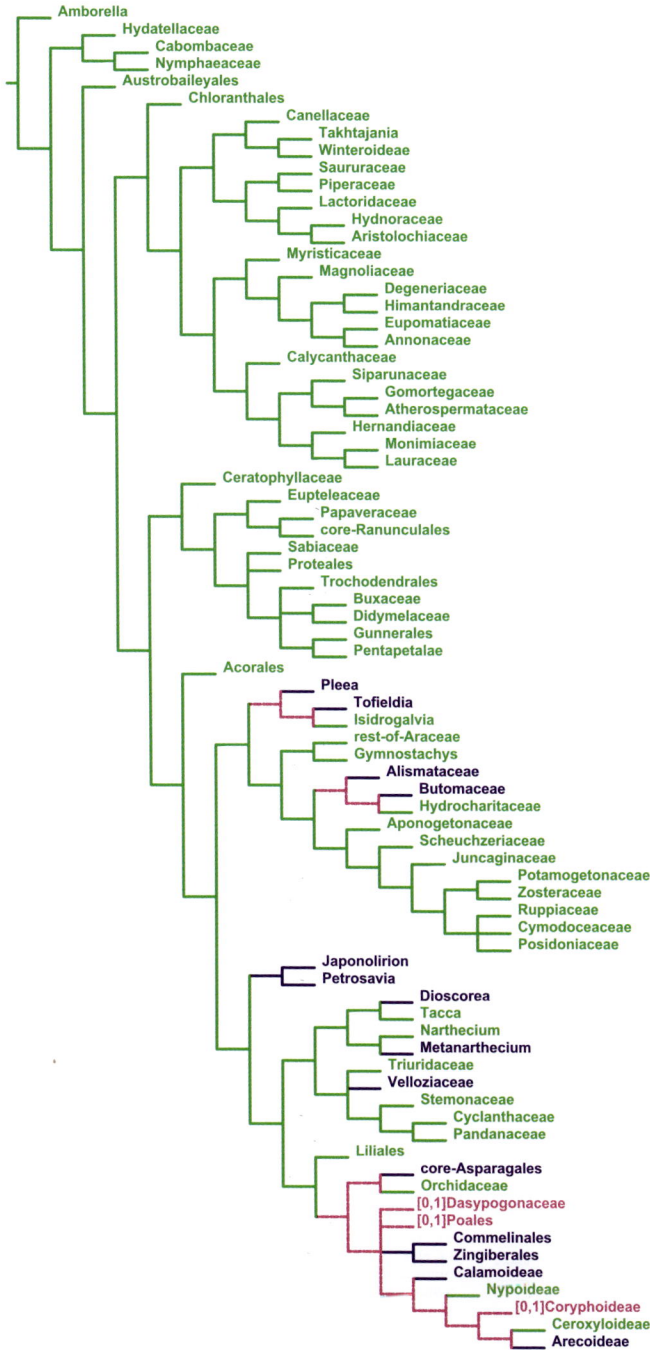

Fig 3.3A Maximum parsimony (MP) optimizations of presence vs. absence of septal (gynopleural) nectaries using tree topology # 3. Green, septal nectaries absent; blue, septal nectaries present; purple, ambiguity. Character coding following Remizowa et al. (2010).

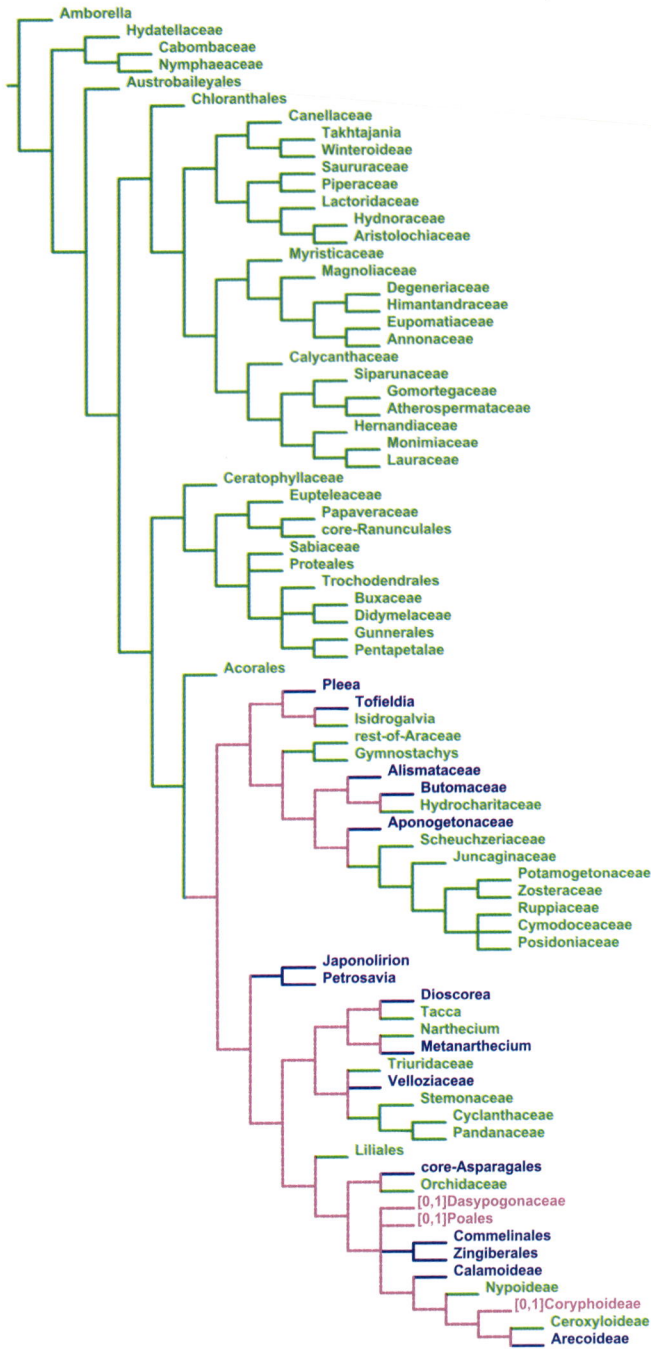

Fig 3.3B The same; Aponogetonaceae are interpreted as possessing reduced septal nectaries.

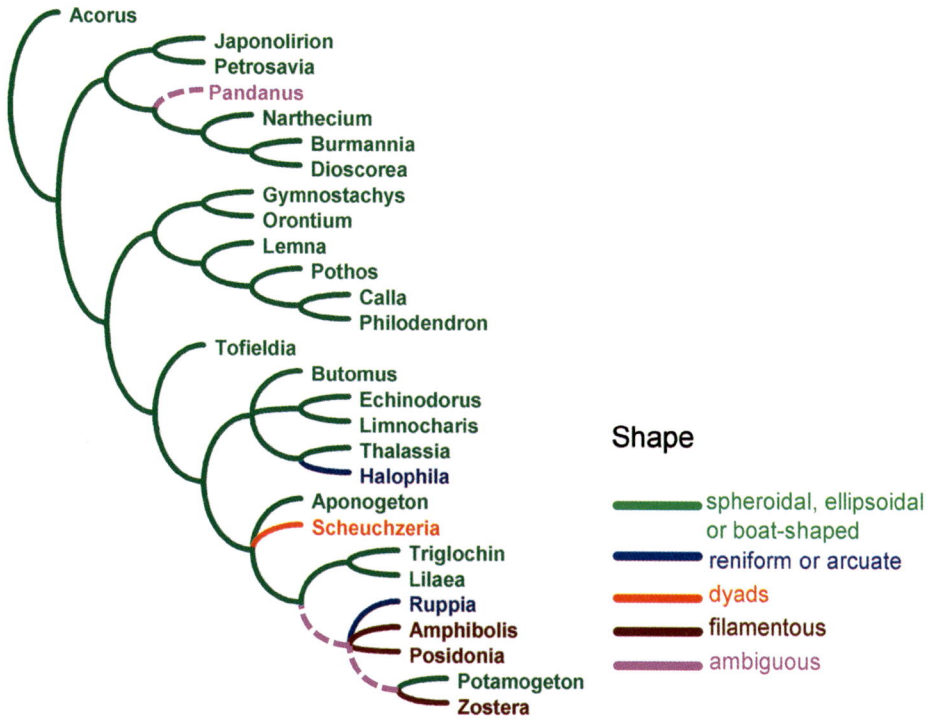

Fig 4.1 Pollen shape.

Figs 4.1–4.6 Pollen and tapetum characters optimized onto a phylogenetic tree of early-divergent monocots using Winclada (Nixon 2002) showing unambiguous changes.

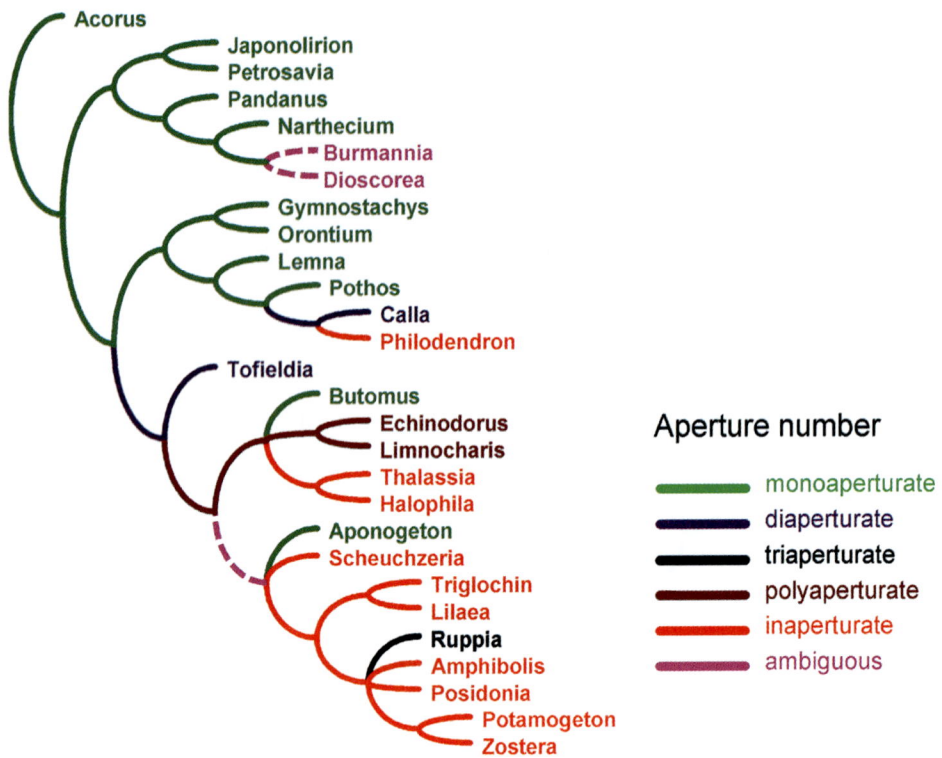

Fig 4.2 Pollen aperture number.

Figs 4.1–4.6 (*cont.*)

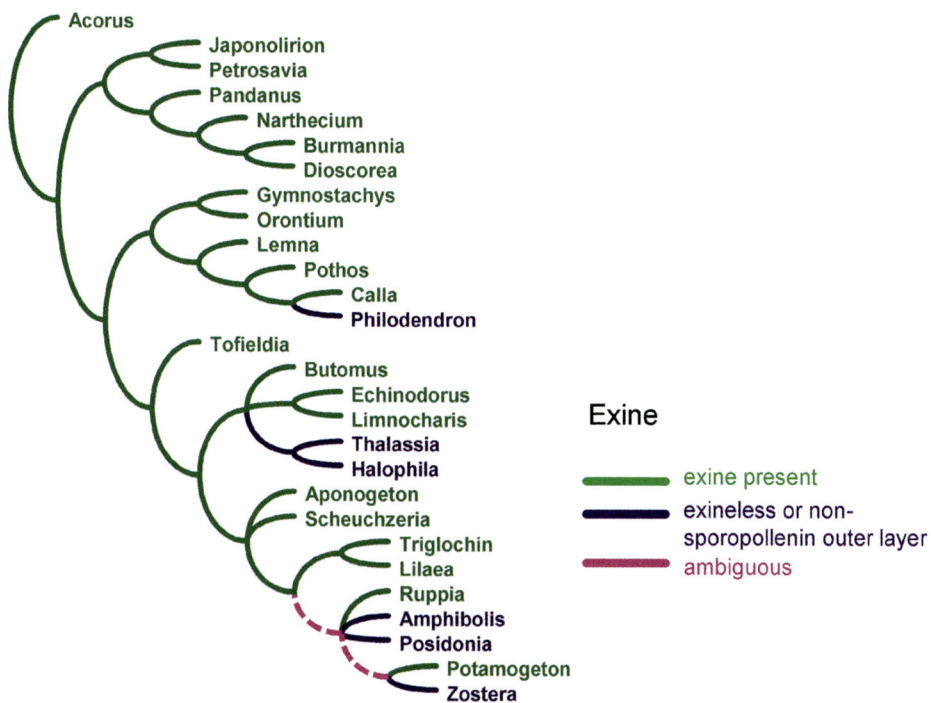

Fig 4.3 Pollen exine.

Figs 4.1–4.6 (*cont.*)

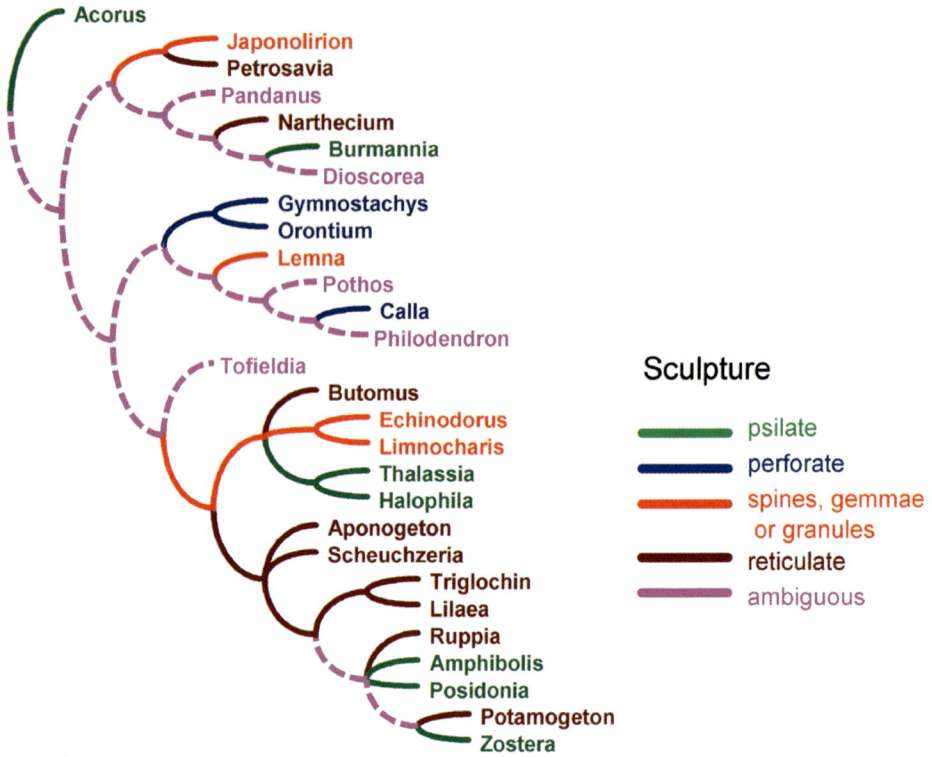

Fig 4.4 Pollen sculpture.

Figs 4.1–4.6 (*cont.*)

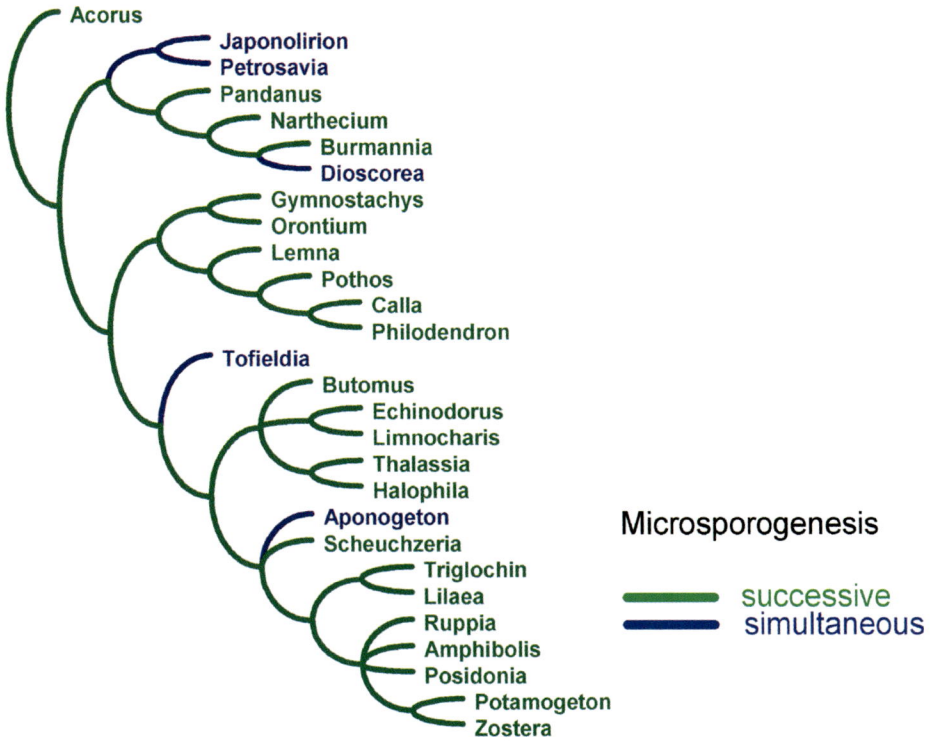

Fig 4.5 Microsporogenesis type.

Figs 4.1–4.6 (*cont.*)

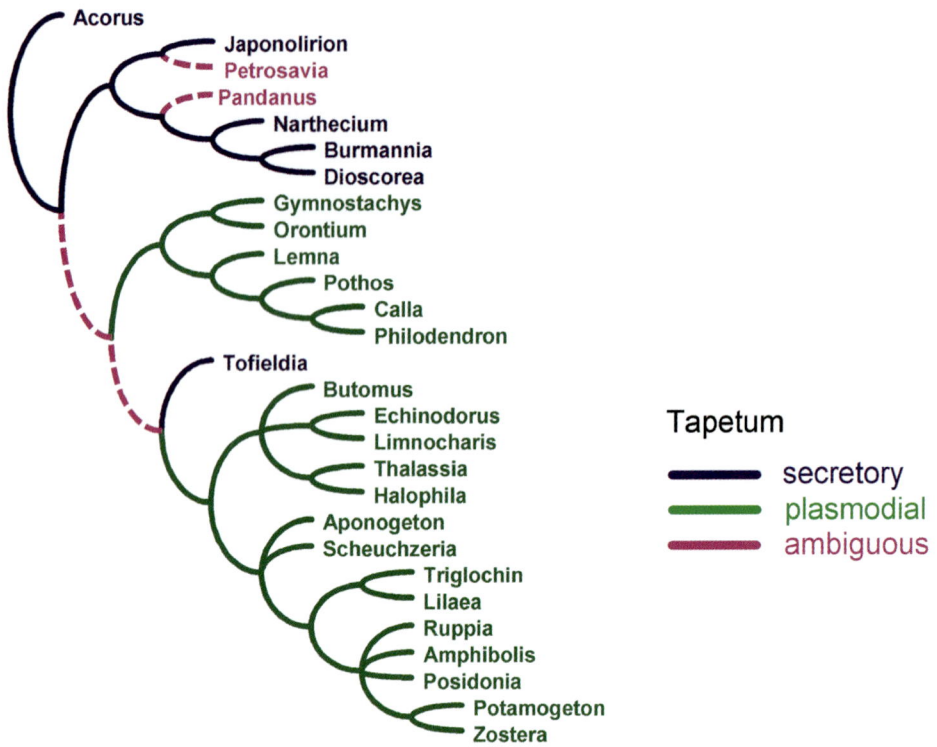

Fig 4.6 Tapetum type. (Reproduced with permission from Furness and Banks, 2010.)

Figs 4.1–4.6 (*cont.*)

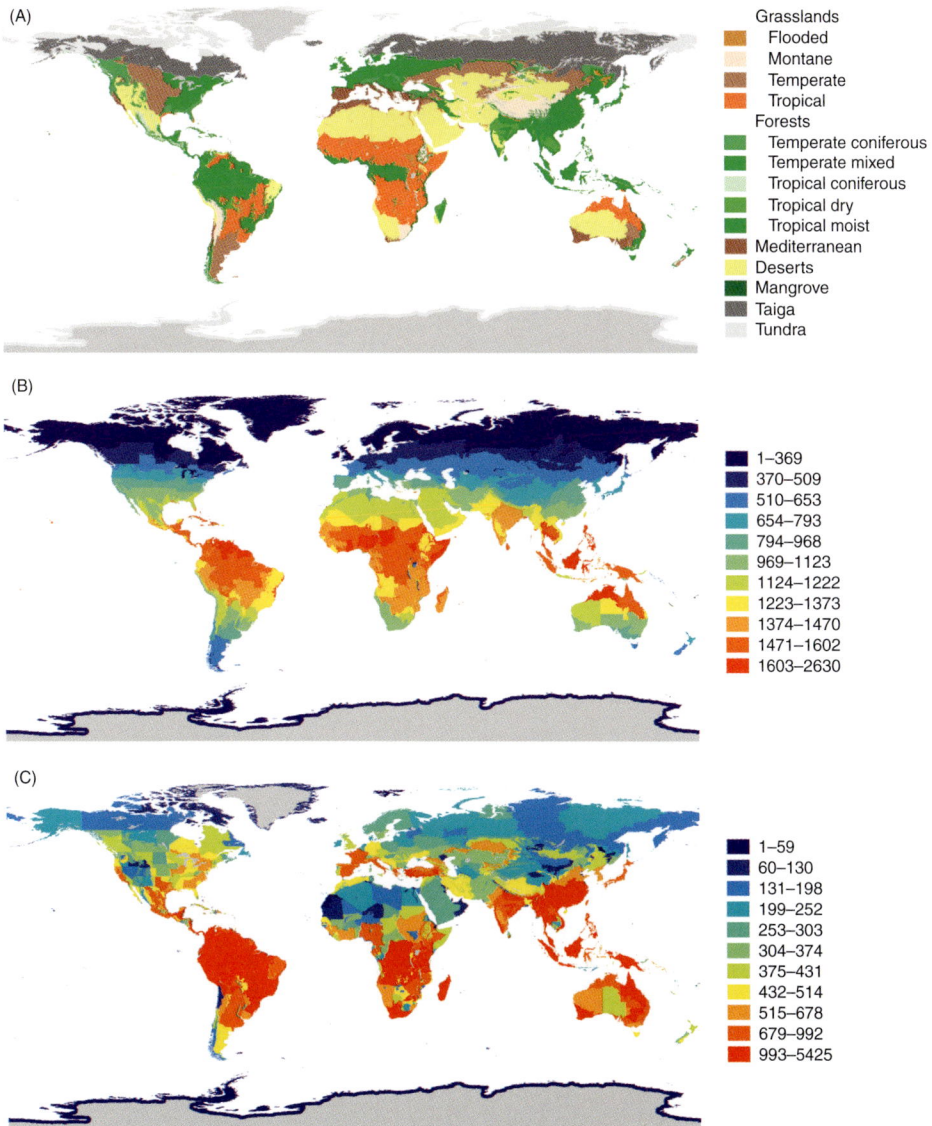

Fig 5.1 Global maps showing the distribution of A: biomes. B: actual evapotranspiration (AET values). C: monocot species richness (numbers of species per spatial region).

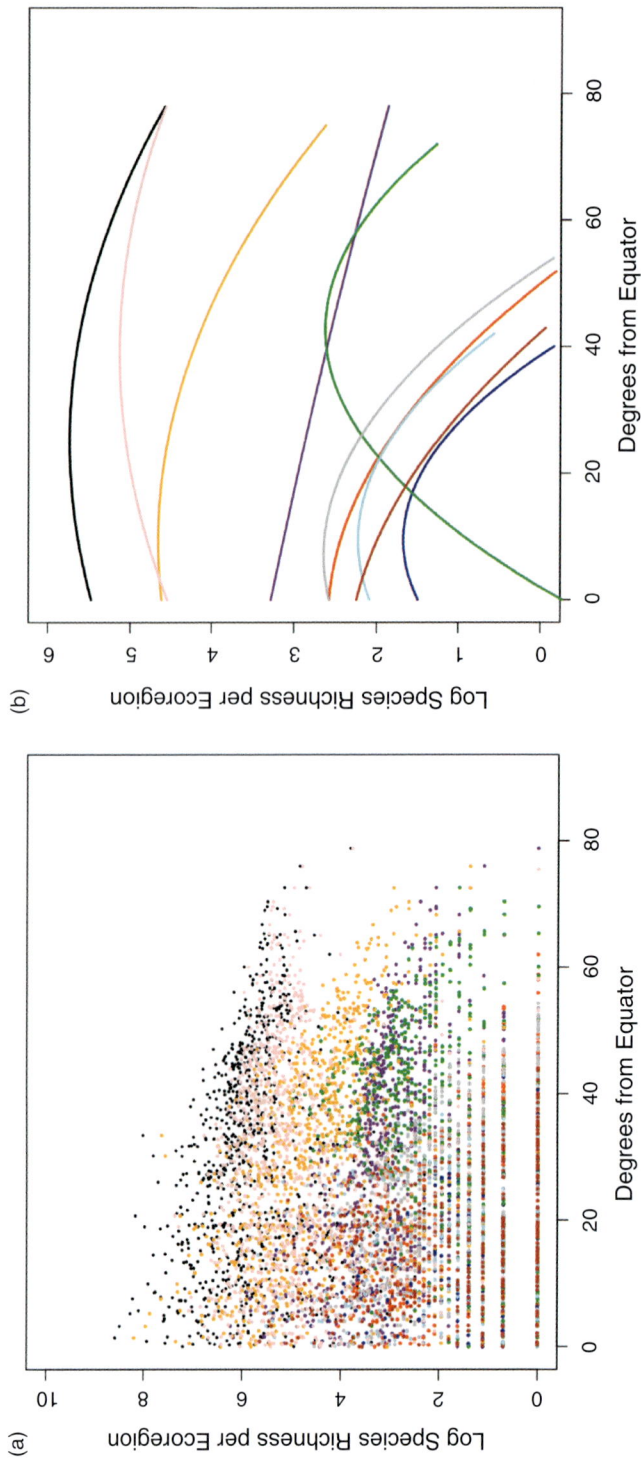

Fig 5.2 Latitudinal gradients in species richness within different monocot orders A: individual data points for orders (see below for colour scheme). B: lines representing the predicted relationship between species number and latitude (degrees from the equator) for orders in A determined using polynomial regression. Black = all monocots (Multiple $R^2 = 0.02$), red = Dioscoreales ($R^2 = 0.31$), blue = Pandanales ($R^2 = 0.08$), green = Liliales ($R^2 = 0.32$), orange = Asparagales ($R^2 = 0.07$), grey = Commelinales ($R^2 = 0.30$), light blue = Zingiberales ($R^2 = 0.04$), pink = Poales ($R^2 = 0.02$), purple = Alismatales ($R^2 = 0.07$), brown = Arecaceae ($R^2 = 0.11$).

Fig 7.1 Flowers of Hydrocharitaceae. a–e: Entomophilous species. a: Male flower of *Egeria densa*. b: Hermaphrodite flower of *Blyxa echinosperma*. c: Hermaphrodite flower of *Ottelia alismoides*. d: Female flower of *Stratiotes aloides*. e: Male flower of *Hydrocharis dubia*. f–g: Anemophilous species: f: Female flower of *Limnobium laevigatum*. g: Male flower of *L. laevigatum*. h–j: Hypohydrophilous species. h: Female flower of *Thalassia hemprichii*. i: Male flower of *T. hemprichii*. j: Female flower of *Halophila ovalis*. k–n: Pollen-epihydrophilous species. k: Spathe of *Hydrilla verticillata* with male flower, which is wrapped in a bubble, probably oxygen produced by photosynthesis, and surfaces due to its buoyancy. l: Male flower of *H. verticillata* on the surface, scattering pollen grains on the water. m: Pollen grains of *H. verticillata* drifting towards female flower on the water surface, being sucked into stigmas between the perianth segments. n: Male flower and pollen grains of *Elodea nuttallii*. o–q: Male flower-epihydrophilous species. o: Male flowers of *Vallisneria asiatica* drifting around a female flower. p: Male flowers of *V. asiatica*. q: Male flowers of *Enhalus acoroides* attached around stigmas of female flower.

Fig 7.1 (*cont.*)

Fig 7.1 (*cont.*)

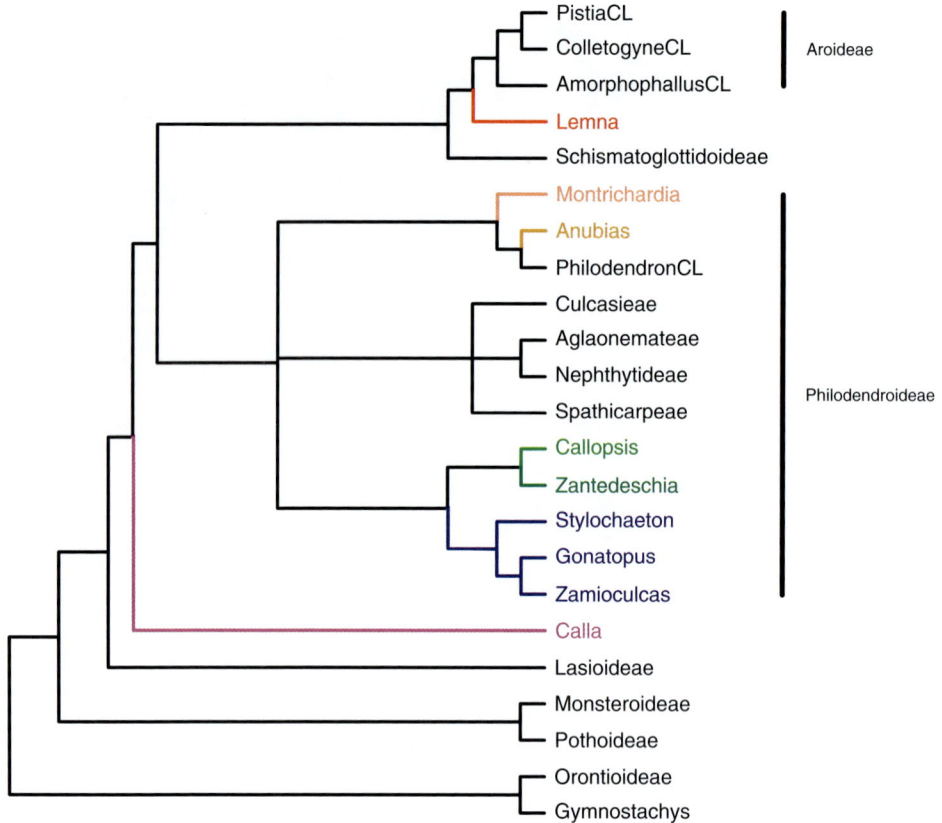

Fig 9.1 Cladogram simplified from French et al. (1995), showing circumscription of subfamilies Aroideae and Philodendroideae as circumscribed by Keating (2003). See Table 9.1 for details of circumscription of suprageneric taxon names. Names ending in '-CL' are abbreviations for '... clade', e.g. PistiaCL ≡ *Pistia* clade. Coloured clades show contrasting positions in Fig 9.2.

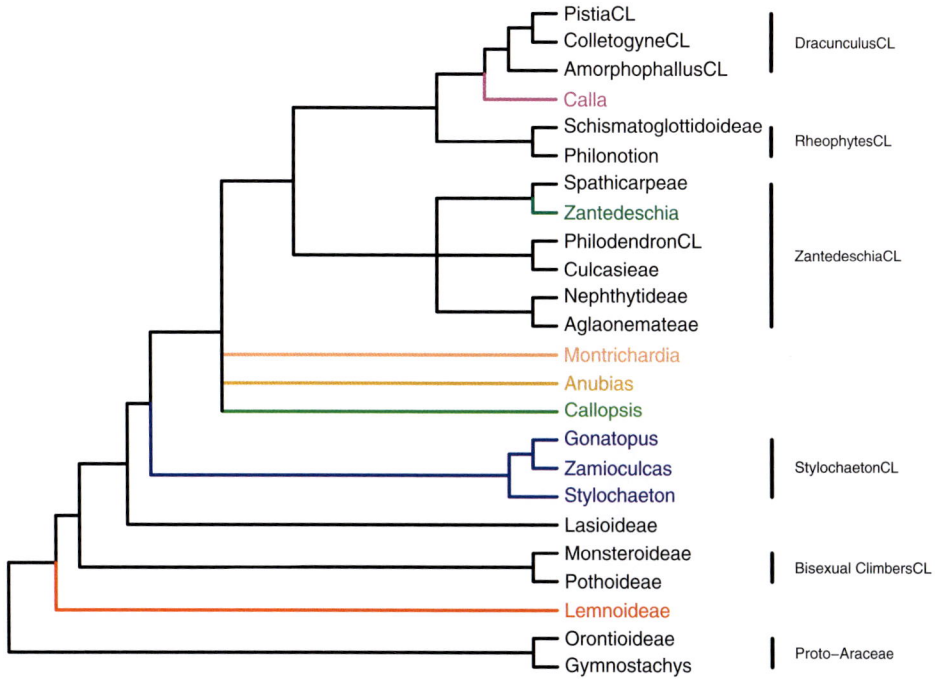

Fig 9.2 Cladogram simplified from Cusimano et al. (2011) showing major changes (coloured clades) from the cladogram in Fig 9.1 (French et al. 1995). See Table 9.1 for details of circumscription of suprageneric taxon names. Names ending in '-CL' are abbreviations for '... clade', e.g. PistiaCL ≡ *Pistia* clade.

Fig 12.1 a: Flower of *Narcissus bulbocodium* showing prominent corona and staminal filaments attached at base of flower. Painted by Rosemary Wise. b: Flower of *Eucharis amazonica* showing staminal filaments fused to form corona.

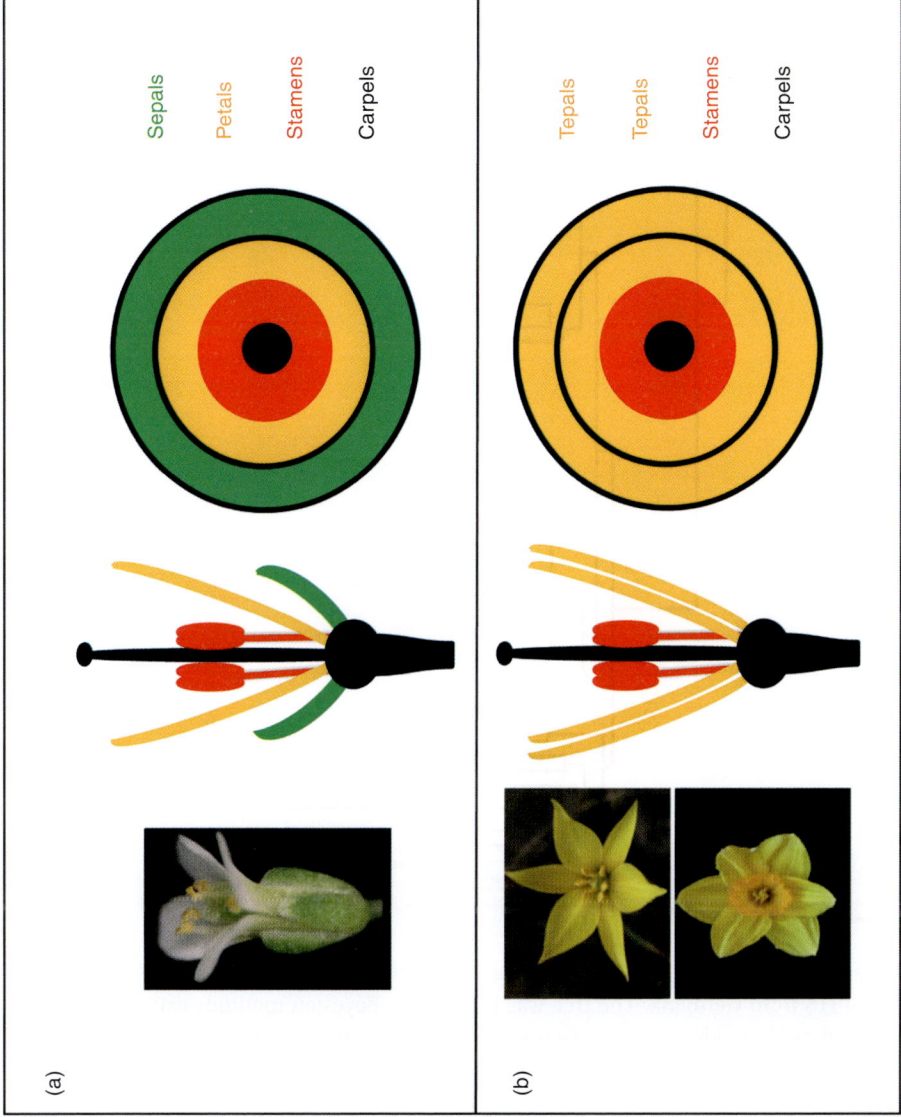

Fig 12.2 a: Flowers of *Arabidopsis thaliana* have four whorls: calyx (sepals), corolla (petals), androecium (stamens), gynoecium (carpels). b: In flowers of petaloid monocots such as tulip and daffodil the two outer whorls have similar identity and are termed tepals. In addition, the daffodil flower has an extra whorl, the corona.

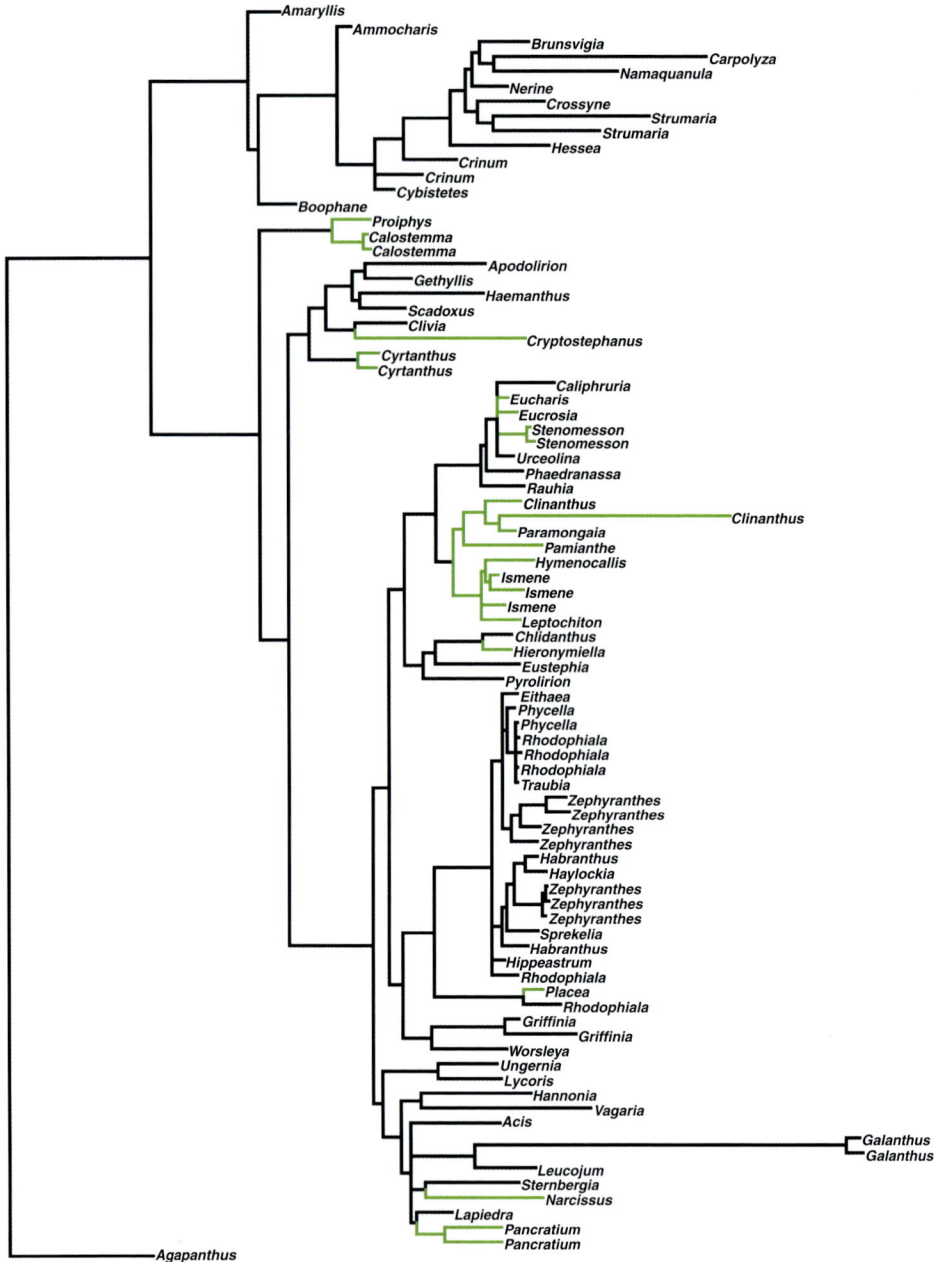

Fig 12.3 Generic-level phylogeny of Amaryllidaceae estimated from available DNA sequences of ITS from GenBank. The tree was inferred by Bayesian methods using the generalized reversible model (nst = 6). Genera with at least some species with a corona are coloured as green terminal branches. The corona has evolved multiple times within Amaryllidaceae.

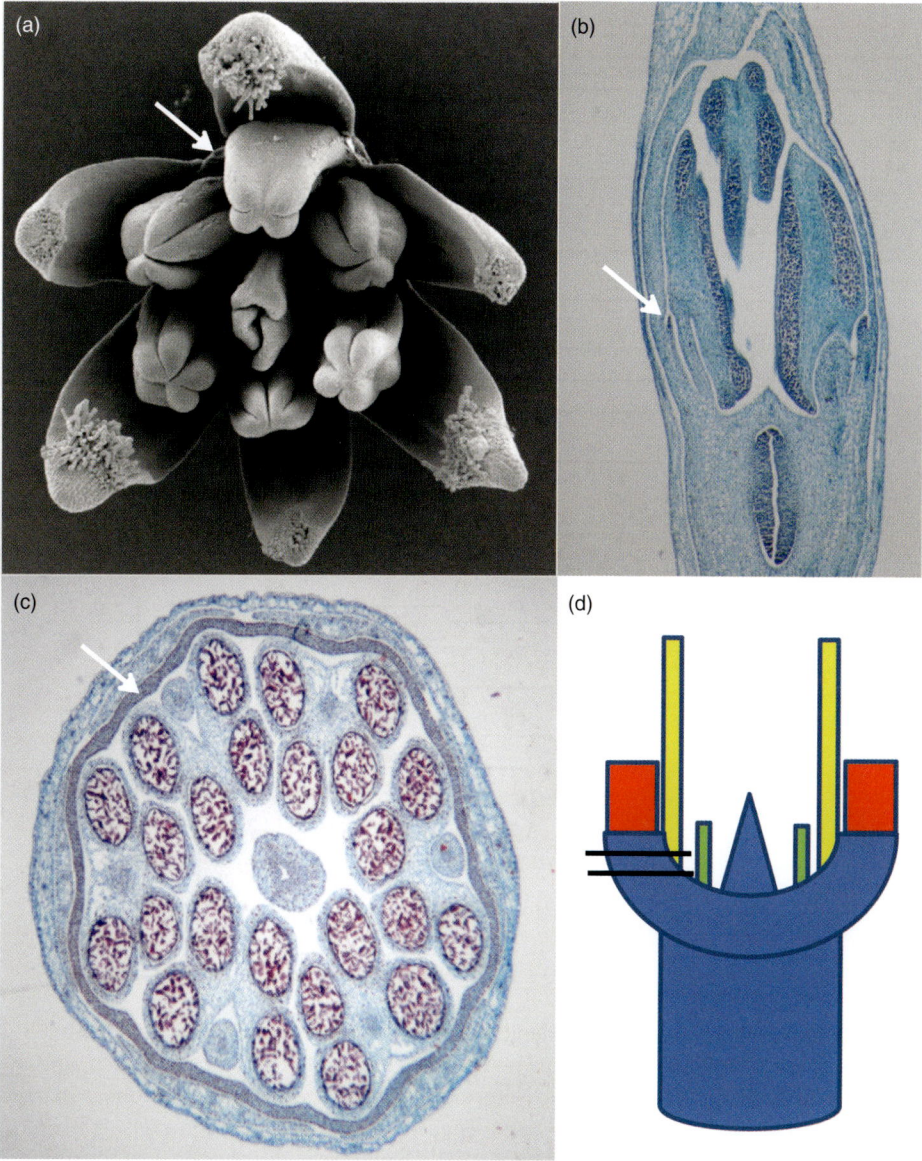

Fig 12.4 a: Very young flower of *Narcissus bulbocodium* with corona initiating late relative to the other floral organs. b: Longitudinal section of young flower of *Narcissus bulbocodium* with initiating corona shown by arrow. c: Transverse section of young flower of *Narcissus bulbocodium* with early stage corona indicated by an arrow. d: Schematic of flower showing carpels (blue), stamens (green), corona (yellow) and tepals (red). The hypanthium region between the stamens and tepals is the point of corona initiation. Black lines indicate the plane of sectioning of Arber (1937).

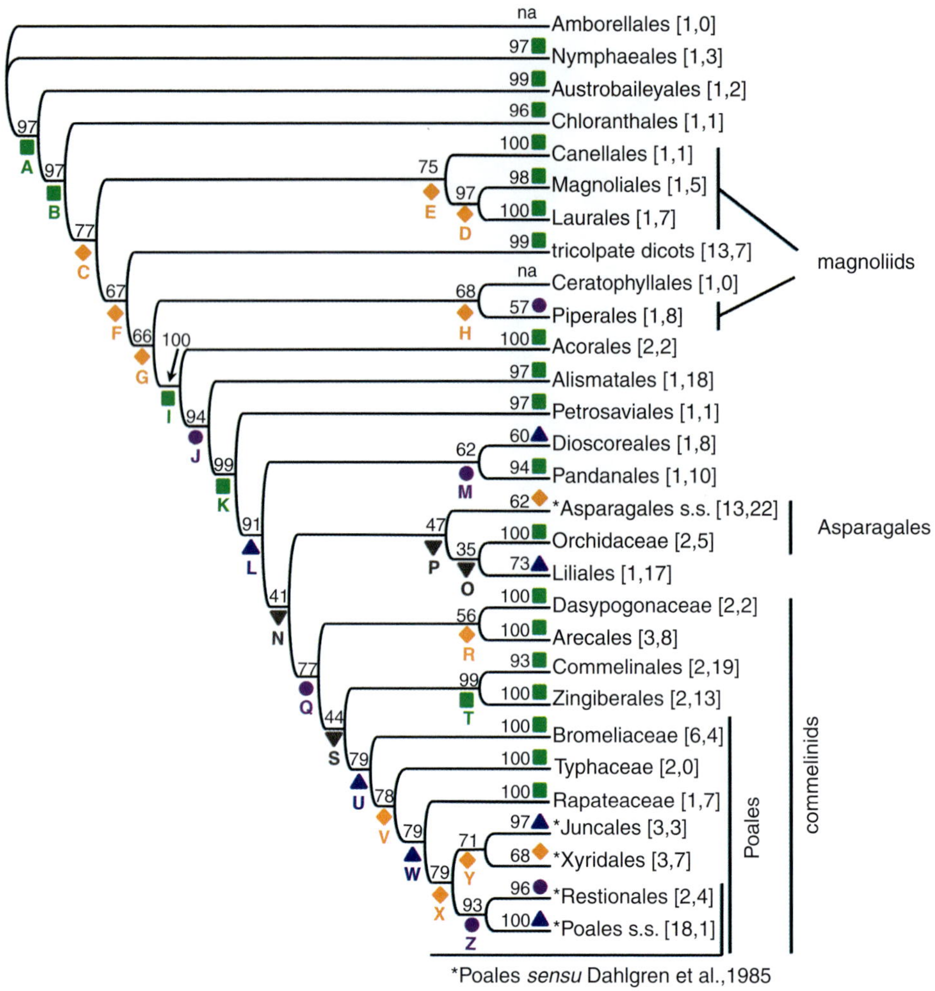

Fig 14.1 Summary consensus tree, depicting principal clades resolved by parsimony analysis of the complete data set, plus two groups represented by one taxon each, and relationships among these 29 groups. Names of groups follow APG III except those marked by asterisks (see text). Following the name of each principal clade, in brackets, is the number of constituent taxa represented by plastid genomes in the data set, and the number represented by *rbc*L and *mat*K sequences (in total, the number of terminals in each of the groups; Appendix). Four higher-level groups are signified by labelled bars at right and bottom. Strict consensus jackknife values are indicated for all principal clades except the two represented by solitary taxa ('na'), and for all relationships among them. Coloured symbols signify five support patterns, as defined in the text and Table 14.3 (green squares = group I; blue triangles, pointed upwards = group II; purple circles = group III; orange diamonds = group IV; grey triangles, pointed downwards = group V).

7.7 References

Cook, C. D. K. (1982). Pollination mechanisms in the Hydrocharitaceae. In *Studies on Aquatic Vascular Plants*, ed. J. J. Symoens, S. S. Hooper and P. Compere. Brussels: Royal Botanical Society of Belgium, pp. 1–15.

Cook, C. D. K. (1985). A revision of the genus *Apalanthe* (Hydrocharitaceae). *Aquatic Botany*, **21**, 157–164.

Cook, C. D. K. and Lüönd, R. (1982a). A revision of the genus *Hydrilla* (Hydrocharitaceae). *Aquatic Botany*, **13**, 485–502.

Cook, C. D. K. and Lüönd, R. (1982b). A revision of the genus *Hydrocharis* (Hydrocharitaceae). *Aquatic Botany*, **14**, 177–204.

Cook, C. D. K. and Lüönd, R. (1982c). A revision of the genus *Nechamandra* (Hydrocharitaceae). *Aquatic Botany*, **13**, 505–513.

Cook, C. D. K. and Triest, L. (1982). *Appertiella* a new genus in the Hydrocharitaceae. In *Studies on Aquatic Vascular Plants*, ed. J. J. Symoens, S. S. Hooper and P. Compere. Brussels: Royal Botanical Society of Belgium, pp. 75–79.

Cook, C. D. K. and Urmi-König, K. (1983a). A revision of the genus *Limnobium* including *Hydromystria* (Hydrocharitaceae). *Aquatic Botany*, **17**, 1–27.

Cook, C. D. K. and Urmi-König, K. (1983b). A revision of the genus *Stratiotes* (Hydrocharitaceae). *Aquatic Botany*, **16**, 213–249.

Cook, C. D. K. and Urmi-König, K. (1984a). A revision of the genus *Egeria* (Hydrocharitaceae). *Aquatic Botany*, **19**, 73–96.

Cook, C. D. K. and Urmi-König, K. (1984b). A revision of the genus *Ottelia* (Hydrocharitaceae). 2. The species of Eurasia, Australasia and America. *Aquatic Botany*, **20**, 131–177.

Cook, C. D. K. and Urmi-König, K. (1985). A revision of the genus *Elodea* (Hydrocharitaceae). *Aquatic Botany*, **21**, 111–156.

Cook, C. D. K., Lüönd, R. and Nair, B. (1981). Floral biology of *Blyxa octandra* (Roxb.) Planchon ex Thwaires (Hydrocharitaceae). *Aquatic Botany*, **10**, 61–68.

Crane, P. R., Friis, E. M. and Pedersen, K. R. (1995). The origin and early diversification of angiosperms. *Nature*, **374**, 27–33.

De Cock, A. W. A. M. (1980). Flowering, pollination and fruiting in *Zostera marina* L. *Aquatic Botany*, **9**, 202–220.

Ducker, S. C., Pettitt, J. M. and Knox, R. B. (1978). Biology of Australian seagrasses: pollen development and submarine pollination in *Amphibolis antarctica* and *Thalassodendron ciliatum* (Cymodoceaceae). *Australian Journal of Botany*, **26**, 265–85.

Furness, C. A. and Rudall, P. J. (1999). Inaperturate pollen in Monocotyledons. *International Journal of Plant Sciences*, **160**, 395–414.

Grayum, M. H. (1992). Comparative external pollen ultrastructure of the Araceae and putatively related taxa. *Monographs in Systematic Botany from the Missouri Botanical Garden*, **43**: 1–167. St Louis, MO: Missouri Botanical Garden.

Guo, Y.-H. and Huang, S.-Q. (1999). Evolution of pollination system and characters of stigmas in Najadales. *Acta Phytotaxonomica Sinica*, **37**, 131–136.

Guo, Y-H., Sperry, R., Cook, C. D. K. and Cox, P. A. (1990). The pollination ecology of *Zannichellia palustris* L. (Zannichelliaceae). *Aquatic Botany*, **38**, 341–356.

Harley, M. M., Kurmann, M. H. and Ferguson, I. K. (1991). Systematic implications of comparative morphology in selected Tertiary and extant pollen from the Palmae and the Sapotaceae. *Pollen et Spores*, **44**, 225–238.

Haynes, R. R. (1977). The Najadaceae in the Southeastern United States. *Journal of the Arnold Arboretum*, **58**, 161–170.

Heslop-Harrison, J. (1976). The adaptive significance of the exine. In *The Evolutionary Significance of the Exine*. Linnean Society Symposium Series Vol. 1, ed. I. K. Ferguson and J. Muller, New York and London: The Linnean Society of London, pp. 27–37.

Heslop-Harrison, J. and Shivanna, K. R. (1977). The receptive surface of the angiosperm stigma. *Annals of Botany*, **41**, 1233–1258.

Heslop-Harrison, Y. (2000). Control gates and micro-ecology: the pollen-stigma interaction in perspective. *Annals of Botany*, **85** (suppl. 1), 5–13.

Hesse, M. (2000). Pollen wall stratification and pollination. *Plant Systematics and Evolution*, **222**, 1–17.

Hesse, M. (2006a). Conventional and novel modes of exine patterning in members of the Araceae - the consequence of ecological paradigm shifts? *Protoplasma* **228**, 145–149.

Hesse, M. (2006b). Pollen wall ultrastructure of Araceae and Lemnaceae in relation to molecular classifications. In: *Monocots: Comparative Biology and Evolution*, ed. J. T. Columbus, E. A. Friar, C. W. Hamilton et al. Claremont, CA: Rancho Santa Ana Botanical Garden, pp. 204–208.

Kress W. J. (1986) Exineless pollen structure and pollination systems of tropical *Heliconia* (Heliconiaceae). In *Pollen and Spores: Form and Function*, ed. S. Blackmore and I. K. Ferguson, London: Academic Press, pp. 329–345.

Kress, W. J., Stone, D. E. and Sellers, S. C. (1978). Ultrastructure of exine-less pollen; *Heliconia* (Heliconiaceae). *American Journal of Botany*, **65**, 1064–1076.

Lee, S. (1978). A factor analysis study of the functional significance of angiosperm pollen. *Systematic Botany*, **3**(1), 1–19.

Les, D. H., Moody, M. L. and Soros, C. L. (2006). A reappraisal of phylogenetic relationships in the monocotyledon family Hydrocharitaceae (Alismatidae). *Aliso*, **22**, 221–230.

Les, D. H., Jacobs, S. W. L., Tippery, N. P. et al. (2008). Systematics of *Vallisneria* (Hydrocharitaceae). *Systematic Botany*, **33**, 49–65.

Lowden, R. M. (1982). An approach to the taxonomy of *Vallisneria* L. (Hydrocharitaceae). *Aquatic Botany*, **13**, 269–298.

Maddison, W. P. and Maddison, D. R. (2003). *MacClade: Analysis of Phylogeny and Character Evolution. Version 4.06*, Sunderland, MA: Sinauer.

Mayo, S. J., Bogner, J. and Boyce, P. C. (1997). *The Genera of Araceae*, Kew: Royal Botanic Gardens.

McConchie, C. A. (1983). Floral development of *Maidenia rubra* Rendle (Hydrocharitaceae). *Australian Journal of Botany*, **31**, 585–603.

McConchie, C. A. and Knox, R. B. (1989). Pollen-stigma interaction in the seagrass *Posidonia australis*. *Annals of Botany*, **63**, 235–248.

Nilsson, S. (1990). Taxonomic and evolutionary significance of pollen morphology in the Apocynaceae. *Plant Systematics and Evolution*, **5** (Suppl.), 91–102.

Pettitt, J. M. (1976). Pollen wall and stigma surface in the marine angiosperms *Thalassia* and *Thalassodendron*. *Micron*, **7**, 21–32.

Pettitt, J. M. (1980). Reproduction in seagrass: nature of the pollen and receptive surface of the stigma in the Hydrocharitaceae. *Annals of Botany*, **45**, 257–271.

Pettitt, J. M. (1981). Reproduction in seagrasses: Pollen development in *Thalassia hemprichii, Halophila stipulacea* and *Thalassodendron ciliatum*. *Annals of Botany*, **48**, 609–622.

Pettitt, J. M. and Jermy, A. C. (1975). Pollen in hydrophilous angiosperms. *Micron*, **5**, 377–405.

Proctor, M., Yeo, P. and Lack, A. (1996). *The Natural History of Pollination*, London: Harper Collins.

Ravikumar, K. and Ganesan, R. (1990). A new subspecies of *Halophila ovalis* (R.Br.) J. D. Hook. (Hydrocharitaceae) from the eastern coast of Peninsular India. *Aquatic Botany*, **36**, 351–358.

Schwanitz, G. (1967). Untersuchungen zur postmeiotischen Mikrosporogenese. II. Vergleichende Analyse der Pollenentwicklung sub- und emers-blühender Arten. *Pollen et Spores*, **9**, 183–209.

Sigrist, M. R. and Sazima, M. (2004). Pollination and reproductive biology of twelve species of neotropical Malpighiaceae: Stigma morphology and its implications for the breeding system. *Annals of Botany*, **94**, 33–41.

Skvarla, J. J. and Rowley, J. R. (1970). The pollen wall of *Canna* and its similarity to the germinal apertures of other pollen. *American Journal of Botany*, **57**, 519–529.

Svedelius, N. (1904). On the life-history of *Enalus acoroides*. *Annals of the Royal Botanic Gardens, Peradeniya*, **2**, 97–267.

Symoens, J. J. and Triest, L. (1983). Monograph of the African genus *Lagarosiphon* Harvey. *Bulletin du Jardin botanique national de Belgique/ Bulletin van de Nationale Plantentuin van België*, **53**, 441–488.

Takahashi, M. (1982). Pollen morphology in North American species of *Trillium*. *American Journal of Botany*, **69**(7), 1185–1195.

Takahashi, M. (1983). Pollen morphology in Asiatic species of *Trillium*. *The Botanical Magazine, Tokyo*, **96**, 377–384.

Takahashi, M. (1987). Development of omniaperturate pollen in *Trillium kamtchaticum* (Liliaceae). *American Journal of Botany*, **74**(12), 1842–1852.

Takahashi, M. (1995). Development of structure-less pollen wall in *Ceratophyllum demersum* L. (Ceratophyllaceae). *Journal of Plant Research*, **108**, 205–208.

Tanaka, N. (2000). Pollination of the genus *Hydrilla* (Hydrocharitaceae) by waterborne pollen grains. *Annals of the Tsukuba Botanical Garden*, **19**, 7–12.

Tanaka, N. (2003). Pollination of the genus *Hydrilla* (Hydrocharitaceae) by waterborne pollen grains: 2. Air bubbles cause the male flower to surface. *Annals of the Tsukuba Botanical Garden*, **22**, 11–13.

Tanaka, N., Setoguchi, H. and Murata, J. (1997). Phylogeny of the family Hydrocharitaceae inferred from *rbc*L and *mat*K gene sequence data. *Journal of Plant Research*, **110**, 329–337.

Tanaka, N., Uehara, K. and Murata, J. (2004). Correlation between pollen morphology and pollination mechanisms in the Hydrocharitaceae. *Journal of Plant Research*, **117**, 265–276.

Tomlinson, P. B. (1982). *Anatomy of the Monocotyledons VII*. Helobiae (Alismatidae). Oxford: Clarendon Press.

Verhoeven, J. T. A. (1979). The ecology of *Ruppia* dominated communities in western Europe. 1. Distribution of *Ruppia* representatives in relation to their autecology. *Aquatic Botany*, **6**, 197–268.

Walker, J. W. (1976). Evolutionary significance of the exine in the pollen of primitive angiosperms. In *The Evolutionary Significance of the Exine*. Linnean Symposium Society Series Volume 1, ed. I. K. Ferguson

and J. Muller, New York and London: Academic Press, pp. 251–308.

Whitehead, D. R. (1968). Wind pollination in the angiosperms: evolutionary and environmental considerations. *Evolution*, **23**, 28–35.

Wodehouse, R. P. (1935). *Pollen Grains*. New York: McGraw-Hill.

Yang, C.-F., Guo, Y.-H., Goturu, R. W. and Sun, S.-G. (2002). Variation in stigma morphology - how does it contribute to pollination adaptation in *Pedicularis* (Orobanchaceae)? *Plant Systematics and Evolution*, **236**, 89–98.

8

Patterns of bract reduction in racemose inflorescences of early-divergent monocots

Margarita V. Remizowa, Dmitry D. Sokoloff
and Paula J. Rudall

8.1 Introduction

Racemose inflorescences (spikes, racemes and spadices) with flower-subtending bracts either absent or superficially inconspicuous are common among early-divergent monocots classified in the orders Alismatales and Acorales. This bract-less condition contrasts with some other monocot orders such as Liliales, Petrosaviales and Dioscoreales, in which flower-subtending bracts are normally present. Two different patterns of bract reduction occur among early-divergent monocots (see also Posluszny and Sattler, 1973; Lieu, 1979; Charlton, 1981; Posluszny, 1981; Buzgo and Endress, 2000; Buzgo, 2001; Remizowa and Sokoloff, 2003; Buzgo et al., 2006; Remizowa et al., 2006): (1) bract suppression leading to formation of a cryptic bract and (2) formation of a single 'hybrid' organ (instead of two distinct abaxial organs situated on the same radius) by overlap of developmental programmes of the flower-subtending bract and the first abaxial organ formed on the floral pedicel. In the latter case, a flower-subtending bract is absent as a separate organ, but its features are partially expressed in the hybrid organ. The second pattern could also be interpreted without involving a hybrid organ concept;

Early Events in Monocot Evolution, eds P. Wilkin and S. J. Mayo. Published by Cambridge University Press. © The Systematics Association 2013.

for example, by assuming that the outer abaxial perianth member is lost and replaced by a flower-subtending bract developing in an unusual position.

The question of bract reduction is of interest in understanding the early evolution of monocots. Earlier investigations (Lieu, 1979; Buzgo and Endress, 2000; Buzgo et al., 2006) have revealed strong similarities in perianth development between *Acorus* (Acoraceae) and *Triglochin maritima* (Juncaginaceae). In both cases, the second pattern of bract reduction was reported. Mature inflorescences of *Acorus* and *Triglochin maritima* appear to be abracteate, but an organ closely resembling a flower-subtending bract is initiated at the earliest developmental stages. In anthetic flowers, this organ could be interpreted as an outer median tepal. Thus, this organ combines features of a flower-subtending bract and a tepal. The similarity between *Acorus* and *Triglochin* is significant because of contrasting results for the phylogenetic placement of *Acorus*. Although most molecular phylogenetic analyses place *Acorus* as sister to all other monocots, with Alismatales as the next branch (e.g. Chase et al., 2000, 2006; Chase, 2004; APG III, 2009), some studies suggest that *Acorus* is embedded within Alismatales (e.g. Qiu et al., 2000; Davis et al., 2004). The latter topology is more congruent with the similarities in perianth development between Acoraceae and Juncaginaceae. However, these speculations are limited by relatively low taxon sampling in detailed developmental studies of early flower development of some key groups. In particular, only two species of Juncaginaceae have previously been studied in detail. The present chapter is aimed at filling this gap by investigating early flower and inflorescence development in some species of Alismatales that have hitherto not been studied in detail.

8.2 Materials and methods

Species examined are listed in Table 8.1. Morphology and development of the inflorescences were examined in three species of *Potamogeton* (Potamogetonaceae), one species of *Tofieldia* (Tofieldiaceae) and four species of *Triglochin* (Juncaginaceae), of which only one (*T. maritima*) was extensively investigated previously (Charlton, 1981; Buzgo et al., 2006). Some data on inflorescence development in *Triglochin barrelieri* were published by Sokoloff et al. (2006) under the name *T. bulbosa* (the same accession was used here and in the previous study). In the present chapter, we follow a recent taxonomic revision of *Triglochin* (Köcke et al., 2010); the taxon here designated as *T. bulbosa* is not the same species as the *T. bulbosa* investigated by Sokoloff et al. (2006). Voucher specimens are deposited at the Herbarium of the Moscow State University (MW) and the Spirit Collection of Department of Higher Plants of Moscow State University.

Plant material was fixed in 70% ethanol. For scanning electron microscopy (SEM), the material examined at Moscow University was dissected in 96% ethanol,

Table 8.1 Collection data of material examined. Voucher specimens are deposited at the Herbarium of Moscow State University (MW) and the Spirit Collection of the Department of Higher Plants of Moscow State University.

Species	Location	Date	Collector
Tofieldia calyculata	Switzerland, Uri, between Wassen and Susten Pass	8 Jul 2002	D. Sokoloff
T. calyculata	Switzerland, Bern, Oberlands, Kandersteg	3 Sep 2005	M. Remizowa
Potamogeton crispus P. lucens P. natans	Russia, Moscow region, Zvenigorod Biological Station of Moscow State University	Jun–Jul 2003–2006	M. Remizowa D. Sokoloff M. Nuraliev S. Majorov
Triglochin barrelieri	Cyprus, Akrotiri	13 Mar 2004	A. Seregin D. Sokoloff
T. bulbosa	South Africa, Cape of Good Hope	Sep 2006	A. Oskolski
T. maritima	Russia, Karelia, White Sea, Pokormezhny Island	7 Jul 2003	M. Remizowa
T. palustris	Russia, Karelia, White Sea Biological Station of Moscow State University	10 Jul 2003	M. Remizowa

dehydrated through absolute acetone and critical-point dried using a Hitachi HCP-2 critical-point dryer (CPD). It was then coated with gold and palladium using an Eiko IB-3 ion-coater, and observed using a Hitachi S-405A, Camscan 4DV or JEOL JSM-6380LA SEM. The material examined at the Royal Botanic Gardens, Kew was dissected in 70% ethanol, dehydrated through absolute ethanol and critical-point dried using a Autosamdri-815B CPD. It was then coated with platinum using an Emitech K550 sputter coater and examined using a Hitachi cold field emission SEM S-4700-II.

8.3 Results

8.3.1 *Tofieldia calyculata*

Flowers are spirally arranged into a terminal many-flowered raceme. Each flower is subtended by a well-developed flower-subtending bract, as in most other *Tofieldia* species except *T. pusilla*, in which the inflorescence is abracteate. Flowers of *Tofieldia* are bisexual, trimerous and pentacyclic, with biseriate tepals and stamens and a gynoecium of three united carpels. Tepals are showy and form a short floral tube that collects nectar secreted by a nectary situated at the gynoecium base.

The flower itself is surrounded by a structure termed a calyculus. There is a short internode between the calyculus and the perianth. The calyculus consists of three united phyllomes. In some specimens of *T. calyculata*, the phyllomes of the calyculus are united to their tips.

Flowers are initiated in an acropetal sequence on the inflorescence axis (Fig 8.1A). A terminal flower is usually absent, and a residual meristem is often present at the inflorescence apex. The subtending bract and flower are initiated simultaneously by two separate primordia (Fig 8.1B, C). These primordia are closely pressed together and divided by a slit. The bract primordium is smaller than the corresponding floral primordium (bract primordia are larger in the lower part of the inflorescence). Calyculus initiation is unidirectional (Fig 8.1D, 8.2A–C). The lateral phyllomes arise as two separate primordia before initiation of the median abaxial phyllome, which is always initiated later as a separate primordium. At later stages, a calyculus tube appears as a result of intercalary growth under the initially free calyculus phyllomes (not shown).

Tepals and stamens are initiated in radial pairs and often (not shown here) as common PA primordia (PA: perianth plus androecium). Tepal–stamen pairs of the same whorl (i.e. all outer tepals plus all outer stamens) are initiated simultaneously (Fig 8.2D). In *T. calyculata*, the tepals and calyculus are considerably delayed in their development compared with other *Tofieldia* species (not shown). Thus, the flower is not protected by perianth and calyculus throughout the development of the stamens and the initiation and early development of the carpels. All protection is provided by the leaves (leaf sheaths) surrounding the developing inflorescence.

8.3.2 Triglochin (*T. barrelieri*, *T. bulbosa*, *T. maritima*, *T. palustris*)

Flowers are spirally arranged into a terminal many-flowered raceme. Flowers are not subtended by bracts. Flowers of *Triglochin* are bisexual, trimerous and hexacyclic with biseriate tepals, stamens and carpels. In *T. maritima*, all six carpels are fertile, whereas the outer carpels are sterile in *T. barrelieri*, *T. bulbosa* and *T. palustris*. Tepals are greenish and inconspicuous. Each stamen is closely associated with its corresponding tepal, resulting in six tepal–stamen pairs.

Flowers are initiated acropetally along the inflorescence axis (Fig 8.3A, B, 8.4A, 8.5). A terminal flower is usually present (not shown).

In *T. maritima*, floral development is unidirectional. Tepals and stamens develop from separate primordia. The first floral organ to be initiated is the outer median abaxial tepal, which cuts off up to one third of the floral meristem (not shown). Primordia of outer median tepals are relatively larger in the lower part of the inflorescence. The outer abaxial tepal grows rapidly and forms a shield-like

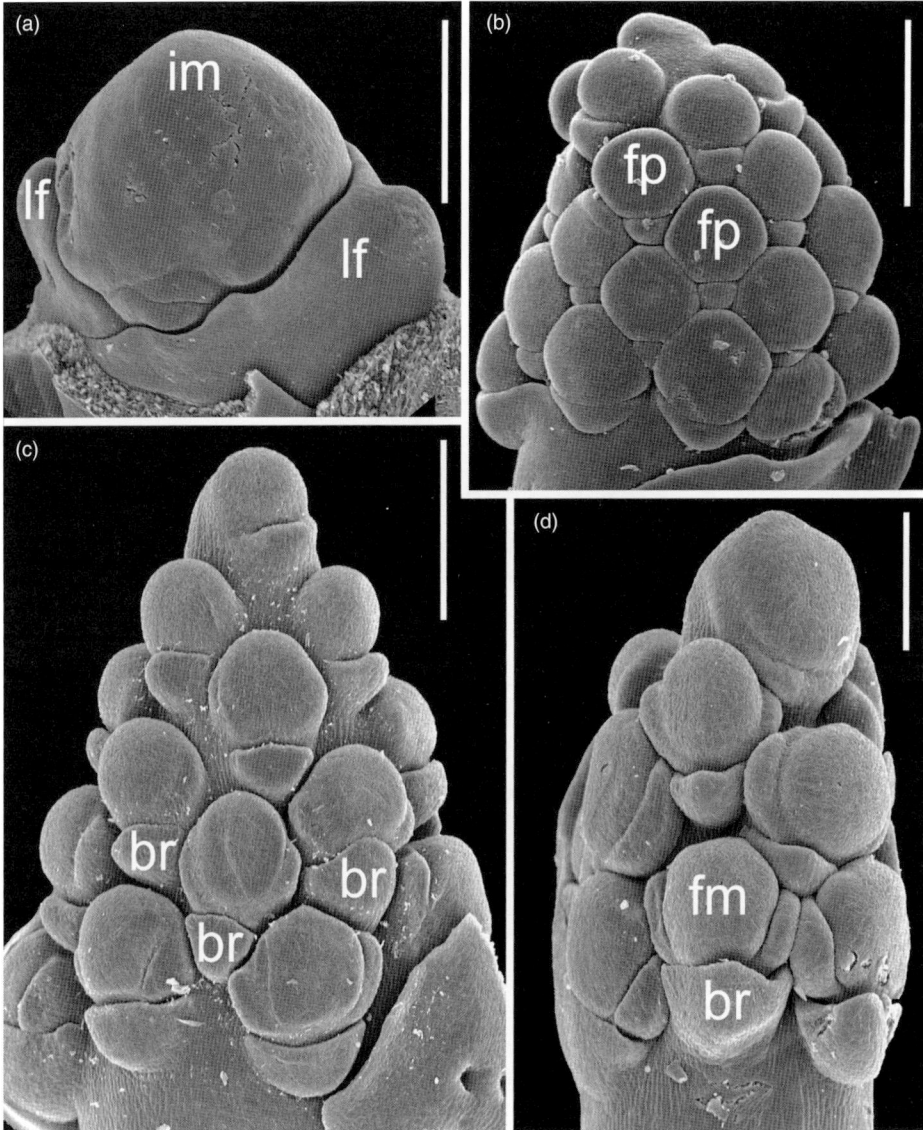

Fig 8.1 Inflorescence development in *Tofieldia calyculata*. A: Lateral view of young inflorescence showing flower initiation. B: Lateral view of young inflorescence at a stage after initiation of subtending bract and flower. C: Lateral view of young inflorescence at an older stage. D. Lateral view of young inflorescence at stage of initiation of lateral calyculus phyllomes. br = primordium of subtending bract; fm = floral meristem; fp = floral primordium; im = inflorescence meristem; lf = the uppermost leaves protecting the developing inflorescence. Scale bars = 300 μm (A, C), 400 μm (B) and 200 μm (D).

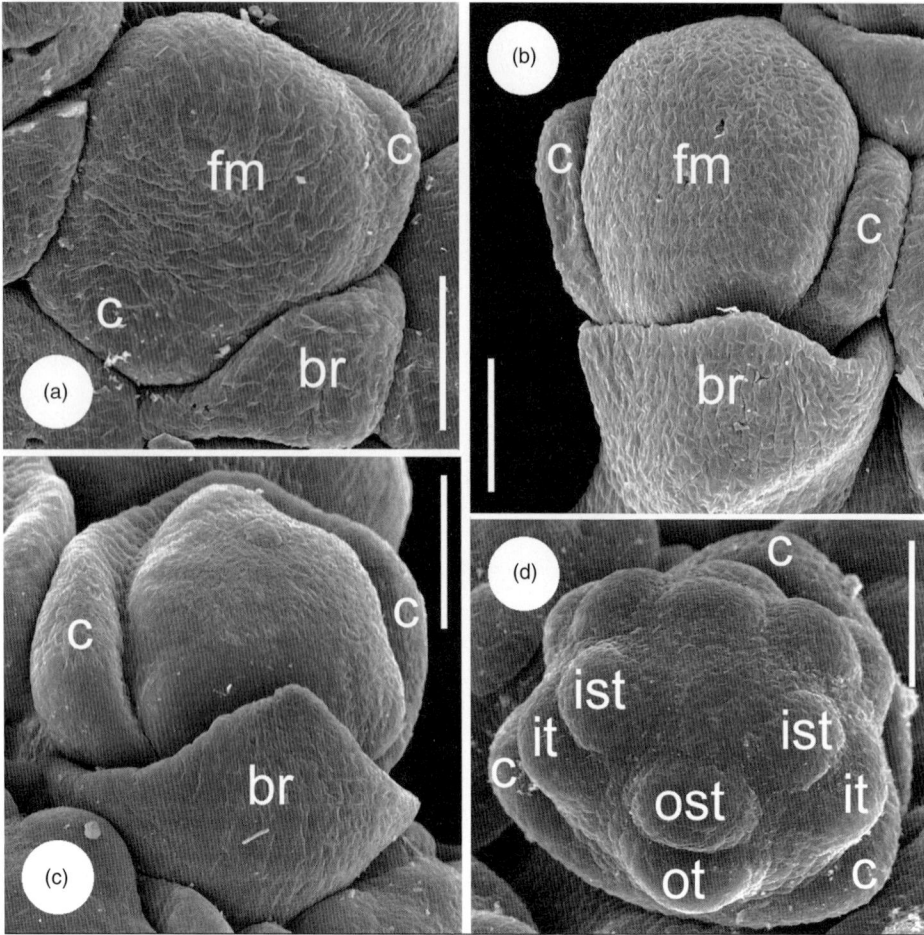

Fig 8.2 Floral development in *Tofieldia calyculata*. A–B: Initiation of lateral calyculus phyllomes. C: Formation of calyculus tube (view from adaxial side). D: Initiation of perianth and androecium (view from adaxial side). br = primordium of subtending bract; c = primordium of calyculus phyllome; fm = floral meristem; ist = primordium of inner stamen; it = primordium of inner tepal; ost = primordium of outer stamen; ot = primordium of outer tepal. Scale bars = 80 μm.

structure that covers the developing flower (Fig 8.3C). The primordia of the outer lateral tepals appear soon after the initiation of the outer abaxial tepal, followed by initiation of the outer lateral stamens. The outer abaxial stamen is initiated simultaneously with the lateral outer ones, or slightly later, depending on the initial size of the abaxial tepal (not shown). The primordia of the inner tepals appear simultaneously in a whorl, followed shortly by initiation of the inner stamens (not shown). The inner tepals are relatively slow in their development; at the stage of gynoecium

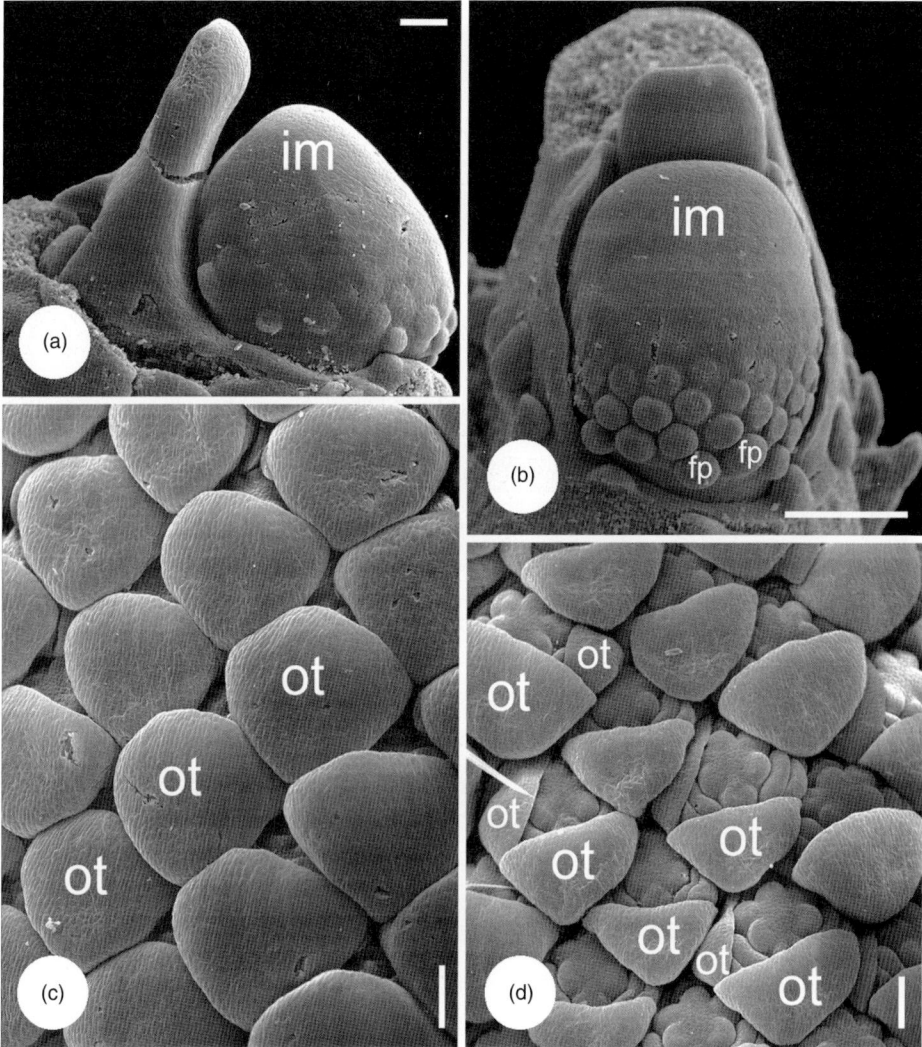

Fig 8.3 Inflorescence development in *Triglochin maritima*. A–B: Young inflorescences showing undifferentiated inflorescence meristem at the top. Note hemispherical flower primordia at the inflorescence base. C: Young inflorescence with flowers covered by median outer tepals that are larger than other tepals at this stage. D: Young inflorescence showing flowers with lateral tepals clearly visible and unprotected floral centre. fp = floral primordium; im = inflorescence meristem; ot = median outer tepal (and lateral outer tepals in D). Scale bars = 100 μm (A, C, D) and 300 μm (B).

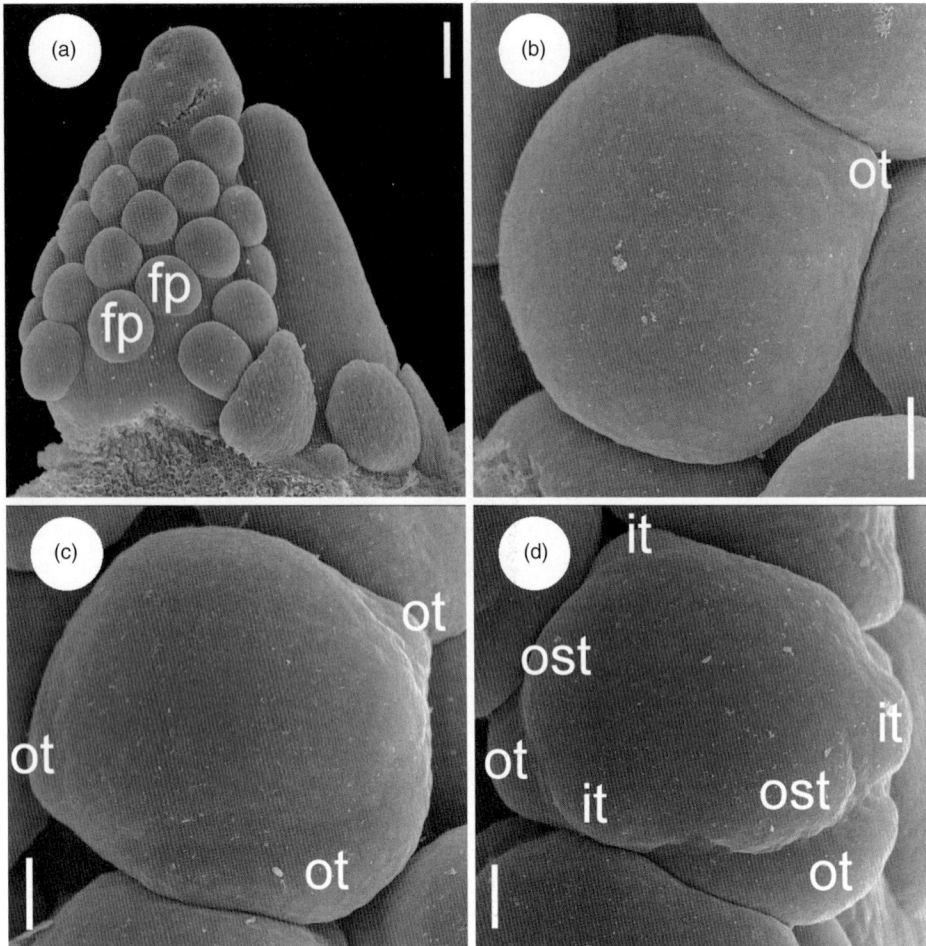

Fig 8.4 Floral development in *Triglochin bulbosa.* A: Acropetal initiation of flowers by hemispherical primordia. B: Initiation of lateral outer tepals. C: Initiation of median outer tepal. Note that the primordium of the median outer tepal is considerably larger than the lateral outer tepal primordia, though the median outer tepal is initiated later. D: Initiation of outer stamens and inner tepals. fp = floral primordium; it = primordium of inner tepal; ost = primordium of outer stamen; ot = primordium of outer tepal. Scale bars = 100 μm (A) and 30 μm (B–D).

initiation they are of the same size as the inner stamens (Fig 8.3D). During early developmental stages, the flower is exclusively protected by the outer abaxial tepal (Fig 8.3C). Later, the lateral outer tepals become more pronounced (Fig 8.3D) and finally take the same shape and size as the abaxial tepal.

In *T. bulbosa* (Fig 8.4), *T. barrelieri* (Fig 8.5) and *T. palustris* (not shown), the outer perianth members are initiated simultaneously within a whorl or their

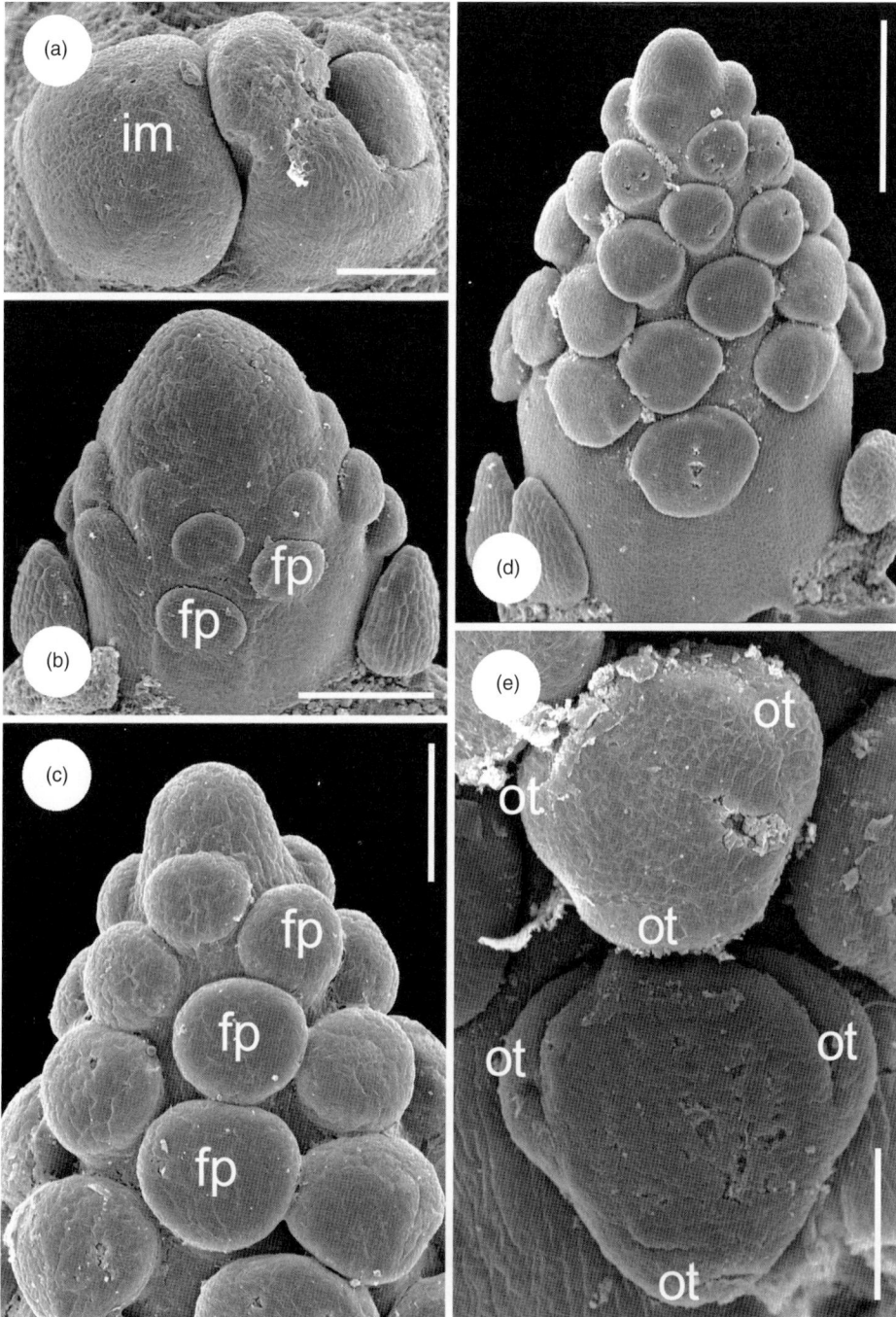

Fig 8.5 Inflorescence development in *Triglochin barrelieri*. A: Undifferentiated inflorescence meristem. B: Acropetal initiation of flowers by hemispherical primordia. C–D: Young inflorescences with all flowers initiated. Note that the lower floral primordia in (D) are almost triangular. E: Initiation of outer tepals. Note that all outer tepal primordia are of the same size. fp = floral primordium; im = inflorescence meristem; ot = primordium of outer tepal. Scale bars = 100 μm (A), 200 μm (B), 150 μm (C) 300 μm (D) and 150 μm (E).

development is slightly unidirectional. In the case of unidirectional development, the two outer lateral tepals are initiated before the abaxial one, which appears later (Fig 8.4B,C). In *T. barrelieri*, both simultaneous (Fig 8.5E) and unidirectional (not shown) types of outer perianth whorl development can co-occur within the same inflorescence (with unidirectional development at the base of the inflorescence). In our material of *T. bulbosa*, the abaxial outer tepal was larger at its inception than the lateral ones (Fig 8.4C), whereas the primordia of the outer tepals are of nearly equal size in both *T. barrelieri* (Fig 8.5E) and *T. palustris* (not shown). The outer stamens and inner tepals in *T. bulbosa* arise almost simultaneously, followed by the inner stamens (Fig 8.4D). As in *T. maritima*, the inner tepals are delayed in their development and the flowers are protected by outer tepals only during early developmental stages (not shown). In addition, the developing inflorescences are protected by the surrounding leaves.

8.3.3 *Potamogeton (P. crispus, P. lucens, P. natans)*

Flowers are arranged into a terminal many-flowered spike. Within the spike, flowers form alternating whorls of two to four flowers, depending on the thickness of the inflorescence axis. In *P. natans* and *P. lucens*, flowers are subtended by small flower-subtending bracts. Inflorescences of *P. crispus* are abracteate.

Flowers of *Potamogeton* are bisexual, tetramerous and tricyclic with uniseriate tepals, stamens and carpels. Tepals are greenish. Stamens and corresponding tepals possess a common stalk and form four tepal–stamen pairs.

Flowers are initiated acropetally on the inflorescence axis (Figs 8.6A, 8.7A, 8.8A). A terminal flower is usually absent and a residual meristem is often visible at the inflorescence apex. In *P. natans* and *P. lucens*, the subtending bract and flower develop from a common meristem. The bract appears simultaneously with the lateral tepals or slightly before them in *P. natans* (Fig 8.6B–D) and simultaneously with, or slightly later than, lateral tepal initiation in *P. lucens* (Fig 8.7B, C). The bract primordium separates from the floral meristem by a transverse slit. The subtending bract is much smaller than the flower from its inception. In *P. crispus*, no sign of a flower-subtending bract is detectable during floral development (Fig 8.8). Tepals and stamens develop as separate primordia, but become united later via formation of a common stalk which appears as a result of intercalary growth under the initially free bases of a tepal and corresponding stamen (not shown). In all species examined, perianth development starts with the inception of the lateral tepals (Figs 8.6B–D, 8.7B, C). Then, the primordia of the median tepals appear. Stamens are initiated after the tepals and their initiation is simultaneous or almost simultaneous. Tepals are delayed in their development until early gynoecium development commences and do not protect inner floral parts (Figs 8.6E, 8.7D, 8.8B). The developing inflorescence is protected by two surrounding leaves.

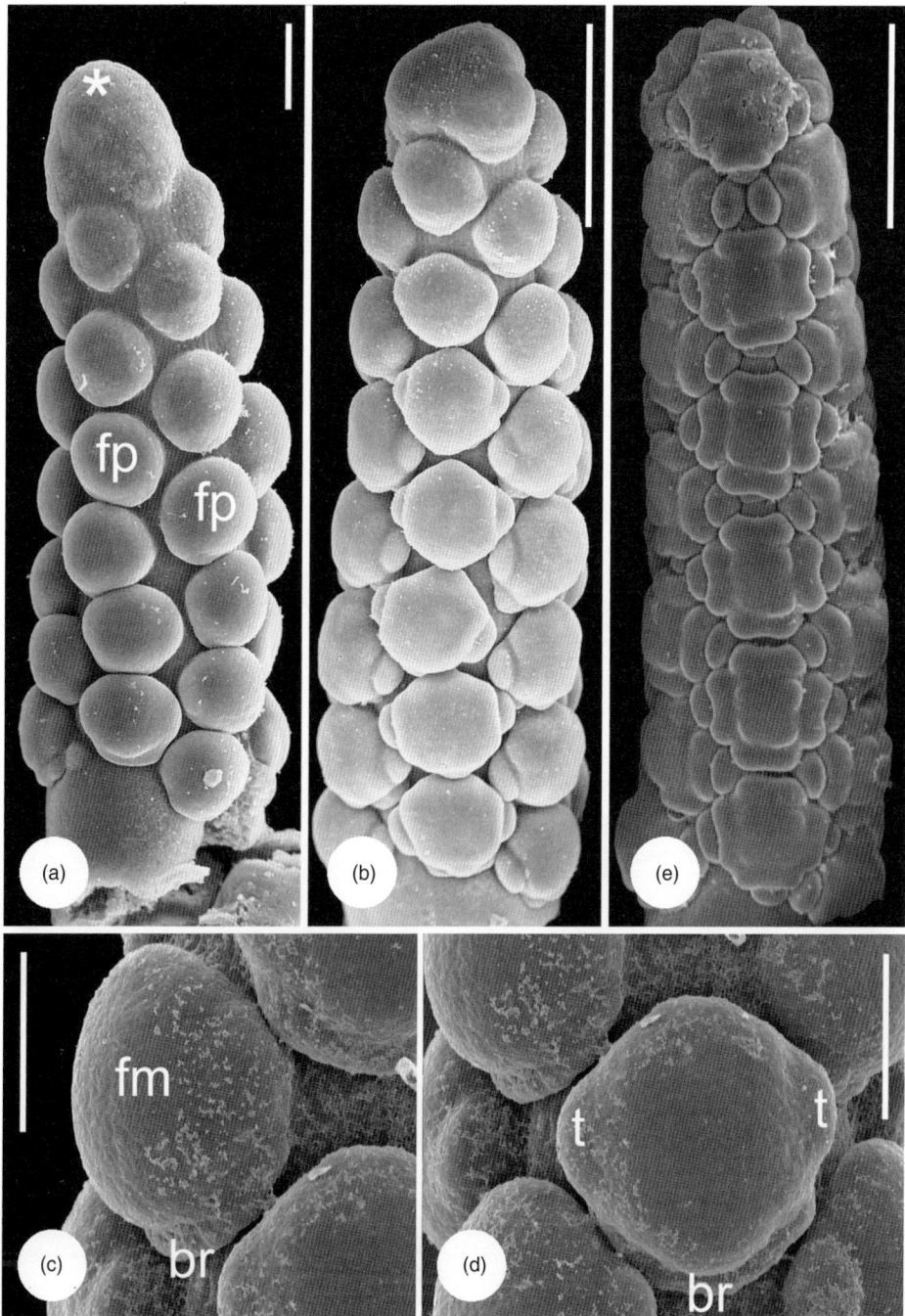

Fig 8.6 Inflorescence development in *Potamogeton natans*. A: Lateral view of young inflorescence showing floral primordia initiated acropetally. B: Lateral view of young inflorescence (terminated by a pseudanthium) at stage of initiation of lateral tepals and subtending bract. In the uppermost flowers the bract is initiated before the perianth or simultaneously with it. C: Initiation of subtending bract. D: Initiation of lateral tepals. E: Lateral view of young inflorescence at stage of gynoecium initiation with perianth and androecium already initiated. At this stage the median tepals are hidden by the stamens and the bracts are not visible. br = primordium of subtending bract; fm = floral meristem; fp = floral primordium; t = tepal primordium; * = residual meristem. Scale bars = 100 μm (A, C, D) and 300 μm (B, E).

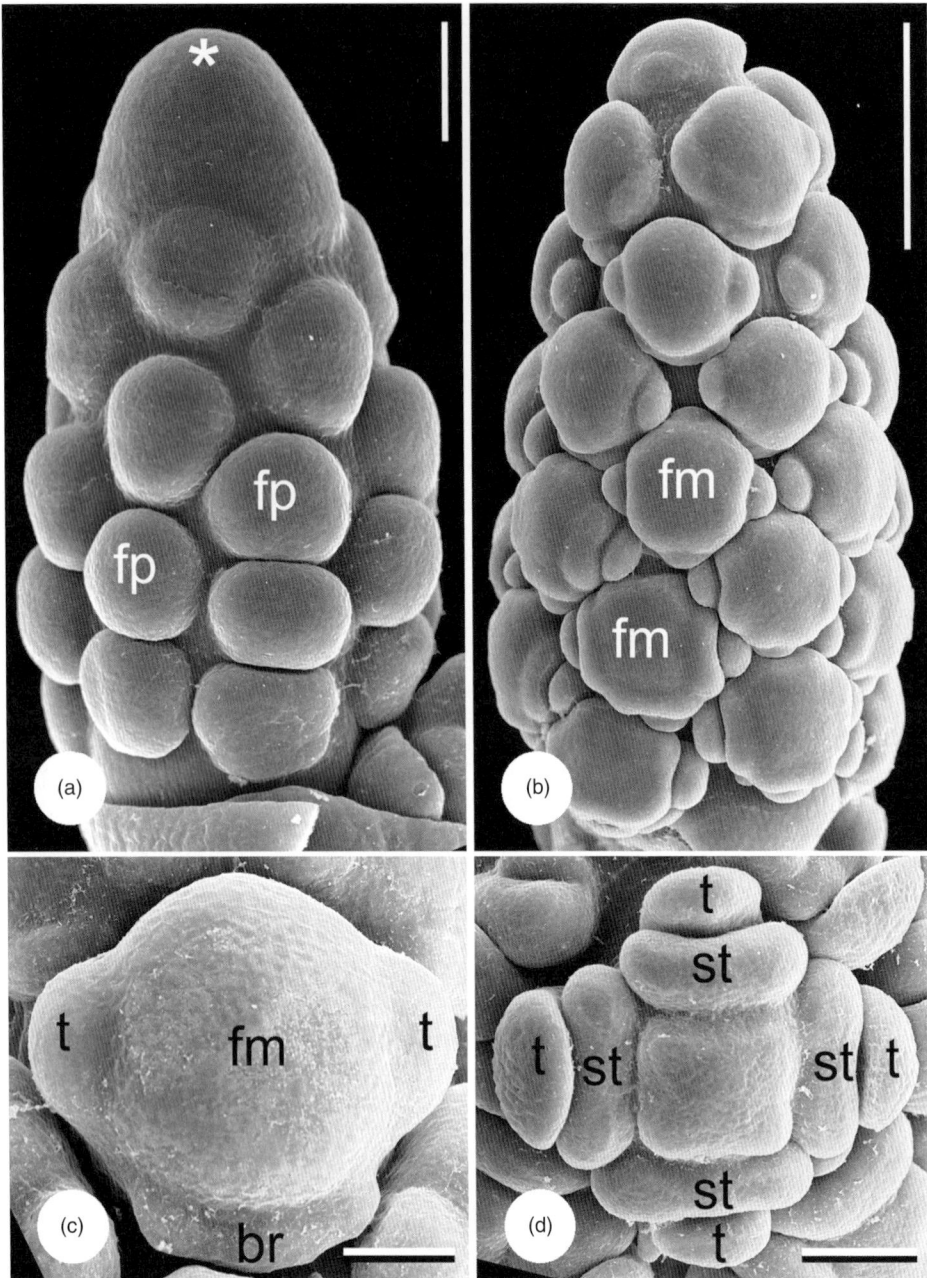

Fig 8.7 Inflorescence and flower development in *Potamogeton lucens*. A: Lateral view of young inflorescence showing floral primordia initiated acropetally. B: Lateral view of young inflorescence (terminated by a pseudanthium) at stage of perianth initiation. Note that the subtending bracts are initiated simultaneously with lateral tepals in the uppermost flowers; they are hidden by flowers in the lower part of the inflorescence. C: A flower at stage of subtending bract and perianth initiation. D: A flower at stage of gynoecium initiation with perianth and androecium already initiated. Subtending bract is not visible. br = primordium of subtending bract; fm = floral meristem; fp = floral primordium; t = tepal primordium; st = stamen primordium; * = residual meristem. Scale bars = 100 μm (A, C, D) and 300 μm (B).

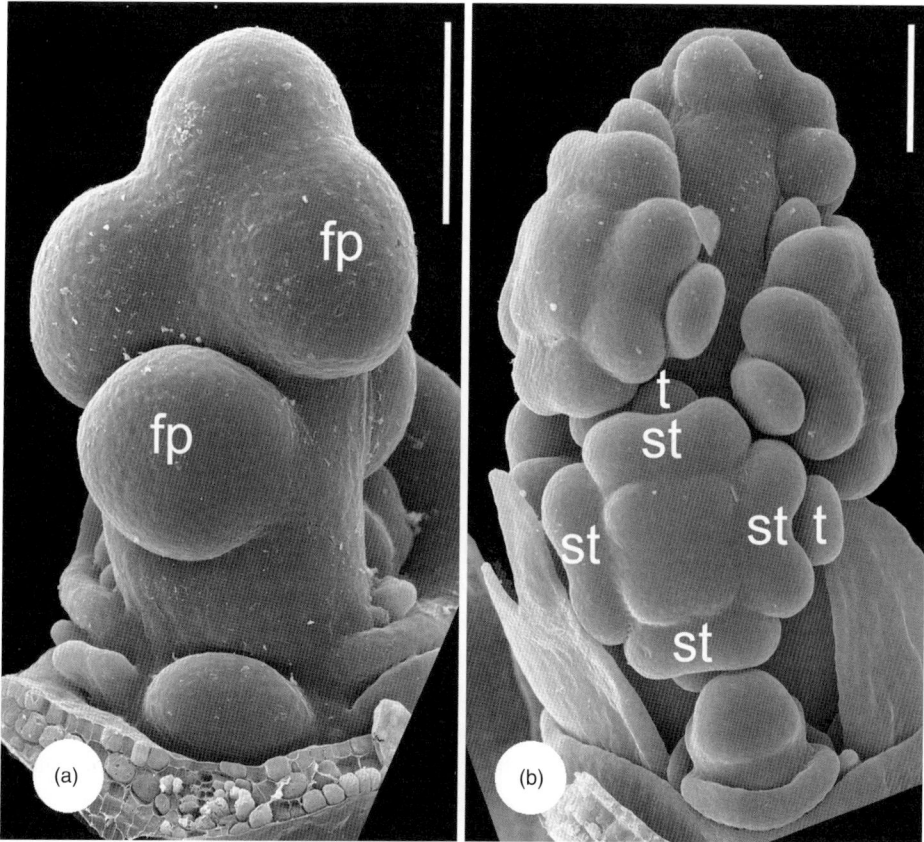

Fig 8.8 Inflorescence development in *Potamogeton crispus*. A: General view of young inflorescence showing floral primordia initiated acropetally. B: General view of young inflorescence at stage of gynoecium initiation with perianth and androecium already initiated. Note that median tepals are hidden by stamens and bracts are absent. fp = floral primordium; t = tepal primordium; st = stamen primordium. Scale bars = 100 μm.

8.4 Discussion

In general, members of Alismatales with racemose inflorescences form two distinct morphological groups: species with well-developed flower-subtending bracts and species that lack subtending bracts, at least superficially. The latter group can be further divided into species possessing a single 'hybrid' organ instead of two separate organs situated on the same radius (the flower-subtending bract and the subsequent abaxial phyllome) and species in which subtending bracts are entirely suppressed. These crucial features of inflorescence architecture are not constant, even at the generic level among tepaloid alismatids. *Scheuchzeria* (Scheuchzeriaceae), *Posidonia* (Posidoniaceae), most *Tofieldia* species

(Tofieldiaceae) and some *Potamogeton* species (Potamogetonaceae) all belong to the group with bracteate inflorescences (Sattler, 1965; Posluszny and Sattler, 1974; Posluszny, 1981, 1983; Tomlinson, 1982; Charlton and Posluszny, 1991; Sun et al., 2000; Remizowa et al., 2006; Sokoloff et al., 2006; Lock et al., 2009; this study). *Tofieldia pusilla*, Araceae, Aponogetonaceae, Juncaginaceae, Ruppiaceae, some Potamogetonaceae and possibly Zosteraceae are characterized by abracteate inflorescences, or at least the flower-subtending bracts are not visible as separate organs (Uhl, 1947; Posluszny and Sattler, 1973; Singh and Sattler, 1977; Lieu, 1979; Charlton and Posluszny, 1991; Soros-Pottruff and Posluszny, 1995a, 1995b; Mayo et al., 1997; Buzgo, 2001; Remizowa and Sokoloff, 2003; Buzgo et al., 2004, 2006; Remizowa et al., 2006).

In most angiosperms, floral primordia arise in the axils of bracts that are already initiated (e.g. Payer, 1857). Among the taxa considered here, this is the case in *Scheuchzeria* (see figures in Posluszny, 1983). In contrast, in some alismatids with racemose inflorescences, the flowers and their subtending bracts are initiated simultaneously, resulting in formation of the flower and its subtending bract by a common primordium in some cases. However, this feature is not always consistent at the family and genus levels; for example, species of *Tofieldia* differ in their patterns of bract initiation. In *T. calyculata*, the flowers and their bracts are initiated by separate primordia (this study), whereas common bract–flower primordia have been reported in *T. coccinea* (Remizowa et al., 2006). In both species, bract and flower initiation are simultaneous. In *Potamogeton*, subtending bracts (where present) arise from a common bract–flower primordium (this study) or are initiated separately (see Posluszny and Sattler, 1974; Posluszny, 1981). In cases of independent bract and flower initiation, the primordium of the flower-subtending bract can be initiated simultaneously with the floral primordium or slightly later, or even later than initiation of the lateral tepals (see figures in Posluszny and Sattler, 1974; Posluszny, 1981; Sun et al., 2000). The presence or absence of a flower-subtending bract does not affect tepal position and sequence of organ initiation. In all cases studied to date, the lateral tepals are initiated before the median one (Posluszny and Sattler, 1974; Posluszny, 1981; Sun et al., 2000). This phenomenon suggests the presence of a cryptic bract in abracteate species of *Potamogeton* such as *P. crispus* (this study).

Cryptic bracts and their contribution to perianth orientation and development have been studied extensively in the model plant *Arabidopsis*, which also possesses a racemose inflorescence (e.g. Baum and Day, 2004; Choob and Penin, 2004; Penin 2008). In *Potamogeton crispus*, as in *Arabidopsis*, we suggest that flower-subtending bracts, though superficially absent, are present physiologically, more precisely at the stage of prepatterning of the floral meristem prior to visible organ development. There is a strong tendency for reduction of the flower-subtending bracts accompanied by loss of their vasculature in *Potamogeton*.

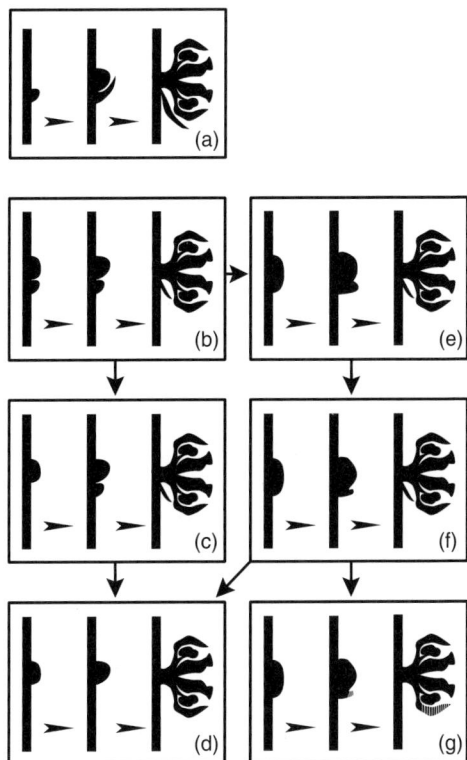

Fig 8.9 Diagram illustrating possible morphological transformations of flower-subtending bracts in *Potamogeton*. a: Hypothetical condition with flower initiated in the axil of a bract that is already well developed. This condition is not found in extant species of *Potamogeton* studied so far. b: Case with small subtending bract and flower initiated simultaneously by separate primordia (not yet reported). c: Case with small bract initiated by separate primordium and later than flower (not yet reported). d: Case with subtending bract entirely suppressed (cryptic bract condition): *Potamogeton crispus*. e–f: Cases with subtending bract and flower initiated by a common meristem (e: *P. natans*, f: *P. lucens*). Cases (e) and (f) differ in the relative size of the bract primordium and time of its formation. g: Case with 'hybrid' (tepal+bract) organ formation (*Potamogeton densus*); 'hybrid' organ dashed. Arrowheads = morphogenetic transformations, arrows = possible evolutionary transformations (see text for explanation).

In species of *Potamogeton* possessing bracteate inflorescences, bracts are nonvascularized (e.g. Uhl, 1947). The same condition occurs in *Posidonia* (Remizowa et al., 2012), in which flowers and their subtending bracts are initiated separately. In *Potamogeton*, either the bract is formed separately as a small bulge below the floral meristem, or the primordium of the subtending bract cuts off a small piece from the floral meristem (Fig 8.9). Subtending bracts are delayed in development and become clearly visible only after anthesis. In younger inflorescences, they are hidden by densely arranged flowers. We propose two contrasting scenarios to illustrate the origin of abracteate inflorescences in *Potamogeton* (Fig 8.9). In scenario (1), during the course of evolution, a stage of bract initiation was entirely eliminated from inflorescence development (Fig 8.9a–b–c–d). This scenario implies that the abracteate condition is achieved directly through inflorescences with separate bract and flower initiation (Fig 8.9a–b–c–d). In this case, an intermediate stage, if present (Fig 8.9c), would have a temporal retardation (heterochronic delay) in subtending bract initiation with respect to flower initiation. Alternatively, scenario (2) implies that a precondition for the evolutionary origin of the abracteate condition is the development of the flower and its subtending bract from a common meristem (Fig 8.9a–b–e–f–d). In turn, a precondition for formation of a common bract–flower primordium is the simultaneous initiation of the flower and its subtending bract (Fig 8.9b). This alternative scenario also implies heterochrony in subtending bract initiation and decrease in size of bract primordium. In both scenarios, a direct pathway to bract reduction (without involving intermediate stages) is also possible.

Regarding the initiation of the flower-subtending bract and flower from a common meristem in some species of *Tofieldia* and *Potamogeton*, there are two possible interpretations, which are impossible to differentiate based on current evidence. Either (1) the flower and its subtending bract are initiated by a common primordium, or (2) the subtending bract is the first organ initiated by the floral meristem (Figs 8.9e, f; 8.10d; 8.11d). Although the second option better reflects the behaviour of the developing organs, it is problematic because it is tempting to interpret the subtending bract and floral parts as belonging to the same lateral shoot and hence to homologize the first median abaxial phyllome initiated by a floral meristem with a perianth part. Regardless of their initiation (in some cases) from a common meristem, by definition a flower and its subtending bract belong to different branching orders – the bract belongs to the inflorescence axis and the floral organs belong to the axillary pedicel. In *T. coccinea*, a species with common bract–flower primordia, the inflorescence vasculature is closely similar to that of other *Tofieldia* species with bracteate inflorescences, in which the flowers and their subtending bracts are initiated separately (Remizowa et al., 2010). In addition, epicaulescence of some pedicels (i.e. their basal congenital fusion with the inflorescence axis) occurs in *T. coccinea*, as also in *T. calyculata*, where subtending bracts and flowers are initiated by separate primordia (Remizowa, 2008).

In common with many other tepaloid alismatids, *Triglochin* lacks visible flower-subtending bracts. Species of *Triglochin* differ from each other in their patterns of perianth initiation (Fig 8.10). In *T. maritima* (Fig 8.10e), the first organ produced by the floral meristem is the median outer tepal, occupying a strictly abaxial position. During early developmental stages, this tepal shows some bract-like features (see also Buzgo et al., 2006). In completely formed flowers, all tepals are

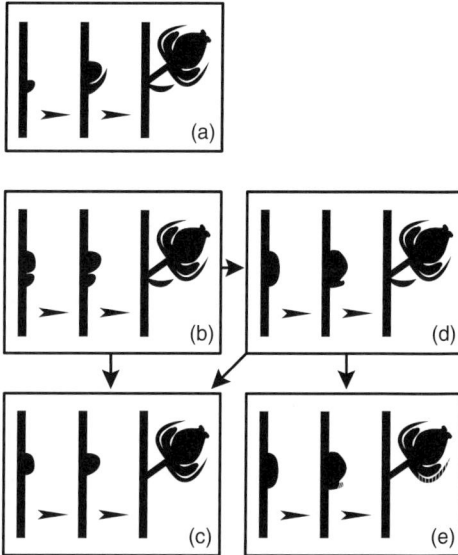

Fig 8.10 Diagram illustrating possible morphological transformations of flower-subtending bracts in *Triglochin*. Conditions (a), (b) and (d) are not found in extant species of *Triglochin* studied so far. a: Hypothetical condition with flower initiated in the axil of a bract that is already well developed. b: Hypothetical case with small subtending bract and flower initiated simultaneously by separate primordia. c: Case with subtending bract entirely suppressed (cryptic bract condition), found in *T. bulbosa, T. palustris, T. barrelieri*. d: Hypothetical case with subtending bract and flower initiated by common meristem. e: Case with 'hybrid' (tepal+bract) organ formation (*T. maritima*); 'hybrid' organ dashed. Arrowheads = morphogenetic transformations, arrows = possible evolutionary transformations.

of the same shape and size. Buzgo et al. (2006) suggested that the median abaxial organ of a *Triglochin* flower is a flower-subtending bract rather than a tepal. More precisely, according to Buzgo et al. (2006) the flower-subtending bract is not extrafloral and is involved in the perianth, i.e. it substitutes the corresponding tepal that has been superficially lost. Despite the long pedicel, the subtending bract is situated at the same level as the remaining five tepals and can be regarded as a perianth part. Recaulescence (a congenital shift of the flower-subtending bract along a pedicel) and loss of the median outer tepal are caused by abbreviation of developmental rates (followed by morphological abbreviation of axillary pedicels) and unidirectional floral development. The tepal-like appearance of the flower-subtending bract is due to shifts in the expression zones of the tepal identity genes into a zone of bract gene expression.

In contrast to Buzgo et al. (2006), we hypothesize that the median abaxial organ in the flower of *Triglochin* is a result of hybridization of the developmental pathways (see Lodkina 1983; Sattler 1988) of the tepal and flower-subtending bract (Fig 8.10e). In other words, the expression zones of the tepal and bract identity genes overlap in the same primordium. Our interpretation implies no shifts of expression zones of genes responsible for perianth development. We agree with Buzgo et al. (2006) that floral development has been abbreviated in *Triglochin* compared with the ancestral condition. *Triglochin* species develop their flowers under extreme environmental conditions; they flower and set seeds during a very short vegetative season, though the growth conditions of *T. maritima* are apparently not the most extreme in the genus. Presumably, in ancestral flowers of Juncaginaceae the median outer tepal and flower-subtending bract were situated on the same radius very close each to other (Fig 8.10a or 8.10b). Increasing the developmental rates resulted in amalgamation of corresponding primordia, while retaining the ancestral prepatterning of the floral meristem (Fig 8.10b–d–e). This new primordium gave rise to a hybrid organ and there is no substitution of the median outer tepal by the flower-subtending bract. The same hybrid organ phenomenon has been found in *T. striata* (see figures in Lieu, 1979), *Potamogeton densus* (Fig 8.9g) (see figures in Posluszny and Sattler, 1973), some Aponogetonaceae (see discussion in Buzgo et al., 2006 – though in most cases *Aponogeton* has no phyllome in the abaxial position) and *Acorus* (Buzgo and Endress, 2000).

We did not find hybrid organs in *Triglochin barrelieri*, *T. bulbosa* or *T. palustris*. Moreover, perianth initiation and development is slightly delayed on the abaxial side of the young flower, suggesting the presence of a cryptic bract, as in abracteate species of *Potamogeton* (*P. crispus*). Flower-subtending bracts are not detectable either in anthetic or developing inflorescences. Morphological reduction of the flower-subtending bracts in the inflorescences of these species is a possible result of the same processes as in species with hybrid tepal–bract organs (Fig 8.10b–c or Fig 8.10b–d–c). In contrast to *T. maritima* and *T. striata*, the bract primordia have been eliminated (as in abracteate *Potamogeton*) instead of being amalgamated with the outer median tepal primordium.

Tofieldia possesses an unusual structure termed a calyculus, which is common to all Tofieldiaceae (Eichler, 1875; Zomlefer, 1997; Remizowa and Sokoloff, 2003; Remizowa et al., 2006, 2010; Takhtajan, 2009). The calyculus consists of three phyllomes. As a consequence of calyculus insertion, *Tofieldia* differs strongly in floral orientation from that of other monocots that possess bracteate racemose inflorescences and lack additional pedicel phyllomes (i.e. bracteoles).

Species of *Tofieldia* (Fig 8.11) differ from each other in calyculus morphology, vasculature and development (Remizowa and Sokoloff, 2003; Remizowa et al., 2006, 2010). In the majority of *Tofieldia* species, the calyculus is inserted just below the flower. There are significant differences between *T. pusilla* and other *Tofieldia* species, both in calyculus morphology, anatomy and development, but

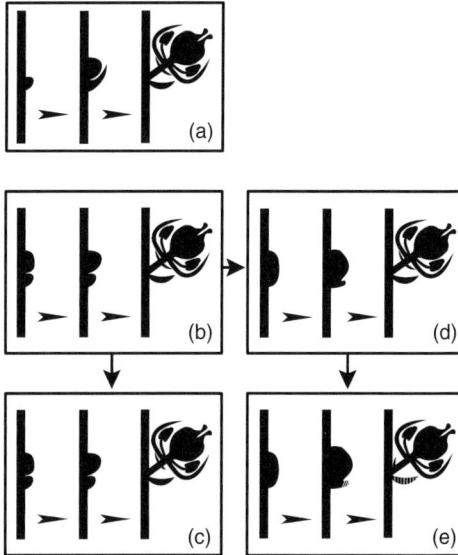

Fig 8.11 Diagram illustrating possible morphological transformations of flower-subtending bracts in *Tofieldia*. a: Hypothetical condition with flower initiated in the axil of a bract that is already well developed. This condition is not found in extant species of *Tofieldia* studied so far. b–c: Cases with small subtending bract and flower initiated simultaneously by separate primordia (*T. calyculata*). Cases (b) and (c) differ only in the relative sizes of the median calyculus phyllome (smaller in c). d: Case with subtending bract and flower initiated by a common meristem (*T. coccinea*). e: Case with 'hybrid' (bract+calyculus phyllome) organ formation (*T. pusilla*); 'hybrid' organ dashed. Arrowheads = morphogenetic transformations, arrows = possible evolutionary transformations.

their calyculi are homologous (Remizowa and Sokoloff, 2003; Remizowa et al., 2005, 2006, 2010). In *T. pusilla*, a flower-subtending bract is apparently absent and the calyculus is inserted at the base of the pedicel. Its tube is open on the adaxial side. The median calyculus phyllome is considerably larger than the lateral phyllomes; it occupies the position of the flower-subtending bract and also shows some bract-like features in vasculature and development (Fig 8.11e). Thus, a hybrid nature has been proposed for the calyculus, or more precisely for the median calyculus phyllome (Remizowa and Sokoloff, 2003).

The formation of hybrid organs that combine characters of the flower-subtending bract and the first median abaxial phyllome on the pedicel is a feature common to *T. pusilla* and some other Alismatales with racemose inflorescences, such as *Triglochin maritima* and *T. striata* (Juncaginaceae) and *Potamogeton densus* (Potamogetonaceae), in addition to the putatively basal monocot, *Acorus*. The formation of similar bract-like organs could represent a homoplastic tendency among basal monocots (Alismatales and Acorales). In *Tofieldia pusilla*, the median calyculus phyllome is the first abaxial phyllome to exhibit bract-like features, whereas in *Acorus*, *Triglochin maritima* and *Potamogeton densus* this role is taken by the outer median tepal. In *Triglochin maritima*, *T. striata*, *Potamogeton densus* and *Acorus*, hybrid organs, which combine features of the flower-subtending bract and the median outer tepal, are initiated in the same way as a flower-subtending bract – i.e. they are initiated before the other floral parts and require more intensive growth to cover the floral meristem. In *Triglochin maritima* and *Acorus* they protect the floral bud during early development, but in mature flowers they resemble tepals in size and shape. Buzgo (2001) and Buzgo et al. (2006) proposed that unidirectional flower development is often correlated with structures resembling subtending bracts (and the absence of a true bract) and with the formation of terminal flower-like structures. However, *Tofieldia* shows no such correlation (Remizowa et al., 2006; this study). All species of *Tofieldia* except *T. pusilla* possess a true flower-subtending bract and lack terminal flower-like structures. In *T. calyculata*, tepals are initiated simultaneously within a whorl. Unidirectional floral development occurs in *Tofieldia coccinea* (Remizowa et al., 2005, 2006). All three species of *Tofieldia* demonstrate unidirectionality in calyculus initiation. In *T. calyculata* and *T. coccinea*, three approximately equal calyculus phyllomes are initiated sequentially. The primordia of the lateral phyllomes appear well before initiation of the median phyllome (see also Remizowa et al., 2006). Delayed development of the median calyculus phyllome is a precondition of reduction of its vasculature (Remizowa et al., 2006, 2010). In *T. pusilla*, the true bract is absent, being replaced by the median organ of the bract-like calyculus, but tepal initiation is simultaneous in each whorl (Remizowa et al., 2005, 2006). Calyculus initiation is also sequential in this species; however, in contrast with other *Tofieldia* species, the first organ to be initiated on the floral meristem is the relatively large primordium of the median calyculus phyllome (Fig 8.11e). The relatively smaller primordia of the lateral phyllomes appear later. The shape of the primordium of the median calyculus phyllome and its timing of initiation are the same as in the flower-subtending bracts of other *Tofieldia* species.

In general, flower-subtending bract features are more pronounced than calyculus features in the hybrid organ of *Tofieldia pusilla*, especially in the vasculature (see Remizowa et al., 2010). In contrast, tepal features are more pronounced than bract features in the hybrid structures of other alismatids that develop such organs. Importantly, vascularization of the hybrid structures in other tepaloid alismatids possessing this feature is the same as in normal tepals of these species, and not what would be expected from a flower-subtending bract (Uhl, 1947; Buzgo and Endress, 2000; Buzgo et al., 2006). Although basal monocots with racemose inflorescences have a general (homoplastic) tendency to bract reduction, the flower-subtending bracts can be lost in two different ways – via morphological reduction (suppression) or via formation of hybrid organs. All reductions within inflorescences of tepaloid alismatids are due to organs that are not directly linked with reproductive success. Both morphological reduction (suppression) and formation of hybrid organs can occur within the same family and even within the same genus (*Triglochin*, *Potamogeton*). Thus, data obtained for a single species of a species-rich family cannot be extrapolated to other representatives of the family. In alismatids, different abracteate conditions co-occur with cases of presence of a true bract, which in turn differ in patterns of relative subtending bract and flower initiation. It remains an unresolved – but highly interesting – question whether the molecular genetic background of bract reduction in Alismatales is similar to that in *Arabidopsis* and other members of Brassicaceae.

8.5 Acknowledgements

We are indebted to Peter Endress for stimulating discussions as well as for his kind help in collecting material of *Tofieldia calyculata* in Switzerland. We thank two anonymous reviewers for their comments on the manuscript. We are grateful to Sergey Majorov, Maxim Nuraliev, Alexei Oskolski and Alexei Seregin for help in collecting plant material. The work of MVR and DDS was supported by RFBR grants 09–04–01155 and 12–04–01070 and the Ministry of Science and Education of Russia (FCP 'Kadry').

8.6 References

APG III (Angiosperm Phylogeny Group III). (2009). An update of the Angiosperm Phylogeny Group classification for the orders and families of flowering plants: APG III. *Botanical Journal of the Linnean Society*, **161**, 105–121.

Baum, D. A. and Day, C. D. (2004). Cryptic bracts exposed. Insights into the regulation of leaf expansion. *Developmental Cell*, **6**, 318–319.

Buzgo, M. (2001). Flower structure and development of Araceae compared with alismatids and Acoraceae. *Botanical Journal of the Linnean Society*, **136**, 393–425.

Buzgo, M. and Endress, P. K. (2000). Floral structure and development of Acoraceae and its systematic relationships with basal angiosperms. *International Journal of Plant Sciences*, **161**, 23–41.

Buzgo, M., Soltis, D. E., Soltis, P. S. and Ma H. (2004). Towards a comprehensive integration of morphological and genetic studies of floral development. *Trends in Plant Science*, **9**, 164–173.

Buzgo, M., Soltis, D. E., Soltis, P. S. et al. (2006). Perianth development in the basal monocot *Triglochin maritima*. *Aliso*, **22**, 107–127.

Charlton, W. A. (1981). Features of the inflorescence of *Triglochin maritima*. *Canadian Journal of Botany*, **59**, 2108–2115.

Charlton, W. A. and Posluszny, U. (1991). Meristic variation in *Potamogeton* flowers. *Botanical Journal of the Linnean Society*, **106**, 265–293.

Chase, M. W. (2004). Monocot relationships: an overview. *American Journal of Botany*, **91**, 1645–1655.

Chase, M. W., Soltis, D. E., Soltis, P. S. et al. (2000). Higher-level systematics of the monocotyledons: An assessment of current knowledge and a new classification. In *Monocots: Systematics and Evolution*, ed. K. L. Wilson and D. A. Morrison. Melbourne: CSIRO, pp. 3–16.

Chase, M. W., Fay, M. F., Devey, D. S. et al. (2006). Multigene analyses of monocot relationships: a summary. *Aliso*, **22**, 63–75.

Choob, V. V. and Penin, A. A. (2004). Structure of flower in *Arabidopsis thaliana*: spatial pattern formation. *Russian Journal of Developmental Biology*, **35**, 224–228.

Davis, J. I., Stevenson, D. W., Petersen, G. et al. (2004). A phylogeny of the monocots, as inferred from *rbc*L and *atp*A sequence variation, and a comparison of methods for calculating jackknife and bootstrap values. *Systematic Botany*, **29**, 467–510.

Eichler, A. W. (1875). *Blüthendiagramme*. Leipzig: W. Engelmann.

Köcke, A. V., von Mering, S., Mucina, L. and Kadereit, J. W. (2010). Revision of the Mediterranean and Southern African *Triglochin bulbosa* complex (Juncaginaceae). *Edinburgh Journal of Botany*, **67**, 353–398.

Lieu, S. M. (1979). Organogenesis in *Triglochin striata*. *Canadian Journal of Botany*, **57**, 1418–1438.

Lock, I. E., Ashurkova, L. D., Belova, O. A. et al. (2009). A continuum between open and closed inflorescences? Inflorescence tip variation in *Potamogeton* (Potamogetonaceae: Alismatales). *Wulfenia*, **16**, 33–50.

Lodkina, M. M. (1983). Features of morphological evolution in plants conditioned by their ontogenesis. *Journal of General Biology*, **44**, 239–253.

Mayo, S. J., Bogner, J. and Boyce, P. C. (1997). *The Genera of Araceae*. Kew: Royal Botanic Gardens.

Payer J.-B. (1857): *Traité d'organogénie comparée de la fleur*. Paris: Victor Masson.

Penin, A. A. (2008). Bract reduction in Cruciferae: possible genetic mechanisms and evolution. *Wulfenia*, **15**, 63–73.

Posluszny, U. (1981). Unicarpellate floral development in *Potamogeton zosteriformis*. *Canadian Journal of Botany*, **59**, 495–504.

Posluszny, U. (1983). Re-evaluation of certain key relationships in the Alismatidae: floral organogenesis of *Scheuchzeria palustris* (Scheuchzeriaceae). *American Journal of Botany*, **70**, 925–933.

Posluszny, U. and Sattler, R. (1973). Floral development of *Potamogeton densus*. *Canadian Journal of Botany*, **51**, 647–656.

Posluszny, U. and Sattler, R. (1974). Floral development of *Potamogeton richardsonii*. *American Journal of Botany*, **61**, 209–216.

Qiu, Y.-L., Lee, J., Bernasconi-Quadroni, F. et al. (2000). Phylogeny of basal angiosperms: analyses of five genes from three genomes. *International Journal of Plant Sciences*, **161** (Suppl.), S3–S27.

Remizowa, M. V. (2008). *Structure, Development and Evolution of Flowers in Some Primitive Monocots*. PhD Thesis. Moscow: Moscow State University.

Remizowa, M. V. and Sokoloff D. D. (2003). Inflorescence and floral morphology in *Tofieldia* (Tofieldiaceae) compared with Araceae, Acoraceae and Alismatales s. str. *Botanische Jahrbücher*, **124**, 255–271.

Remizowa, M. V., Sokoloff, D. D. and Moskvicheva, L. A. (2005). Morphology and development of flower and shoot system in *Tofieldia pusilla* (Tofieldiaceae). *Botanichesky Zhurnal*, **90**, 840–853.

Remizowa, M. V., Sokoloff, D. D. and Rudall, P. J. (2006). Patterns of floral structure and orientation in *Japonolirion, Nartecium*, and *Tofieldia. Aliso*, **22**, 159–171.

Remizowa, M. V., Sokoloff, D. D., Timonin, A. C. and Rudall, P. J. (2010). Floral vasculature in *Tofieldia* (Tofieldiaceae) is correlated with floral morphology and development. In *Diversity, Phylogeny, and Evolution in the Monocotyledons*, ed. O. Seberg, G. Petersen, A. Barfod and J. I. Davis. Aarhus: Aarhus University Press, pp. 81–99.

Remizowa, M. V., Sokoloff, D. D., Calvo, S., Thomasello, A. and Rudall, P. J. (2012). Flowers and inflorescences of the seagrass *Posidonia* (Posidoniaceae Alismatales). *American Journal of Botany*, **99**, 1592–1608.

Sattler, R. (1965). Perianth development of *Potamogeton richardsonii. American Journal of Botany*, **52**, 35–41.

Sattler, R. (1988). Homeosis in plants. *American Journal of Botany*, **75**, 1606–1617.

Singh, V. and Sattler, R. (1977). Floral development of *Aponogeton natans* and *A. undulatus. Canadian Journal of Botany*, **55**, 1106–1120.

Sokoloff, D. D., Rudall P. J. and Remizowa, M. V. (2006). Flower-like terminal structures in racemose inflorescences: a tool in morphogenetic and evolutionary research. *Journal of Experimental Botany*, **57**, 3517–3530.

Soros-Pottruff, C. L. and Posluszny, U. (1995a). Developmental morphology of reproductive structures of *Zostera* and a reconsideration of *Heterozostera* (Zosteraceae). *International Journal of Plant Sciences*, **156**, 143–158.

Soros-Pottruff, C. L. and Posluszny, U. (1995b). Developmental morphology of reproductive structures of *Phyllospadix* (Zosteraceae). *International Journal of Plant Sciences*, **155**, 405–420.

Sun, K., Zhang, Z.-Y. and Chen, J.-K. (2000). Floral organogenesis of *Potamogeton distinctus* A. Benn. (Potamogetonaceae). *Acta Phytotaxonomica Sinica*, **38**, 528–531.

Takhtajan, A. (2009). *Flowering Plants*, 2nd edn. New York: Springer.

Tomlinson, P. B. (1982). Helobiae (Alismatidae), including the seagrasses. In *Anatomy of Monocotyledons, Volume 3*, ed. C. R. Metcalfe. Oxford: Clarendon Press.

Uhl, N. (1947). *Studies in the Floral Anatomy and Morphology of Certain Members of the Helobiae*. PhD Thesis. Ithaca, New York: Cornell University.

Zomlefer, W. B. (1997). The genera of Tofieldiaceae in the southeastern United States. *Harvard Papers in Botany*, **2**, 179-194.

9

Recent progress in the phylogenetics and classification of Araceae

SIMON J. MAYO, JOSEF BOGNER AND NATALIE CUSIMANO

9.1 Introduction

The aim of this paper is to review progress in phylogenetic research of Araceae during the period since publication of the first major molecular study by French et al. (1995). This, the first cladogram of the whole family inferred from DNA molecular data (Fig 9.1), resulting from research by J.C. French, M. Chung and Y. Hur, was based on chloroplast restriction site data (RFLPs or restriction fragment length polymorphisms). Their paper was highly significant and marked the beginning of the modern era of molecular phylogenetics of Araceae, nowadays based on DNA sequence data (e.g. Cabrera et al., 2008; Cusimano et al., 2011; Nauheimer et al., 2012b). It was innovative for Araceae in other ways as well, being the first family-scale cladistic analysis using computer algorithms and the first published cladogram for the family as a whole using genera as the ultimate operational taxonomic units (OTUs).

No attempt has been made in the present chapter to discuss in detail the work of the previous 25 years during which many significant advances in systematic knowledge of the family were made, both in morphological taxonomy, but also in other fields such as cytology, palynology, phytochemistry, anatomy, fossil aroids, pollination biology and seedling morphology. Reviews of this literature have been provided by various authors, including Petersen (1989), Grayum (1990),

Early Events in Monocot Evolution, eds P. Wilkin and S. J. Mayo. Published by Cambridge University Press. © The Systematics Association 2013.

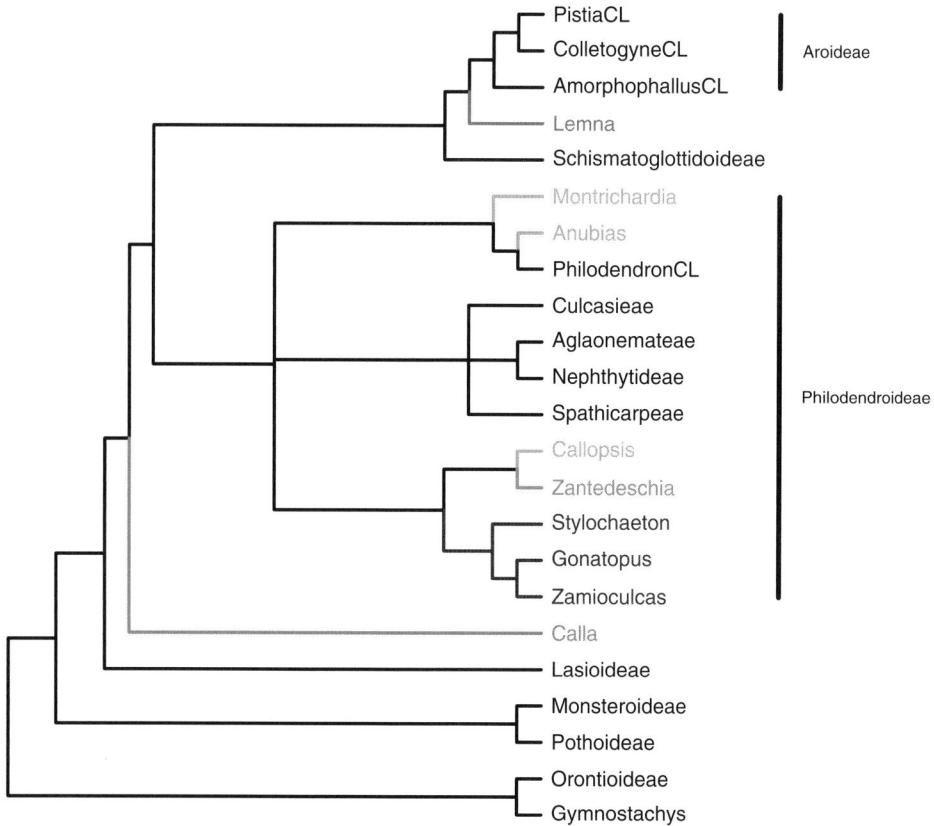

Fig 9.1 Cladogram simplified from French et al. (1995), showing circumscription of subfamilies Aroideae and Philodendroideae as circumscribed by Keating (2003). See Table 9.1 for details of circumscription of suprageneric taxon names. Names ending in '-CL' are abbreviations for '... clade', e.g. PistiaCL ≡ *Pistia* clade. Coloured clades show contrasting positions in Fig 9.2. Colour version to be found in colour plate section.

Mayo et al. (1997) and Keating (2003), and Nicolson (1960) and Croat (1990) summarized classifications published in this earlier period.

From the beginning of the 1980s, J.C. French, initially in collaboration with P.B. Tomlinson, produced a steady stream of anatomical studies (see review and references in French, 1997) that threw doubt on a number of aspects of the then-prevailing Englerian classification as exemplified by the synopsis published by Bogner (1979). A critical watershed then came with the doctoral thesis of M.H. Grayum (Grayum, 1984), a comprehensive survey of aroid systematic literature combined with the first family-wide SEM survey of pollen surface structure, which coincided with the more widespread use of cladistic methods in botanical taxonomy. Grayum's powerfully argued case for the removal of *Acorus* from Araceae (Grayum, 1987) had a wide impact and set the scene for what was to

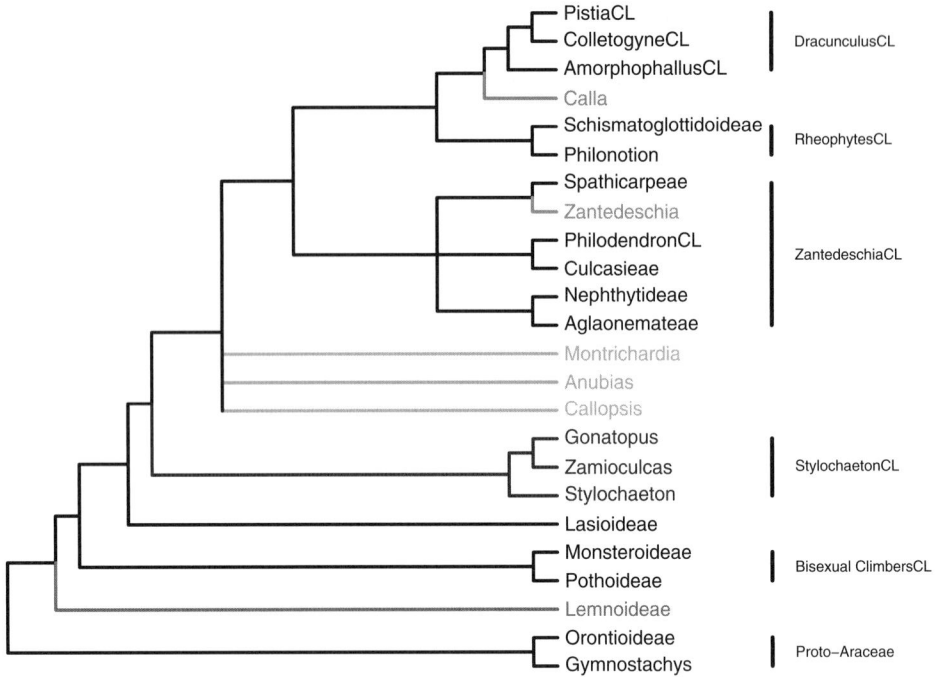

Fig 9.2 Cladogram simplified from Cusimano et al. (2011) showing major changes (coloured clades) from the cladogram in Fig 9.1 (French et al. 1995). See Table 9.1 for details of circumscription of suprageneric taxon names. Names ending in '-CL' are abbreviations for '... clade', e.g. PistiaCL ≡ *Pistia* clade. Colour version to be found in colour plate section.

come. A few years later, with his general papers on the evolution, phylogeny and SEM palynology of Araceae, Grayum (1990, 1992) effectively brought an end to the century-long pre-eminence of the Engler system, proposing a new classification in its place. In the meantime, J. Bogner and D.H. Nicolson had been collaborating over a number of years to update and modify Engler's system, incorporating results from French's anatomical work, and with the additional aim of producing a diagnostic key to all the genera; this work was eventually published in 1991 (Bogner and Nicolson, 1991) and represents the last version of the Englerian classification.

The comparison of new data between the Bogner and Nicolson, and Grayum systems then became a stimulus for new hypotheses by other workers and it was in this context that the cladogram of French et al. (1995) appeared. It presented many features that were substantially different from previous classifications but subsequent work has confirmed many of their novel results. The present review takes in turn the main clades of the French et al. (1995) cladogram (Fig 9.1) and discusses the results of more recent work (Fig 9.2) in relation to each. This shows that the molecular view of Araceae phylogenetics is broadly consistent and that the basis for a new classification exists for which there is a reasonable expectation of stability in general outline, although much detailed work remains to be done and certain genera do not yet seem to have found a stable phylogenetic position.

Fossil Araceae have become an increasingly important focus for research in the past decade and apart from their intrinsic interest as direct evidence of ancient Araceae taxa, have a vital role to play in providing calibration for dating phylogenetic trees. Recent palaeobotanical work is here discussed in the context of the clades to which they have been ascribed. In particular, Nauheimer et al. (2012b) have published a comprehensive treatment of the global history of the family based on fossil dating and a new molecular analysis of all genera.

9.2 A note on names

The various classifications and phylogenetic analyses of recent decades have resulted in different taxon concepts for many suprageneric names. To deal with this, suprageneric names mentioned in the text, unless referenced otherwise, refer to the taxon concepts given here in alphabetical order as Table 9.1; this is based on the list of well-supported clades presented by Cusimano et al. (2011), with additional taxa referenced usually to the last formal synopsis published by Bogner and Petersen (2007). The informal term 'lemnoid' refers to the taxa of duckweeds, i.e. subfamily Lemnoideae (previously Lemnaceae). The informal term 'aroid' is today ambiguous in meaning, since it may refer to all Araceae (i.e. including lemnoids), Araceae minus lemnoids or just the taxa belonging to subfamily Aroideae; where used in this paper, it means Araceae minus lemnoids.

9.3 *Gymnostachys*

This highly distinctive Australian endemic was grouped with *Acorus* by Engler (1920). Bogner and Nicolson (1991) placed it at the beginning of their classification as the only genus of a new subfamily Gymnostachydoideae. Grayum (1990: 669) noted interesting phenotypic similarities between *Gymnostachys* and Orontioideae (assigned by Grayum, 1990: 688 to his subfamily Lasioideae), but despite this he placed the genus in his broadly circumscribed subfamily Pothoideae as the first taxon of his classification. French et al. (1995) found that *Gymnostachys* was indeed closely related to Orontioideae and that this well-supported clade was placed as the most basal branch of their cladogram. Duvall et al. (1993), using plastid *rbc*L sequence data, had previously recorded a sister-group relationship between *Gymnostachys* and *Symplocarpus* (Orontioideae) but the significance of this finding was not then apparent as their analysis included only nine genera of Araceae.

Table 9.1 Clades, clade numbers and clade names based on the circumscriptions of Cusimano et al. (2011).

Taxon	Reference	Clade number	Genera included
Aglaonemateae	Cusimano et al. 2011	9	*Aglaodorum, Aglaonema*
Alocasia clade	Cusimano et al. 2011	30	*Alocasia, Arisaema, Arum, Biarum, Dracunculus, Eminium, Helicodiceros, Lazarum, Pinellia, Sauromatum, Theriophonum, Typhonium*
Ambrosina clade	Cusimano et al. 2011	36	*Alocasia, Ambrosina, Ariopsis, Arisaema, Arisarum, Arophyton, Arum, Biarum, Carlephyton, Colletogyne, Colocasia, Dracunculus, Eminium, Helicodiceros, Lazarum, Peltandra, Pinellia, Pistia, Protarum, Remusatia, Sauromatum, Steudnera, Theriophonum, Typhonium, Typhonodorum*
Amorphophallus clade	Cusimano et al. 2011	35	*Amorphophallus, Caladium, Chlorospatha, Filarum, Hapaline, Jasarum, Pseudodracontium, Scaphispatha, Syngonium, Ulearum, Xanthosoma, Zomicarpa, Zomicarpella*
Anadendreae	Bogner and Petersen 2007		*Anadendrum*
Anchomanes clade	Cusimano et al. 2011	26	*Aglaodorum, Aglaonema, Anchomanes, Nephthytis, Pseudohydrosme*
Araceae	Cusimano et al. 2011	44	*Alloschemone, Alocasia, Ambrosina, Amorphophallus, Amydrium, Anadendrum, Anaphyllopsis, Anaphyllum, Anthurium, Apoballis, Aridarum, Ariopsis, Arisaema, Arisarum, Arophyton, Arum, Bakoa, Biarum, Bucephalandra, Caladium, Calla, Carlephyton, Chlorospatha, Colletogyne, Colocasia, Cryptocoryne, Cyrtosperma, Dracontioides, Dracontium, Dracunculus, Eminium, Epipremnum, Filarum, Gonatopus, Gymnostachys, Hapaline, Helicodiceros, Hestia, Heteropsis, Holochlamys, Jasarum, Lagenandra, Landoltia, Lasia, Lasimorpha, Lazarum, Lemna, Lysichiton, Monstera, Ooia, Orontium, Pedicellarum, Peltandra, Philonotion, Phymatarum, Pichinia, Pinellia, Piptospatha, Pistia, Podolasia, Pothoidium, Pothos,*

			Genera
			Protarum, Pseudodracontium, Remusatia, Rhaphidophora, Rhodospatha, Sauromatum, Scaphispatha, Schismatoglottis, Schottariella, Scindapsus, Spathiphyllum, Spirodela, Stenospermation, Steudnera, Stylochaeton, Symplocarpus, Syngonium, Theriophonum, Typhonium, Typhonodorum, Ulearum, Urospatha, Wolffia, Wolffiella, Xanthosoma, Zamioculcas, Zomicarpa, Zomicarpella
Areae	21	Cusimano et al. 2011	*Arum, Biarum, Dracunculus, Eminium, Helicodiceros, Lazarum, Sauromatum, Theriophonum, Typhonium*
Arisareae	18	Cusimano et al. 2011	*Ambrosina, Arisarum*
Aroideae	39	Cusimano et al. 2011	*Aglaodorum, Aglaonema, Alocasia, Ambrosina, Amorphophallus, Anchomanes, Anubias, Apoballis, Aridarum, Ariopsis, Arisaema, Arisarum, Arophyton, Arum, Asterostigma, Bakoa, Biarum, Bognera, Bucephalandra, Caladium, Calla, Callopsis, Carlephyton, Cercestis, Chlorospatha, Colletogyne, Colocasia, Croatiella, Cryptocoryne, Culcasia, Dieffenbachia, Dracunculus, Eminium, Filarum, Furtadoa, Gearum, Gorgonidium, Hapaline, Helicodiceros, Hestia, Homalomena, Incarum, Jasarum, Lagenandra, Lazarum, Mangonia, Montrichardia, Nephthytis, Ooia, Peltandra, Philodendron, Philonotion, Phymatarum, Pichinia, Pinellia, Piptospatha, Pistia, Protarum, Pseudodracontium, Pseudohydrosme, Remusatia, Sauromatum, Scaphispatha, Schismatoglottis, Schottariella, Spathantheum, Spathicarpa, Steudnera, Synandrospadix, Syngonium, Taccarum, Theriophonum, Typhonium, Typhonodorum, Ulearum, Xanthosoma, Zantedeschia, Zomicarpa, Zomicarpella*
Arophyteae	19	Cusimano et al. 2011	*Arophyton, Carlephyton, Colletogyne*
Bisexual Climbers clade	31	Cusimano et al. 2011	*Alloschemone, Amydrium, Anadendrum, Anthurium, Epipremnum, Heteropsis, Holochlamys, Monstera, Pedicellarum, Pothoidium, Pothos, Rhaphidophora, Rhodospatha, Scindapsus, Spathiphyllum, Stenospermation*

Table 9.1 (*cont.*)

Taxon	Reference	Clade number	Genera included
Caladieae	Cusimano et al. 2011	17	*Caladium, Chlorospatha, Filarum, Hapaline, Jasarum, Scaphispatha, Syngonium, Ulearum, Xanthosoma, Zomicarpa, Zomicarpella*
Colletogyne clade	Cusimano et al. 2011	33	*Ambrosina, Arisarum, Arophyton, Carlephyton, Colletogyne, Peltandra, Typhonodorum*
Colocasia clade	Cusimano et al. 2011	20	*Ariopsis, Colocasia, Remusatia, Steudnera*
Colocasieae	Bogner and Petersen 2007		*Alocasia, Ariopsis, Colocasia, Protarum, Remusatia, Steudnera*
Cryptocoryneae	Cusimano et al. 2011	14	*Cryptocoryne, Lagenandra*
Culcasieae	Cusimano et al. 2011	11	*Cercestis, Culcasia*
Dieffenbachieae	Bogner and Petersen 2007		*Bognera, Dieffenbachia*
Dracunculus clade	Cusimano et al. 2011	37	*Alocasia, Ambrosina, Amorphophallus, Ariopsis, Arisaema, Arisarum, Arophyton, Arum, Biarum, Caladium, Carlephyton, Chlorospatha, Colletogyne, Colocasia, Dracunculus, Eminium, Filarum, Hapaline, Helicodiceros, Jasarum, Lazarum, Peltandra, Pinellia, Pistia, Protarum, Pseudodracontium, Remusatia, Sauromatum, Scaphispatha, Steudnera, Syngonium, Theriophonum, Typhonium, Typhonodorum, Ulearum, Xanthosoma, Zomicarpa, Zomicarpella*
Gymnostachydoideae	Bogner and Petersen 2007		*Gymnostachys*

Heteropsideae	Bogner and Petersen 2007		*Heteropsis*
Heteropsis clade	Cusimano et al. 2011	4	*Alloschemone, Heteropsis, Rhodospatha, Stenospermation*
Homalomena clade	Cusimano et al. 2011	27	*Cercestis, Culcasia, Furtadoa, Homalomena, Philodendron*
Homalomeneae	Bogner and Petersen 2007		*Furtadoa, Homalomena*
Lasioideae	Cusimano et al. 2011	7	*Anaphyllopsis, Anaphyllum, Cyrtosperma, Dracontioides, Dracontium, Lasia, Lasimorpha, Podolasia, Pycnospatha, Urospatha*
Lemnoideae	Cusimano et al. 2011	2	*Landoltia, Lemna, Spirodela, Wolffia, Wolffiella*
Monstereae	Bogner and Petersen 2007		*Alloschemone, Amydrium, Epipremnum, Monstera, Rhaphidophora, Rhodospatha, Scindapsus, Stenospermation*
Monsteroideae	Cusimano et al. 2011	24	*Alloschemone, Amydrium, Anadendrum, Heteropsis, Holochlamys, Epipremnum, Monstera, Rhaphidophora, Rhodospatha, Scindapsus, Spathiphyllum, Stenospermation*
Nephthytideae	Cusimano et al. 2011	10	*Anchomanes, Nephthytis, Pseudohydrosme*
Orontioideae	Cusimano et al. 2011	1	*Lysichiton, Orontium, Symplocarpus*
Peltandreae	Bogner and Petersen 2007		*Peltandra, Typhonodorum*

Table 9.1 (*cont.*)

Taxon	Reference	Clade number	Genera included
Philodendreae	Bogner and Petersen 2007		*Philodendron*
Philodendron clade	Cusimano et al. 2011	12	*Furtadoa, Homalomena, Philodendron*
Philonotion clade	Cusimano et al. 2011	38	*Alocasia, Ambrosina, Amorphophallus, Apoballis, Aridarum, Ariopsis, Arisaema, Arisarum, Arophyton, Arum, Bakoa, Biarum, Bucephalandra, Caladium, Calla, Carlephyton, Chlorospatha, Colletogyne, Colocasia, Cryptocoryne, Dracunculus, Eminium, Filarum, Hapaline, Helicodiceros, Hestia, Jasarum, Lagenandra, Lazarum, Ooia, Peltandra, Philonotion, Phymatarum, Pichinia, Pinellia, Piptospatha, Pistia, Protarum, Pseudodracontium, Remusatia, Sauromatum, Scaphispatha, Schismatoglottis, Schottariella, Steudnera, Syngonium, Theriophonum, Typhonium, Typhonodorum, Ulearum, Xanthosoma, Zomicarpa, Zomicarpella*
Pistia clade	Cusimano et al. 2011	34	*Alocasia, Ariopsis, Arisaema, Arum, Biarum, Colocasia, Dracunculus, Eminium, Helicodiceros, Lazarum, Pinellia, Pistia, Protarum, Remusatia, Sauromatum, Steudnera, Theriophonum, Typhonium*
Podolasia clade	Cusimano et al. 2011	41	*Alocasia, Ambrosina, Amorphophallus, Anaphyllopsis, Anaphyllum, Apoballis, Aridarum, Ariopsis, Arisaema, Arisarum, Arophyton, Arum, Bakoa, Biarum, Bucephalandra, Caladium, Calla, Carlephyton, Chlorospatha, Colletogyne, Colocasia, Cryptocoryne, Cyrtosperma, Dracontioides, Dracontium, Dracunculus, Eminium, Filarum, Gonatopus, Hapaline, Helicodiceros, Hestia, Jasarum, Lagenandra, Lasia, Lasimorpha, Lazarum, Ooia, Peltandra, Philonotion, Phymatarum, Pichinia, Pinellia, Piptospatha, Pistia, Podolasia, Protarum, Pseudodracontium, Pycnospatha, Remusatia, Sauromatum,*

			Scaphispatha, Schismatoglottis, Schottariella, Steudnera, Stylochaeton, Syngonium, Theriophonum, Typhonium, Typhonodorum, Ulearum, Urospatha, Xanthosoma, Zamioculcas, Zomicarpa, Zomicarpella
Potheae	Cusimano et al. 2011	3	*Pedicellarum, Pothoidium, Pothos*
Pothoideae	Cusimano et al. 2011	23	*Anthurium, Pedicellarum, Pothoidium, Pothos*
Proto-Araceae	Cusimano et al. 2011	22	*Gymnostachys, Lysichiton, Orontium, Symplocarpus*
Rhaphidophora clade	Cusimano et al. 2011	6	*Amydrium, Anadendrum, Epipremnum, Monstera, Rhaphidophora, Scindapsus*
Rheophytes clade	Cusimano et al. 2011	28	*Apoballis, Aridarum, Bakoa, Bucephalandra, Cryptocoryne, Hestia, Lagenandra, Ooia, Philonotion, Phymatarum, Pichinia, Piptospatha, Schismatoglottis, Schottariella*
Schismatoglottideae	Cusimano et al. 2011	15	*Apoballis, Aridarum, Bakoa, Bucephalandra, Hestia, Ooia, Phymatarum, Pichinia, Piptospatha, Schismatoglottis, Schottariella*
Spathicarpeae	Cusimano et al. 2011	13	*Asterostigma, Bognera, Croatiella, Dieffenbachia, Gearum, Gorgonidium, Incarum, Mangonia, Spathantheum, Spathicarpa, Synandrospadix, Taccarum*
Spathiphylleae	Cusimano et al. 2011	5	*Holochlamys, Spathiphyllum*
Spirodela clade	Cusimano et al. 2011	43	*Alloschemone, Alocasia, Ambrosina, Amorphophallus, Amydrium, Anadendrum, Anaphyllopsis, Anaphyllum, Anthurium, Apoballis, Aridarum, Ariopsis, Arisaema, Arisarum, Arophyton, Arum, Bakoa, Biarum, Bucephalandra, Caladium, Calla, Carlephyton, Chlorospatha, Colletogyne,*

Table 9.1 (cont.)

Taxon	Reference	Clade number	Genera included
			Colocasia, Cryptocoryne, Cyrtosperma, Dracontioides, Dracontium, Dracunculus, Eminium, Epipremnum, Filarum, Gonatopus, Hapaline, Helicodiceros, Hestia, Heteropsis, Holochlamys, Jasarum, Lagenandra, Landoltia, Lasia, Lasimorpha, Lazarum, Lemna, Monstera, Ooia, Pedicellarum, Peltandra, Philonotion, Phymatarum, Pichinia, Pinellia, Piptospatha, Pistia, Podolasia, Pothoidium, Pothos, Protarum, Pseudodracontium, Pycnospatha, Remusatia, Rhaphidophora, Rhodospatha, Sauromatum, Scaphispatha, Schismatoglottis, Schottariella, Scindapsus, Spathiphyllum, Spirodela, Stenospermation, Steudnera, Stylochaeton, Syngonium, Theriophonum, Typhonium, Typhonodorum, Ulearum, Urospatha, Wolffia, Wolffiella, Xanthosoma, Zamioculcas, Zomicarpa, Zomicarpella
Stylochaeton clade	Cusimano et al. 2011	25	Gonatopus, Stylochaeton, Zamioculcas
Thomsonieae	Cusimano et al. 2011	16	Amorphophallus, Pseudodracontium
True Araceae clade	Cusimano et al. 2011	42	Alloschemone, Alocasia, Ambrosina, Amorphophallus, Amydrium, Anadendrum, Anaphyllopsis, Anaphyllum, Anthurium, Apoballis, Aridarum, Ariopsis, Arisaema, Arisarum, Arophyton, Arum, Bakoa, Biarum, Bucephalandra, Caladium, Calla, Carlephyton, Chlorospatha, Colletogyne, Colocasia, Cryptocoryne, Cyrtosperma, Dracontioides, Dracontium, Dracunculus, Eminium, Epipremnum, Filarum, Gonatopus, Hapaline, Helicodiceros, Hestia, Heteropsis, Holochlamys, Jasarum, Lagenandra, Lasia, Lasimorpha, Lazarum, Monstera, Ooia, Pedicellarum, Peltandra, Philonotion, Phymatarum, Pichinia, Pinellia, Piptospatha, Pistia, Podolasia, Pothoidium, Pothos, Protarum, Pseudodracontium, Remusatia, Rhaphidophora,

Clade	Reference	No.	Genera
			Rhodospatha, Sauromatum, Scaphispatha, Schismatoglottis, Schottariella, Scindapsus, Spathiphyllum, Stenospermation, Steudnera, Stylochaeton, Syngonium, Theriophonum, Typhonodorum, Typhonium, Ulearum, Urospatha, Xanthosoma, Zamioculcas, Zomicarpella
Typhonodorum clade	Cusimano et al. 2011	29	*Arophyton, Carlephyton, Colletogyne, Peltandra, Typhonodorum*
Unisexual Flowers clade	Cusimano et al. 2011	40	*Alocasia, Ambrosina, Amorphophallus, Apoballis, Aridarum, Ariopsis, Arisaema, Arisarum, Arophyton, Arum, Bakoa, Biarum, Bucephalandra, Caladium, Calla, Carlephyton, Chlorospatha, Colletogyne, Colocasia, Cryptocoryne, Dracunculus, Eminium, Filarum, Gonatopus, Hapaline, Helicodiceros, Hestia, Jasarum, Lagenandra, Lazarum, Ooia, Peltandra, Philodendron, Philonotion, Phymatarum, Pichinia, Pinellia, Piptospatha, Pistia, Protarum, Pseudodracontium, Remusatia, Sauromatum, Scaphispatha, Schismatoglottis, Schottariella, Steudnera, Stylochaeton, Syngonium, Theriophonum, Typhonium, Typhonodorum, Ulearum, Xanthosoma, Zamioculcas, Zomicarpa, Zomicarpella*
Zamioculcadoideae	Cusimano et al. 2011	8	*Gonatopus, Zamioculcas*
Zantedeschia clade	Cusimano et al. 2011	32	*Aglaodorum, Aglaonema, Anchomanes, Asterostigma, Bognera, Cercestis, Croatiella, Culcasia, Dieffenbachia, Furtadoa, Gearum, Gorgonidium, Homalomena, Incarum, Mangonia, Nephthytis, Philodendron, Pseudohydrosme, Spathantheum, Spathicarpa, Synandrospadix, Taccarum, Zantedeschia*
Zantedeschieae	Bogner and Petersen 2007		*Zantedeschia*

As noted by French et al. (1995), their demonstration of the cladistic association of *Gymnostachys* with Orontioideae was novel. Prior to their study the three genera of Orontioideae had a varied taxonomic history. Engler (1887–1889, 1920) grouped them with *Calla* in his subfamily Calloideae, but Grayum (1990), and Bogner and Nicolson (1991) grouped them instead in their respective circumscriptions of subfamily Lasioideae. Mayo et al. (1997), having also obtained the pairing of *Gymnostachys* with the Orontioideae in a morphological cladistic analysis, named the clade informally as 'proto-Araceae', but nevertheless recognized Gymnostachydoideae and Orontioideae as distinct subfamilies. Rothwell et al. (2004) found a sister-group relation between *Gymnostachys* and *Symplocarpus* in their analysis of 27 genera of Araceae. Tam et al. (2004), in a study of the DNA sequences of the plastid *trnL-F* spacer/intron region of 42 Araceae genera, confirmed that *Gymnostachys* and all three genera of Orontioideae formed a robust clade and advocated formal recognition of proto-Araceae as a subfamily. Cabrera et al. (2008), in a comprehensive and detailed study of 102 of the 110 genera then accepted, used sequences from five regions of coding and noncoding plastid DNA and their result provided further confirmation, as did Cusimano et al. (2011) more recently.

9.4 Orontioideae

This taxon has been the object of several specific phylogenetic studies in addition to those already mentioned in the wider context of family phylogenetics. An early cladistic paper was by Barabé and Forget (1987), who used morphological characters to show that the orontioids and *Calla* did not form a natural group when placed together in the Englerian subfamily Calloideae. Wen et al. (1996) reported on the biogeography of the genus *Symplocarpus*, using plastid DNA restriction site data, and this was followed by papers by Kitano et al. (2005) on *Symplocarpus* in Japan using plastid DNA sequences and Nie et al. (2006) on the Orontioideae as a whole. The last paper described a study based on DNA sequences from *trnL-F* and *ndh*F regions of the plastid genome which sampled six species of Orontioideae and eight other genera of Araceae, including *Gymnostachys* and *Calla*. In their cladograms, *Gymnostachys* grouped with the orontioids as found by other authors of molecular studies, whereas *Calla* paired with *Philodendron* within the Unisexual Flowers clade (Table 9.1).

Nie et al. (2006) made use of the recent discovery of a Cretaceous orontioid fossil, *Albertarum pueri* (Bogner et al., 2005) to date their tree. *Albertarum pueri* is a remarkable fossil infructescence from Late Cretaceous deposits of Alberta, Canada which Bogner et al. (2005) placed near *Symplocarpus* although it is distinct within the Orontioideae due to its trilocular ovary, uniovulate locules and seeds with a ribbed testa. Leaf fossils of Orontioideae were reported by Bogner et al. (2007), comprising four species. The three North American species, from the Late Cretaceous and Paleogene, are new (*Orontium wolfei, O. mackii, Symplocarpus hoffmaniae*). *Lysichiton austriacus*, from the Late Cretaceous of Austria, was transferred from the fossil genus *Araciphyllites*.

Nauheimer et al. (2012b) report the existence of an orontioid fossil leaf from Ceará in Northeast Brazil.

9.5 Lemnoideae

The most recent cladistic analyses of Araceae (Cabrera et al., 2008; Cusimano et al., 2011; Nauheimer et al., 2012b) placed the five genera of Lemnoideae (*Spirodela, Lemna, Landoltia, Wolffia* and *Wolffiella*) near the base of the cladogram just above the proto-Araceae, but prior to these studies lemnoid relationships were not clear.

French et al. (1995) included a single species of Lemnoideae (duckweeds) in their analysis and found that it grouped within the unisexual Araceae — but not close to *Pistia* — as sister to Keating's (2003, 2004) subfamily Aroideae (curiously enough the same position occupied by *Calla* in the cladograms of Cabrera et al., 2008 and Cusimano et al., 2011). Duvall et al.'s (1993) early *rbc*L analysis also placed *Lemna* high up in the Araceae clade, but their sample of taxa was small (nine Araceae genera).

According to Mayo et al. (1995, 1997), the weight of evidence favoured inclusion of the lemnoid genera within Araceae, but these authors did not do so in their classification. Keating (2003), on the other hand, basing his classification on the cladogram of French et al. (1995), formally recognized subfamily Lemnoideae. Barabé et al. (2002), in a phylogenetic analysis of 34 Araceae genera including one lemnoid species (*Lemna* sp.) and using the plastid *trn*L and *trn*L-*F* DNA regions, were the first authors to place the lemnoids in their currently recognized near-basal position.

Because of a primary focus on ingroup topology, this result did not emerge in Les et al.'s (2002) comprehensive phylogenetic analysis of Lemnoideae (as Lemnaceae), which included all 38 accepted species and used five plastid genes (*rbc*L, *mat*K, 5′ *trn*K spacer, 3′ *trn*K spacer, *rpl16* intron). Having reviewed the existing evidence, these authors concluded that the most probable position for Lemnaceae was sister to unisexual Araceae. This formed the basis for their choice of *Pistia* and seven other unspecified Araceae genera as outgroups.

The topology for lemnoids found by Barabé et al. (2002) emerged again in the study by Tam et al. (2004), which placed the single lemnoid taxon sampled (*Lemna* sp.) above the four genera of proto-Araceae and sister to the 37 sampled genera of the True Araceae clade (*sensu* Mayo et al. 1997: 70).

Rothwell et al. (2004), in an analysis of the *trnL-trnF* region of the plastid genome in 27 Araceae genera, showed the five lemnoid genera in a quite distinct position from *Pistia*, the latter associated with genera from Keating's (2003) subfamily Aroideae. This study built on an earlier paper by Stockey et al. (1997) on the Late Cretaceous lemnoid fossil genus *Limnobiophyllum*, in which, by contrast, *Pistia* had emerged as sister to the lemnoids in a morphological cladistic analysis of 11 Araceae genera. Kvaček (1995) had previously highlighted the importance of *Limnobiophyllum* for the relationship of Lemnoideae and Araceae. Rothwell et al. (2004) concluded that the floating habit had evolved twice independently in extant Araceae and if the fossil record were also considered then the floating habit would have evolved within the Araceae clade perhaps three times independently since the Cretaceous. Stockey et al. (2007) reinforced this by demonstrating that the Late Cretaceous fossil *Cobbania corrugata* may be related to Araceae, although its position within the family is unclear.

To settle the question of lemnoid relationships a general analysis of Araceae as a whole was needed, and this was presented by Cabrera et al. (2008). Unlike previous workers, these authors analysed not only all lemnoid genera but also most of the aroid genera (93%) recognized at that time as well. Their result confirmed the pattern reported by Barabé et al. (2002), but with a wider range of genes and a near-comprehensive generic sampling. Cusimano et al.'s (2011) analysis of an augmented version of the same molecular matrix added further support for the conclusion that lemnoids are an early (but not the earliest) branch of the Araceae clade. Bogner (2009) has provided a well-illustrated survey of the fossil and extant floating Araceae.

9.6 Pothoideae and Monsteroideae

French et al.'s (1995) cladogram grouped together into a single clade the genera of subfamilies Pothoideae and Monsteroideae, confirming the views of Hotta (1970) and Keating (2003), who grouped them together formally as the single subfamily Pothoideae. Carvell's detailed study of the floral anatomy of Pothoideae and Monsteroideae also supported this grouping (Carvell 1989). Grayum's (1990) cladogram and classification did not associate these two groups exclusively, but presented a wider concept of the Pothoideae, which also included Zamioculcadeae (our Zamioculcadoideae) and *Gymnostachys*. Bogner and Nicolson (1991) reduced Engler's Pothoideae to consist only of the three genera of tribe Potheae and recognized subfamily Monsteroideae as distinct with the four tribes Monstereae, Spathiphylleae, Anadendreae and Heteropsideae, the latter two transferred from Engler's Pothoideae. Mayo et al. (1997), and Bogner and Petersen (2007) maintained this position with the difference that they included *Anthurium* in Pothoideae.

In all later molecular analyses (Barabé et al., 2002; Tam et al., 2004; Cabrera et al., 2008; Cusimano et al., 2011) there is a single clade consisting of two major subclades corresponding respectively to Pothoideae (*Pothos, Pothoidium, Pedicellarum, Anthurium*) and Monsteroideae (*Alloschemone, Amydrium, Anadendrum, Epipremnum, Heteropsis, Holochlamys, Monstera, Rhaphidophora, Rhodospatha, Scindapsus, Spathiphyllum, Stenospermation*).

The latter three studies, which have a complete representation of the monsteroid genera, reveal a new topology in which *Heteropsis* and *Anadendrum* are embedded in two different well-supported clades, but the position of tribe Spathiphylleae is less clear. Tam et al. (2004) first showed that the genera *Rhodospatha, Stenospermation, Alloschemone* and *Heteropsis* formed a clade, and this was confirmed with better support by Cabrera et al. (2008) and Cusimano et al. (2011). The last two studies also showed equally strong support for a clade formed by six other genera (*Monstera, Epipremnum, Scindapsus, Rhaphidophora, Anadendrum, Amydrium*), but in Tam et al.'s (2004) cladogram, with a much larger sampling of monsteroid species, this clade had no support greater than 50%.

The ambiguous position of Spathiphylleae vis-à-vis these other two internal clades of Monsteroideae (*Heteropsis* clade and *Rhaphidophora* clade) is interesting. Although Spathiphylleae have been regarded as phenotypically plesiomorphic in many respects (Grayum, 1990: 670), there is as yet little molecular support for its sister-group position to a clade of the other monsteroid genera, a topology which their morphology would suggest. Both Spathiphylleae and the *Rhaphidophora* clade have remarkable amphi-Pacific geographical distributions (Grayum, 1990: 670–671; Mayo, 1993). Tarasevich (1988) discovered that the striate pollen of *Spathiphyllum*, hitherto thought to be inaperturate, was actually multiaperturate. Carvell (1989) discovered vestigial tepal-like structures in the flowers of species of *Alloschemone, Monstera, Rhaphidophora* and *Rhodospatha*, thus weakening still further the morphological distinction between these three clades.

At the population level, I.M. Andrade and colleagues studied species of *Anthurium* (Pothoideae) and *Monstera* (Monsteroideae) in northeastern Brazil (Andrade et al., 2007, 2009; Andrade and Mayo, 2010) using AFLP molecular markers to gather evidence of population and species differentiation in isolated humid forest fragments.

Fossil Monsteroideae have been studied by various workers. An early Cretaceous pollen fossil from Portugal, *Mayoa portugallica,* was discovered and identified as a member of the tribe Spathiphylleae by Friis et al. (2004), but doubt has been thrown on this assignment more recently by Hofmann and Zetter (2010). The tribe Monstereae is well established from the Late Cretaceous (Santonian?, Campanian to Maastrichtian) of Portugal by an *Epipremnum*-like inflorescence with stamens and pollen *in situ*, as well as seeds (E.-M. Friis, pers. comm.). A viny axis, *Rhodospathodendron tomlinsonii*, from the Late Cretaceous (Late Maastrichtian) was described from India and assigned to Monstereae by Bonde (2000). Hesse and Zetter (2007) reinterpreted certain *Ephedripites* forms from the Late Cretaceous and Palaeogene as *Spathiphyllum (S. vanegensis, S. elsikii)*. A zona-aperturate pollen fossil of *Monstera* or *Gonatopus* type has been reported from the Lower Eocene of Austria (Zetter et al., 2001; Hesse and Zetter, 2007), similar to the fossil *Proxapertites operculatus*. Wilde et al. (2005) described the leaf fossil *Araciphyllites tertiarius* from the Middle Eocene Messel Formation of Germany and ascribed it to the tribe Monstereae. More recent are fossil seeds of *Epipremnites* and *Scindapsites* of tribe Monstereae (Gregor and Bogner, 1984, 1989) from European Tertiary deposits. *Teichosperma spadiciflorum* (Upper Eocene to Lower Oligocene) is known from infructescences, fruits and seeds resembling those of *Epipremnum* in the Monstereae (Renner, 1907; Kräusel and Stromer, 1924, Tiffney, pers. comm.).

In the Pothoideae, *Anthurium* has been the object of recent molecular studies by L. Temponi (2006) and M. Carlsen (pers. comm.). Temponi's study focussed mainly on Brazilian taxa and indicated that most Brazilian species belonged to a single clade which appears to have evolved independently of the rest of the genus in the eastern Atlantic forest of South America. Carlsen's study is a broader survey of this, the largest genus of Araceae (pers. comm.).

Herrera et al. (2008) described the fossil genus *Petrocardium* with two species from the Paleocene of Colombia and showed that this taxon is more similar in its leaf morphology to *Anthurium* than to any other genus. Wilde et al. (2005) emended the circumscription of the fossil leaf genus *Araceophyllum*, based on *A. engleri* Kräusel from the Neogene of Sumatra, and placed it in the tribe Potheae because of the leaf venation type.

9.7 Lasioideae

Lasioideae have been a problematic group since their first description by Engler (1876), and it was only with French et al.'s (1995) molecular cladogram that a clearer concept of this taxon was achieved. It was Grayum (1990: 672–673) who first clearly targeted the artificiality of Engler's Lasioideae, aided by new data on pollen structure and anatomy. In his original concept of the subfamily, Engler relied on vegetative characters to group the genera and in doing so brought together taxa with diverse reproductive structures, as discussed by Grayum (1990: 672–673). It seems likely that Engler was misled by a mistaken belief that the bisexual-flowered tribe Lasieae ($\equiv$ our Lasioideae) possessed laticifers (Mayo and Bogner, Chapter 10, this volume), a question that was only settled in recent times by French (1988) and Keating (2003).

Although he removed the genera of Engler's unisexual-flowered tribes Nephthytideae, Amorphophalleae (= Thomsonieae) and Montrichardieae to other subfamilies, Grayum (1990) introduced the three genera of Orontioideae and unisexual-flowered *Stylochaeton* into his Lasioideae, which makes its circumscription different from the current one. Bogner and Nicolson (1991) followed Grayum by including the Orontioideae and removing the genera of Thomsonieae, but they included bisexual-flowered *Anthurium* and various unisexual-flowered genera: *Montrichardia*, *Cercestis*, *Culcasia*, *Nephthytis*, *Anchomanes*, *Pseudohydrosme*, *Callopsis* and the Zamioculcadeae (= Zamioculcadoideae). In contrast, the morphological cladistic analysis of Mayo et al. (1997) obtained a result similar to that of French et al. (1995) in which the genera of the bisexual-flowered clade tribe Lasieae formed a group distinct from all the others with which they had previously been associated. The agreement of the two analyses on this point led Mayo et al. (1997) to formally recognize what is now the currently accepted circumscription for Lasioideae, corresponding to the Englerian tribe Lasieae (Engler 1920). The systematics of this group were studied in detail by Hay (1986, 1988, 1992). Hay and Mabberley (1991) published a wide-ranging essay on the phenotypic evolution of this group that included consideration of homoeotic saltation, a topic later taken up by Barabé et al. (2008).

Keating (2003, 2004), basing his classification on the molecular analysis by French et al. (1995), proposed the same concept of Lasioideae, and all subsequent molecular analyses have upheld its distinctness (Barabé et al., 2002; Tam et al., 2004; Cabrera et al., 2008; Cusimano et al., 2011). However, in the latest molecular analyses, the relationships of the Lasioideae clade to the rest of the family are still not secure. The studies of Cabrera et al. (2008) and Cusimano et al. (2011) agreed in placing Lasioideae above the Pothoideae–Monsteroideae clade, but the interrelationships of the main unisexual-flowered clade (Aroideae) to Lasioideae and the *Stylochaeton* clade are not well supported. Cabrera et al. (2008) showed two different topologies for these three clades, and in Cusimano et al. (2011) the position for Lasioideae below the *Stylochaeton* clade and Aroideae is only weakly supported by molecular data.

Lasioideae pollen fossils from the Upper Cretaceous Timerdyakh Formation in Vilui Basin (Siberia, Russia) were recently described by Hofmann and Zetter (2010) as *Lasioideaecidites hessei* and *L. bogneri* and represent the oldest evidence for the subfamily so far discovered. Smith and Stockey (2003) presented a detailed description of *Keratosperma allenbyense*, the seeds of a fossil genus of Lasioideae from the Middle Eocene of British Columbia in Canada. This study, based on earlier work by Cevallos-Ferriz and Stockey (1988), confirmed that *Keratosperma* is a distinct lasioid genus. More recent palaeobotanical evidence for Lasioideae is provided by the seed fossil taxon *Urospathites* from Eurasian mid-Tertiary (Gregor and Bogner, 1984, 1989).

9.8 Monoecious aroids form a monophyletic lineage

Perhaps the most striking of all the results presented by French et al.'s (1995) cladogram was the grouping together into a single large clade of all the unisexual-flowered genera. In proposing his phylogenetic classification of Araceae, Engler (1876, 1920) had explicitly rejected as artificial the primary division of the family into bisexual-flowered and unisexual-flowered taxa put forward by his predecessor H.W. Schott (1860), emphasizing instead patterns of vegetative anatomy and morphology. One result of this was the formation of two subfamilies, Pothoideae and Lasioideae, in which bisexual- and unisexual-flowered genera were brought together and within which Engler envisaged parallel evolution from the former to the latter condition. This unparsimonious idea persisted in most Araceae classifications (e.g. Grayum, 1990; Bogner and Nicolson, 1991; Hay and Mabberley, 1991) until the advent of molecular systematics.

Mayo et al.'s (1997) morphological analysis obtained a similar result to French et al. (1995) in this respect and they consequently formalized the unisexual clade as a new and much broader concept of subfamily Aroideae, since it represented a major new classificatory feature supported by congruence of independent data sets. Keating (2003), however, argued that the four distinct subclades within this large group shown by French et al.'s cladogram were supported by anatomical data and on this basis he recognized a formal division of the unisexual clade into subfamilies Philodendroideae, Schismatoglottidoideae, Lemnoideae and Aroideae.

Leaving aside the question of the circumscription of subfamilies, subsequent molecular analyses have in general tended to group the unisexual-flowered genera together. Barabé et al. (2002) grouped those they sampled into one subclade of a trichotomy with unisexual-flowered *Zamioculcas* and the bisexual-flowered Lasioideae. Tam et al. (2004) resolved the position of the Zamioculcadoideae as sister to a clade containing all other unisexual-flowered genera they sampled.

As discussed in the previous section, in the comprehensive analyses by Cabrera et al. (2008) and Cusimano et al. (2011) the validity of a single origin for the unisexual-flowered genera remains equivocal due to the lack of strong support for the sister-group relationship of the *Stylochaeton* clade to the other unisexual genera. This still leaves open the possibility that Lasioideae might be the sister group for the main Araceae clade and that unisexuality evolved independently in the *Stylochaeton* clade. Cusimano et al. (2011) provide strong support for the exclusion of the *Stylochaeton* clade from Aroideae, which is noteworthy since these three genera were included in the philodendroid clade by French et al. (1995) and Keating (2003, 2004).

A further complicating factor in the concept of the bisexual–unisexual transition as a strict synapomorphy in Araceae is provided by *Calla*, which in the most recent and most complete molecular analyses (Cabrera et al., 2008; Cusimano et al., 2011) is embedded within the Aroideae. *Calla* flowers have bisexual morphology (although the upper part of the spadix is functionally male), and there are other important characters (particularly of the pollen) discussed in more detail by Cabrera et al. (2008) and Cusimano et al. (2011), which make its position within the unisexual Aroideae clade seemingly anomalous.

Friis et al. (2010) report fragments of an inflorescence interpreted as bearing staminate, naked flowers from the Early Cretaceous of Portugal ('Araceae fossil sp. A') noting that this fossil has aperturate, semi-tectate reticulate pollen. Other fossils that have been tentatively ascribed to the monoecious (unisexual-flowered) Araceae clade include *Cobbania corrugata* (see below) and *Cobbanicarpites amurensis* (Krassilov and Kodrul, 2009).

9.9 The *Stylochaeton* clade: Zamioculcadoideae + *Stylochaeton*

Engler's tribe Zamioculcadeae (*Zamioculcas, Gonatopus*) was raised to subfamily rank as Zamioculcadoideae by Bogner and Hesse (2005) and is currently treated as such by Bogner and Petersen (2007). The association of this group with *Stylochaeton* was yet another innovation of the French et al. (1995) analysis. Previous classifications had kept them separate. Grayum (1990) and Hay and Mabberley (1991) kept Zamioculcadeae in subfamily Pothoideae, where Engler had originally placed them, while *Stylochaeton* was transferred to the Lasioideae. Bogner and Nicolson (1991) on the other hand moved Zamioculcadeae into the Lasioideae, while keeping *Stylochaeton* in its own tribe in the Aroideae, as had Engler (1920).

French et al.'s (1995) analysis grouped the three genera as a distinct clade within the large unisexual Aroideae clade, with *Stylochaeton* as sister to the two zamio-culcad genera. Mayo et al.'s (1997) morphological analysis placed Zamioculcadeae and then *Stylochaeton* as successive sister groups to Aroideae. The plesiomorphies of absence of laticifers and presence of a perigon in these three genera seemed to qualify them uniquely as transitional genera at the base of the unisexual Aroideae in a paraphyletic arrangement, and these authors dubbed them informally as the 'perigoniate Aroideae'. Keating (2003, 2004) maintained them as separate but adjacent tribes in his subfamily Philodendroideae following French et al.'s (1995) topology, pointing out the distinct morphology and anatomy of the two taxa.

Hesse et al. (2001) focussed on the palynology of the three genera of 'perigoniate Aroideae' and showed, among other things, that both Zamioculcadeae and *Stylochaeton* differed significantly from Aroideae in pollen structure, but also that each had peculiarities that mirrored their differences in macro-morphology. In a later paper Bogner and Hesse (2005) separated the zamioculcads as a subfamily and recommended that *Stylochaeton* be regarded as the basal element of the Aroideae clade, by virtue of its transitional pollen and floral characters.

Neither Barabé et al. (2002) nor Tam et al. (2004) sampled both taxa, and it was not until Cabrera et al.'s (2008) analysis that French et al.'s (1995) result was examined again with molecular data. Their result was unequivocal support for a monophyletic 'perigoniate Aroideae' and these authors argued that subfamily Zamioculcadoideae should be expanded to include *Stylochaeton*, despite the clear phenotypic divergence. Cusimano et al.'s (2011, *Stylochaeton* clade) results lent further support to this view.

As noted by Cabrera et al. (2008), these three genera are African endemics and geophytes, each exhibiting peculiar morphology. The difficulties thus posed for phylogenetic interpretation based on the phenotype suggest that many pieces of this particular evolutionary jigsaw puzzle became extinct during the Tertiary aridification of Africa (Raven and Axelrod, 1974). As with the lemnoid genera, this is probably a case where molecular evidence has special value in providing insights.

9.10 Progress in establishing the internal structure of subfamily Aroideae

One of the great strengths of Engler's classification was its subfamily structure, i.e. the reduction of the complexity of Schott's many tribes into a small number of larger groups, most of which seemed to have a more-or-less recognizable 'facies'. In the words of Mayo et al. (1997: 73):

> In a rough and ready manner subsequent specialists of the family learned to recognize subfamily *Pothoideae* by their complete lack of laticifers, subfamily *Monsteroideae* by their trichosclereids and mostly aperigoniate bisexual flowers, subfamily *Calloideae* by their temperate Northern hemisphere distribution and preference for swampy habitats, subfamily *Lasioideae* by their frequent possession of deeply sagittate or dracontioid leaves, subfamily *Philodendroideae* by their unisexual flowers and parallel-pinnate leaf venation, subfamily *Colocasioideae* by their unisexual flowers, anastomosing laticifers and special type of leaf venation ('colocasioid' ...) and subfamily *Aroideae* by their unisexual flowers, mostly geophytic habit and frequent possession of a smooth terminal spadix appendix.

All subsequent authors have maintained the classification of Araceae in subfamilies, and there is no doubt that this facilitates conceptual thinking about the classification. The formalization of the large unisexual clade as the subfamily Aroideae by Mayo et al. (1997) violated this tradition, since the Aroideae consequently contained well over half the genera. However, these authors were led to this conclusion not only because of the striking and well-supported simplification thus introduced into the family classification and phylogenetic structure, but also because their morpho-anatomical analyses showed an inconsistent internal topology for Aroideae. Their cladistic results did not support the circumscription of any of the unisexual subfamilies of Engler (1920), Bogner and Nicolson (1991), Grayum (1990) or Hay and Mabberley (1991). Many tribal circumscriptions recognized by these authors were upheld, but the intertribal relationships advocated previously were mostly not supported. Mayo et al. (1997), seeking congruence between molecular and morphological patterns as the preferred basis for formal circumscriptions of supra-generic taxa, felt compelled to leave the classification of their large subfamily Aroideae largely unresolved above tribal rank: 'The internal topology of our subfamily *Aroideae* concept remains largely unresolved above the tribal level and this is a problem to which future phylogenetic studies should be devoted.' (Mayo et al., 1997: 70).

In contrast, as is now easier to see, French et al.'s (1995) large unisexual clade (Aroideae of Mayo et al., 1997) revealed three major clades in its internal topology if the positions of Lemnoideae, Zamioculcadoideae and *Stylochaeton* are ignored (see previous sections). These have turned out to be the foundation for a more consistent view of this part of the family's phylogenetic relationships. Keating (2003) presented a major new study of vegetative anatomical data that fitted this topology and on this basis he formalized the three clades as subfamilies Philodendroideae, Schismatoglottidoideae and Aroideae. This structure has been largely confirmed by later studies, which together represent substantial progress towards a better understanding of the phylogeny of this complex group.

The studies of Cabrera et al. (2008) and Cusimano et al. (2011), based largely on the same data but using somewhat different methods of analysis, both yielded the same three main aroid clades, if *Calla*, *Callopsis*, *Montrichardia* and *Anubias* are ignored (see below).

9.10.1 The *Zantedeschia* clade ('philodendroids')

This clade, still the least robust of the three major unisexual subclades, brings together the genera of tribes Aglaonemateae, Culcasieae, Dieffenbachieae, Homalomeneae, Nephthytideae, Philodendreae, Spathicarpeae and Zantedeschieae, as circumscribed in Bogner and Petersen (2007). It largely mirrors the earlier Philodendroideae clade of French et al. (1995) and Keating (2003). French et al.'s (1995) cladogram did not present bootstrap support, and even cautious recognition of the clade had to be tempered at that time by the lack of agreement with Mayo et al.'s (1997) morphological analyses and inconsistency of subsequent sequence analyses. In the earlier DNA sequence studies of Barabé et al. (2002) and Tam et al. (2004) this clade did not emerge, instead the various components sampled by them formed polytomies. It was not until the comprehensive analyses by Cabrera et al. (2008) and Cusimano et al. (2011) that this clade was revealed as a well-supported unit.

The most important phylogenetic studies undertaken within this clade to date are those of Gonçalves (2002) and Gonçalves et al. (2007) on Spathicarpeae, and Gauthier et al. (2008) on *Philodendron*. French et al.'s (1995) molecular cladogram resuscitated a relationship between *Dieffenbachia* and the genera of the Spathi-carpeae (in the older, narrower sense of Bogner and Petersen, 2007), which had been proposed much earlier by Schott (1860) on the basis of similarity in floral characters. Eduardo Gonçalves (Gonçalves, 2002; Gonçalves et al., 2007) tested this with a comprehensive analysis of the genera of Spathicarpeae that included the rare genera *Mangonia*, *Bognera* and *Gearum*. He used as outgroups genera (*Aglaonema*, *Cercestis*) representing the two sister groups present in French et al.'s polytomous topology and confirmed that tribe Spathicarpeae should be augmented to include both *Dieffenbachia* and *Bognera*, using both plastid DNA sequences (*mat*K, *trnL-F*) and phenotypic data.

Gauthier et al. (2008) focussed on an investigation of the phylogeny of the large Neotropical genus *Philodendron*, using DNA sequences of two nuclear genes (ITS, ETS) and one plastid intron (*rpl*16) and a sample of 72 species of *Philodendron*, nine species of *Homalomena* and two outgroup genera, *Anchomanes* and *Culcasia*. Their choice of outgroups was guided by the earlier molecular study of Barabé et al. (2002), but is even better justified in the more recent cladogram of Cusimano et al. (2011). Their results showed, amongst other things, that the subgenera of *Philodendron* are monophyletic and confirmed the close relationship between (at least the Neotropical species of) *Homalomena* and *Philodendron*, which had been recognized since Engler (1920) and highlighted by French et al. (1995) and unpub-lished molecular studies by Cassia Sakuragui (1998 and pers. comm.).

Studies of population genetics and fingerprinting have been carried out in this clade. Cuartas-Hernández and Núñez-Farfán (2006) studied genetic variability in fragmented populations of *Dieffenbachia seguine* in Mexico using allozymes. Molecular fingerprinting of horticultural cultivars has been studied in several genera, which has provided important information for future studies of wild populations: *Aglaonema* (Chen et al., 2004a), *Dieffenbachia* (Chen et al., 2001) and *Philodendron* (Devanand et al., 2004).

Wilde et al. (2005) described the leaf fossil *Araciphyllites schaarschmidtii* and ascribed it to either tribe Homalomeneae (*Philodendron* clade) or tribe Aglaonemateae, i.e. from the *Zantedeschia* clade.

9.10.2 The rheophytes clade

French et al. (1995) found a close sister-group relationship between the two tribes Schismatoglottideae and Cryptocoryneae, which had previously been placed widely apart in different subfamilies by Engler (1920), Grayum (1990), Bogner and Nicolson (1991) and Hay and Mabberley (1991). Mayo et al. (1997) recognized this group informally as the *Schismatoglottis* alliance, based on French et al.'s result, and later Keating formalized it as subfamily Schismatoglottidoideae. The more recent global analyses of Cabrera et al. (2008) and Cusimano et al. (2011) have confirmed this clade as robust; the latter authors have provided it with the informal name 'rheophytes clade' now that the genus *Philonotion* has been resurrected (see below).

Taxa of the rheophytes clade have long been an especially interesting focus of taxonomic attention. The work of N. Jacobsen and colleagues (e.g. Jacobsen, 1977; Othman et al., 2009; Bastmeijer et al., 2010) has made the tribe Cryptocoryneae cytologically and taxonomically one of the best-studied genera of Araceae. Othman (1997) carried out a detailed molecular study of 25 species of *Cryptocoryne*, using *Lagenandra* as outgroup. This study is notable for its wide-ranging methodological approach, which tested a variety of molecular markers, including nuclear (ITS), plastid (RFLP and sequenced *trn*K and *mat*K) and RAPDs. This study provided insights into the historical biogeography of the species. Ipor et al. (2010) used DNA fingerprinting with M13 universal primer to investigate hybrids in *Cryptocoryne purpurea*.

Peter Boyce, S.Y. Wong and colleagues are currently carrying out wide-ranging phylogenetic, taxonomic and ecological studies of the Schismatoglottideae (e.g. Boyce and Wong, 2008, 2009; Wong and Boyce, 2010a, 2010b, 2010c). Wong et al. (2010) published a major phylogenetic study of the whole clade with 77 taxa, using plastid markers (*mat*K, *trn*K, *trnL-F*) and the genera *Aglaonema, Anchomanes, Dieffenbachia, Hapaline* and *Homalomena* as outgroups. This revealed an unexpected phylogenetic insight in that the American species previously regarded as *Schismatoglottis americana* came out as sister to the two tribes combined. On this basis these authors resurrected the Schottian genus *Philonotion* and described a new tribe Philonotieae. The new phylogenetic tree also showed that the concept of an amphi-Pacific distribution for *Schismatoglottis* (e.g. Mayo, 1993) no longer held and sheds a sceptical light on the comparable distributions of *Homalomena* and the Spathiphylleae, the former now under active investigation by P. Boyce (pers. comm.).

9.10.3 The *Dracunculus* clade (Aroideae *sensu* Keating 2002)

This clade was first identified by French et al.'s (1995) RFLP study and has recently been confirmed by the DNA sequence analyses of Cabrera et al. (2008) and Cusimano et al. (2011). The earlier and less comprehensive sequence studies of Barabé et al. (2002) and Tam et al. (2004) each showed a clade that is compatible with this group.

In comparison to earlier classifications, the key new features of this part of the French et al. topology were: (1) a clade (*Amorphophallus* clade) consisting of the

Thomsonieae and Caladieae, the latter augmented by its merging with the Zomicarpeae, (2) a clade (*Colletogyne* clade) consisting of the pairing of *Ambrosina* and *Arisarum* and their association with a subclade consisting of *Peltandra*, *Typhonodorum* and the endemic Madagascan tribe Arophyteae and (3) a clade (*Pistia* clade) in which *Pistia* and *Protarum* attach to the basal nodes of a subclade grouping the *Colocasia* clade (including *Ariopsis*), *Alocasia* (a separate branch), *Arisaema*, *Pinellia* and the Areae. This radically different pattern represented a new understanding of the relationships of the most derived groups of the family. Previously Thomsonieae had not been associated with Caladieae, and the latter had been grouped with Colocasieae (in the sense of Bogner and Petersen, 2007) by almost all authors as subfamily Colocasioideae. The positions of *Pistia*, *Ambrosina* and *Protarum* had never been clear. *Pistia*, in particular, because of uniquely reduced floral structures and unusual vegetative morphology, had been treated by most previous authors as a separate tribe or subfamily. This was further reinforced by H.-D. Behnke's detailed survey of Araceae sieve element plastid ultrastructure (Behnke, 1995), which found that *Pistia*, uniquely in the family, has S-type plastids.

Molecular sequence studies have been carried out in the *Amorphophallus* clade by Grob et al. (2002) using 48 species from Thomsonieae with outgroups from six genera, *Filarum*, *Hapaline*, *Anchomanes*, *Arisaema*, *Sauromatum* and *Gonatopus*, based on plastid markers *mat*K and *trn*L. This work was aimed primarily at testing the generic and sectional subdivision of Thomsonieae and resulted in the recognition of five major clades that represent a new understanding of the phylogeny and classification of the group, including the monophyly of the large group of African species and the reduction of *Pseudodracontium* to a subclade of *Amorphophallus*.

Loh et al. (2000), following an earlier fingerprinting study of *Caladium* cultivars (Loh et al., 1999), applied the technique of AFLP marker analysis to investigate relationships between five species of *Caladium* and three species of *Xanthosoma* and showed that these two genera could thus be distinguished.

The topology of the the *Colletogyne* clade, the second subclade of the *Dracunculus* clade, has since been confirmed by Cabrera et al. (2008), Mansion et al. (2008) and Cusimano et al. (2010). These results differ in whether *Peltandra* and *Typhonodorum* form their own subclade (corresponding to tribe Peltandreae in the sense of Bogner and Petersen, 2007) or whether they are successively sister to the three genera of Arophyteae. This is of biogeographic interest because of the fact that *Typhonodorum* and Arophyteae are Madagascan, whereas *Peltandra* occurs in eastern North America. However, the pairing of *Ambrosina-Arisarum* and their sister-group status to the clade comprising the former genera is well supported in all these studies.

Cobbania corrugata (Stockey et al., 2007) is a fossil from the Late Cretaceous of Alberta, Canada and the Amur region in the Russian Far East that as yet cannot be ruled out of consideration as a unisexual-flowered aroid, though this remains highly speculative. The Palaeocene-Eocene leaf fossils of *Nitophyllites*, on the other hand, are more certainly ascribable to the *Dracunculus* clade. Wilde et al. (2005) have given a recent survey of this genus, which includes three species (*N. zaisa-nicus, N. limnestis, N. bohemicus*). These authors place *Nitophyllites* in this clade, relating it either to tribe Peltandreae or tribe Arophyteae.

The most intensively studied of the groups found by French et al. (1995) is the third and most derived, the *Pistia* clade (*sensu* Cusimano et al., 2011). Renner and Zhang (2004), who first dubbed the clade with this informal name, carried out a study of 37 species from 16 genera of the *Pistia* clade, with outgroup species from the genera *Caladium, Peltandra, Typhonodorum* and *Xanthosoma*, representing the other two subclades of the *Dracunculus* clade. These authors used DNA sequence data from three plastid regions (*trn*L-*trn*F, *rpl20-rps12, trn*L) and one mitochondrial region (*nad1* b/c) both to test earlier phylogenetic results and investigate the historical biogeography of *Pistia* with the help of fossils. This study was important in confirming the phylogenetic relationships of *Pistia*, previously never clear, but also reinforced the robustness of the *Pistia* clade and established its internal topology as consisting of five components above the basal nodes with *Pistia* and *Protarum*, successively the *Colocasia* clade (with *Ariopsis* and without *Alocasia*), *Alocasia* and finally a trichotomy of the tribe Areae, *Arisaema* and *Pinellia*, the last two genera not forming a monophyletic group as previously supposed by all authors, including French et al. (1995).

Chen et al. (2004b) used AFLP fingerprinting of 23 cultivars from 17 species of *Alocasia*, but did not include other genera of the *Pistia* clade in their study, which was aimed at investigating the potential for horticultural hybrid development. Nauheimer et al. (2012a) published an important study of 71 species of *Alocasia* and 25 species of other genera from the *Pistia* clade, based on plastid and nuclear DNA sequence data. They confirmed that *Alocasia* is a monophyletic group and showed that its sister group is *Colocasia gigantea* (Blume) Hook.f., which cannot now be considered to belong to the genus *Colocasia*. This study used extensive data to investigate the historical biogeography of the genus since the Miocene and was able to trace the origin of domesticated *Alocasia macrorrhizos* (Giant Taro) to the Philippines.

Mansion et al. (2008) used a phylogenetic tree of genera from the *Colletogyne* and *Pistia* clades to make a detailed and fascinating study of the historical biogeography of the western Mediterranean region. They sampled 54 species as 88 OTUs, including as outgroups *Amorphophallus, Caladium* and *Xanthosoma* from the *Amorphophallus* clade and the genus *Calla*, which is sister to the entire clade (*Dracunculus* clade) sampled. They used data from six regions of the plastid genome (*trn*L intron, *trn*L-*trn*F spacer, part of the *trn*K intron, *mat*K, *rbc*L and *rps16*) and obtained a result which is compatible with the topologies of Cusimano et al. (2010, 2011); some species of *Typhonium* are now species of *Sauromatum*. The only significant disagreement is the grouping of *Colocasia esculenta* with *Alocasia* in a clade apart from that of *Ariopsis*, *Remusatia* and *Steudnera*. *Arisaema* groups with the Areae rather than *Pinellia*, but this sister relationship has only weak support and may be considered equivalent to the polytomy found between these three groups by Renner and Zhang (2004) and Cusimano et al. (2011).

Renner and Zhang's (2004) result indicated that *Typhonium* was paraphyletic, and this has been further investigated in two more recent studies. Cusimano et al. (2010), in a study of 86 of the total 153 species of the Areae, including 53 species of *Typhonium* (in the broad sense), used both plastid (*rpl20-rps12*, *trn*K) and nuclear (*Phy*C) genome regions to analyse all relevant genera, using *Arisaema* as outgroup. Their results confirmed that the taxon concept of *Typhonium* hitherto used was paraphyletic, but showed that if *Sauromatum* were resurrected and the Australian species recognized as a distinct genus, all other current genera would then emerge as well-supported groups.

Ohi-Toma et al. (2010) used six plastid genome regions (3′ *trn*L-*trn*F, *rpl20*-5′ *rps12*, *psb*B-*psb*H, *trn*G intron, *rpo*C2-*rps2* and *trn*K 3′ intron) to analyse 17 species of *Typhonium* and 8 species from other genera of Areae (*Arum*, *Biarum*, *Helicodiceros*, *Theriophonum*), using outgroup taxa from *Alocasia*, *Colocasia*, *Remusatia*, *Arisaema* and *Pinellia*. Their study originated from a pioneer molecular (RFLP) paper by Sriboonma et al. (1993), one of the first to be specifically aimed at Araceae systematics. The differences between this study and that of Cusimano et al. (2010) appear to concern mainly the topology above *Theriophonum*. Cusimano et al. (2010) obtained two main clades in this part of the cladogram, one corresponding to *Sauromatum* and the other to the Mediterranean clade (*Arum*, *Biarum*, *Dracunculus*, *Eminium*, *Helicodiceros*). Ohi-Toma et al. (2010), with a much smaller number of species from Areae and from *Typhonium s.l.* (five species), obtained a polytomy of four monophyletic elements, one of which corresponds to the Mediterranean clade. From the other branches they described three new monospecific genera, *Hirsutiarum*, *Diversiarum* and *Pedatyphonium*. A consequence of both papers was the resurrection of *Sauromatum*.

Renner et al. (2004) studied 77 species of the large and widespread genus *Arisaema* using plastid DNA regions *trn*L intron, *trn*L-*trn*F spacer and *rpl20*-*rps12* spacer, with outgroups *Pinellia* and *Typhonium s.l.* (including *Sauromatum giganteum*, *S. hirsutum*, *S. horsfieldii*, *S. venosum*). Their aim was to test the monophyly of the genus and to elucidate the historical biogeography of the genus, which is represented today in Africa, Asia and North America. They found that the sister-group relationships of the North American species involved East Asian species in patterns comparable to those revealed in Orontioideae by Nie et al. (2006), discussed earlier. They also did not find a sister-group relation with

Pinellia, which had been the general view of taxonomists hitherto. Genetic structure studies of *Arisaema* using allozymes have been carried out by Boles (1996) and Maki and Murata (2001).

Mansion et al. (2008) sampled 20 of the 30 currently recognized species of *Arum*, but two yet more recent studies focussed exclusively on the genus *Arum*. Espíndola et al. (2010) sampled 28 species of *Arum*, using *Dracunculus* and *Biarum* as outgroups using DNA sequences from the four plastid regions *3'rps16–5'trnK*, *ndhA* intron, *psbD-trnT* and *rpl32-trnL*. These authors traced selected phenotypic characters, such as ploidy, tuber morphology, flower type and ratio of spadix length to spathe length, and carried out a historical biogeographic analysis, including dating and dispersal-vicariance analysis. Their molecular cladogram only agreed partially with the infrageneric classification proposed by Boyce (1993, 2006); the two subgenera *Arum* and *Gymnomesium* (the latter consisting of just *Arum pictum*) emerged, but these authors recommended replacing the current two sections of subgenera *Arum* by five sections corresponding to the main subclades obtained in their analysis.

Linz et al. (2010) analysed the species of *Arum* with molecular data from three plastid (*matK*, *rbcL*, *trnL*) regions, one nuclear (ITS1, 5.8S, ITS2) region and AFLPs to produce a cladogram of 24 species, with *Dracunculus* as outgroup. Besides the elucidation of a robust and well-resolved cladogram of the genus, their objective also included the establishment of a phylogenetic framework for interpreting the evolution of pollination syndromes within the genus. The cladistic result was in broad agreement with the classification of Boyce (1993, 2006) and allowed these authors to hypothesize that dung mimicry and hence pollination by coprophilous flies and beetles is the oldest pollination syndrome in the genus.

There are a number of fossils known from the *Pistia* clade. Wilde et al. (2005) placed the leaf fossil genus *Caladiosoma* possibly in Colocasieae or near to *Alocasia*, but also with equal probablility in the Caladieae (*Amorphophallus* clade). Their assignment was based especially on the leaf venation of their new species (*C. messelense*) from the Middle Eocene Messel deposits of Germany. *Pistia* fruits and seeds are known from Siberia (*P. sibirica*, Oligocene, Kvaček and Bogner, 2008).

As discussed by Cabrera et al. (2008) and Cusimano et al. (2011), phylogenetic studies have so far not been able to establish stable positions for the genera *Zantedeschia*, *Montrichardia*, *Anubias* and *Callopsis*. These taxa have always been considered isolated taxonomically and biogeographically, and many authors have placed them individually in monogeneric tribes. Recent molecular studies (Cabrera et al., 2008; Cusimano et al., 2011; Nauheimer et al., 2012b) have grouped *Calla* within the unisexual Araceae clade, despite some major phenotypic differences, as previously discussed; this placement requires further testing by future phylogenetic analyses. Herrera et al. (2008) have reported the first reliably authenticated fossil of *Montrichardia* from the Palaeocene of Colombia, *M. aquatica*.

9.11 Conclusions

There has been substantial progress in the phylogenetic understanding of Araceae since 1995, not only with the aid of molecular cladistics but also through palaeo-botanical studies. It is becoming increasingly clear that further comparative morphological studies are needed to link these two data fields. Although it is surely now indisputable that comparative morphology has been largely superseded by molecular studies for establishing the phylogenetic framework of extant taxa, at least above species level, the same cannot be said when it comes to finding the optimal phylogenetic placement for the increasingly abundant and carefully worked fossil taxa. To make this procedure more precise it is indispensable to establish more reliable and easily accessible data from the morphological data fields most commonly represented in fossils (e.g. pollen, seeds and leaves) and to optimize the distributions of these characters on molecular cladograms to serve as a more reliable basis for palaeobotanical interpretations. Excellent contributions have been made in recent times in relevant fields, e.g. M. Hesse and colleagues for pollen (Hesse and Zetter, 2007), E. Seubert for seed structure (Seubert, 1993, 1997a, 1997b) and V. Wilde and colleagues for leaf venation (Wilde et al., 2005). The further development of such studies will help the dating of phylogenetic trees and the further progress of historical biogeographical studies, such as the important recent review with new analyses by Nauheimer et al. (2012b).

9.12 References

Andrade, I.M. and Mayo, S.J. (2010). Molecular and morphometric patterns in Araceae from fragmented Northeast Brazilian forests. In *Diversity, Phylogeny and Evolution in the Monocotyledons*, ed. O. Seberg, G. Petersen A.S. Barfod and J.R. Davis. Aarhus: Aarhus University Press, pp. 115–128.

Andrade, I.M., Mayo, S.J., Van den Berg, C. et al. (2007). A preliminary study of genetic variation in populations of *Monstera adansonii* var. *klotzschiana* (Araceae) from North-East Brazil, estimated with AFLP molecular markers. *Annals of Botany*, **100**, 1143–1154.

Andrade, I.M., Mayo, S.J., Van den Berg, C. et al. (2009). Genetic variation in natural populations of *Anthurium sinuatum* and *A. pentaphyllum* var. *pentaphyllum* (Araceae) from north-east Brazil using AFLP molecular markers. *Botanical Journal of the Linnean Society*, **159**, 88–105.

Barabé, D. and Forget, S. (1987). Analyse phylogénique des Calloideae (Araceae). *Naturaliste can. (Rev. Écol. Syst.)* **114**, 487–494.

Barabé, D., Bruneau, A. Forest, F. and Lacroix, C. (2002). The correlation between development of atypical bisexual flowers and phylogeny in the Aroideae (Araceae). *Plant Systematics and Evolution*, **232**, 1–19.

Barabé, D., Lacroix, C. and Jeune, B. (2008). Quantitative developmental analysis of homeotic changes in the inflorescence of *Philodendron* (Araceae). *Annals of Botany*, **101**, 1027–1034.

Bastmeijer, J.D., Idei, T., Jacobsen, N., Ramsdal, A.M. and Sookchaloem, D. (2010). Notes on Cryptocoryne (Araceae) of Thailand, including a new species from Loei Province. *Thai Forest Bulletin (Bot.)*, **38**, 179–183.

Behnke, H.-D. (1995). P-type sieve-element plastids and the systematics of the *Arales* (sensu Cronquist 1988) – with S-type plastids in *Pistia*. *Plant Systematics and Evolution* **195**(1–2), 87–119.

Bogner, J. (1979). A critical list of the Aroid genera. *Aroideana*, **1**(3), 63–73.

Bogner, J. (2009). The free-floating Aroids (Araceae) – Living and fossil. *Zitteliana*, **A48–49**, 113–128.

Bogner, J. and Hesse, M. (2005). Zamioculcadoideae, a new subfamily of Araceae. *Aroideana*, **28**, 3–20.

Bogner, J. and Nicolson, D.H. (1991). A revised classification of Araceae with dichotomous keys. *Willdenowia*, **21**, 35–50.

Bogner, J. and Petersen, G. (2007). The chromosome numbers of the Aroid genera. *Aroideana*, **30**, 82–90.

Bogner, J., Hoffman, G.L. and Aulenback, K.R. (2005). A fossilized aroid infructescence, *Albertarum pueri* gen. nov. et sp.nov., of Late Cretaceous (Late Campanian) age from the Horseshoe Canyon Formation of southern Alberta, Canada. *Canadian Journal of Botany*, **83**, 591–598.

Bogner, J., Johnson, K.R., Kvaček, Z. and Upchurch Jr., G.R. (2007). New fossil leaves of Araceae from the Late Cretaceous and Paleogene of western North America. *Zitteliana*, A**47**, 133–147.

Boles, R. (1996). *Genetic Structure and Sexual Reproduction Potential of the Perennial Aroid* Arisaema dracontium.

M.Sc. Thesis, Windsor, Ontario, Canada: University of Windsor.

Bonde, S.D. (2000). *Rhodospathodendron tomlinsonii* gen. et sp. nov., an araceous viny axis from the Nawargaon intertrappean beds of India. *The Palaeobotanist*, **49**, 85–92.

Boyce, P.C. (1993). *The Genus Arum*.Kew: Royal Botanic Gardens.

Boyce, P.C. (2006). *Arum* – a decade of change. *Aroideana*, **29**, 132–139.

Boyce, P.C. and Wong, S.Y. (2008). Studies on Schismatoglottideae (Araceae) of Borneo VII: *Schottarum* and *Bakoa*, two new genera from Sarawak, Malaysian Borneo. *Botanical Studies (Taiwan)*, **49**, 393–404.

Boyce, P.C. and Wong, S.Y. (2009). *Schottariella mirifica* P.C. Boyce and S.Y. Wong: a new name for *Schottarum sarikeense* (Araceae: Schismatoglottideae). *Botanical Studies (Taiwan)*, **50**, 269–271.

Cabrera, L.I., Salazar, G.A., Chase, M.W. et al. (2008). Phylogenetic relationships of aroids and duckweeds (Araceae) inferred from coding and noncoding plastid DNA. *American Journal of Botany*, **95**, 1153–1165.

Carvell, W.N. (1989). *Floral Anatomy of the Pothoideae and Monsteroideae (Araceae)*. Ph.D. Dissertation, Oxford, OH: Miami University.

Cevallos-Ferriz, S.R.S. and Stockey, R.A. (1988). Permineralized fruits and seeds from the Princeton chert (Middle Eocene) of British Columbia: Araceae. *American Journal of Botany*, **75**, 1099–1113.

Chen, J., Henny, R.J., Norman, D.J., Devanand, P.S. and Chao, C.T. (2001). Analysis of genetic relatedness of *Dieffenbachia* cultivars using AFLP markers. *Journal of the American Society of Horticultural Science*, **129**, 81–87.

Chen, J., Devanand, P.S., Henny, R.J., Norman, D.J. and Chao, C.T. (2004a). Genetic relationships of *Aglaonema* cultivars inferred from AFLP markers. *Annals of Botany* **93**, 157–166.

Chen, J., Devanand, P.S., Henny, R.J., Norman, D.J. and Chao, C.T. (2004b). Interspecific relationships of *Alocasia* revealed by AFLP analysis. *Journal of Horticultural Science and Biotechnology*, **79**, 582–586.

Croat, T.B. (1990). A comparison of Aroid classification systems. *Aroideana*, **13(1–4)**, 44–63.

Cuartas-Hernández, S. and Núñez-Farfán, J. (2006). The genetic structure of the tropical understory herb *Dieffenbachia seguine* L. before and after forest fragmentation. *Evolutionary Ecology Research*, **8**, 1061–1075.

Cusimano, N., Barrett, M.D., Hetterscheid, W.L.A. and Renner, S.S. (2010). A phylogeny of the Areae (Araceae) implies that *Typhonium*, *Sauromatum*, and the Australian species of *Typhonium* are distinct clades. *Taxon* **59**, 439–447.

Cusimano, N., Bogner, J., Mayo, S.J. et al. (2011). Relationships within the Araceae: Comparison of morphological patterns with molecular phylogenies. *American Journal of Botany*, **98**, 654–668.

Devanand, P.S., Chen, J., Henny, R.J. and Chao, C.-C.T. (2004). Assessment of genetic relationships among *Philodendron* cultivars using AFLP markers. *Journal of the American Society for Horticultural Science*, **129**, 690–697.

Duvall, M.R., Learn Jr., G.H., Eguiarte, L.E. and Clegg, M.T. (1993). Phylogenetic analysis of *rbc*L sequences identifies *Acorus calamus* as the primal extant monocotyledon. *Proceedings of the National Academy of Sciences USA*, **90**, 4641–4644.

Engler, A. (1876). Vergleichende Untersuchungen über die morphologischen Verhältnisse der Araceae. I. Theil. Natürliches System der Araceae. *Nova Act. Leop.-Carol. Akad. Naturf.*, **39(3)**, 133–155.

Engler, A. (1887–1889). Araceae. In *Die natürlichen Pflanzenfamilien*, ed. A. Engler and K. Prantl. Leipzig: W. Engelmann, Volume **II. 3**, pp. 102–153.

Engler, A. (1920). Araceae. Pars generalis et Index familiae generalis. In *Das Pflanzenreich*, ed. A. Engler. Leipzig: W. Engelmann, Volume **74** (IV.23A), pp. 1–71.

Espíndola, A., Buerki, S., Bedalov, M., Küpfer, P. and Alvarez, N. (2010). New insights into the phylogenies and biogeography of *Arum* (Araceae): unravelling its evolutionary history. *Botanical Journal of the Linnean Society*, **163**, 14–32.

French, J.C. (1988). Systematic occurrence of anastomosing laticifers in Araceae. *Botanical Gazette*, **149**, 71–81.

French, J.C. (1997). Vegetative Anatomy. In *The Genera of Araceae*, ed. S.J. Mayo, J. Bogner and P.C. Boyce. Richmond: Royal Botanic Gardens, Kew, pp. 9–24.

French, J.C., Chung, M.G. and Hur, Y.K. (1995). Chloroplast DNA phylogeny of the Ariflorae. In *Monocotyledons: Systematics and Evolution, Volume 1*, ed. P.J. Rudall, P.J. Cribb, D.F. Cutler and C.J. Humphries. Richmond: Royal Botanic Gardens, Kew, pp. 255–275.

Friis, E.M., Pedersen, K.R. and Crane, P.R. (2004). Araceae from the Early Cretaceous of Portugal: evidence on the emergence of monocotyledons. *Proceedings of the National Academy of Sciences USA*, **101(47)**, 16565–16570.

Friis, E.M., Pedersen, K.R. and Crane, P.R. (2010). Diversity in obscurity: fossil flowers and the early history of angiosperms. *Philosophical Transactions of the Royal Society B*, **365**, 369–382.

Gauthier, M.-P.L., Barabé, D. and Bruneau, A. (2008). Molecular phylogeny of the genus *Philodendron* (Araceae): delimitation and infrageneric classification. *Botanical Journal of the Linnean Society*, **156**, 13–27.

Gonçalves, E.G. (2002). *Sistemática e evolução da tribo Spathicarpeae (Araceae)*. Ph.D. Thesis, São Paulo: Instituto de Biociências da Universidade de São Paulo.

Gonçalves, E. G. Mayo, S. J., Van Sluys, M.-A. and Salatino, A. (2007). Combined genotypic-phenotypic phylogeny of the tribe Spathicarpeae (Araceae) with reference to independent events of invasion to Andean regions. *Molecular Phylogenetics and Evolution*, **43**, 1023–1039.

Grayum, M.H. (1984). *Palynology and Phylogeny of the Araceae*. Ph.D. thesis, 852 pp., Univ. Massachusetts (Amherst).

Grayum, M.H. (1987). A summary of evidence and arguments supporting the removal of *Acorus* from the Araceae. *Taxon*, **36**, 723–729.

Grayum, M.H. (1990). Evolution and phylogeny of the Araceae. *Annals of the Missouri Botanical Garden*, **77**, 628–697.

Grayum, M.H. (1992). Comparative external pollen ultrastructure of the Araceae and putatively related taxa. *Monographs in Systematic Botany from the Missouri Botanical Garden*, **43**, 1–167.

Gregor, H.J. and Bogner, J. (1984). Fossile Araceen Mitteleuropas und ihren rezenten Vergleichsformen. *Documenta Naturae*, **19**, 1–12.

Gregor, H.J. and Bogner, J. (1989). Neue Untersuchungen an tertiären Araceen II. *Documenta Naturae*, **49**, 12–22.

Grob, G.B.J., Gravendeel, B., Eurlings, M.C.M. and Hetterscheid, W.L.A. (2002). Phylogeny of the tribe Thomsonieae (Araceae) based on chloroplast *matK* and *trnL* intron sequences. *Systematic Botany*, **27**, 453–467.

Hay, A. (1986). Cyrtosperma *Griff. and the Origin of the Aroids*. Ph.D. thesis. Oxford: Oxford University.

Hay, A. (1988). *Cyrtosperma* (Araceae) and its Old World allies. *Blumea*, **33**, 427–469.

Hay, A. (1992). Tribal and subtribal delimitation and circumscription of the genera of Araceae tribe *Lasieae*. *Annals of the Missouri Botanical Garden*, **79**, 184–205.

Hay, A. and Mabberley, D.J. (1991). 'Transference of Function' and the origin of aroids: their significance in early angiosperm evolution. *Botanische Jahrbücher*, **113**, 339–428.

Herrera, F.A., Jaramillo, C.A., Dilcher, D.L., Wing, S.L. and Gómez-N. C. (2008). Fossil Araceae from a Paleocene neotropical rainforest in Colombia. *American Journal of Botany*, **95**, 1569–1583.

Hesse, M. and Zetter, R. (2007). The fossil pollen record of Araceae. *Plant Systematics and Evolution*, **263**, 93–115.

Hesse, M., Bogner, J., Halbritter, H.-M. and Weber, M. (2001). Palynology of the perigoniate Aroideae: *Zamioculcas*, *Gonatopus* and *Stylochaeton* (Araceae). *Grana*, **40**: 26–34.

Hofmann, Chr.-Ch. and Zetter, R. (2010). Upper Cretaceous sulcate pollen from the Timerdyakh Formation, Vilui Basin (Siberia). *Grana*, **49**, 170–193.

Hotta, M. (1970). A system of the family Araceae in Japan and adjacent areas. *Memoirs of the Faculty of Science, Kyoto Imperial University, Series Biology*, **4**, 72–96.

Ipor, I., Tawan, C., Jacobsen, N. et al. (2010). Genotyping natural hybrids of *Cryptocoryne purpurea* in Borneo and Peninsular Malaysia. *Journal of Tropical Biology and Conservation*, 7, 81–86.

Jacobsen, N. (1977). Chromosome numbers and taxonomy in *Cryptocoryne* (Araceae). *Botaniska Notiser*, 130, 71–87.

Keating, R.C. (2003). Acoraceae and Araceae. In *Anatomy of the Monocotyledons, Volume 9*, ed. M. Gregory and D.F. Cutler. Oxford: Oxford University Press, pp.1–327.

Keating, R.C. (2004). Vegetative anatomical data and its relationship to a revised classification of the genera of Araceae. *Annals of the Missouri Botanical Garden*, 91, 485–494.

Kitano, S., Otsuka, K., Uesugi, R. and Goka, K. (2005). Molecular phylogenetic analysis of the genus *Symplocarpus* (Araceae) from Japan based on chloroplast DNA sequences. *Journal of Japanese Botany*, 80, 334–339.

Krassilov, V. and Kodrul, T. (2009). Reproductive structures associated with *Cobbania*, a floating monocot from the Late Cretaceous of the Amur region, Russian Far East. *Acta Palaeobotanica*, 49, 233–251.

Kräusel, R. and Stromer, E. (1924). Die fossilen Floren Ägyptens. Ergebnisse der Forschungsreisen Prof. E. Stromers in den Wüsten Ägyptens. *Abhandlungen der Bayerischen Akademie der Wissenschaften, Mathematisch-naturwissenschaftliche Abteilung*, 30, 1–48.

Kvaček, Z. (1995). *Limnobiophyllum* Krassilov – a fossil link between the Araceae and the Lemnaceae. *Aquatic Botany*, 50, 49–61.

Kvaček, Z. and Bogner, J. (2008). Twenty-million-year-old fruits and seeds of *Pistia* (Araceae) from Central Europe. *Aroideana*, 31, 90–97.

Les, D.H., Crawford, D.J., Landolt, E., Gabel, J.D. and Kimball, R.T. (2002). Phylogeny and systematics of Lemnaceae, the duckweed family. *Systematic Botany*, 27, 221–240.

Linz, J., Stökl, J., Urru, I. et al. (2010). Molecular phylogeny of the genus *Arum* (Araceae) inferred from multi-locus sequence data and AFLPs. *Taxon*, 59, 405–415.

Loh, J.P., Kiew, R., Kee, A., Ganb, L.H. and Gan, Y.-Y. (1999). Amplified Fragment Length Polymorphism (AFLP) provides molecular markers for the identification of *Caladium bicolor* cultivars. *Annals of Botany*, 84, 155–161.

Loh, J.P., Kiew, R., Hay, A. et al. (2000). Intergeneric and interspecific relationships in Araceae tribe Caladieae and development of molecular markers using Amplified Fragment Length Polymorphism (AFLP). *Annals of Botany*, 85, 371–378.

Maki, M. and Murata, J. (2001). Allozyme analysis of the hybrid origin of *Arisaema ehimense* (Araceae), *Heredity*, 86, 87–93.

Mansion, G., Rosenbaum, G., Schoenenberger, N. et al. (2008). Phylogenetic analysis informed by geological history supports multiple, sequential invasions of the Mediterranean Basin by the Angiosperm family Araceae. *Systematic Biology*, 57, 269–285.

Mayo, S.J. (1993). Aspects of aroid geography. In *The Africa-South America Connection*, ed. W. George and R. Lavocat. Oxford: Clarendon Press, pp. 44–58.

Mayo, S.J., Bogner, J. and Boyce, P.C. (1995). The Arales. In *Monocotyledons: Systematics and Evolution, Volume 1*, ed. P.J. Rudall, P.J. Cribb, D.C. Cutler and C.J. Humphries.Kew: Royal Botanic Gardens, pp. 277–286.

Mayo, S.J., Bogner, J. and Boyce, P.C. (1997). *The Genera of Araceae*. Kew: Royal Botanic Gardens.

Nauheimer, L., Boyce, P.C. and Renner, S.S. (2012a). Giant taro and its relatives: a phylogeny of the large genus *Alocasia* (Araceae) sheds light on Miocene floristic exchange in the Malesian region. *Molecular Phylogenetics and Evolution*, **63**, 43–51.

Nauheimer, L., Metzler, D. and Renner, S.S. (2012b). Global history of the ancient monocot family Araceae inferred with models accounting for past continental positions and previous ranges based on fossils. *New Phytologist,* **195**, 938–950.

Nicolson, D.H. (1960). A brief review of classifications in the Araceae. *Baileya*, **8**, 62–67.

Nie, Z.-L., Sun, H., Li, H. and Wen, J. (2006). Intercontinental biogeography of subfamily Orontioideae (*Symplocarpus*, *Lysichiton*, and *Orontium*) of Araceae in eastern Asia and North America. *Molecular Phylogenetics and Evolution*, **40**, 155–165.

Ohi-Toma, T., Wu, S., Yadav, S.R., Murata, H. and Murata, J. (2010). Molecular phylogeny of *Typhonium* sensu lato and its allied genera in the tribe Areae of the subfamily Aroideae (Araceae) based on sequences of six chloroplast regions. *Systematic Botany*, **35**, 244–251.

Othman, A.S. (1997). *Molecular Systematics of the Tropical Aquatic Plant Genus* Cryptocoryne *Fischer ex Wydler (Araceae)*. Ph.D. Thesis. St. Andrews: University of St Andrews.

Othman, A.S., Jacobsen, N. and Mansor, M. (2009). Cryptocoryne *of Peninsular Malaysia*. 102 p. Pulau Pinang: Penerbit Universiti Sains Malaysia.

Petersen, G. (1989). Cytology and systematics of Araceae. *Nordic Journal of Botany*, **9**, 119–166.

Raven, P.H. and Axelrod, D.I. (1974). Angiosperm biogeography and past continental movements. *Annals of the Missouri Botanical Garden*, **61**, 539–673.

Renner, O. (1907). *Teichosperma*, eine Monocotylenfrucht aus dem Tertiärs Ägyptens. *Beiträge zur Paläontologie und Geologie Österreich-Ungarns und des Orients*, **20**, 217–220.

Renner, S.S. and Zhang, L.-B. (2004). Biogeography of the *Pistia* clade (Araceae): based on chloroplast and mitochondrial DNA sequences and Bayesian divergence time inference. *Systematic Biology*, **53**, 422–432.

Renner, S.S., Zhang, L.-B. and Murata, J. (2004). A chloroplast phylogeny of *Arisaema* (Araceae) illustrates Tertiary floristic links between Asia, North America, and East Africa. *American Journal of Botany*, **91**, 881–888.

Rothwell, G.W., Van Atta, M.R., Ballard Jr., H.E. and Stockey, R.A. (2004). Molecular phylogenetic relationships among Lemnaceae and Araceae using the chloroplast *trn*L-*trn*F intergenic spacer. *Molecular Phylogenetics and Evolution*, **30**, 378–385.

Sakuragui, C.M. (1998). *Taxonomia e filogenia das espécies de* Philodendron *Seção* Calostigma *(Schott) Pfeiffer no Brasil*. Ph.D. Thesis. São Paulo: Instituto de Biociências da Universidade de São Paulo.

Schott, H.W. (1860). *Prodromus systematis Aroidearum*. Vienna: Typis congregationis mechitharisticae.

Seubert, E. (1993). *Die Samen der Araceen: Die Samenmerkmale der Araceen und ihre Bedeutung für die Gliederung der Familie*. PhD. Dissertation, University of Kaiserslautern. Königstein: Koeltz Scientific Books.

Seubert, E. (1997a). Sclereids of Araceae. *Flora*, **192**, 31–37.

Seubert, E. (1997b). A comparative study of the seeds of Lasieae (Araceae). *Botanische Jahrbücher*, **119**, 407–426.

Smith, S.Y. and Stockey, R.A. (2003). Aroid seeds from the Middle Eocene Princeton Chert (*Keratosperma allenbyense*, Araceae): comparisons with extant Lasioideae. *International Journal of Plant Sciences*, **164**, 239–250.

Sriboonma, D., Hasebe M., Murakami, N., Murata, J. and Iwatsuki, K. (1993). Phylogeny of *Typhonium* (Araceae) inferred from restriction fragment analysis of chloroplast DNA. *Journal of Plant Research*, **106**, 11–14.

Stockey, R.A., Hoffman, G.L. and Rothwell, G.W. (1997). The fossil monocot *Limnobiophyllum scutatum*: resolving the phylogeny of Lemnaceae. *American Journal of Botany*, **84**, 355–368.

Stockey, R.A., Rothwell, G.W. and Johnson, K.R. (2007). *Cobbania corrugata gen. et comb. nov.* (Araceae): a floating aquatic monocot from the Upper Cretaceous of western North America. *American Journal of Botany*, **94**, 609–624.

Tam, S.-M., Boyce, P.C., Upson, T.M. et al. (2004). Intergeneric and infrafamilial phylogeny of subfamily Monsteroideae (Araceae) revealed by chloroplast *trnL-F* sequences. *American Journal of Botany*, **91**, 490–498.

Tarasevich V.F. (1988). Peculiarities of morphology in ridged pollen grains in some representatives of Araceae. In *Palynology in the USSR, Papers of the Soviet Palynologists to the VII International Palynological Congress Brisbane, Australia, 1988*, ed. A.F. Chlonova. Novosibirsk: Nauka, pp. 58–61.

Temponi, L.G. (2006). *Sistemática de Anthurium Sect. Urospadix (Araceae).*

Ph.D. Thesis, Instituto de Biociências da Universidade de São Paulo, São Paulo.

Wen, J., Jansen, R.K. and Kilgore, K. (1996). Evolution of the Eastern Asian and Eastern North American disjunct genus *Symplocarpus* (Araceae): insights from chloroplast DNA restriction site data. *Biochemical Systematics and Ecology*, **24**, 735–747.

Wilde, V., Kvaček, Z. and Bogner, J. (2005). Fossil leaves of the Araceae from the European Eocene and notes on other aroid fossils. *International Journal of Plant Sciences*, **166**, 157–183.

Wong, S.Y. and Boyce, P.C. (2010a). Studies on Schismatoglottideae (Araceae) of Borneo IX: A new genus, *Hestia*, and resurrection of *Apoballis*. *Botanical Studies*, **51**, 249–255.

Wong, S.Y. and Boyce, P.C. (2010b). Studies on Schismatoglottideae (Araceae) of Borneo X. *Pichinia*, a new genus from Sarawak, Malaysian Borneo. *Gardens' Bulletin Singapore*, **61**, 541–548.

Wong, S.Y. and Boyce, P.C. (2010c). Studies on Schismatoglottideae (Araceae) of Borneo XI: *Ooia*, a new genus, and a new generic delimitation for *Piptospatha*. *Botanical Studies*, **51**, 543–552.

Wong, S.Y., Boyce, P.C., bin Othman, A.S. and Pin, L.C. (2010). Molecular phylogeny of tribe Schismatoglottideae (Araceae) based on two plastid markers and recognition of a new tribe *Philonotieae*, from the neotropics. *Taxon*, **59**, 117–124.

Zetter, R., Hesse, M. and Frosch-Radivo, A. (2001). Early Eocene zona-aperturate pollen grains of the *Proxapertites* type with affinity to Araceae. *Review of Palaeobotany and Palynology*, **117**, 267–279.

The first evolutionary classification of Araceae: A. Engler's Natural System

SIMON J. MAYO AND JOSEF BOGNER

10.1 Introduction

Adolf Engler's scientific interest in Araceae began when he worked at the Munich Botanical Institute between 1871 and 1878 and remained strong throughout his life (Diels, 1931; Lack, 2000). Besides many subsidiary papers, he published three major taxonomic monographs of the family, in A. and C. de Candolle's series *Monographiae Phanerogamarum* (Engler 1879a), in *Die natürlichen Pflanzenfamilien* (Engler, 1887 – 1889) and in the *Das Pflanzenreich* series in partnership with K. Krause (Engler, 1905, 1908, 1911, 1912, 1915, 1920a, 1920b, 1920c; Engler and Krause, 1908, 1920; Krause, 1908, 1913); these are all different versions of the same basic concept (Nicolson, 1983, 1988). His classification dominated aroid taxonomy for over 100 years and was not displaced until Michael Grayum's comprehensive taxonomic overhaul of the family (Grayum, 1984, 1990), which coincided with the advent of cladistic methods and molecular systematics.

Engler's classification of Araceae was the first to be based on phylogenetic principles and succeeded that of Heinrich Wilhelm Schott whose species monograph (Schott, 1860) was published only 16 years before Engler's first paper on aroid taxonomy (Engler, 1876a). Schott's work was in the mould of the pre-Darwinian Natural System (Stevens, 1994) and Engler's new view was revolutionary, at least as much as that brought about by molecular methods (French et al., 1995).[1] He had completed his doctoral studies on the genus *Saxifraga* in 1866,

Early Events in Monocot Evolution, eds P. Wilkin and S. J. Mayo. Published by Cambridge University Press. © The Systematics Association 2013.

when Darwinism was having its first impact on European botany (Diels, 1931; Junker, 1989, 2009) and it must have been this context that led him, right at the beginning of his professional taxonomic career, to propose classifications explicitly aimed at tracing the evolutionary connections within families (Lack, 2000).

He had already published monographs and explanatory papers on other plant families prior to working on aroids, but his Araceae classification was far more extensive and based on a wider range of new observations. We will discuss these earlier essays first because they provide some insight into his methods as he grappled with the task of working out taxonomic revisions within a phylogenetic framework.

Almost all of Engler's publications were written in German or Latin and most have not yet been translated into English. Dan Nicolson provided a number of English translations, both unpublished and published (Nicolson, 1983) and explanations of Engler's classification (Nicolson, 1988). The most important similar contribution in recent years is a complete illustrated English translation of Engler's monograph on shoot architecture (Engler, 1877) by Ray and Renner (1990), which includes an extensive commentary on Engler's work and a glossary of the German morphological terms he used. Over a period of some 50 years, Engler not only wrote the single greatest opus of formal taxonomy on Araceae but also many other discursive explanatory papers, some of which have been a focus of this study. Tom Croat (1983) gave a short description of these achievements. The English translations presented in this paper and on the CATE website (at http://cate-araceae.myspecies. info/content/translations-classic-araceae-studies/) are literal and unpolished. We have frequently given the original German word or phrase in square brackets ('[...]') alongside the translation, especially in the case of terms such as *Entwicklungsgeschichte* and *Verwandtschaft*, where the context is important for Engler's meaning, and where we have done so, the German words are left in their original grammatical cases. Where it was felt to be necessary to add English text for clarification these additions are in curly brackets ('{...}'). We have been rather liberal in the number and length of the quotations made in the text, but we ask the readers' forbearance because this work is unfamiliar to many aroid botanists and because the goal of our study is to understand Engler's intellectual approach to his phylogenetic studies of aroids. This involves not only tackling translations, but also straddling a major gulf in understanding between modern views of evolution and those that surrounded Engler in the 1870s, when the world of plant systematics was still reeling from the impact of Darwin's *Origin* and most of the fundamental principles of modern evolutionary theory would not be firmly established for many years to come.

Engler's classification of Araceae was strongly influenced by the evolutionary ideas of C.W. Nägeli, a major figure in nineteenth-century German botany (Junker, 1989). Nägeli was especially influential from 1871, when Engler moved to Munich to take up the position of Acting Curator of the State Botanical Collections, of which Nägeli was Director and thus his formal superior (Diels, 1931: VI). Nägeli's orthogenetic *Vervollkommnungstheorie* (Nägeli, 1884) influenced Engler's (1884) theoretical justification of his classification of Araceae. However, in his extensive appreciation of Nägeli's career, Engler (1917) reveals that they did not agree on all points. Various comments made by Engler discussed further below show divergence in their views, notably in regard to Engler's conclusion that a reduction trend in floral structure predominated in the evolution of Araceae.

10.2 Earliest phylogenetic interpretations: Saxifragaceae, Rutaceae, Simaroubaceae, Burseraceae and Ochnaceae

In his doctoral studies (Engler, 1867), Engler was fairly negative about the usefulness of Darwin's theory for explaining the distribution and origin of *Saxifraga* species, but by the time his later monograph was published (Engler, 1872), he was convinced of the fact of evolution (Diels, 1931; Junker, 1989). Even here, however, he applied evolutionary concepts to taxonomy only tentatively, using Types (Ground Plans) to interpret biogeographical patterns. By considering the morphological characters of the species, he concluded that the present-day geographical distribution of *Saxifraga* could be interpreted as the result of diversification of six basic Types (viewed as ancestral forms) which existed at the end of the Tertiary period (Engler, 1872: 67) and from which he derived the taxonomic sections he recognized.

Engler's main commitment in taking up his post at Munich was the preparation of various families for the *Flora Brasiliensis* under A.W. Eichler's editorship (Diels, 1931; Ray and Renner, 1990; Lack, 2000). Engler seems to have resolved to take the opportunity that these taxonomic treatments provided to express phylogeny more explicitly in classification and the resulting studies are among the earliest such works to have been published in botany (Stevens, 1994: 240). The papers he wrote alongside the publication of the *Flora* monographs explain and justify his conclusions.

In his paper on the Rutaceae and allied families (Engler, 1874), the opening paragraph suggests that he was self-consciously branding himself a Darwinist, at least in regard to evolution, though not as it turned out later (Junker, 1989), so much to the theory of natural selection:-

> Now that the Theory of Descent has entered the field of systematic botany – and it is thanks to Darwin's theory, at first feared and rejected by the majority of descriptive botanists and zoologists, that systematics has once again returned to high esteem, though long treated shabbily and neglected by the most prominent researchers of recent decades – the following investigations into the systematic position and phylogenetic relationships of certain plant groups with which the most important systematists and plant experts thoroughly concerned themselves

during the years 1820–1825, [Engler cites studies for this period by A.P. de Candolle, Nees von Esenbeck and Martius, A. St. Hilaire and A. de Jussieu] may also perhaps find favour with the botanical public.

(Engler, 1874: 3)

This paper provides a key early insight into the methodology which Engler developed early in the new era of phylogenetic taxonomy. His main objective was to conceptualize natural higher-level groups – he often accepted as natural the genera and species of his predecessors (e.g. Engler, 1876a: 99) and sought to extend 'naturalness' to suprageneric levels in the classification – and link them together in a phylogenetic scheme that would express their relative levels of advancement (Engler, 1876b: 139). Confronted with a confusing range of patterns in floral morphology he used vegetative anatomy and morphology wherever he could as a source of characters to group genera into higher taxa. He also used the geographical distributions of the taxa so formed as a separate source of evidence for phylogenetic reasoning.

The scope of the 1874 Rutaceae paper concerns primarily the problem of whether the three families Rutaceae, Simaroubaceae and Burseraceae are distinct and how they are related to each other in a phylogenetic sense. He states that since he had undertaken treatments for the *Flora Brasiliensis*,

I was impelled to obtain an insight into the phylogenetic relationships [*verwandtschaftlichen Beziehungen*] of those families.

(Engler, 1874: 5)

Later he makes another statement which highlights his interest in expressing phylogeny in taxonomy:

When it is not simply a question of identifying plants and placing them in a particular position in the herbarium, but rather to establish as precisely as possible the phylogenetic relationships [*verwandtschaftlichen Beziehungen*] of genera and higher groups [*Gruppen*] then it is an essential scientific requirement to seek after a constant character which belongs to a group of plant forms that are also linked together by numerous intermediate forms. *If such characters are present, then they indicate a common descent and allow us to treat the groups, with greater probability, as members of one and the same Type* {our emphasis}; however if such characters are lacking, then one is not justified in interpreting such a plant group as a unique Type …

(Engler, 1874: 6)

He sets out three main questions to answer: (1) Can the families he is studying be derived from same Basic Type [*Grundtypus*]?, (2) Do the tribes grouped by Hooker (1862) into the families Rutaceae, Simaroubaceae and Burseraceae really form natural taxa?, (3) Are there general diagnostic characters which justify the families as distinct or have these been lost through the appearance of intermediates which arose by later fusion of lineages originating from the same Basic Type?

In an extensive footnote on page 8 he clarifies the meaning of the third question by proposing a developmental current [*Entwicklungsstrom*] as an alternative meta-phor to an evolutionary tree (Diels, 1931 refers to this):

> As is well known, it is customary to employ the metaphor of a genealogical tree [*Stammbaum*] for the elucidation of how the morphological development of the forms of a genus, family or a still larger taxon [*Formencomplexes*] took place in nature or, often, simply as imagined by authors. The setting up of such genea-logical trees may be challenged on the basis of the often very incomplete knowledge of living forms, being even more defective in the case of extinct taxa, and it may be contended that other trees can easily be constructed which are just as justifiable on the basis of known facts; nevertheless, this kind of representa-tion at least still has the benefit that it shows us in the most concise and lucid manner the course of evolutionary development [*Entwicklungsgang*] which each respective author believes to be best justified according to his investigations. It seems to me that there is another metaphor still more suitable for the same purpose, namely that of a current, from which lesser currents branch in various directions and which can endlessly branch again and again, sometimes here and there joining together again, and sometimes, once a course is taken, following it without ever meeting any other smaller branches.
>
> (Engler, 1874: 8)

He affirms that his families all belong to the same Basic Type, the Geraniales of Bentham and Hooker, based on floral morphology, and then embarks on a discussion of whether the tribal composition of each of the families Rutaceae, Simaroubaceae and Burseraceae as proposed by Hooker (1862) can be justified as natural groupings. The conclusion is that floral morphology is too variable across the families, but clearer distinctions can be found in their vegetative anatomy, e.g. the presence of stone cells in the phloem, punctate pellucid glands in the leaves and oil-ducts in the hypodermis of the stem. These characters are based on his own original anatomical observations and illustrated in the paper. His own words provide a guide to his later approach to the higher classification of Araceae:

> The circumstance that the Rutaceae in the widest sense are characterized by the aroma of their leaves and fruits, the Simarubaceae by the great bitterness of their bark, and the Burseraceae by their rich resin content decisively indicates that in spite of the impossibility of characterizing the families on floral structure, never-theless there must be certain characteristics inherent to the three groups which perhaps are of greater constancy than floral and fruit characters. I therefore set myself the task of investigating anatomically all genera available to me which were assigned to those families ...
>
> (Engler, 1874: 28)

Engler, however, takes care to stress that the importance of these anatomical characters also lies in the fact that they reinforce a general, but not crisply defined, commonality in the traditional reproductive characters:

> If the groups characterized by these characters were not also characterized by other inner connections between their constituent taxa then these anatomical diagnostic characters would be of no value. Since, however, these characters are the only constant {ones} and thus are very important, it can be presumed on these grounds that the plant forms which we now call Rutaceae, Simarubaceae and Burseraceae have indeed developed from forms of the same Basic Type, but that the developmental currents of each of these groups have spread out and divided amongst each other [*dass aber die Entwicklungsströme jeder dieser Gruppen sich für sich ausgebreitet und vertheilt haben*].
>
> (Engler, 1874: 35)

In the geographical discussion Engler expresses his conviction that biogeography has evolutionary significance:

> In the discussion of the genealogical relationships [*verwandtschaftliche Beziehungen*] of the Rutaceae, their geographical distribution should not be omitted from consideration since this contributes in a high degree to the illumination of the genealogical relationships [*verwandtschaftlichen Beziehungen*] of the individual groups, and to demonstrate as quite natural the systematic grouping which we have proposed.
>
> (Engler, 1874: 40)

In a similar paper on Ochnaceae (Engler, 1875) published shortly afterwards, he once again poses three questions: (1) What is the delimitation of the family?, (2) What is the Basic Type [*Grundtypus*] and its higher relationships within Dicotyledons?, (3) What approximate phylogeny can be derived from available information?

Answers to the first two questions are based mostly on floral character data with some support from leaf morphology and venation. The many floral formulae and floral diagrams of the genera provided (this paper appeared just before publication of the first volume of A.W. Eichler's *Blütendiagramme*) represent the basis for discussing the family's Type:

> When it is a question of determining the Basic Type, then it is essential to establish the limits within which the development of the individual floral structures vary.
>
> (Engler, 1875: 15–16)

He determines that the 'Stammform' of the Ochnaceae, i.e. the Type, is characterized by the following floral structure: calyx and corolla: 5-merous whorl each; stamens: 3 whorls of indeterminate mery; gynoecium: 2–3 five-merous whorls of free carpels. As regards inferring the phylogeny:

> As soon as we are clear about the Type of the family, we will be able to indicate which genera agree most with this Type and which diverge most from it; with consideration of this state of affairs as regards the primordial Type and the

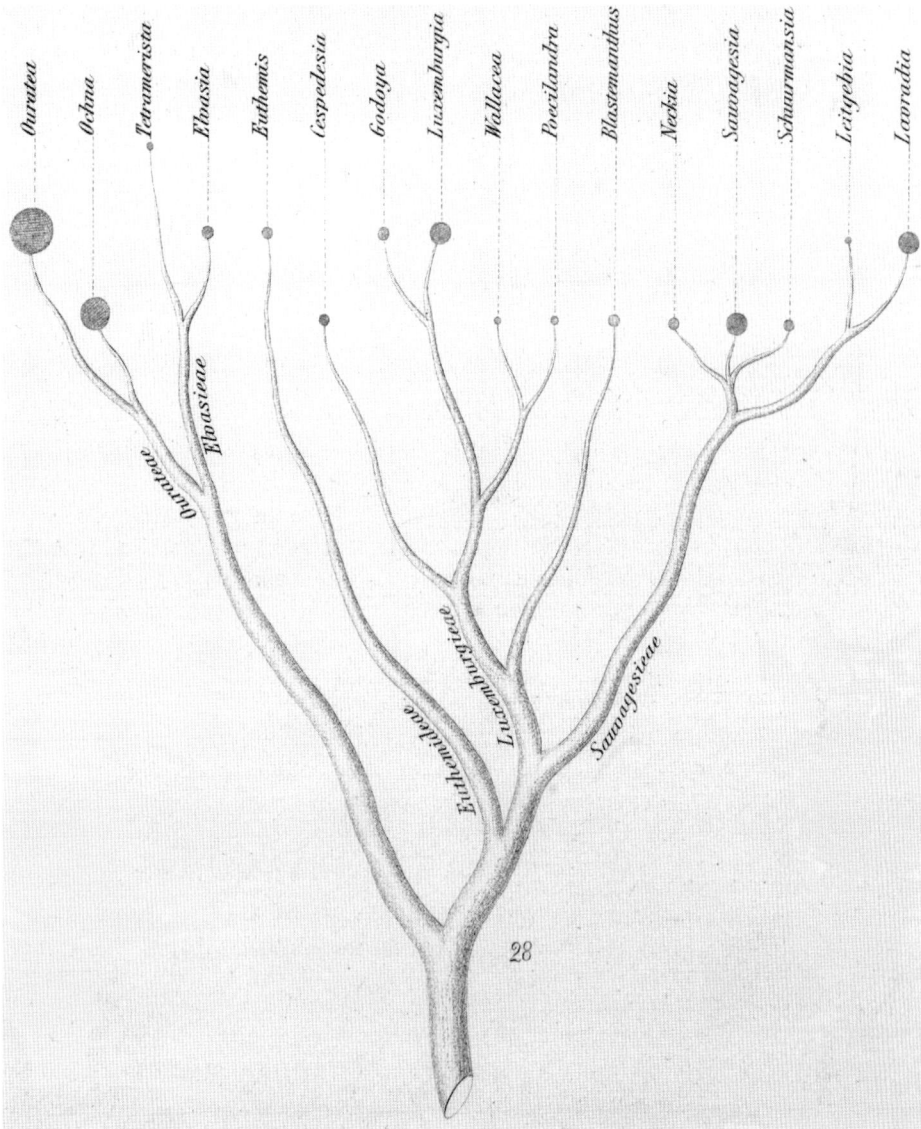

Fig 10.1 Engler's phylogenetic tree of the Ochnaceae (Engler 1875: Plate XIII). Reproduced by kind permission of the Board of Trustees of the Royal Botanic Gardens, Kew.

individual genera, we can construct a picture of the evolution [*Entwicklung*] of the family, as I have tried to do in the following ...

(Engler, 1875: 17)

Then follows a verbal description of the branching structure of the phylogeny from the Type to the extant genera, which is also illustrated (Plate XIII) as a branching

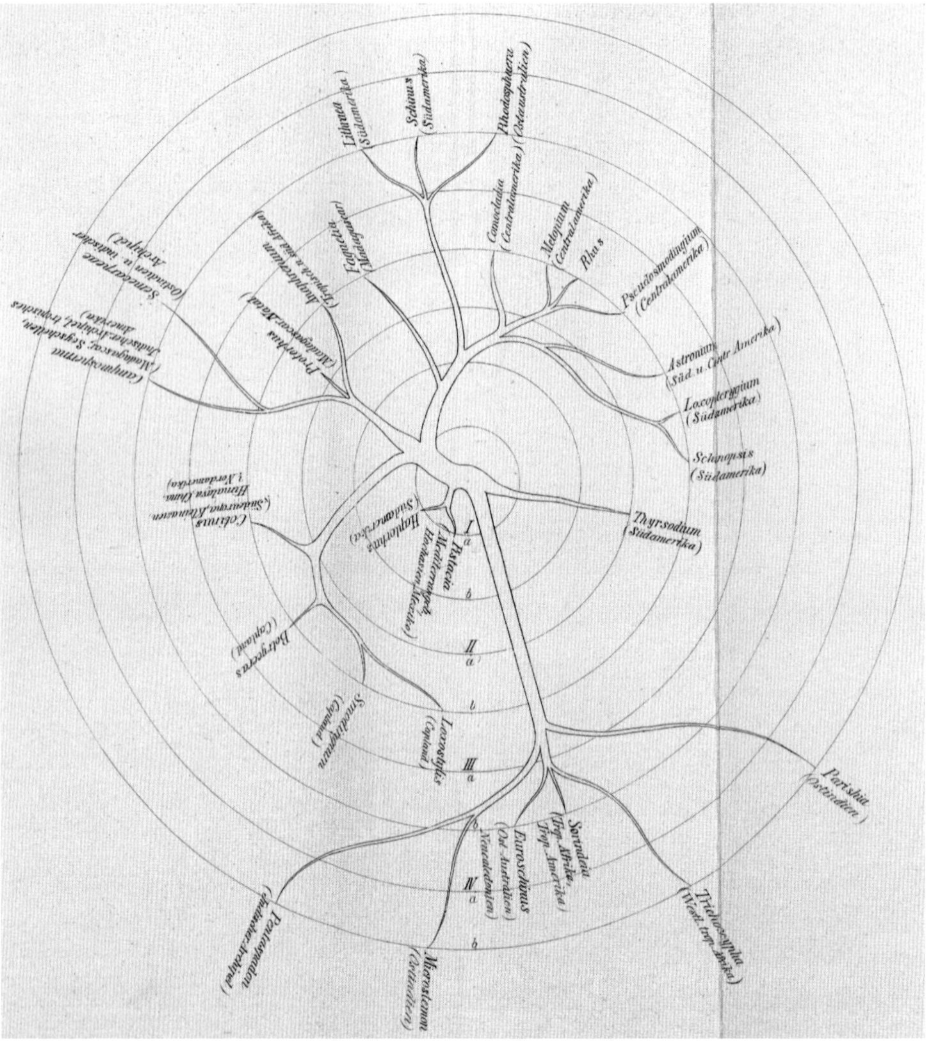

Fig 10.2 Engler's phylogenetic diagram of the Anacardiaceae (Engler 1881: Plate IV). Reproduced by kind permission of the Board of Trustees of the Royal Botanic Gardens, Kew.

diagram (our Fig 10.1, also reproduced by Stevens 1994: his Fig 22 and dust jacket). He ends the paper with a discussion of the biogeography of the family in an evolutionary context. Here he lays down some principles to guide such interpretations:

> If we wish to utilize these geographical facts together with the morphological facts for the purposes of evolutionary explanation [*für die Descendenz*], we must above all not leave out of consideration the following generally recognized or at least established basic principles.

(1) Species-rich genera, whose forms are distributed over all parts of the Old and New Worlds, are older than those whose numerous forms are united in a restricted open region which offers few major obstacles to migration.

(2) Genera which occur with a greater number of closely related genera in a restricted region are of more recent origin.

(3) Genera of isolated systematic position are either the remnants of previously more strongly developed Types if they have a wide distribution, or forms which have recently appeared and are localized in a small region if they have a small distribution range.

(Engler, 1875: 24–25)

In a later paper on the evolution of Anacardiaceae, Engler (1881) presented in a different and remarkable style an evolutionary tree[2] (reproduced in our Fig 10.2) of the genera of a family developing from a common stem.

10.3 Engler's 'Natural Classification' of Araceae

Engler's classification of Araceae (Engler, 1876b), published the year after his Ochnaceae paper, was the first application of phylogenetic ideas to aroid taxonomy and was innovative, not only in seeking a detailed explanation of the evolution of the family, but in generating a large body of new data to support his proposals. An important difference from his previous studies was that Engler was here working on a plant group for which he had extensive living materials available in botanic gardens at Munich, Vienna and elsewhere. This made it possible for him to widen the scope of his comparative morphological and anatomical studies. It was the work of a young and outstandingly energetic botanist already making his mark among leading figures such as H.R. Göppert, A.W. Eichler and C.W. Nägeli, and on the way to extraordinary later achievements (Fig 10.3). His Araceae classification was important not only for aroid taxonomy, but also as a key formative stage in the development of his system for the plant kingdom as a whole (Barabé and Vieth, 1990: 405). His work on aroids was by far the most extensive of the various systematic studies he embarked on during his happy and productive Munich years (Lack, 2000) and it seems that Engler, ambitious from the start, used Araceae as a vehicle to experiment with the large-scale application of phylogeny to higher plant classification.

The classification emerged in a series of publications that followed one another in rapid succession (Engler, 1876a, 1876b, 1877, 1878, 1879a, 1884; see Diels, 1931 for a full bibliography of Engler's works) and which are considered here as the fruit of a single sustained study; they are discussed chronologically and we have tried to highlight text which reveals aspects of Engler's thought on the relationship between his taxonomic proposals and his phylogenetic ideas.

Fig 10.3 Adolf Engler in 1873, during his years in Munich. From H.W. Lack (2000). Reproduced with the kind permission of the Botanischer Garten und Botanisches Museum Berlin-Dahlem.

The first of these publications (Engler, 1876a) was an article published in two successive numbers of the *Botanische Zeitung* in February 1876 entitled *Zur Morphologie der Araceae* (On the morphology of Araceae). In this paper Engler presents a preliminary summary of the main results of his comparative investigations on shoot organization, leaf development, stem and leaf anatomy and floral morphology, based on 85 genera, of which about 30 were available as living plants. In view of this broad representation:

> ... I could well presume that the results obtained from these comparative studies would be valid for the whole family and could thus hope to acquire the foundations for a natural system of the Araceae and knowledge of the course of evolution [*des Entwicklungsgangs*] of this most interesting family.
>
> (Engler, 1876a: 81)

The results are presented as numbered paragraphs, of which the first 17 deal with shoot organization and leaf arrangement. Paragraphs 18 and 20 explain and compare shoot organization in *Pistia* and Lemnaceae, the latter in some detail. With hindsight, the most important result reported in this part is the occurrence of a particular pattern of sympodial branching in which the reiteration of the shoot arises in the axil of the penultimate leaf before the spathe – the so-called '$(n - 1)$' sympodium. Paragraph 19 points out that this pattern is very rare otherwise in Monocotyledons.[3] The second part of the article discusses mainly the classification and floral morphology and here Engler clarifies why he prioritized vegetative anatomy in working out the major subdivisions of the family and shows how he is applying the same approach used previously for Rutaceae:

> Already, my study of leaf position and shoot architecture [*Sprossverhältnisse*] had shown me that some genera which belong to distantly related divisions [*Abthei-lungen*] in Schott's system, which is based exclusively on the construction of the flowers, show a close relationship, for example the *Dracontioninae* and *Amorphophallinae*; I therefore examined these genera with a view to their anatomy and found complete agreement. Since, furthermore, the floral differences which exist between the main divisions of Schott's {system}, the Monoclines and the Diclines, as well as those between their subdivisions, are of a kind which must have formed by abortion or reduction of individual parts of the flowers, I attempted to make use of the anatomy [*anatomische Beschaffenheit*] of the stem and leaves as the first criterion of classification [*als ersten Eintheilungsgrund*]; in this way I obtained better morphological series than by the primary consideration of the floral structure.
>
> 22. Since in very closely related genera the floral structure may be different, so it follows that floral structure is more variable in the family Araceae while the anatomical structure is of greater constancy. The classification of the Araceae based primarily on anatomy [*auf die anatomische Beschaffenheit*] will thus correspond better to the evolution [*der natürlichen Entwickelung*] of this family.
>
> (Engler, 1876a: 97–98)

It is possible that Engler's interest in anatomy as a tool for systematics was influenced by L. Radlkofer, who also worked in Munich, although we have found no direct indication of this. In a well-known lecture to the Bavarian Academy of Sciences advocating the wider use of anatomy in systematics, Radlkofer (1883: 30) stated, 'Perhaps the *Monographie der Sapindaceen Gattung Serjania* published by our Academy in 1875 may be designated as one of the first studies in which the anatomical method in the sense advocated here was employed in the search for anatomical data for the aims of systematics . . .'. It is at least interesting to note that at this very time in the same place, Engler was busy studying the vegetative anatomy of various plant families for just these aims; it is another indication of Engler's forward-looking and innovative approach to research at this period.

Engler's preference for anatomical characters over floral ones led him ultimately to propose two subfamilies, Pothoideae and Lasioideae, which more recent anatomical, palynological and molecular studies have shown to be highly unnatural taxa. The Pothoideae were defined primarily on the absence of both trichosclereids and laticifers; in a later discussion of the Pothoideae he states:

> In the subfamilies [*Gruppen*] already dealt with, it was not difficult to ascertain an inner coherence between the genera; within the subfamily of the Pothoideae

{however} *which show no outstanding histological peculiarity* {our italics}, this {coherence} is more difficult to demonstrate.

(Engler, 1884: 327)

The great importance that Engler laid on the absence of these anatomical characters led him to make assignments of genera to the Pothoideae which are very surprising in the light of both Schott's classification and modern ideas.

The definition of the Lasioideae[4] was more complex and obscure, involving a confusion over the presence of laticifers and the strong but superficial resemblance of the highly divided leaves of *Dracontium, Anchomanes* and *Amorphophallus* (see Hay, 1992 and Hay and Mabberley, 1991 for illuminating discussions). In this subfamily Engler plainly commits himself to an over-riding predominance of vegetative over floral characters as group (Type) -defining criteria.

This earliest classification was intended as only a preliminary statement (the subtitle is '*vorläufige Mittheilung*'). It was proposed in February 1876 but differs considerably from his much more detailed presentation published a few months later. There are three subfamilies, based on anatomical characters: (1) the *Pothoideae* (*Acorus, Heteropsis, Anthurium, Lasimorpha, Cyrtosperma, Lasia, Anaphyllum, Symplocarpus, Pothos, Pothoidium, Gymnostachys*), which lack any special anatomical characteristics, (2) the *Monsteroideae* (*Anepsias, Spathiphyllum, Rhaphidophora, Rhodospatha, Atimeta, Monstera, Tornelia, Epipremnum, Scindapsus, Anadendron, Cuscuaria*), characterized by bisexual flowers, usually distichous leaves and presence of trichosclereids, which at that time he understood as H-shaped sclerenchymatic tube cells, and (3) the *Aroideae*, comprising all other genera recognized by Schott, together with *Pistia* and Lemnaceae, characterized by the presence of laticifers in the phloem and flowers,

in which in an unusual fashion, a retrogression from more highly developed forms with complete flowers to lower simpler forms with completely reduced flowers is revealed.

(Engler, 1876a: 99)

This latter phrase suggests that Engler was using an idea from C.W. Nägeli (see below) which envisaged an orthogenetic progression from simple flowers of less integrated structure to more complex and consistently constructed ones. On this hypothesis, bisexual flowers with stable numbers of floral elements would be more advanced than unisexual ones with varying numbers of elements. This is probably why Engler writes that his proposed evolutionary progression from bisexual-flowered genera to unisexual-flowered ones is 'an unusual ... retrogression' (In later, more general, work (e.g. Engler, 1919: XIX), Engler took the view that bisexual-flowered taxa were in general phylogenetically older than comparable taxa with unisexual flowers and that this derivation had happened many times in the Angiosperms).

In paragraph 24, Engler states that:

> Within the major divisions there are several very natural groups of genera, which are closely related to one another. For the most part these were proposed by Schott, who exactly described and illustrated their floral characters in a manner which was masterful for its time, but who did not employ the comparative method [*die vergleichende Methode*] and thus did not recognize the genealogical relationships [*verwandtschaftichen Beziehungen*].
>
> (Engler, 1876a: 99)

Engler was thus carrying out a reorganization of relationships between lower-level elements (genera, species) which he took up largely unaltered from Schott (1858, 1860) and he saw his main task as establishing higher-level natural groups (his subfamilies) and the phylogenetic connections between them. The remainder of the paper is taken up with examples of morphological series which show progressions of evolutionary reduction.

Many of the most innovative aspects of Engler's classification are mentioned here in this first work. The most fundamental change was to place the bisexual-flowered genera first, precisely the inverse of the order of Schott's system. Although Engler does not say why he did this in the '*Zur Morphologie ...*', he states in the general part of his monograph for A. and C. de Candolle's *Monographiae Phanerogamarum* what is in effect an outgroup comparison as a justification:

> Among the subfamilies of Araceae distinguished by me, the Pothoideae come nearest to the normal Monocotyledon Type in every respect: they differ from the Liliaceae essentially only by the fleshy outer seed integument ...
>
> (Engler, 1879a: 48)

Engler here uses a similar approach as he employed in the Ochnaceae, using the more general Monocot Type from which to derive Araceae phylogenetically. This contrasts with Nägeli's theory, which would imply that the bisexual-flowered taxa were the most advanced and would fit better with Schott's sequence. Despite this, as we will see later on, Engler was very influenced by Nägeli's views on the evolution of plants (Diels, 1931), like many other botanists at this time (Junker, 1989).

A more detailed and much revised presentation of Engler's classification appeared just a few months later (Engler, 1876b) as Part 1 of a two-part monograph entitled *Comparative Studies on the Morphology of the Araceae* [*Vergleichende Untersuchungen über die morphologischen Verhältnisse der Araceae*]. The first part is entitled *Natural System of the Araceae* [*Natürliches System der Araceae*] and is a summary of the entire classification with diagnoses of all taxa above genus rank; the genera are assigned to their respective subtribes, tribes and subfamilies and synonymy of Schott's genera is given where relevant. A notable feature is the

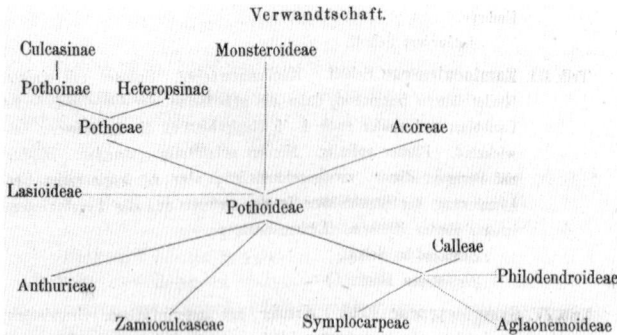

Fig 10.4 Phylogenetic diagram of Araceae subfamily Pothoideae (Engler 1876a: Figure on p. 142, 1879a: Figure on p. 64). Reproduced by kind permission of the Board of Trustees of the Royal Botanic Gardens, Kew.

provision of branching diagrams for each subfamily showing the evolutionary connections between the constituent tribes (Fig 10.4 shows the diagram for Pothoideae).

The foreword provides some details of Engler's approach. He discusses sceptically the question of whether there are classes of characters which are 'systematically essential' or not. He concludes that:

> The only correct way known to proceed in this regard has been to group the different elements [*die Formen*] of a group of related taxa [*eines Verwandtschaftskreises*] according to their greater or lesser similarity and then use the results to reach a conclusion on the systematic value of individual characters.
>
> (Engler, 1876b: 135)

This statement implies that Engler first undertook a phase of intuitive grouping by general similarity, followed by a more explicit reassessment of the taxonomic value of different characters given his initial groupings. He goes on to point out that vegetative characters have hitherto been undervalued:

> It is an even more deep-rooted axiom that the floral parts alone have systematic importance and that the nature of the leaves, leaf arrangement and shoot architecture, as well as the anatomical structure of individual organs are of a much lesser degree of importance.
>
> (Engler, 1876b: 135)

Vegetative anatomy and shoot organization are all too often neglected and thus:

> On these grounds, some groupings {with}in families where investigation of floral organs is very complete, are still to be regarded as unnatural. This is the case in the family Araceae ...
>
> (Engler, 1876b: 136)

This is an indirect reference to Schott's primary division of the family into the unisexually flowered Diclines and the bisexually flowered Monoclines. He warns that anatomy can be similar in groups that are not closely related:

> It should be noted here that anatomical characters should only be applied to classification with great care, since many comparative anatomical studies have shown that the same or similar anatomical conditions occur in plant groups which systematically certainly do not belong together and could absolutely not be connected by relationship [*welche systematisch zweifellos nicht zusammen gehö-ren und durchaus nicht in verwandtschaftlicher Beziehung stehen können*]. This is particularly relevant to the distribution of the fibrovascular and phloem strands. In fact the earlier comparative anatomical studies carried out by van Tieghem {van Tieghem 1867} on the Araceae have shown that the groups of Araceae based on the distribution of fibrovascular bundles do not correlate with the groups of Schott's system of the Araceae and that genera of the most profound natural relationship [*natürlichen Verwandschaft*] belong to different Types regarding their anatomical structure.
>
> (Engler, 1876b: 137)

But nevertheless some anatomical characters do have systematic value:

> Yet there are certain histological characters in the Araceae which I have shown to be constant and characteristic for distinct groups as a result of my extensive studies of fresh and dried material. I have been guided by these {characters} alone no more than by the leaf arrangement, the shoot architecture and the floral structure. If however they appear united with other morphological characters, I have granted them due attention and so I am now in a position to determine at least to which major group a plant must belong for any Araceae otherwise unknown to me, using {only} a small piece of its petiole or stem and with the help of a piece of the leaf blade.
>
> (Engler, 1876b: 137)

The introduction to the synoptical treatment of the classification has some key statements regarding his phylogenetic intentions:

> This system allows it to be very clearly perceived that reduction of the floral parts must have occurred in several [*verschiedenen*] {Engler's emphasis} subfamilies [*Gruppen*] and thus that Schott's classification, in which the major divisions are based on floral structure, is unnatural.
>
> (Engler, 1876b: 139)

In particular he states here that:

> My system of the Araceae would have been [*ausgefallen*] quite different if I had had the aim of providing a means of identification for botanists less familiar with the family. My purpose is different: in the following system all phylogenetic relationships [*verwandtschaftlichen Beziehungen*] between individual subfamilies

[*einzelnen Gruppen*] should be presented as clearly as possible, and hence the number of subordinate groups is larger than perhaps seems necessary at first sight.

(Engler, 1876b: 139)

Part 2 of his monograph on comparative morphology of Araceae is the important and widely known work, *On Leaf Placement and Shoot Organization of Araceae* [*Ueber Blattstellung und Sprossverhältnisse der Araceae*] (Engler, 1877), which has been translated into English in its entirety by Ray and Renner (1990). This is a detailed exposition of shoot architecture in Araceae, restudied and expanded a century later by Ray (e.g. Ray, 1990). Among many other findings, the most important result was the discovery of the near-universal occurrence in Araceae of '(*n* – *1*)' sympodial architecture (see earlier comments). Engler's analysis also revealed that in *Acorus*, *Gymnostachys* and the genera currently recognized as the subfamily Orontioideae (*Orontium*, *Lysichiton* and *Symplocarpus*), reiteration of the sympodium takes place in the axil of the last foliage leaf (*n* morphology), and furthermore that some highly specialized lianescent genera (*Pothos*, *Pothoidium*, *Heteropsis*) have monopodial architecture. The monograph also includes an important morphological analysis of *Pistia* and its relationship to the Lemnaceae.

In the following year, Engler (1878) published a treatment of Araceae for *Flora Brasiliensis* which includes, in addition to taxonomic treatments of the Brazilian species of Araceae, the most complete illustrated version of Engler's own vegetative anatomical studies (pp. 32–37, Table II–V); one of the anatomical plates consists of photographs of transverse sections of stems and petioles, the other three are his own drawings. There are shorter essays on affinities, floral structure, geographical distribution and uses. The synoptical presentation of the genera (pp. 40–48, *Conspectus generum in Brasilia provenientium*) is a summary version of the 1876 classification in which non-Brazilian taxa are included in their respective positions, but with no diagnoses.

This rather extensive material, additional to the flora treatment, was republished and expanded further, but this time in German, in the general part of Engler's species-level monograph for the *Monographiae Phanerogamarum* (Engler, 1879a), published a few months later. All key points regarding the basis for Engler's classification are addressed in this 55-page exposition. This essay is especially important for the long section on vegetative anatomy, which is a German version of the corresponding essay in *Flora Brasiliensis* and which refers to the illustrations in the latter. A summary of his 1877 findings on shoot organization and leaf arrangement is also given, as well as sections on leaf shape, floral structure, pollination, seed and seedling morphology. A description of the supposed evolution of Araceae forms the last part of the discussion of geographical distribution and is significant as Engler's first major published statement on the evolution of the family. The synoptical presentation of the classification (pp. 62–78) corresponds in almost every respect (including the branching diagrams, e.g. Fig 10.4) with the 1876 classification. The monograph is an outstanding *tour de force*, when one considers that he managed to completely revise the species taxonomy, as well as carry out a total reconstruction of the classification backed up with detailed original treatments of the most important characters he used as its basis. The monograph is incomplete only in the omission of species treatments for the Lemnoideae, which as he explained in a footnote on p. 77, had been presented in detail by Hegelmaier (1868) only a decade earlier. Engler (1879a) published very few novel taxa; most new binomials refer to combinations made as a result of changed generic or species limits. This monograph, in contrast to Schott's *Prodromus Systematis Aroidearum* (Schott, 1860), was not the fruit of 40 years' taxonomic study (as Schott's work was described by Engler in his review of the *Aroideae Maximilianae*, Engler, 1879b, English translation available at http://cate-araceae.myspecies.info/content/translations-classic-araceae-studies), but rather a thorough revision of existing knowledge of the species taxonomy. Indeed it represented the starting point for a further 40 years of new discoveries and revisions that only ended with the completion of Engler's monograph for *Das Pflanzenreich* in 1920 (Engler, 1905–1920).

10.4 Engler's subfamilies

The most crucial difference between the taxonomies of Schott (1860) and Engler (1876b) consists in the primary divisions, the former consisting of two rankless taxa corresponding to subfamilies and the latter of ten explicitly ranked subfamilies. Schott's two 'subfamilies' have no diagnoses, but their definitions are implicit in their names *Diclines* (i.e. diclinous or unisexual flowers) and *Monoclines* (bisexual or hermaphrodite flowers). This simplicity, though supported by modern molecular phylogenetic studies (bisexual flowers ≡ plesiomorphy, unisexual flowers ≡ synapomorphy), must have been a prime motive for Engler's view of it as artificial, because no other supporting characters are proposed by Schott. By contrast, Engler based his major subdivisions, the subfamilies, on a broad range of characters from vegetative anatomy (trichosclereids, tannin cells, laticifers, collenchyma patterns), shoot organization, phyllotaxis, leaf venation, leaf-blade shape, life form and stem morphology, presence of prickles, spathe morphology, spadix morphology, floral sexuality, floral diagram data, presence of a perigon, connation of androecium, ovule morphology and presence of endosperm in the seed. Engler's subfamilies are not as sharply defined as Schott's two major groups, the implicit characters of which essentially function as the leads of a synoptical key.

Engler based his subfamilies on combinations of characters common to broad groups of genera. He interpreted the different states of these characters as

evolutionary transformation series. He attempted to justify the evolutionary presumption of these series and their direction using a variety of arguments, refined
especially in his 1884 paper (see later) where he determined morphological
'Progressions', following the precepts of Nägeli (1884). It is in this stage of the
process that we sense most forcefully the difference that 140 years of progress in
evolutionary biology and genetics has made between modern views and the
intellectual milieu in which Engler worked. The ten subfamily delimitations stand
on combination of characters which he states to be predominantly or universally
present in each, although the characters may not be unique to a given subfamily.
Since Engler was influenced by Nägeli's theory of evolution, it is questionable
whether the phylogenetic or genealogical lines he draws between the subfamilies
in his branching diagrams are directly comparable to the stems of a cladogram. It
seems more likely that he regarded the subfamilies as stages in the evolution of
greater vegetative complexity and progressive floral reduction which lower-level
taxa (e.g. genera) have undergone. The genera proceed phylogenetically through
this scheme of linked stages on a particular path (close to a Lamarckian idea of
evolution, see Stevens, 1994), but there is a contingency element in that not all taxa
pass through every Type on their way to higher forms, e.g. the Colocasioid genera
of today did not pass through the Monsteroid Type (Engler, 1884: 331). Engler's
Araceae dendrograms represent a summation of separate lineages which each
began separately from the Pothoid Type, itself a level of organization which is seen
as only a little further advanced from a general monocot Type.

10.5 The 1884 paper

Five years after his 1879 monograph, Engler completed his taxonomic edifice in a
110-page illustrated paper (*Über den Entwicklungsgang in der Familie der Araceen
und über die Blütenmorphologie derselben*, Engler, 1884; for an English translation
see http://cate-araceae.myspecies.info/content/translations-classic-araceae-studies/),
where he laid out in more detail the theoretical basis for his phylogenetic
classification. Nicolson (1983, 1988) and Grayum (1984, 1990) have previously
drawn attention to this work. Engler must have been somewhat sensitive to the
need to justify his classification since in a footnote he remarks that 'A valued
colleague', who had made a thorough study of Engler's work on the family
'… often declared that he was still waiting for the detailed evidence of my
expressed assertions' (Engler, 1884: 143); it seems likely that this would have
been his director, Carl von Nägeli.

   The introduction contains a number of interesting general statements by Engler
which shed light on his approach to constructing taxonomies that reflected phylogeny. In the next part of this long essay, Engler sets out the various states of organ

systems which he used to assemble the subfamilies and to connect them phylo-genetically. The key feature is that he recognizes within each organ system a progressive series of stages [*Stufen*] which are treated as morphological or ana-tomical transformation series [*Progressionen*]. These transformations are not always interpreted as simple linear series; sometimes they are seen as branched sequences (for example in the section on tissue development [*Ausbildung der Gewebe*], p. 147). However, they are interpreted by him essentially as a time series.

The exposition of these Progressions is accompanied by extensive commentary and explanation. Discussion is particularly elaborate with regard to the flowers (pp. 160–173). He begins this section by setting out eight general propositions on the morphological interpretation of floral structure and these are followed by nine pages of discussion, with examples not only from Araceae, but also from other plant families. This part of the essay seems to be the fruit both of Engler's lively interest in contemporary theoretical morphology (Diels, 1931) and the very recent publication of Nägeli's *Abstammungslehre* (Nägeli, 1884), which is cited many times.

In the third part of the treatise, Engler discusses each of his subfamilies separ-ately, beginning with the Lasioideae, noting that:

> The sequence in which I present the individual subfamilies has no significance in a phylogenetic sense; I begin with a group of related taxa which I name the Lasioideae, on account of the fact that this group shows the most diverse forms of development of the inflorescence and thus immediately provides the reader with an opportunity to become familiar with the most important {morphological} phenomena in the family Araceae.
>
> (Engler, 1884: 173)

The separate subfamilial discussions focus mainly, but not exclusively, on repro-ductive structures, and each terminates with a table summarizing the characters of their constituent tribes. The reason for this focus on floral characters is probably due to Engler's concern to demonstrate the step-wise progression of the varied floral structure within his subfamilies, in the delimitation of which vegetative anatomy and morphology play the dominant role. The discussions reveal the extent to which Engler attempted to interpret varying floral structures as either more ancestral or more derived, within and between genera, e.g. his discussion of *Cyrtosperma* gynoecia beginning on p. 174.

At the end of the paper, there is a short paragraph on the relationship of the subfamilies to one another. The previously published trees of each subfamily (Engler, 1876b, 1879a) are here replaced by a single tree showing only the inter-relations of the subfamilies themselves (Fig 10.5), and this is introduced with these words:

> Whether one recognizes the above discussed groups of Araceae as subfamilies or as tribes, is rather unimportant, as in any case they are linked together by

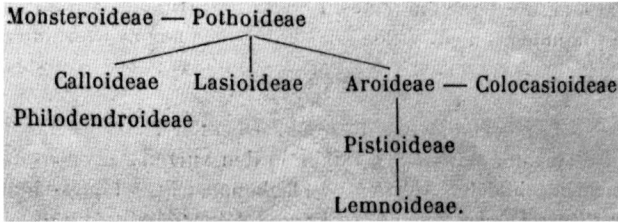

Fig 10.5 Phylogenetic diagram of the subfamilies of Araceae from Engler (1884: 331) and Engler (1887–1989: 111). Reproduced by kind permission of the Board of Trustees of the Royal Botanic Gardens, Kew.

> genealogical relationships [*verwandtschaftlicher Beziehung*]; only in the case of the Lemnoideae can any doubt be warranted. According to the principles expressed in the first sections the subfamilies must stand together in the following relationship . . . {dendrogram follows}
>
> (Engler, 1884: 331)

Engler was clearly thinking largely in Nägelian terms about phylogeny – i.e. advanced forms are ancient in relation to the rest – since he comments finally:

> This arrangement gives absolutely no evidence for the age of the genera within the individual subfamilies [*Gruppen*], it suggests *only the relative age of the subfamilies* {Engler's emphasis}, the genera of the Philodendroideae, Aroideae, Pistioideae which now exist could even be older than the extant genera of the Pothoideae, because they are phylogenetically already the furthest advanced.
>
> (Engler, 1884: 331)

This extensive paper is exceptionally interesting as a discussion of the thought that lay behind Engler's Araceae classification. It reveals the extent to which he was influenced by Nägeli, and also his intense interest in the detailed application of evolutionary ideas to taxonomy within the context of comparative morphology.

He published two further versions of his Araceae classification, for *Die natürlichen Pflanzenfamilien* (Engler, 1887–1889) and for *Das Pflanzenreich* (Engler, 1905–1920; Engler and Krause, 1908, 1920; Krause, 1908, 1913). In the *Pflanzenreich* volumes Engler also wrote extended essays for each subfamily which provided an updated and more detailed exposition of his taxonomic system. Although there were some notable changes in these later works, e.g. the treatment of the Lemnaceae as a separate family and substantial changes in the genera recognized, the overall structure of the classification remained remarkably constant. The monograph for *Das Pflanzenreich* (summarized in Engler, 1920b), the last version of Engler's system, was the standard classification for Araceae and has not yet been surpassed as a single statement of Araceae taxonomy. The high-level subfamily classification was replaced only with the advent of modern phylogenetic systematics based on cladistic principles and in particular the work of Mike Grayum (1984, 1990). After this, aroid classification underwent many changes and the modern view is represented by the molecular phylogenies of Cabrera et al. (2008) and Cusimano et al. (2011). The most recent formal synopsis of the classification of Araceae was published by Bogner and Petersen (2007).

10.6 Engler's phylogenetic method and the influence of C.W. Nägeli's theory of evolution

In his evolutionary thinking Engler seems always to have been relatively conservative,[5] but what perhaps marks him out especially was his determination to carry out in detail the construction of large-scale taxonomic systems based on the evolutionary paradigm, even if his version of phylogenetic classification did not find universal favour (see e.g. Mägdefrau, 1973). He was strongly influenced by C.W. Nägeli and as a result his understanding of phylogeny was very divergent from that of today, but not from that of many other contemporary botanists of late nineteenth-century Germany (Junker, 1989). Engler's Araceae classification was born in the mid 1870s and revised over a 45-year period during which evolutionary ideas underwent constant and much-debated change, with widely differing views current at any one time on such basic questions as the mechanism of heredity and the likelihood of a monophyletic origin of plants. It is hardly surprising, therefore, that to modern botanists his method of constructing his phylogenetic classification is difficult to comprehend clearly in all details.

In attempting to understand Engler's classification of Araceae it is useful to consider the context in which he worked, i.e. the evolutionary debate in German botany during the second half of the nineteenth century. Thomas Junker (e.g. 1989, 2002, 2009) has provided an impressive analysis of the impact of Darwin's theory on the German botanical community and he shows that by the 1870s, most botanists, especially of Engler's generation (e.g. E. Strasburger), had accepted evolution as the explanation for morphological diversity and the mutability of species. On the other hand, natural selection had not been widely accepted as the prime mechanism and controversy over this was to continue for many decades to come (Junker, 1989). During the early years of this period, Julius Sachs's highly successful textbook of botany (e.g. 4th edition, Sachs, 1874) was significant in increasing the acceptance of evolutionary thought among botanists.

Carl Wilhelm von Nägeli (Fig 10.6), like Sachs, was a highly influential voice and especially for Engler during his Munich years in the 1870s, when he was under Nägeli's directorship.[6] Nägeli was one of the major thinkers on questions of evolution and heredity in the German-speaking botanical community (Junker, 1989) and soon after Darwin's *Origin* appeared had published an influential orthogenetic

Fig 10.6 Carl Wilhelm von Nägeli (1817–1891). Professor of Botany at Munich University 1857–1889 and Engler's chief in Munich between 1871 and 1878. Nägeli was a major figure in plant evolutionary thinking in Germany prior to rediscovery of G. Mendel's genetics research. Reproduced from C. Cramer (1896), by kind permission of the Board of Trustees of the Royal Botanic Gardens, Kew.

evolutionary theory (Nägeli, 1865; Junker, 2009). However, Nägeli (1884) further developed his theory in ways which brought it into conflict with Darwinism, primarily because of its orthogenetic basis. According to Nägeli's theory, evolution was mainly the result not of chance factors (as in Selection Theory), but of an inherent deterministic process of morphological complexification and structural stabilization, hence the name Perfection Theory [*Vervollkommnungstheorie*]. From this Nägeli argued that spontaneous generation, inheritance of acquired characters and polyphyly were necessary consequences of what he explicitly intended as a nonvitalistic, mechanistic [*mechanische*] view of evolution; that is, if evolution of morphological forms were no more than the consequence of deterministic physical and chemical processes that are still the same today, then it follows that spontaneous generation of the simplest living forms has been happening throughout the history of life. It follows in turn from this that the diversity of morphological forms present today is more likely to be the product of many concurrent evolutionary trees rather than just one. Although he postulated that the constant action of spontaneous generation would result in lineages, none of which could be the same in their constitution, he nevertheless argued that orthogenetic evolution from simple to complex organisms in similar conditions will lead to similar morphologies in higher organisms. This follows from his uniformitarianist view regarding the fundamental physical and chemical basis of morphological development.

Engler (1879a: 48–55)[7] presented his earliest discussion of the evolution of Araceae as part of a discussion of the geographical distribution of the family.

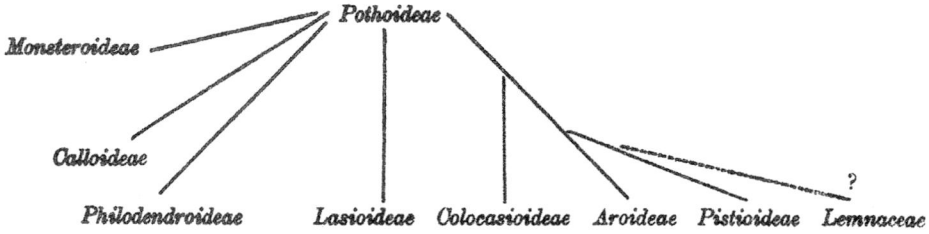

Fig 10.7 Phylogenetic diagram of the subfamilies of Araceae from Engler (1920b: 63). Reproduced by kind permission of the Board of Trustees of the Royal Botanic Gardens, Kew.

He based his subfamilies on correspondence in vegetative and reproductive characters and linked them together by phylogenetic connections in the form of trends of increasing complexity in vegetative characters and reduction trends in floral morphology. In these early treatments he did not explicitly specify evolutionary stages and progressions, but focussed on grouping genera and tribes into more natural higher taxa:[8]

> Natural grouping does not consist of arranging the genera in such a way as to facilitate identification or accessibility for nonspecialists, but rather the establishment of an arrangement which demonstrates evolution [*den Entwickelungsgang*] in the family most clearly. *Since the representation of ancestors is thus sought* {our emphasis}, which is out reach of direct observation, it is obvious that even the best 'natural arrangement' must suffer shortcomings, . . .
>
> (Engler, 1884: 141–142)

His dendrograms, however, are more circumspect and although he published many in the early works (Engler, 1876b, 1879a), from 1884 onwards he restricted himself just to a single diagram (Figs 10.5, 10.7) showing evolutionary connections between the subfamilies (Engler, 1884, 1887–1889, 1920b).

Perhaps by 1884 Engler had modified his views on monophyly of major groups under Nägeli's influence. He was unsure regarding the difficult problem of determining relatedness:

> In fact, today, {even} with our direct aim of establishing the genetic connections within a family, we also get no further than the recognition that some genera are more closely 'related' {Engler's quotes} and others stand further apart. However, the establishment of real genetic connections is {in this way} only very weakly made and in fact one must often be satisfied if it is possible to state with confidence that one genus belongs to a phylogenetically older 'Type' {Engler's quotes} and another to a phylogenetically younger one. Furthermore, the phylogenetic trees [*Stammbäume*] that have been constructed here and there, which were set up only extremely rarely with the presumption that they should exactly express the evolutionary development [*Entwicklung*] that actually happened, serve rather to show that one genus belongs to a phylogenetically older stage

and another to one phylogenetically younger. In fact, however, there is often a lack of clarity about what is to be regarded as an ancient [*ältere*] structure [*Bildung*] and what a more recent one. Moreover, more recent investigations have often showed that not rarely, external correspondences {of characters between taxa}, which lead erroneously to the acceptance of relationship, are no more than adaptations brought about by the same cause at different times.

(Engler, 1884: 144)

Nägeli (1884) discusses extensively the relation between evolutionary theory and systematics (in Chapter 9 – *Morphology and Systematics as Phylogenetic Sciences*) and he there sets out a detailed series of evolutionary stages [*Stufen*] within each major organ system of the Angiosperms: shoot system, leaves, flower and floral parts. These stages are postulated as transformations undergone as the plants pass from simpler, less tightly organized forms to more complex and more stable 'higher' forms under his Perfection Theory.

Engler used Nägeli's method in his 1884 paper (on page 145, Nägeli's book is cited with the date 1883) as the framework for his argumentation.[9] It was here that Engler first published his Progressions (republished later in Engler, 1887–1889, 1920b). But there are other aspects of Nägeli's theory; belief in an extreme level of polyphyly, absence of genetic connections between even taxa that are very similar, inheritance of acquired characters and an inner orthogenetic force of evolutionary complexification, immune to the influence of external conditions. Nägeli was by no means the only botanist who had such ideas about evolution; as Junker (1989) discusses in detail, J. Sachs also held very similar views.

Engler's ambiguous attitude regarding the monophyly/polyphyly of his Araceae subfamilies is important to bear in mind. It seems that his morphological and anatomical Progressions occurred in parallel in his separate subfamilies via some kind of orthogenetic mechanism. It seems surprising to us, for example, that Engler, after his preliminary paper (Engler, 1876a), did not use the presence of laticifers as a character to group those subfamilies in which they are present, despite the apparently obvious correlation with unisexual flowers and the clear fit with Schott's unisexual-flowered higher taxon Diclines. But this must reflect the influence of Nägeli's theory, in which polyphyly was fundamental. Sometimes a lineage acquires a unique character (e.g. trichosclereids in the Monsteroideae), but often many subfamily lineages will independently (polyphyletically) attain the same stage, as, for example, the acquisition of laticifers and unisexual flowers, and the loss of the perigon in Philodendroideae, Aroideae, Colocasioideae, Pistioideae, and within Lasioideae. For Engler they are indications of parallel evolution rather than monophyly, as they are interpreted today (Cusimano et al., 2011). Engler used these character transformations as tools for establishing relationships, if in a different theoretical context to today's, and modern studies (e.g. Keating, 2002; Cabrera et al., 2008) show just how many remain important in Araceae phylogenetics at some level.

Engler's system of Araceae, in its essentials, is now over 130 years old and many of the theoretical concepts underlying his explanation of the family's phylogeny and evolution have been subsequently shown to be wrong. It is nevertheless right to recognize the magnitude of his achievement and the influence that his classification continues to exert, not through evolutionary theory but through his taxonomic discoveries. His anatomy and morphology in particular, provide modern aroid systematists with important tools to interpret molecular phylogenies and continue to pose key biological problems. Why, for example, should '$n - 1$' sympodial architecture have been such an important step in the early evolution of Araceae (Cabrera et al., 2008; Cusimano et al., 2011)? What is the biological significance of the evolution of laticifer systems in aroids, and why did this occur along with the evolution of unisexuality? Engler assembled an enormous amount of new taxonomic data on aroids in a relatively short time. He constructed his evolutionary taxonomy with the theoretical tools then most widely accepted. He explained in detail why he did what he had done. Perhaps most important of all, he provided the taxonomic framework on which the flowering of modern aroid systematics was built. In achieving this, he made huge strides in increasing our knowledge of the biology of Araceae and left to future generations a whole range of fascinating evolutionary problems, many of which remain to be investigated.

This paper is dedicated to Dan H. Nicolson by the authors with esteem, gratitude and affection.

10.7 References

Barabé, D. and Vieth, J. (1990). Les principes de systématique chez Engler. *Taxon*, **39**, 394–408.

Bogner, J. and Petersen, G. (2007). The chromosome numbers of the aroid genera. *Aroideana*, **30**, 82–90.

Cabrera, L.I., Salazar, G.A., Chase, M.W. et al. (2008). Phylogenetic relationships of aroids and duckweeds (Araceae) inferred from coding and noncoding plastid DNA. *American Journal of Botany*, **95**, 1153–1165.

Candolle, A.-P. de. (1813). *Théorie Élémentaire de la Botanique*. Paris: Déterville.

Cramer, C. (1896). *Leben und Wirken von Carl Wilhelm von Nägeli*. Zürich: F. Schulthess.

Croat, T.B. (1983). Heinrich Gustav Adolph Engler: A prodigious Aroid worker. *Aroideana*, **6**(3), 68–70.

Cusimano, N., Bogner, J., Mayo, S.J. et al. (2011). Relationships within the Araceae: comparison of morphological patterns with molecular phylogenies. *American Journal of Botany* **98**, 654–668.

Diels, L. (1931). Zum Gedächtnis von Adolf Engler. *Botanische Jahrbücher*, **64**, I–LVI.

Engler, A. (1867). Beiträge zur Naturgeschichte des Genus *Saxifraga* L. *Linnaea*, **35**, 1–124.

Engler, A. (1872). *Monographie der Gattung Saxifraga L. mit besonderer Berücksichtigung der geographischen Verhältnisse*. Breslau: J.U. Kern's Verlag.

Engler, A. (1874). Studien über die Verwandtschaftsverhältnisse der Rutaceae, Simarubaceae und Burseraceae nebst Beiträgen zur Anatomie und Systematik dieser Familien. *Abhandlungen der naturforschenden Gesellschaft zu Halle*, **13**, 1–50, plates XII–XIII.

Engler, A. (1875). Ueber Begrenzung und systematische Stellung der natürlichen Familie der Ochnaceae. *Nova Acta Kaiserlich Leopoldinisch-Carolinischen Deutschen Akademie der Naturforscher*, **37**(2), 1–28, plates XII–XIII.

Engler, A. (1876a). Zur Morphologie der Araceae: Vorläufige Mittheilung. *Botanische Zeitung*, **34**(6–7), 81–90, 97–105.

Engler, A. (1876b). Vergleichende Untersuchungen über die morphologischen Verhältnisse der Araceae. I. Theil. Natürliches System der Araceae. *Nova Acta Academiae Caesareae Leopoldino-Carolinae Germanicae Naturae Curiosorum* **39**(3), 133–155.

Engler, A. (1877). Vergleichende Untersuchungen über die morphologischen Verhältnisse der Araceae. II. Ueber Blattstellung und Sprossverhältnisse der Araceae. *Nova Acta Academiae Caesareae Leopoldino-Carolinae Germanicae Naturae Curiosorum* **39**(4), 157–232, plates 1–6 (VIII–XIII).

Engler, A. (1878). Araceae. In *Flora Brasiliensis, Volume 3(2)*, ed. C.F.P. Martius. Munich: Typographia Regia C. Wolf et fil, pp. 25–224, plates 2–52.

Engler, A. (1879a). Araceae. In *Monographiae Phanerogamarum, Volume 2*, ed. A.. and C. de Candolle. Paris: G. Masson, pp. 1–681.

Engler, A. (1879b). Aroideae Maximilianae. *Botanische Zeitung (Leipzig)*, **37**, 853–856.

Engler, A. (1881). Ueber die morphologischen Verhältnisse und die geographische Verbreitung der Gattung *Rhus*, sowie der mit ihr verwandten, lebenden und ausgestorbenen *Anacardiaceae. Botanische Jahrbücher*, **1**, 365–426, plate IV.

Engler, A. (1884). Beiträge zur Kenntnis der Araceae V. 12. Über den Entwicklungsgang in der Familie der Araceen und über die Blütenmorphologie derselben. *Botanische Jahrbücher*, **5**, 141–188, 287–336, plates I–V.

Engler, A. (1887–1889). Araceae. In *Die natürlichen Pflanzenfamilien, Volume II, Part 3*, ed. A. Engler and K. Prantl. Leipzig: W. Engelmann, pp. 102–153.

Engler, A. (1905). Araceae – Pothoideae. In *Das Pflanzenreich*, Heft **21** (IV.23B), ed. A. Engler. Leipzig: W. Engelmann, pp. 1–330.

Engler, A. (1908). Additamentum ad Araceas – Pothoideas. In *Das Pflanzenreich*, Heft **37** (IV.23B), ed. A. Engler. Leipzig: W. Engelmann, pp. 1–3.

Engler, A. (1911). Araceae – Lasioideae. In *Das Pflanzenreich*, Heft **48** (IV.23C), ed. A. Engler. Leipzig: W. Engelmann, pp. 1–130.

Engler, A. (1912). Araceae – Philodendroideae – Philodendreae. Allgemeiner Teil, Homalomeninae und Schismatoglottidinae. In *Das Pflanzenreich*, Heft **55** (IV.23Da), ed. A. Engler. Leipzig: W. Engelmann, pp. 1–134.

Engler, A. (1915). Araceae – Philodendroideae – Anubiadeae, Aglaonemateae, Dieffenbachieae, Zantedeschieae, Typhonodoreae, Peltandreae. In *Das Pflanzenreich*, Heft **64** (IV.23Dc), ed. A. Engler, Leipzig: W. Engelmann, pp. 1–78.

Engler, A. (1917). Karl Wilhelm von Nägeli. *Internationale Monatsschrift für Wissenschaft, Kunst und Technik*, **12** (1917–1918), 63–84.

Engler, A. (1919). *Prinzipien der systematischen Anordnung.* In *Syllabus der Pflanzenfamilien*, 8th edn., ed. A. Engler and E. Gilg. Berlin: Gebrüder Borntraeger, pp. VIII–XXV.

Engler, A. (1920a). Araceae – Aroideae und Araceae – Pistioideae. In *Das Pflanzenreich*, Heft **73** (IV.23F), ed. A. Engler. Leipzig: W. Engelmann, pp. 1–274.

Engler, A. (1920b). Araceae – Pars generalis et index familiae generalis. In *Das Pflanzenreich*, Heft **74** (IV.23A), ed. A. Engler. Leipzig: W. Engelmann, pp. 1–71.

Engler, A. (1920c). Additamentum ad Araceas – Philodendroideas. In *Das Pflanzenreich*, Heft **71** (IV.23E), ed. A. Engler. Leipzig: W. Engelmann, pp. 1–2.

Engler, A. and Krause, K. (1908). Araceae – Monsteroideae. In *Das Pflanzenreich*, Heft **37** (IV.23B), ed. A. Engler. Leipzig: W. Engelmann, pp. 4–139.

Engler, A. and Krause, K. (1920). Araceae – Colocasioideae. In *Das Pflanzenreich*, Heft **71** (IV.23E), ed. A. Engler. Leipzig: W. Engelmann, pp. 3–132.

French, J.C. (1988). Systematic occurrence of anastomosing laticifers in Araceae. *Botanical Gazette*, **149**, 71–81.

French, J.C., Chung, M. and Hur, Y. (1995). Chloroplast DNA phylogeny of the Ariflorae. In *Monocotyledons: Systematics and Evolution, Volume 1*, ed. P.J. Rudall, P.J. Cribb, D.C. Cutler and C.J. Humphries. Kew: Royal Botanic Gardens, pp. 255–275.

Grayum, M.H. (1984). *Palynology and Phylogeny of the Araceae.* Ph.D. thesis. Amherst, MA: University of Massachusetts.

Grayum, M.H. (1990). Evolution and phylogeny of the Araceae. *Annals of the Missouri Botanical Garden*, **77**, 628–697.

Haeckel, E. (1866). *Generelle Morphologie der Organismen.* 2 Volumess. Berlin: G. Reimer.

Hay, A. (1992). Tribal and subtribal delimitation and circumscription of the genera of Araceae tribe Lasieae. *Annals of the Missouri Botanical Garden* **79**, 184–205.

Hay, A. and Mabberley, D.J. (1991). 'Transference of Function' and the origin of aroids: their significance in early angiosperm evolution. *Botanische Jahrbücher*, **113**, 339–428.

Hegelmaier, F. (1868). *Die Lemnaceen.* Leipzig: W. Engelmann.

Hooker J.D. (1862). Ordo XXXIX. Rutaceae – Ordo XLII. Burseraceae. In *Genera Plantarum, Volume 1(1)*, ed. G. Bentham and J.D. Hooker. Kew: A. Black, pp. 278–327.

Junker, T. (1989). *Darwinismus und Botanik: Rezeption, Kritik und theoretische Alternativen im Deutschland des 19. Jahrhunderts.* Stuttgart: Deutscher Apotheker Verlag.

Junker, T. (2002). Carl Nägeli und der Anti-Darwinismus. Von der Vervollkommnungstheorie zur Makroevolution. In *Pratum floridum: Festschrift für Brigitte Hoppe*, ed. M. Folkerts, S. Kirschner and A. Kühne. Augsburg: Rauner, pp. 205–219.

Junker, T. (2009). Charles Darwin, Carl Nägeli und das Rätsel der 'neutralen Merkmale'. In *Darwin und die Botanik*, ed. J. Stöcklin and E. Höxtermann. Rangsdorf: Basilisken-Presse, pp. 192–211.

Jussieu, A.-L. de. (1789). *Genera Plantarum.* Paris: Hérissant et Barrois, Paris.

Keating, R.C. (2002). IX. Acoraceae and Araceae. In *Anatomy of the Monocotyledons*, ed. M. Gregory and D.F. Cutler. Oxford: Clarendon Press, pp. 1–327.

Krause, K. (1908). Araceae – Calloideae. In *Das Pflanzenreich*, Heft **37** (IV.23B), ed. A. Engler. Leipzig: W. Engelmann, pp. 140–155.

Krause, K. (1913). Araceae – Philodendroideae – Philodendreae – Philodendrinae. In *Das Pflanzenreich*, Heft **60** (IV.23Db), ed. A. Engler. Leipzig: W. Engelmann, pp. 1–143.

Lack, H.W. (2000). *Botanisches Museum Berlin. Adolf Engler – Die Welt in einem Garten*. München: Prestel.

Mägdefrau, K. (1973). *Geschichte der Botanik: Leben und Leistung grosser Forscher*. Stuttgart: Gustav Fischer Verlag.

Nägeli, C.W. (1865). *Entstehung und Begriff der naturhistorischen Art*. München: Königliche Akademie.

Nägeli, C.W. (1884). *Mechanisch-physiologische Theorie der Abstammungslehre*. München: R. Oldenbourg.

Nicolson, D.H. (1983). Translation of Engler's classification of Araceae with updating. *Aroideana* **5**, 67–88.

Nicolson, D.H. (1988, '1987'). History of aroid systematics. *Aroideana* **10**(4), 23–30.

Radlkofer, L. (1883). *Ueber die Methoden in der botanischen Systematik, insbesondere die anatomische Methode*. Festrede zur Vorfeier des Allerhöchsten Geburts- und Namensfestes Seiner Majestät des Königs Ludwig II. gehalten in der öffentlichen Sitzung der k. Akademie der Wissenschaften zu München am 25. Juli 1883. München:

Verlag der königlichen bayerischen Akademie.

Ray, T.S. (1990). Metamorphosis in the Araceae. *American Journal of Botany*, **77**, 1599–1609.

Ray, T.S. and Renner, S.S. (1990). A. Engler, Comparative studies on the morphology of the Araceae Part II. On leaf placement and shoot organization of Araceae. *Englera*, **12**, 1–140, 6 plates.

Sachs, J. (1874). *Lehrbuch der Botanik*, 4th edn. Leipzig: W. Engelmann.

Schott, H.W. (1858). *Genera Aroidearum*, 98 plates. Vienna: C. Ueberreuter.

Schott, H.W. (1860). *Prodromus Systematis Aroidearum*. Vienna: Typis congregationis mechitharisticae.

Stevens, P.F. (1994). *The Development of Biological Systematics: Antoine-Laurent de Jussieu, Nature, and the Natural System*. New York: Columbia University Press.

van Tieghem, P. (1867). Recherches sur la structure des Aroidées. *Annales des Sciences Naturelles série 5, Botanique*, **6**, 72–210, plates 1–10.

Wagner Jr., W.H. (1980). Origin and philosophy of the groundplan-divergence method of cladistics. *Systematic Botany*, **5**, 173–193.

Wettstein, R. (1935). *Handbuch der systematischen Botanik*, 4th edn. Leipzig: F. Deuticke.

Endnotes

1 Schott's classification was conceived and written in the classic period that followed Jussieu (1789) and de Candolle (1813) when the Natural Method displaced the Linnean approach to classification and the aim of virtually all botanists was to trace the 'natural' pattern of relationships between and within plant families (see Stevens, 1994 for a very detailed account of this period). Schott's classification was 'natural' in that classic sense, relying on the correlation of mainly reproductive characters to form groups, following the precepts of Jussieu and de Candolle, but without recourse to evolutionary explanations.

Engler admitted that he found most of Schott's lower-ranking taxa (tribes, subtribes, genera) to be natural: 'A large part of the tribes and smaller groups set up by Schott were however so natural that with a few alterations they could be maintained.' (e.g. Engler, 1884: 141) and so it seems that the artificiality lay for Engler at the higher level, and especially in Schott's primary division of the Araceae into unisexual- (Diclines) and bisexual- (Monoclines) flowered taxa. Schott arranged his classification without any reference to an evolutionary scenario, although this is difficult to verify as Schott was extremely sparing in any commentary on his classification. Engler on the other hand was prolix; as Diels notes:

> Engler has discussed more extensively and more frequently than many of his forerunners the theoretical grounds of his system, so that we have extensive literature to hand when we wish to assess his systematic ideas.
>
> (Diels, 1931: XVI)

2 This example of Engler's earlier style of phylogenetic explanation, which owes more to Haeckel's monophyletic viewpoint (Haeckel, 1866) than Nägeli's polyphyletic one, is given in a passage from his study of Anacardiaceae systematics and evolution which accompanies and explains the dendrogram in plate IV, itself seemingly a forerunner of W.H. Wagner's (Wagner, 1980) groundplan divergence dendrograms (Engler, 1881: 398–401, plate IV). See http://cate-araceae.myspecies.info/content/translations-classic-araceae-studies for an English translation of this passage.

3 Modern molecular phylogenies show that this character is a synapomorphy near the base of the family phylogeny (Cusimano et al., 2011).

4 In his 1876a paper, genera of subtribe *Lasinae* (*Lasimorpha, Cyrtosperma, Lasia, Anaphyllum*) are assigned to the Pothoideae because of their lack of special anatomical characteristics. However, he mentions earlier in this paper that if the floral dissimilarities are not taken into account, the *Dracontioninae* have a close relationship and the same anatomical characteristics as the *Amorphophallinae*, which can only mean in this context that in his opinion laticifers were present in the *Dracontioninae* – sometimes written as *Dracontieae* in this paper – (*Godwinia, Echidnium, Ophione* and *Dracontium*). In the *Monographiae Phanerogamarum* Engler states that

> The most common manifestation {of laticifers} is that the latex-containing cells form straight series on both sides of the phloem, more rarely also in the middle, and later form tubes by fusing together, as in *Lasia, Dracontium* (Engl. l.c. [*Flora Brasiliensis*] t. III. f. 12, 12a) ...
>
> (Engler, 1879a: 8)

For *Lasia* he adds the footnote

> In *Lasia* the occurrence of laticifers {latex tubes} [*Milchsaftröhren*] is denied both by Karsten (Monatsber. D. Berl. Akad. 1857 p. 253) and by De Bary (Vergl. Anatomie p. 451). But they are in fact present, although here lying on the inner side of the phloem immediately adjacent to the xylem.
>
> (Engler, 1879a: 8)

Engler's two figures of *Dracontium polyphyllum* in the *Flora Brasiliensis* (Engler, 1878: 217 (legend) t. III, f. 12, 12a) are transverse and longitudinal sections of the petiole showing laticifers as shaded cells, and in the case of the LS, as a continuous tube with no transverse cell walls; these figures were republished in Engler (1887–1889: 105, Fig 73). Engler's observation is not mentioned in the survey of laticifer anatomy of Araceae by French (1988), who found that no member of the modern Lasioideae (= Lasieae of Engler) contained laticifers. Engler's erroneous observation (as it seems) of laticifers in *Lasia* and *Dracontium* must be one of the primary reasons which led Engler to set up his highly heterogeneous Lasioideae, combining bisexual- and unisexual-flowered genera, a circumscription which, along with Engler's Pothoideae, has been decisively rejected by findings from modern molecular systematics (see also Grayum, 1990).

5 Diels writes as follows on Engler's mature views on evolution and systematics:

> Engler avowed himself to be an evolutionist, but at the same time he was convinced that for the present we are not yet in a position to reproduce in our system, so to speak, the real evolutionary pathway of the plant kingdom. He granted only the possibility to express the genetic interconnections in the system for the ultimate branches of the stems . . . But he had many misgivings in regard to attempting this for all plants. For he thought it more likely that the main lineages of the plant kingdom were separate phylogenetically [*eine phyletische Getrenntheit der Hauptstämme*] than that they were monophyletic. . . . Thus a double task is indicated to systematists: to determine the constancy of characters for the purpose of comprehending the groups, and to investigate the progressions in order to arrange their members logically. Whether these progressions correspond to historical reality, we cannot decide, because the palaeontological evidence is inadequate for this purpose. We remain reliant on the subjective appraisal of the given status of the morphology. . . . In summary it may be said of Engler that his scientific ideal is the phylogenetic system. But he regards it as impossible at the present time to reach this ideal. In order to come closer to it in the meanwhile, he sees the main task to be the amassing of new facts. . .
>
> (Diels, 1931: XVI–XVII)

Mägdefrau (1973) in a brief mention of Engler's work emphasizes this view of Engler's approach to phylogeny:

> Engler was of the view that systematics and phylogeny should be kept apart and that the system should be built on similarity in characters, still [*nach wie vor*] without considering phylogenetic interpretations. In other words: Systematics or Taxonomy organizes the diversity of forms and Phylogeny explains it; the means of representing systematics is a unidimensional system and that of representing phylogeny is the two dimensional phylogenetic tree [*Stammbaum*].
>
> (Mägdefrau, 1973: 194)

Richard von Wettstein, author of an influential phylogenetic classification of plants (*Handbuch der systematischen Botanik*, 4th edn., Wettstein, 1935), also took the view that phylogenetic interpretations should only be included in classifications with caution:

> Since a complete construction of the system on a phylogenetic basis can hardly be achieved, we must be content if the system, as far as possible, gives a reflection of our phylogenetic knowledge and we must reckon with the possible need to avoid expressing empirical knowledge of phylogeny in the form of the system itself.
>
> (Wettstein, 1935: 2).

6 Engler published an extensive appreciation of Nägeli (Engler, 1917) and Diels (1931) provided the following assessment of Nägeli's influence on Engler's thinking:

The exchange of views with Nägeli and the insight he {Engler} gained from him in thought and productive activity must have been of lasting influence for his intellectual development. This great scholar was very different in mentality from the type of teacher Engler had in Breslau; this contrast could not fail to have its effect. In Nägeli's productive output and his restless diligence Engler found similar tendencies to his own character; however, in his {Nägeli's} often deductive research approach addressing the most general questions, he {Engler} encountered something quite new. These intellectual factors took full effect in the extraordinary clarity of his {Nägeli's} discourse 'without any rhetorical decoration' and made it fruitful for every listener who could follow it. 'Whoever worked in Nägeli's laboratory', said Engler in a centenary essay about his former director {Engler, 1917}, 'or who spoke with him about scientific matters, had to prepare themselves for sharp criticism'. Here we obviously hear the memory of his own experience. For scientifically, Engler's problems offered close contact with the questions with which Nägeli was dealing in the {Eighteen} Seventies as he worked on speciation [*Formenbildung*] in *Hieracium* on the largest scale and indeed using methods which Engler had also used in his study of *Saxifraga* and applied as widely as his limited means allowed. Thus both in actual discussion, and even more in thought, Engler had come to terms with Nägeli's ideas on relationships and descent of organisms and one has the impression that {Engler's} views, for example on phylogenetic systematics, had been significantly influenced by them.

(Diels, 1931: VII)

7 Translation of Engler's (Engler, 1879a) first substantial published statement on the evolution of Araceae, following a description of the family's biogeography (he published an updated and amplified version of this essay in *Das Pflanzenreich* (Engler, 1920b):

I may now be allowed to add a theoretical reflection on the evolutionary development [*Entwicklung*] of the Araceae, by means of which some relations can be considered which could not be expressed in the tables by numbers {only}.

In the first place it should be borne in mind that in all Araceae the viability of the seeds is very short, and that as a result their long-range migration destroys the germinability of the seeds. It is furthermore to be considered that with the exception of the tuberous Araceae and the floating *Pistioideae* and *Lemnoideae*, the great majority are incapable of colonizing exposed terrain and that on the contrary a different vegetation must be already present which offers support and shelter to climbing or epiphytic Araceae.

Among the subfamilies of Araceae distinguished by me, the Pothoideae come nearest to the normal Monocotyledon Type in every respect: they differ from the Liliaceae essentially only by the fleshy outer seed integument. Admittedly the shoot organization is usually different, but in this regard there is a great diversity in the Araceae itself and the rule which holds for all other Araceae, that the sympodial reiteration shoot arises in the axil of the penultimate leaf before the spathe, suffers some exceptions in the Pothoideae. There are even in the Pothoideae some genera (*Pothos, Heteropsis*) in which the flowering branches are axillary to a leafy main axis. The Pothoideae also show no conspicuous histological peculiarities, neither '*Intercellularhaare*' {trichosclereids} as in the Monsteroideae, nor any laticifers [*Milchsaftschläuche*] bound to the phloem strands; however it can easily be supposed that among extinct related groups, trichosclereids or laticifers of a particular arrangement {must have} existed; the tannin-containing cells scattered irregularly through the tissues of Pothoideae seem to be somewhat related [*verwandt*] to the laticifers of other Araceae. It is however also the geographical distribution of the Pothoideae that suggests that in them we see the oldest subfamily of the Araceae. They reach to the farthest limits of the whole area which is occupied by the Araceae. In Australia occurs the extremely divergent monotypic genus *Gymnostachys* and at the northern limits of the range of the Araceae we find the four monotypic genera *Calla, Orontium, Symplocarpus, Lysichitum,* among which *Calla* is distributed from eastern North America to east Siberia while *Orontium* occurs only in eastern North America, *Symplocarpus* in East America [*Ostamerika* {≡ eastern North America}] and East Asia and *Lysichitum* in West America [*Westamerika* {≡ western North America}] and East Asia. Also the genus *Acorus*, which is related to these monotypic genera, particularly to *Gymnostachys*, is distributed over a large part of the northern and southern hemispheres, and thus occurs between the furthest outposts of the Araceae-Pothoideae. The remaining Pothoideae show a great similarity in their floral structure, but significant differences in their ecology [*Wachstumsverhältnissen*], so that it is necessary to presume a great number of extinct intermediates, a presumption which is well justified for a family today most diverse still only in the tropics. The Monsteroideae must have descended from shrubby forms with the habit of the genus *Anadendron*; the Lasioideae and most of the other subfamilies must be derived from other sympodial forms like those represented today in *Amydrium* and *Anthurium*. That the Pothoideae were previously more diverse than today is indicated by the many monotypic and in no way closely related genera {of which it is composed}. The most diverse genus of the Pothoideae is *Anthurium* with around 160 species between 25° N and 30° S; the species are in part very closely related, and in spite of the diversity of leaf shape which we find in this genus as in *Philodendron*, the individual sections of the genus are not sharply separated from one another. All this shows that the genus *Anthurium* is at the height of its {evolutionary} development [*Entwicklung*] and is relatively younger than most other Pothoideae, which in large part are monotypic; it can well be presumed, in view of the restriction of such a species-rich genus as *Anthurium* to America, that its evolutionary development occurred at a more recent period in which the present-day division of ocean and continents impeded to a great extent an exchange of forms between the Old and New Worlds.

Next to *Anthurium*, *Pothos* itself is the most species-rich genus of the Pothoideae, of which the great majority of species are scattered through the Monsoon Region [*Monsumgebiet*], {with} one however occurring in Madagascar.

The genera of the Monsteroideae, which are rather equally distributed in the Monsoon Region and in Central and South America, are more closely related to one another [*stehen unter einander in näherer Verwandtschaft*] than the genera of the Pothoideae. The genera are restricted either to the Monsoon Region or to America; only *Spathiphyllum* is an exception; while 17 species occur in America, we find one, *Sp. commutatum* Schott, in the Celebes and the Philippines, although it is also rather closely related to a Brazilian species. This example shows how little it is justifiable to presume that the place of strongest development {i.e. most species diversity} of a genus is also its centre of origin [*Ausgangspunkt*]. Had this single species in the Celebes {and the Philippines} also by chance become extinct with the others that formerly may have existed in the Monsoon Region, the home range [*Heimath*] of the genus *Spathiphyllum* would be transferred to Central America; it is however much more probable that the original range [*Heimath*] of this genus, and indeed of the Monsteroideae, whose genera are so closely related to one another, is to be sought in the eastern part of the Monsoon Region. From here the Monsteroideae may then have spread out to the West and East; {although} they do not seem to have reached Sudan {= Africa between the Sahara and the Cape}.

The situation is quite different in the Lasioideae, which alongside a number of heterogeneous forms also includes closely related forms which occur in regions which at the present day are separated by wide zones of ocean. This indicates that this subfamily is probably of more ancient origin than the Monsteroideae. First there is *Cyrtosperma*, the genus which comes closest to the original morphological Type [*morphologischen Urtypus*] of this subfamily, with two species in the Monsoon Region, two in Sudan and two in the Hylaea and Cisaequatorial America. In America the tuberous and still bisexual-flowered Dracontioninae evolved [*entwickelten sich*] from the rhizomatous Lasinae to which, apart from *Cyrtosperma*, *Urospatha* also belongs, as well as the arborescent *Montrichardia*, which still has the prickles of the Lasinae. *Syngonium* is connected to the Lasioideae on account of its embryological structure; however in its anatomy {it is closer} to the Colocasioideae, so that I am still somewhat doubtful about the natural position of this genus. The African and Asiatic Amorphophalleae are intimately related to the American Dracontioninae, {the former being} no more than Dracontioninae in which the monoclinous flowers have become diclinous. Whether the Amorphophalleae are directly descended from the American Dracontioninae or from extinct Dracontioninae of Africa and Asia cannot of course be definitely asserted. It is probable that the Amorphophalleae represent a more recent development in the Old World, which were likewise preceded by Dracontioninae. In America similar groups to the Amorphophalleae may evolve [*sich entwickeln*] from the Dracontioninae still present today. In the African genus *Anchomanes*, which also still shows the prickles of the Lasinae, the spadix is still covered to the apex with fertile flowers, and thus stands closer to the *Dracontium* – Type than the true Amorphophallinae, whose upper flowers atrophy in their development and together form the well-known spadix appendix. The fact that the genera of the Amorphophallinae are somewhat numerous but very closely related [*sehr nahe verwandt sind*], argues that they are younger than the Lasinae and Dracontioninae.

The Colocasioideae are such a natural subfamily that this taxon has long been recognized. They represent a decidedly later development, on the one hand because their sympodium is shortly erect or tuberous and on the other because their laticifers reveal a more advanced formation, branching and anastomosing, while those of the Lasioideae form only straight rows, but above all because, as in the Amorphophalleae, dicliny is completely established and in the male flowers the formation of synandria occurs, likewise a later development. The only plant which because of the staminodes present in the female flowers, still calls to mind the ancestral Type [*Urform*] is *Steudnera colocasiaefolia* from Burmah. The other genera are very similar to one another in habit and were placed in *Caladium*, after the Linnean procedure of denoting almost all Araceae as *Arum* had been abandoned, until Schott brought clarity to these relationships and distinguished the various very natural genera in which we recognize these plants today. The Colocasioideae subfamily is rather uniformly developed in the Monsoon Region and in tropical America, but is lacking in Sudan; its history may thereby be similar to that of the Monsteroideae. *Xanthosoma* and *Caladium* in America and on the other hand *Alocasia* in the Monsoon Region have numerous closely related species at the present time, so that there can be no doubt that these genera are of more recent origin than *Steudnera*, *Gonatanthus* and *Remusatia*.

As regards the Philodendroideae, various considerations suggest that this subfamily is older than the Monsteroideae and Colocasioideae, {and} somewhat similar in age to the Lasioideae. Apart from the tribe [*Abtheilung*] Philodendreae, which is so richly developed in the Monsoon Region and in tropical America, there are four genera, *Richardia* in the Cape, *Peltandra* in Virginia, *Anubias* in West Africa {and} *Typhonodorum* in Madagascar, which are isolated both in their floral structure and in their geographical ranges, but which nevertheless are closest to the Philodendroideae {than to any other subfamily}. The Philodendreae are in any case younger than these genera; it follows from the close relationship [*aus der nahen Verwandtschaft*] that connections between the Monsoon Region and South America must have existed; and particularly from the fact that *Homalomena* has 10 species in the Monsoon Region, but also 5 in Costa Rica and New Grenada, {the latter} to be sure, belonging to a different Section. *Philodendron*, like *Anthurium*, is at the height of its evolutionary development; the individual Sections are however more sharply separated and distributed within {more} restricted regions {than in *Anthurium*}. Thus the arborescent Sections *Meconostigma* and *Sphincterostigma* are restricted to Brazil, while *Polytomium* is represented chiefly in Central America and the West Indies.

The subfamilies Aglaonemoideae and Staurostigmoideae are of less interest; they also belong both to the eastern and the western hemispheres.

More interesting is the subfamily of the genuine Aroideae. In order to remove all doubt about the great age of these morphologically reduced forms it is enough to refer to *Arisaema*, a genus which is also interesting because there, alone among the Araceae, dioecy of the sexes occurs. The genus is distributed throughout the Monsoon Region; but it is also found in Amurland and in the Chinese-Japanese region, where the related genus *Pinellia* occurs with it; furthermore there are four species in eastern North America, in Texas and in Mexico, two species in Arabia and two in Abyssinia. This is a distributional range which could have come about only before the present-day separation of the western continents from the eastern ones. Otherwise the genera of the Aroideae are very scattered, *Spathicarpa* in Brazil, *Spathantheum* in the Andes, *Sauromatum* in Nubia, *Stylochiton* in Natal and in Sudan; however by far the majority of Aroideae belong to the Monsoon Region and to the Mediterranean region. The connections which in other respects exist between the Azores, Canary Islands and the Mediterranean region also appear here, in that the genera of the Mediterranean region occur on these islands with only a few divergent forms. There are also in the Steppe region a few Aroideae, and so the two main regions of occurrence of the subfamily, the Mediterranean and the Monsoon Region, are linked together. Certainly in the latter other genera occur, particularly *Typhonium* and *Theriophonum*, but these are related to those of the Mediterranean region. *Cryptocoryne* and *Lagenandra* however are connected to *Ambrosinia* and to *Arisarum*. The conditions under which *Cryptocoryne* and *Lagenandra* grow on the river banks of the Monsoon region are found both in tropical Africa and in tropical America. The fruit ripens partly under water, as *Stylochiton* below ground. Like any other Araceae fruit, those of *Cryptocoryne* and *Lagenandra* would have to be transported to distant regions by water and there would be able to germinate if they find open ground on the sea shore ['*Meeresstande*' !sic {this is a misunderstanding as well as a misspelling (for *Meeresstrande*): *C. ciliata* grows in brackish water but no species does so in permanently salt water}]. Since however these genera are not encountered outside the Monsoon region, it can be presumed that the evolutionary development [Entwicklung] of these genera took place at a later time than those forms of the Monsoon region which were able to reach tropical America. Although therefore the Aroideae subfamily, like the other subfamilies, must have developed early {in time}, nevertheless many genera evolved later than others, particularly {later} than *Arisaema*; the great age of the latter {genus} is implied by the fact that some of its species have already attained dioecy, which has not happened in the other genera. In regard to the descent of the Aroideae from other subfamilies [*Gruppen*] it may be remarked that they are closest to the Lasioideae, at least anatomically and in leaf venation; however, they are all distinguished from the Lasioideae by the seeds containing endosperm and so they may have evolved [*sich entwickelt*] from early times in parallel with the later development of the Lasioideae. It is of great interest that there exists for us an ancestral form of the Aroideae Type, represented by *Stylochiton*, a genus in which we already find diclinous flowers but with the normal number of elements (6 stamens in the male flower and 3 carpels in the female) and still surrounded by a perigon. The other genera have much more reduced flowers; in most, as in the reduced Lasioideae and the reduced Colocasioideae, the upper part of the inflorescence does not develop normally {but} turns into a so-called appendix. The Pistioideae and Lemnoideae are those subfamilies of the Araceae in which the reduction has reached the furthest point. Although the development of *Pistia* and thus also of the Lemnoideae has become comprehensible to me only by {consideration of} the *Aroideae, Lagenandra* and *Cryptocoryne*, I nevertheless cannot see how *Pistia* could be derived from these genera; various morphological peculiarities indicate that the Pistioideae should have a position within the Araceae of equal rank to the other subfamilies. Although we recognize only a single species represented by a number of varieties, I nevertheless do not hesitate to regard this as the representative of a subfamily whose more diverse forms must long since have disappeared. *Pistia* must be one of the oldest reduced forms of the Araceae. Its wide distribution in the freshwaters of the tropics and sporadically in the extratropical region{s} is only partly explained by the facile dispersal of detached shoots; the seeds, just as in those of other Araceae, soon lose their viability, so that dispersal across vast stretches of ocean is not easily possible.

The best evidence, of course, for the great age of the genus *Pistia* is that fossils are found already in the oldest strata of the Tertiary period, in the Flandrian stage in Fuveau at the mouth of the Rhone; this is *P. Mazelii* Sap. et Mar. Unfortunately the Araceae are so unsuitable for preservation in a fossil state that we may expect little corroboration in this regard of the views expressed earlier; so much the more must one seek to study the morphological relations in depth, because they can well provide an idea of the history of the family or of the genera. The wide distribution and extreme reduction of the Lemnoideae also implies that their evolution [*Entwicklung*], like the Pistioideae, must be of very ancient date. (Engler,1879a: 48 – 55).

8 Junker (1989) explains that during 19th century botany, there were two kinds of Type concept, an idealistic spiritual one, resembling a Platonic essence, which is the source of reproductions of itself (e.g. individuals of a species) each of which bore no material

genealogical connection to the others, and one that emerged with the evolutionary paradigm as the common structural form of a set of descendants, all of which bore its stamp, a Bauplan which indicated ancestral character combinations representing a particular stage of evolution. The key change that took place with the acceptance of evolutionary theory by many botanists after 1859 was a more general reinterpretation of the previously idealistic Type concept as evidence of ancestral characters passed on to descendants by heredity; e.g. Strasburger, 1874 (as cited and discussed by Junker, 1989: 37 – 39). According to Junker, the general mode of thought in the 19th century as regards the latter interpretation was that progressions or transformations from one Type to another did not happen, but phylogenetic diversification within Types was accepted.

9 Engler (1884), citing the *Abstammungslehre* (Nägeli, 1884), shows his commitment to Nägeli's evolutionary theory:

> 'It is of the greatest significance for Morphology and Systematics that von Nägeli ... standing unequalled among botanists for his logical rigour, has undertaken to elucidate the phylogenetic laws of the plant kingdom and particularly to emphasize the features brought about through adaptation (as a result of external stimuli) in contrast to those forms of organization produced by inherent causes. The inner causes, just as outer ones, produce form changes in different parts of the plant. In a plant, the form of one part can be phylogenetically more advanced [*phylogenetisch vorgeschrittener*] and the form of another part more retarded [*phylogenetisch mehr zurückgeblieben*] than in another plant. It is thus immediately apparent that in comparing numerous plants of a higher stage of evolutionary development [*höheren Entwicklungsstufe*] the establishment of the phylogeny [*der phylogenetische Entwicklung*] encounters considerable difficulties on account of the large number of parts to be compared. Nevertheless, the phylogenetic pathway [*der phylogenetische Entwicklungsgang*] is more likely to be discovered in those families in which a great diversity of forms occurs, particularly of the reproductive organs, at least in broad outline, rather than in those that present a general uniformity in floral structure. The ascertainment of the ontogenies [*Ontogenieen*] is of great significance for the description of phylogenetic advancement [*phylogenetischen Fortschrittes*], or simply, progression, and especially interesting are those cases in which the development stages of less advanced plants can be recognized in the ontogeny of an advanced plant. The Araceae family is however rather rich in such forms and hence the establishment of the resulting progressions is here and there facilitated.'
>
> (Engler, 1884: 145).

11

Aroid floral morphogenesis in relation to phylogeny

Denis Barabé

11.1 Introduction

What is the link between morphogenesis and phylogeny? This question was addressed by Haeckel, in the nineteenth century, when he formulated his controversial biogenetic law, stating that ontogeny is a short and rapid recapitulation of phylogeny (Haeckel, 1866). Since that time, many zoological and botanical studies discussing the idea of the usefulness of ontogeny in determining phylogeny have been published (e.g. Gould, 1977; Nelson, 1978). Ontogenetic features have been used, for example, to determine the phylogenetic relationships of Saururaceae and Piperaceae (Tucker et al., 1993). On the other hand, Mishler (1986) considered that an independent phylogeny should be established to adequately interpret the evolution of ontogenetic characters. In the present chapter, I will address this general question by using the floral morphogenesis of Araceae as a case study.

The Araceae family comprises 117 genera and nearly 3300 species (Haigh et al., 2010). In recent phylogenies, Araceae belong to the Alismatales and are positioned as a sister group of the rest of the order (Stevens, 2001). The family includes eight subfamilies: Gymnostachydoideae, Orontioideae, Lemnoideae, Pothoideae, Monsteroideae, Lasioideae, Zamioculcadoideae and Aroideae, if we accept the recently proposed Zamioculcadoideae (Bogner and Hesse, 2005, Cusimano et al., 2011).

The floral developmental morphology of several species with unisexual flowers (subfamily Aroideae) has been investigated in the context of inflorescence

Early Events in Monocot Evolution, eds P. Wilkin and S. J. Mayo. Published by Cambridge University Press. © The Systematics Association 2013.

structure (Uhlarz, 1982, 1986; Buzgo, 1994), homeosis (Barabé et al., 2004a, 2008), morphogenetic gradient (Barabé and Lacroix, 1999) and phylogeny (Barabé et al., 2002, 2004b). Until the work of Buzgo (1999, 2001) on basal Araceae (*Gymnostachys*, *Orontium*, *Lysichiton*, *Pothos*, *Pothoidium* and *Spathiphyllum*), relatively few genera belonging to subfamilies with bisexual flowers (Calloideae, Gymnostachydoideae, Lasioideae, Orontoideae, Pothoideae, Monsteroideae) had been investigated from a developmental point of view, with the exception of studies of the phyllotactic patterns of *Monstera deliciosa* (Fujita, 1942) and the development of the inflorescence and flowers of *Symplocarpus foetidus* (Barabé et al., 1987b; Barabé, 1994). More recently, this type of developmental study has been applied to the study of *Anthurium* (Barabé and Lacroix, 2008a) and the subfamily Lasioideae (Barabé and Lacroix, 2008b; Poisson and Barabé, 2011).

The aroid flower presents us with a great amount of morphological variability which highlights some important developmental problems with regard to phylogeny. These include, for example, the developmental relationships between bisexual and unisexual flowers, the morphological nature of sterile floral appendages present in many genera and the mode of development of homeotic atypical bisexual flowers in the Aroideae. However, floral development has so far been studied in less than 20% of aroid genera. This absence of floral developmental studies may be explained by the practical difficulty of obtaining many samples of inflorescences at different stages of development to conduct complete developmental studies, particularly in the case of geophytic plants. Nevertheless, sufficient documentation is available in print to allow us to create a portrait of the floral developmental morphology of Araceae in the context of the recent molecular phylogenies published by Cabrera et al. (2008) and Cusimano et al. (2011).

The specific goals of this short survey of the morphogenesis of the aroid flower are: (1) to determine the type of relationship that exists between phylogeny and floral development and (2) to analyse the developmental constraints on morphogenesis.

11.2 Unidirectionality of floral development in basal Araceae

Buzgo and Endress (2000) have shown clearly how floral development may be linked to the phylogenetic position of *Acorus*. In the past, *Acorus* was included among Araceae. However, over the last few decades, molecular studies revealed *Acorus* as the sister to all other monocots (Stevens, 2001). Among important developmental features of *Acorus*, Buzgo and Endress (2000) found an abaxially median tepal that is initiated first and is similar to a flower-subtending bract, and a strongly unidirectional early floral development with an inversion of the organ initiation sequence in the second tepal whorl. A similar mode of formation of the floral appendages leading to an inversion in unidirectional flowers was observed in other monocots and in Piperales (Buzgo and Endress, 2000). However, such a unidirectional development does not occur in aroids. Furthermore, the same authors have also shown that the mature gynoecium of *Acorus* is largely synascidiate, while the early development of the carpels is plicate and the apocarpous portion persists until anthesis. This mode of development has also been reported for *Dracontium*, a basal aroid. *Dracontium* appears to be the only genus of Araceae showing an apocarpous (plicate) upper ovarian portion during early development (Poisson and Barabé, 2011). The presence of this character in a basal aroid remains difficult to explain.

In Araceae, Buzgo (2001) reported a slightly unidirectional development of the flower (tepals and stamens are larger on the abaxial side) in *Gymnostachys*, *Lysichiton* and *Orontium*. This unidirectionality has not been observed in Pothoideae and Lasioideae. In basal Araceae, although tepals and stamens of the same sectors are closely associated, they develop on clearly distinct whorls and not from a common primordium. However, notable exceptions are *Symplocarpus* and *Lysichiton*, in which tepal and stamen of the same sector are formed from a common primordium (Barabé, 1994; Buzgo, 2001).

The mode of floral development observed in *Anthurium* is comparable to that reported for *Gymnostachys* (Buzgo, 2001), which also has a tetramerous flower. In both genera, the lateral tepals and stamens develop first. However, a unidirectional mode of floral development has not been observed in *A. jenmanii* (Barabé and Lacroix, 2008a) as it has in *Gymnostachys* (Buzgo, 2001). The inner whorls of the perianth and androecium develop equally on the abaxial and adaxial sides. Contrary to *Gymnostachys*, the lateral stamens of *A. jenmanii* are the first to release pollen.

The mode of floral development in *Anthurium* supports the idea that there is a correlation between the absence of both unidirectional development and a flower-subtending bract or tepal in Araceae. In *Anthurium*, compaction of the primordial flower is so strong that the development of the floral parts begins before the complete individuation of the floral primordium. The floral primordium has a tetragonal shape even prior to the initiation of tepals (Barabé and Lacroix, 2008a). Consequently, there is no space for a subtending bract, and no unidirectional development. Here, we can hypothesize that the extreme compaction of floral primordia would not have been possible if the flower had a bract.

11.3 Lasioideae and merosity

The subfamily Lasioideae, which contains 10 genera (Hay, 1992), offers a phylogenetic and developmental framework to study changes in the number of floral parts (merosity) that can occur in closely related taxa. With respect to this question, the floral developmental morphology of *Anaphyllopsis*, *Dracontium* and

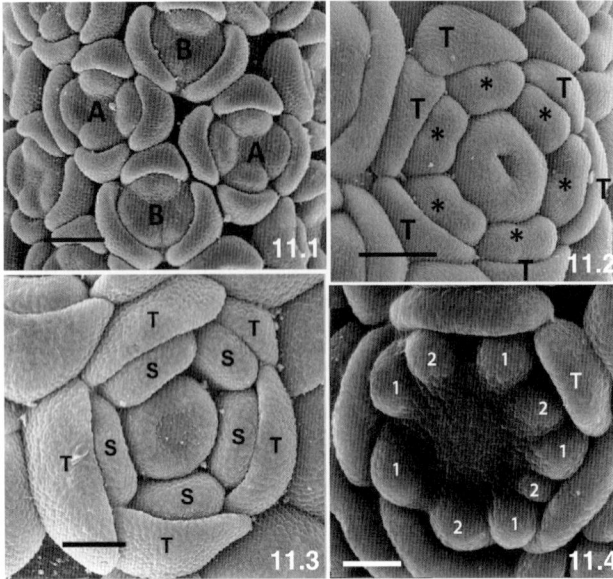

Fig 11.1 *Urospatha sagittifolia.* Variation in number of floral organs: flowers A are tetramerous; flowers B have three tepals and four stamen primordia. Bar = 75µm. From Barabé et al. (2011).

Fig 11.2 *Urospatha sagittifolia.* Initiation of a ring-shaped gynoecial structure centrally in a hexamerous flower. Stamens (*), Tepals (T). Bar = 75µm. From Barabé et al. (2011).

Fig 11.3 *Anaphyllopsis americana.* Pentamerous flower consisting of five tepals (T) and five stamens (S). Bar = 75µm. From Barabé and Lacroix (2008b).

Fig 11.4 *Dracontium polyphyllum.* Early stages of development showing flower with 5 tepals (T) two whorls of stamens (1–2) surrounding the site of the future location of the gynoecial primordium. Bar = 50 µm. From Poisson and Barabé (2011).

Urospatha is particularly illuminating (Barabé and Lacroix, 2008b; Barabé et al., 2011; Poisson and Barabé, 2011). *Urospatha* and *Anaphyllopsis* (Figs 11.1–11.3) are characterized by tetramerous flowers, normally comprising four tepals and four stamens surrounding a unilocular (or bilocular) gyneocium. However, both genera also bear pentamerous flowers and hexamerous flowers have recently been observed in the inflorescence of *Urospatha* (Barabé et al., 2011). The number of floral parts in *Dracontium* (Fig 11.4) is highly variable. Flowers can have five to nine tepals and six to twelve stamens surrounding a gyneocium with one to six carpels.

In all types of flowers of *Anaphyllopsis* and *Urospatha*, the stamens are closely associated with and opposite to the facing tepals. In both genera, pentamerous flowers are formed by the addition of a sector comprising a stamen and tepal, that is, a stamen–tepal unit additional to the basal tetramerous flower (Figs 11.2, 11.3). In the case of hexamerous flowers, two sectors are added, each representing a stamen–tepal complex.

In *Dracontium*, in contrast to *Anaphyllopsis* and *Urospatha*, one observes a structural dissociation between the positions of stamens and tepals. The first whorl of stamens is initiated in an alternate position with respect to the first whorl of tepals. The second whorl of stamens develops in an alternate position to the first whorl of stamens (Fig 11.4). There is no visible sectorial addition of a stamen–tepal complex (Poisson and Barabé, 2011). The number of whorls of stamens increases and they develop in a more-or-less alternate position not linked to the number of tepals. The mode of formation of polymerous flowers in *Dracontium* would represent a developmental novelty in the subfamily Lasioideae, an evolution from the basal tetramerous type occurring in *Anaphyllopsis* and *Urospatha*. The only way to allow for the presence of a greater number of stamens and tepals in alternate positions in *Dracontium* would have been for a modification to take place in the mode of development underlying the basal pattern of Lasioideae.

In the phylogeny of Lasioideae (Cabrera et al., 2008; Cusimano et al., 2011) there was a shift from a regular sectorial development characterized by a stable number of tepals and stamens (four to six) to a more variable development linked to a greater number of floral parts inserted in a more-or-less alternate fashion. If we consider the number of floral parts in Lasioideae, the shift present in *Dracontium* occurred only once in the phylogeny of Lasioideae. Perhaps the way to develop more than six floral parts was by a sudden evolutionary change in the mode of development that, at the molecular level, could be reflected in the mutation of a single regulatory gene. This change could be linked with two different morphologies observed in the subfamily. The first one occurs in *Dracontium*, where we observe an increase in the number of stamens and tepals, and a more-or-less disorganized developmental pattern resulting in the alternate position of certain tepals. The other pattern is represented by *Pycnospatha*. The flower of this genus has no tepals, but the number of stamens can be greater than six (Hay, 1992). In this situation, the absence of tepals is another morphological shift in the subfamily Lasioideae associated with an increase in stamen number (greater than six).

The basal flower type in the clade (Cusimano et al., 2011) comprising *Urospatha*, *Dracontium*, *Anaphyllopsis* and *Dracontioides* is tetramerous, with a potential for pentamery and hexamery in the same inflorescence. We can hypothesize that in this clade there was a shift from typical tetramerous flowers (*Anaphyllopsis* and *Dracontioides*) to polymerous flowers (*Dracontium*). Therefore, the presence of polymerous flowers in *Dracontium* would constitute an apomorphic character within Lasioideae. This interpretation is also supported by the phylogeny of Cabrera et al. (2008), given that *Urospatha* is the sister group of all other Lasioideae.

Given that in basal Araceae dimerous whorls occur in certain genera (*Anthurium*, *Gymnostachys*), one may hypothesize that tetramerous whorls occurring in basal Lasioideae are homologous to two dimerous whorls from an evolutionary point of view. Therefore, the pentamerous or hexamerous flowers appearing in *Urospatha* and *Anaphyllopsis* would be a modification of the typical dimerous flower by the addition of one or two stamen–tepal sectors.

11.4 The intermediate position of Zamioculcadoideae

The subfamily Zamioculcadoideae (Bogner and Hesse, 2005; *Stylochaeton* clade in Cusimano et al., 2011) forms a monophyletic group inserted between basal Araceae (bisexual flowers) and Aroideae (unisexual flowers), and constitutes the sister group of Aroideae. In the inflorescence of *Zamioculcas*, the female and male zones are separated by a short constricted zone bearing sterile flowers. Male flowers consist of four tepals, four free stamens and one clavate pistillode. In female flowers there are four tepals and one bilocular gynoecium with one ovule per locule. Each sterile flower consists of four tepals surrounding a clavate pistillode. The vascularization of the sterile flowers is comparable to that of the fertile female flowers. In the sterile gynoecium of male and sterile flowers, one can observe cavities corresponding to aborted locules (Barabé et al., 2002).

The position of *Zamioculcas* in molecular phylogenies (Cabrera et al., 2008; Cusimano et al., 2011) is supported by the anatomy and developmental morphology of the flower. In *Zamioculcas*, the sterile flowers located in the intermediate portion of the inflorescence are derived from female flowers and do not correspond to atypical bisexual flowers found in Aroideae. The sterile flowers of the intermediate zone in *Zamioculcas* and Aroideae are not homologous (Barabé et al., 2002). However, the functionally male flowers appear to correspond to sterile bisexual flowers, because they consist of a whorl of fertile stamens enclosing a sterile gynoecium. In this case, the female and male appendages are inserted on different whorls. From a morphological point of view, the flowers of *Zamioculcas* seem to represent an intermediate stage between the bisexual flowers of basal subfamilies of Araceae and the typical unisexual flowers of Aroideae. Therefore there appears to be agreement between anatomical and molecular data in suggesting that the genus *Zamioculcas* should be related to Aroideae but in the separate subfamily Zamioculcadoideae.

11.5 Atypical bisexual flowers in Aroideae

11.5.1 *Cercestis* and *Philodendron* types

The subfamily Aroideae is characterized by the presence of unisexual flowers, except in the genus *Calla*. However, flowers with male and female characteristics (called '*monströse Blüten*' by Engler and Krause, 1912: 16) are often found in the intermediate zone of the inflorescence, between the male and female zones in

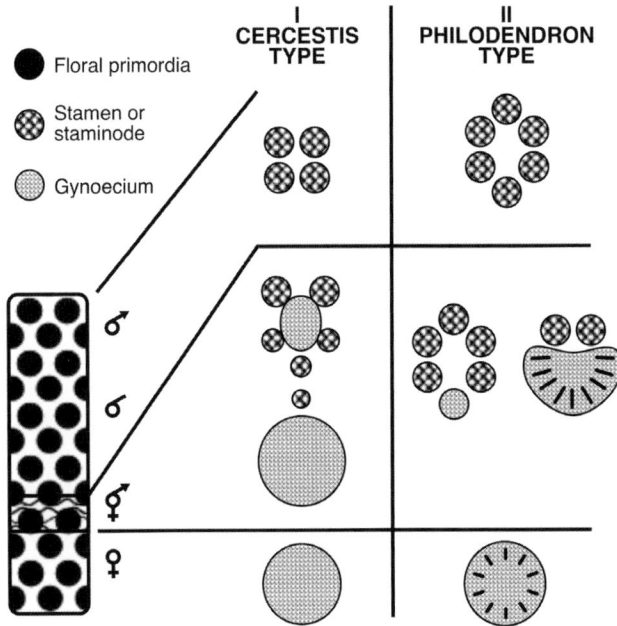

Fig 11.5 Diagrammatic representation of floral types corresponding to male, sterile male, atypical bisexual and female portions of the inflorescence axis of *Cercestis* – Type I (left column) and *Philodendron* –Type II (right column). From Barabé and Lacroix (1999).

many genera. This phenomenon has been well documented, particularly in the genera *Cercestis* (Barabé and Bertrand, 1996) and *Philodendron* (Barabé and Lacroix, 1999; Barabé et al., 2000). Two main types of developmental sequences of atypical bisexual flowers (ABFs) have been recognized: the *Cercestis* type and the *Philodendron* type (Fig 11.5). In the *Cercestis* type, the bisexual flowers are characterized by a functional or nonfunctional gynoecium surrounded by a few (one to five) vestigial stamens (Fig 11.6). In the *Philodendron* type, bisexual flowers generally consist of functional carpels and staminodes inserted on the same whorl (Fig 11.7). Basically these two morphologies may correspond to a 'vestigial' type vs. a 'homeotic' type that can be explained by different molecular regulatory pathways (Zluvova et al., 2006). Let us note, however, that these bisexual flowers are not true functional bisexual flowers. Even though the pistillate portion is fertile, the staminate portion consists of staminodes. Therefore, from a functional reproductive point of view, these flowers remain unisexual. However, from a developmental point of view, we can consider these flowers as bisexual because pistillate and staminate primordia are initiated on the same flower, even if the staminate primordia later develop into staminodes.

Atypical bisexual flowers (ABFs) have also been reported in other aroid genera. Early stages of development of the inflorescence of *Montrichardia* reveal the

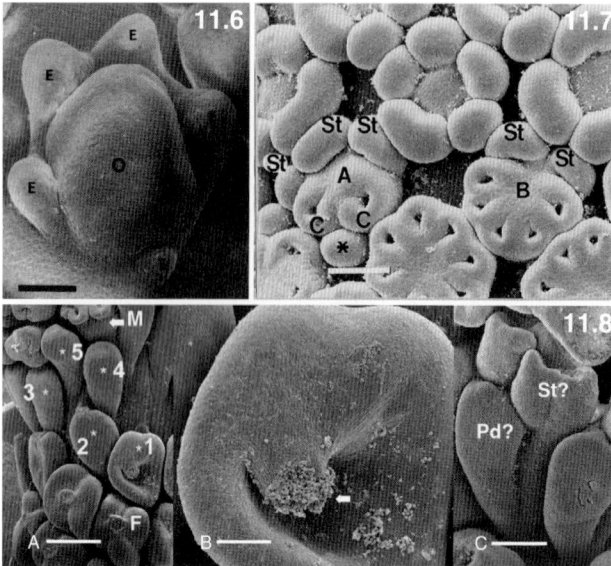

Fig 11.6 *Cercestis stigmaticus*. Early stage of development of an atypical bisexual flower showing an aborted gynoecium (O) surrounded by a whorl of staminodes (E). Bar = 100 µm. From Barabé and Bertrand (1996).

Fig 11.7 *Philodendron fragrantissimum*. Early stages of development of bisexual flowers (A and B) showing carpels (C) and staminodes (St) inserted on the same whorl. Some bisexual flowers have staminodes (*) facing the female zone rather than the male zone. Bar = 150 µm. From Barabé et al. (2000).

Fig 11.8 *Schismatoglottis calyptrata*. Variety of morphologies in the transition zone. A, General view of the transition zone showing organs (*) intermediate between typical female flowers (F) and male flowers (M). Bar = 750 µm. B, Close-up of an undefined flower. Arrow = stigmatic tissue. Bar = 150 µm. C, Fused stamen-like (St?) and pistillode-like (Pd?) structures. Bar = 300 µm. From Barabé et al. (2004b).

presence of ABFs between the female and male zones. These atypical flowers are not recognizable on mature inflorescences and appear to correspond to the *Philodendron* type (Barabé and Lacroix, 2001).

In *Caladium*, the sterile, morphologically male flowers are depressed–obpyramidal synandrodes. However, in contrast to *Philodendron*, there is no broad ring of ABFs, but during the early stages of the development of the inflorescence a few ABFs are recognizable between the female zone and the male-sterile zone (Barabé and Lacroix, 2002). These atypical flowers have a staminal portion, the rudimentary gynoecium belongs to the same meristematic unit and can be viewed as a single whorl. From a morphological point of view, the structure of the atypical flowers in the inflorescences of *Caladium* and *Montrichardia* is much closer to that of the *Philodendron* type than to the *Cercestis* type.

In *Arisaema*, there is a sex dimorphism based on plant size (Ewing and Klein, 1982). Individuals of *Arisaema triphyllum*, for example, can be of three types, depending on the sex of the inflorescence: male, female or bisexual (less frequent; Barriault et al., 2010). Fukai (2004) reported the presence of few bisexual flowers on the border between the male and female zones on monoecious spadices. These bisexual flowers had an anther and a pistil. However as this author did not observe the early stage of development of bisexual flowers, it is not possible to determine whether they represent the *Philodendron* or *Cercestis* type.

In the case of *Pistia*, there are no traces of ABFs. However, a sterile annulus surrounding the spadix below the male flowers and a unilateral horizontal scale-like structure located above the gynoecium (Buzgo, 1994) could be interpreted as modified staminodes. On the other hand, based on available information on the development of the flower of *Pistia*, it is not possible to determine with certainty whether or not this unilateral scale and the gynoecium belong to two different whorls.

To date, the *Philodendron* floral type has only been observed in the genera *Caladium*, *Philodendron* and *Montrichardia*. This may be due in part to a lack of developmental studies of flowers in the subfamily Aroideae. The *Cercestis* type of bisexual flower (atypical flower) occurs in the genera *Cercestis*, *Culcasia*, *Dieffen-bachia* and *Spathicarpa* (Barabé et al., 2002). The presence of ABFs of the *Philodendron* type in the genus *Caladium* shows that this type of ABF appears in two different clades (Caladieae and Philodendreae) that do not form a monophyletic group in the cladogram of Cusimano et al. (2011). The same holds true for ABFs of the *Cercestis* type present in Spathicarpeae, Dieffenbachieae, Aglaononematae and Culcasieae.

With regard to the phylogenetic tree of Cusimano et al. (2011), there is no close correspondence between organogenesis of ABFs and molecular phylogeny. These results show that the *Cercestis*-type and *Philodendron*-type developmental patterns of ABFs in Aroideae do not correspond to two separate evolutionary lineages within clades, but are homoplastic. Given the great diversity of ABFs that seem to occur in Aroideae, it is very difficult to predict the type of ABFs present in genera where the mode of floral development is unknown. However, the presence of two whorls of floral parts in ABFs of *Cercestis* indicates that this character could represent a symplesiomorphy in Aroideae. On the other hand, the presence of the *Philodendron*-type in Aroideae would constitute a homoplasy.

11.5.2 Other types of atypical flowers

Although *Philodendron*- and *Cercestis*-types of ABF can be recognized in different genera of Aroideae, there are other types of atypical flowers that do not fit into these two categories.

In the inflorescence of *Schismatoglottis*, the intermediate zone located between the female and male zones consists of atypical flowers that display a wide variety of forms ranging from more-or-less typical female flowers to male flowers (Barabé et al., 2004b). In *S. calyptrata*, atypical flowers located above typical female flowers are generally flattened. Some have a prominent cavity in the centre, with a residual flattened, triangular structure also topped with a residual stigma (Fig 11.8). Other atypical flowers located near the male zone have a clavate form and resemble interpistillar pistillodes. Near the male zone, there are usually a few atypical flowers with two different appendages joined at their base. In Fig 11.8c, one of these appendages resembles an interpistillar staminode (Pd?) and the other a stamen (St?).

In *Arum italicum*, the female zone located at the very base of the inflorescence and the male zone represent approximately 45% of the total length of the inflorescence. The zone of bristle-like staminodes located above the male zone occupies only 5% of the total length of the inflorescence. The staminate floral primordia are arranged irregularly on the surface of the inflorescence. The female primordia on the other hand form a more-or-less regular lattice on the surface of the inflorescence. The rest of the inflorescence (approx. 50%) consists exclusively of a sterile appendix located directly above the male. During early stages of development, there is no free space between the different zones of the inflorescence (Barabé et al., 2003).

In different species of *Arum* (Boyce, 1993), a morphological transition between typical female flowers and pistillodes can be observed. The overall morphology of transitional forms of bristles indicates that the sterile flowers located between the female and male zones could correspond to under-developed female flowers. However, there are no atypical flowers that are morphologically intermediate between male and female flowers as in other genera of the subfamily Aroideae (e.g. *Cercestis*, *Philodendron*, *Schismatoglottis*). Additionally, no residual ovary or stigma has been observed in these structures as in atypical flowers of *Schismatoglottis*. From a morphological point of view, the structure of the bristles on the inflorescences of *Arum* does not correspond to any type of atypical flower (unisexual or bisexual) that has been analysed previously (Barabé et al., 2003). This shows that developmental constraints (genetic or physiological) experienced by the floral primordia in the intermediate zones are not exactly the same for all genera of Aroideae.

The presence of different types of atypical flowers in Aroideae is linked directly to the morphogenesis of normal unisexual flowers bordering these particular structures in the intermediate zone of the inflorescence. Normal functional flowers represent a developmental constraint canalizing the nature of atypical organs that can develop.

11.6 Homeosis and developmental constraints

In *Philodendron*, atypical bisexual flowers generally consist of a combination of carpels and staminodes inserted on the same whorl arising through the homeotic replacement of carpels by staminodes. The existence of sterile stamens (staminodes) and carpels on the same whorl indicates that the identity of organs that will be formed is independent of the floral whorl on which they arise. The nature of the floral organs that will be formed depends on their position along the axis of the inflorescence.The staminodes of the ABF will tend to be formed on the side facing the sterile male zone of the inflorescence, while the carpels will be located on the side facing the female zone (Barabé and Lacroix, 1999).

The number of carpels involved in a homeotic transformation varies between species, and the number of staminodes is linked to the number of carpels in ABFs. Even though quantitative analyses cannot predict the precise number of homeotic transformations that will take place in a given ABF, they show that there is regularity in the homeotic transformation involving *P. melinonii, P. pedatum, P. solimoesense, P. insigne* and *P. squamiferum* (Barabé et al., 2004a, 2008). The number of carpels present in female flowers acts as a border condition or constraint on the number of appendages that develop in ABFs (Barabé et al., 2004a). In this system, the number of carpels in normal flowers constitutes a developmental constraint (primordium size vs. number of organs) acting on morphological characteristics of ABFs.

In the zone of ABFs, flowers located in the restricted space of the intermediate zone experience the influence of the female flowers on one side and that of the male flowers on the other side. The hypothesis of a hormonal gradient has been formulated to explain the presence of ABFs in the inflorescences of both *Philodendron* (Barabé et al., 2000) and *Cercestis* (Barabé and Bertrand, 1996). Based on this interpretation, it is plausible that the ABFs, which are subjected to male and female influences simultaneously along a hormonal gradient, would form a whorl of stamens if there were enough space between floral primordia, as is the case in *Culcasia* and *Cercestis*. However, in genera such as *Caladium, Montrichardia* and *Philodendron*, where the floral primordia are packed very tightly on the surface of the inflorescence, male and female organs appear on the same whorl. Therefore, due to a lack of space for a whorl of staminodes to form, an ABF of the *Philodendron* type can only develop through a homeotic substitution where stamens replace carpels. The presence of ABFs of the *Philodendron* type in different clades of Araceae may thus represent a homoplastic tendency due to a design limitation (Wake, 1991; Bowman et al., 1999).

11.7 *Calla* and pseudomonomery

The monospecific genus *Calla* is characterized by bisexual flowers lacking a perigon with a unilocular ovary surrounded by 10–12 stamens (Figs 11.9–11.10). In the recent molecular phylogeny of Cusimano et al. (2011), *Calla* is placed in the Aroideae as the sister group of a large clade consisting of the *Amorphophallus,*

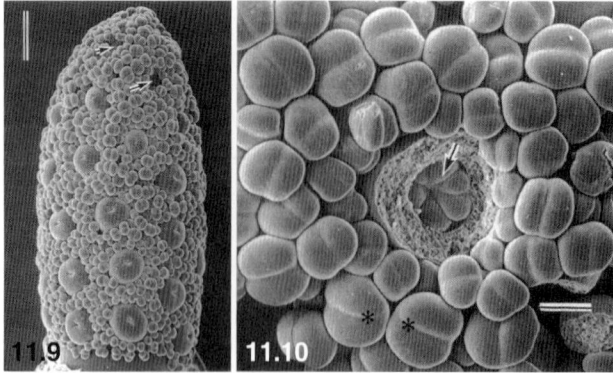

Fig 11.9 *Calla palustris*. Side view of entire spadix with fully formed flowers; arrows indicate failure of gynoecium to develop, producing male flowers. Bar = 500 μm. From Scribailo and Tomlinson (1992).
Fig 11.10 Flower with top of gynoecium removed to show basal placentae (arrow), each of which will produce an ovule. Bar = 100 μm. From Scribailo and Tomlinson (1992).

Colletogyne and *Pistia* clades. The position of bisexual *Calla* in the Aroideae is not in accordance with the distribution of morphological, anatomical and palyno-logical characters. For Cusimano et al. (2011: 12): '... The phenotype of *Calla* suggests a position in the "transition zone" between the bisexual taxa and uni-sexual clades similar to that of the *Stylochaeton* clade.'

Based on anatomical data, Barabé and Labrecque (1983) considered that 5-6 of the 10–12 stamens of *Calla* correspond to modified tepals, and interpreted the gynoecium as pseudomonomerous (unilocular ovary composed of more than one carpel) with three fused carpels. After studying development and vascularization in the same species, Scribailo and Tomlinson (1992) rejected the interpretation of pseudomonomery and modified tepals. They did not find developmental traces of distinct carpel primordia or modified tepals. On the other hand, during the early stages of development, Lehmann and Sattler (1992) found irregular lobes on top of the ovary which they interpreted as distinct carpels, supporting the interpretation of pseudomonomery. Although the position of stamen primordia on flower buds of *Calla* is variable and erratic, Lehmann and Sattler (1992) observed the presence of outer and inner stamen morphs that can be identified throughout development. They concluded that the outer stamens may correspond to a perigon transformed by a process of homeosis. However, this assumption depends on the phylogenetic position of *Calla* within Araceae.

The interpretation of a shift from tepals to stamens by the process of homeosis was formulated when the genus *Calla* was included in the subfamily Calloideae, which is characterized by bisexual flowers with a perianth. In this systematic context, the interpretation of homeosis based on anatomical data (Barabé and Labrecque, 1983) and developmental data (Lehmann and Sattler 1992) is plausible. With regard to recent molecular phylogenies (Cabrera et al., 2008; Cusimano et al., 2011), the absence of tepals in *Calla* could result from the replacement of a part of the perigon by stamens or by a loss of tepals followed by an increase in stamen number. However, it remains difficult to interpret the presence of a second whorl by invoking a homeotic transformation, since *Calla* is in the subfamily Aroideae, whose members lack a perianth. A novel interpretation of the flower in the framework of the molecular phylogeny of Aroideae is required.

The morphological nature of the gynoecium of *Calla* raises the problem of pseudomonomery in subfamilies of Araceae. Pseudomonomery is defined from an anatomical (vascularization) or developmental point of view (vestigial carpel): a unilocular gynoecium comprising more than one carpel (Barabé et al., 1987a). The presence of two or three locules (pluricarpellate gynoecium) would be the original state in some Araceae and a reduction from several locules appears to have occurred several times in the family (Grayum, 1990; Buzgo, 2001). In *Gymnostachys*, for example, Buzgo (2001) states that the apical, median placentation, and the gynoecium's ascidiate carpel appearance indicate a truly unicarpellate gynoecium. On the other hand, the unusual vasculature of the gynoecium base in *Gymnostachys* gives no clue as to the number of carpels involved. In the Orontioideae, the unilocular gynoecium of *Symplocarpus* has been interpreted as pseudomonomerous. In *Lysichiton*, the frequent presence of unilocular gynoecia, corresponding to pseudomonomerous gynoecia has been reported (Barabé et al., 1987a). In basal Araceae, Buzgo (2001) notes that a unicarpellate gynoecium can form in different ways: by fusion of locules or carpels (*Lysichiton*) or by suppression of septa. In *Pothos*, Buzgo (2001) considers that unilocular gynoecia may be truly monomerous with a secondary multiplication of vascular bundles. This author also notes that the recognition of more than one carpel is difficult in unilocular Lasioideae and unicarpellate conditions are likely to occur. Lasioideae would certainly represent a very interesting taxon on which to perform an anatomical and developmental study of the evolution of a unilocular ovary in closely related taxa.

Until now, pseudomonomery has been studied mainly by using anatomical data (Eckardt, 1937; Eyde et al., 1967; Barabé et al., 1987a). However, can morphogenesis help to determine if a gynoecium is monocarpellate or not? How could we determine the mode of reduction of the number of carpels in aroids from a developmental perspective? We have seen in *Calla* that developmental data may be as controversial as anatomical data. Therefore, the problem of determining the number of carpels involved in aroid pseudomonomerous gynoecia remains an open question which will be very difficult to answer by using only developmental data.

11.8 Lemnoideae and neoteny

Lemnoideae are highly reduced plants compared to other aroid genera. Although the phylogenetic position of Lemnoideae is well resolved, it remains difficult to interpret the anatomy and morphogenesis of the flower.

Recent molecular phylogenies showed that Lemnoideae appears among the basal subfamilies of Araceae, between the Proto-Araceae (*Gymnostachys* + Orontioideae) and Pothoideae (Cabrera et al., 2008; Cusimano et al., 2011). For Bogner (2009), Lemnoideae, particularly Wolffieae, represents a case of extreme neoteny. Fossils of putative ancestors of Lemnoideae (with the genus *Limnobiophyllum*) go back to the late Cretaceous (Stockey et al., 1997; Bogner, 2009), and their position among basal Araceae indicates clear phylogenetic relationships with genera having bisexual flowers. Many authors follow the interpretation of Shih (1979), who considered the flower of *Lemna* as bisexual. Bogner (2009), on the other hand, interpreted the inflorescence of Lemnoideae 'as one bisexual flower in the Wolffieae and as one bisexual flower (with a single stamen) and one male flower (with a single stamen) in the Lemneae.' The interpretation of Bogner (2010) is based on the fact that one stamen matures prior to the other. However, this phenomenon also occurs in some genera with bisexual flowers, for example *Anthurium*, *Gymnostachys* and *Anaphyllopsis*. Unfortunately, the two opposite interpretations are based on Shih's developmental study (Shih, 1979), which does not document early stages of development. Perhaps a comparison of early stages of development of other genera of Lemnoideae with basal Araceae would help to solve this problem, which should be analysed in the general context of the distinction between flowers and inflorescences in basal angiosperms (Rudall, 2003).

11.9 Conclusions

The floral development of *Cercestis* and *Philodendron* shows that in closely related taxa, morphogenetic processes are not necessarily similar. This is also the case in Lasioideae (e.g. *Dracontium*, *Urospatha*), where changes in the number of floral parts are governed by different processes. The floral development of atypical flowers in the subfamily Aroideae indicates that morphogenesis is subordinated to the structure of the mature normal flower that is the evolving unit experiencing the action of selection. Morphogenetic patterns are the consequence of the types of mature structures that have been selected during evolution.

The spadix of Araceae, and of the subfamily Aroideae in particular, presents a great diversity of developmental features and a unique system to study the transition of different floral types within the same inflorescence. Does the morphological gradient observed in genera with unisexual flowers also occur at the physiological or molecular level? This remains an open question. A study of molecular developmental genetics would certainly be able to provide new insights into the nature of aroid floral organs in a phylogenetic context.

In general, floral development helps us to understand the nature of structures that are not clearly recognizable on mature inflorescences. However, until now it has not been very useful in solving phylogenetic problems in the aroid family. On the contrary, to interpret the morphogenesis of the aroid flower one needs a well-resolved phylogeny based on mature structures or molecular data.

11.10 Acknowledgements

I would like to thank Dr. Christian Lacroix for his comments on the manuscript. The comments of an anonymous reviewer were greatly appreciated. This research was supported in part by an operating grant from the Natural Sciences and Engineering Research Council of Canada and by the project '2ID' from the CNRS Amazonie Program (Conseil national de la recherche scientifique, France).

11.11 References

Barabé, D. (1994). Développement et phyllotaxie de l'inflorescence du *Symplocarpus foetidus* (L.) Nutt. (Araceae). *Canadian Journal of Botany*, **72**, 715–725.

Barabé, D. and Bertrand, C. (1996). Organogénie florale des genres *Culcasia* et *Cercestis* (Araceae). *Canadian Journal of Botany*, **74**, 898–908.

Barabé, D. and Labrecque, M. (1983). Vascularisation de la fleur de *Calla palustris* (Araceae). *Canadian Journal of Botany*, **61**, 1718–1726.

Barabé, D. and Lacroix, C. (1999). Homeosis, morphogenetic gradient and the determination of floral identity in the inflorescences of *Philodendron solimoesense* (Araceae). *Plant Systematics and Evolution*, **219**, 243–261.

Barabé, D. and Lacroix, C. (2001). The developmental morphology of the flower of *Montrichardia arborescens* (Araceae) revisited. *Botanical Journal of the Linnean Society*, **135**, 413–420.

Barabé, D. and Lacroix, C. (2002). Aspects of floral development in *Caladium bicolor* (Araceae). *Canadian Journal of Botany*, **80**, 899–905.

Barabé, D. and Lacroix, C. (2008a). Developmental morphology of the flower of *Anthurium jenmanii*: a new element in our understanding of basal Araceae. *Botany*, **86**, 45–52.

Barabé, D. and Lacroix, C. (2008b). Developmental morphology of the flower of *Anaphyllopsis americana* and its relevance to our understanding of basal Araceae. *Botany*, **86**, 1467–1473.

Barabé, D., Chrétien, L. and Forget, S. (1987a). On the pseudomonomerous gynoecia of the Araceae. *Phytomorphology*, **37**, 139–143.

Barabé, D., Forget, S. and Chrétien, L. (1987b). Organogénèse de la fleur de

Symplocarpus foetidus (Araceae). Canadian Journal of Botany, **65**, 446–455.

Barabé, D., Lacroix, C. and Jeune, B. (2000). Development of the inflorescence and flower of *Philodendron fragrantissimum* (Araceae): a qualitative and a quantitative study. *Canadian Journal of Botany*, **78**, 557–576.

Barabé, D., Bruneau, A., Forest, F. and Lacroix, C. (2002). The correlation between development of atypical bisexual flowers and phylogeny in the Aroideae (Araceae). *Plant Systematics and Evolution*, **232**, 1–19.

Barabé, D., Lacroix, C. and Gibernau, M. (2003). Development of the flower and inflorescence of *Arum italicum* (Araceae). *Canadian Journal of Botany*, **81**, 622–632.

Barabé, D., Lacroix, C. and Jeune, B. (2004a). The game of numbers in homeotic flowers of *Philodendron* (Araceae). *Canadian Journal of Botany*, **82**, 1459–1467.

Barabé, D., Lacroix, C., Bruneau, A., Archambault, A. and Gibernau, M. (2004b). Floral development and phylogenetic position of *Schismatoglottis* (Araceae). *International Journal of Plant Sciences*, **165**, 173–189.

Barabé, D., Lacroix, C. and Jeune, B. (2008). Quantitative developmental analysis of homeotic changes in the inflorescence of *Philodendron* (Araceae). *Annals of Botany*, **101**, 1027–1034.

Barabé, D., Lacroix, C. and Gibernau, M. (2011). Floral development of *Urospatha*: merosity and phylogeny in the Lasioideae (Araceae). *Plant Systematics and Evolution*, **296**, 41–50.

Barriault, I., Barabé, D., Cloutier, L. and Gibernau, M. (2010). Pollination ecology and reproductive success in

Jack-in-the-Pulpit (*Ariseama triphyllum*) in Québec. *Plant Biology*, **12**, 161–171.

Bogner, J. (2009). The free-floating aroids (Araceae) – living and fossil. *Zitteliana*, **A48/49**, 113–128.

Bogner, J. (2010). Are the flowers of the Duckweeds (Araceae-Lemnoideae) bisexual or unisexual? *Aroideana*, **33**, 178–182.

Bogner, J. and Hesse, M. (2005). Zamioculcadoideae, a new subfamily of Araceae. *Aroideana*, **28**, 3–20.

Bowman J. L., Brüggemann, H., Lee, J. Y. and Mummenhoff, K. (1999). Evolutionary changes in floral structure within *Lepidium* L. (Brassicaceae). *International Journal of Plant Science*, **160**, 917–929.

Boyce, P. C. (1993). *The Genus Arum*. Kew: Royal Botanic Gardens.

Buzgo, M. (1994). Inflorescence development of *Pistia stratiotes* (Araceae). *Botanische Jahrbücher*, **115**, 557–570.

Buzgo, M. (1999). *Flower Structure and Development of Acoraceae and Basal Araceae and their Systematic Position among Basal Monocotyledons*. Thesis, Zürich University (Universität Zürich), Switzerland.

Buzgo, M. (2001). Flower structure and development of Araceae compared with alismatids and Acoraceae. *Botanical Journal of the Linnean Society*, **136**, 393–425.

Buzgo, M. and Endress P. K. (2000). Floral structure and development of Acoraceae and its systematic relationships with basal Angiosperms. *International Journal of Plant Sciences*, **161**, 23–41.

Cabrera, L. I., Salazar, G. A., Chase, M. W. et al.(2008). Phylogenetic relationships of Aroids and Duckweeds (Araceae) inferred from coding and non-coding plastid DNA. *American Journal of Botany*, **95**, 1153–1165.

Cusimano, N., Bogner, J., Mayo, S. J. et al. (2011). Relationships within the Araceae: comparison of morphological patterns with molecular phylogenies. *American Journal of Botany*, **98**, 654–668.

Eckardt, T. (1937). Untersuchungen über Morphologie, Entwicklungsgeschichte und systematische Bedeutung pseudomonomeren Gynoeceums. *Nova Acta Leopoldina*, **5**, 3–112.

Engler, A. and Krause, K. (1912). Araceae-Philodendroideae-Philodendreae. In *Das Pflanzenreich*, IV, 23Da. Heft **55**, ed. A. Engler. Leipzig: Engelmann, Reprinted 1966 (J. Cramer), pp. 1–134.

Ewing, J. W. and Klein R. M. (1982). Sex expression in Jack-in-the-Pulpit. *Bulletin of the Torrey Botanical Club*, **109**, 47–50.

Eyde, R. H., Nicolson, D. H. and Sherwin, P. (1967). A survey of floral anatomy in Araceae. *American Journal of Botany*, **54**, 478–479.

Fujita, T. (1942). Zur Kenntnis der Organstellungen im Pflanzenreich. *Japanese Journal of Botany*, **12**, 1–55.

Fukai, S. (2004). Floral initiation and development of the sex-changing plant *Arisaema sikokianum* (Araceae). *International Journal of Plant Sciences*, **165**, 739–744.

Gould, S. J. (1977). *Ontogeny and Phylogeny*. Cambridge, MA: Harvard University Press.

Grayum, M. H. (1990). Evolution and phylogeny of the Araceae. *Annals of the Missouri Botanical Garden*, **77**, 628–697.

Haeckel, E. (1866). *Generelle Morphologie der Organismen*, Berlin: Georg Reinser.

Haigh, A., Clark, B., Reynolds, L. et al. (2010). *CATE Araceae*.http://www.cate-araceae.org (accessed October 19, 2010).

Hay, A. (1992). Tribal and subtribal delimitation and circumscription of the genera of Araceae tribe *Lasieae*. *Annals of the Missouri Botanical Garden*, **79**, 184–205.

Lehmann, N. L. and Sattler, R. (1992). Irregular floral development in *Calla palustris* (Araceae) and the concept of homeosis. *American Journal of Botany*, **79**, 1145–1157.

Mishler, B. D. (1986). Ontogeny and phylogeny in *Tortulla* (Musci: Pottiaceae). *Systematic Botany*, **11**, 189–208.

Nelson, G. (1978). Ontogeny, phylogeny, paleontology and the biogenetic law. *Systematic Zoology*, **27**, 324–345.

Poisson, G. and Barabé, D. (2011). Developmental morphology of the flower of *Dracontium polyphyllum* in the context of the phylogeny of the Araceae. *Kew Bulletin* **66**: 537–543.

Rudall, P. J. (2003). Monocot pseudanthia revisited: floral structure of mycoheterotrophic family Triuridaceae. *International Journal of Plant Sciences*, **164** (5 Suppl.), S307 –S320.

Scribailo, R. W. and Tomlinson, P. B. (1992). Shoot and floral development in *Calla palustris* (Araceae-Calloideae). *International Journal of Plant Sciences*, **153**, 1–13.

Shih, C. Y. (1979). SEM studies of the flowering duckweed *Lemna perpusilla*. *Journal of Scanning Electron Microscopy*, **3**, 479–486.

Stevens, P. F. (2001). *Angiosperm Phylogeny Website*. Version 9, June 2008. http://www.mobot.org/MOBOT/research/APweb/.

Stockey, R. A., Hoffman, G. L. and Rothwell, G. W. (1997). The fossil monocot *Limnobiophyllum scutatum*: resolving the phylogeny of Lemnaceae. *American Journal of Botany*, **84**, 355–368.

Tucker, S. C., Douglas, A. W. and Liang, H.-X. (1993). Utility of ontogenic and conventional characters in determining phylogenetic relationships of Saururaceae and Piperaceae (Piperales). *Systematic Botany*, **18**, 614–641.

Uhlarz, H. (1982). Typologische und ontogenetische Untersuchungen an *Spathicarpa sagittifolia* Schott (Araceae): Wuchsform und Infloreszenz. *Beiträge zur Biologie der Pflanzen*, **57**, 389–429.

Uhlarz, H. (1986). Zum Problem des 'blattlosen Sprosses': Morphologie und Anatomie der Infloreszenz von *Pinella tripartita* (Blume) Schott (Araceae, Aroideae). *Beiträge zur Biologie der Pflanzen* **61**: 241–282.

Wake, D. B. (1991). Homoplasy: the result of natural selection, or evidence of design limitations. *American Naturalist*, **138**, 534–567.

Zluvova, J., Nicolas, M., Berger, A., Negrutiu, I. and Monéger, F. (2006). Premature arrest of the male flower meristem precedes sexual dimorphism in the dioecious plant *Silene latifolia*. *Proceedings of the National Academy of Sciences*, **103**, 18854–18859.

Some observations on the homology of the daffodil corona

ROBERT W. SCOTLAND

12.1 Introduction

Flowers of the genus *Narcissus* (the daffodils) contain a corona, a cup-shaped structure, positioned between the stamens and tepals (Fig 12.1a). The identity and homology of this structure has been a source of debate and confusion for more than a century (Doell, 1857; Masters, 1865; Smith, 1866; Eichler, 1875; Pax, 1888; Celakovsky, 1898; Velenovsky, 1907, 1910; Goebel, 1933; Arber, 1937; Singh, 1972). Here I provide a short review of this issue, which explains the context for an ongoing developmental genetics research study attempting to ascertain the homology of the daffodil corona.

12.1.1 Historical perspective

Understanding and interpreting floral anatomy across the full range of angiosperms largely comes from a combination of positional homology and identity of floral organs. The expectation that flowers have four whorls consisting of a calyx (sepals), corolla (petals), androecium (stamens) and gynoecium (carpels) allows interpretation of a wide range of angiosperm flowers (Fig 12.2). However, there are many exceptions to this basic pattern not only in monoecious and dioecious flowers, but also due to a range of processes including reduction, fusion, novelty, homeosis, duplication etc. One such difference is that found in many petaloid monocots in which the two outer whorls of the flower have similar

Early Events in Monocot Evolution, eds P. Wilkin and S. J. Mayo. Published by Cambridge University Press. © The Systematics Association 2013.

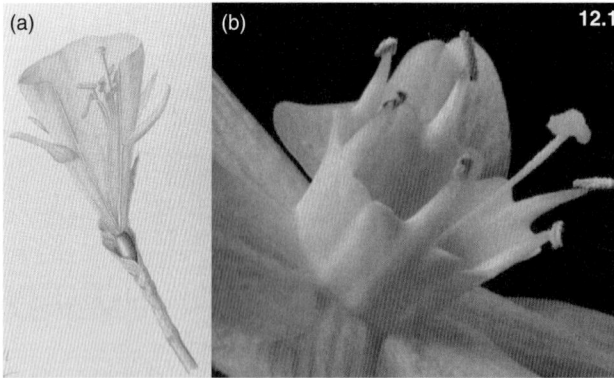

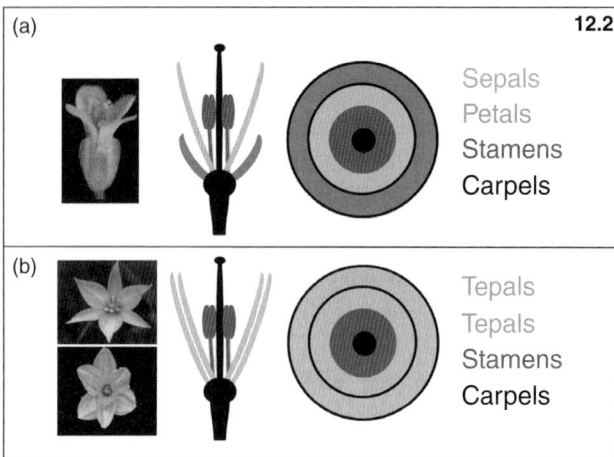

Fig 12.1 a: Flower of *Narcissus bulbocodium* showing prominent corona and staminal filaments attached at base of flower. Painted by Rosemary Wise. b: Flower of *Eucharis amazonica* showing staminal filaments fused to form corona.

Fig 12.2 a: Flowers of *Arabidopsis thaliana* have four whorls: calyx (sepals), corolla (petals), androecium (stamens), gynoecium (carpels). b: In flowers of petaloid monocots such as tulip and daffodil the two outer whorls have similar identity and are termed tepals. In addition, the daffodil flower has an extra whorl, the corona. Colour versions to be found in colour plate section.

identity and are collectively termed tepals rather than sepals and petals (Fig 12.2b). Many common ornamental plants including tulips and daffodils have this type of flower. The genus *Narcissus* is characterized by flowers that apparently contain a distinct fifth structure within the flower. These flowers comprise two whorls of three tepals, a whorl of six stamens and a whorl of three fused carpels. In addition, however, flowers of the 70 or so species of the genus *Narcissus* (with one exception *N. cavanillesii*) contain an additional structure between the tepals and the stamens (Fig 12.1a). This structure, termed the corona, is a cup-shaped or crown-like structure that gives these flowers a very characteristic appearance. The prevailing framework for interpreting the daffodil corona has been to view it as a modification of established floral whorls such as stamens or tepals (Doell, 1857; Masters, 1865; Smith, 1866; Eichler, 1875; Pax, 1888; Celakovsky, 1898; Velenovsky, 1907, 1910; Goebel, 1933; Arber, 1937; Singh, 1972). For example, some have interpreted the corona as an outgrowth of tepals equivalent to ligules of leaves (Doell, 1857) or confluent petal stipules (Smith, 1866) and others as a modification of stamens (Masters, 1865). In an influential paper Agnes Arber stated that 'the corona of *Narcissus* has no connexion with the stamens, but is split off from the inner surface of the perianth tube' (Arber, 1937: 300–301).

The comparative context for interpreting the corona of daffodil flowers has been partly established by the presence of a corona in several other genera of Amaryllidaceae including *Eucharis*, *Hymenocallis* and *Pancratium*. These three and several other genera of Amaryllidaceae have what are termed staminal coronas. This type of corona has been interpreted as being formed from fusion of staminal filaments because the anthers are positioned at the apex of the corona (Fig 12.1b). The confusion and disagreement surrounding the daffodil corona partly stems from the observation that the anther filaments have an apparently different positional relationship to the corona as they are attached to its base rather than forming the corona itself (Fig 12.1a). This has led to the distinction between staminal and perianthal coronas within Amaryllidaceae. Figure 12.3 is a generic-level phylogeny of Amaryllidaceae. Genera with at least some species with a corona are represented as green terminal branches. The phylogeny shows that the corona has a homoplastic distribution and is therefore likely to have evolved independently multiple times within Amaryllidaceae. Nevertheless, the question of whether the corona of daffodils is modified from stamens, as in other genera of Amaryllidaceae or is a modification of tepals or something novel, remains open.

12.1.2 Daffodil corona is derived from tepals

An important paper entitled 'On the "Corona" and Androecium in Certain Amaryllidaceae' by Agnes Arber in 1937, makes the case that the corona of *Narcissus* has no connection with the stamens but is a derivative of the perianth. Arber (1937) sectioned mature flowers of *Hymenocallis*, *Pancratium* and *Narcissus*. With regard to *Hymenocallis festalis* and *Pancratium illyricum* she stated that 'the stamen-cup is not only uniform in its obvious derivation from the stamens, it is also uniformly non-vascular, except for a single massive bundle belonging to each filament' (Arber, 1937: 300). In contrast, in relation to *Narcissus* she showed that the corona 'tends to be richly vascular; it is traversed by numerous bundles and went onto state that the stamen cup of the Pancratieae and the corona of *Narcissus* are fundamentally different' (Arber, 1937: 301).

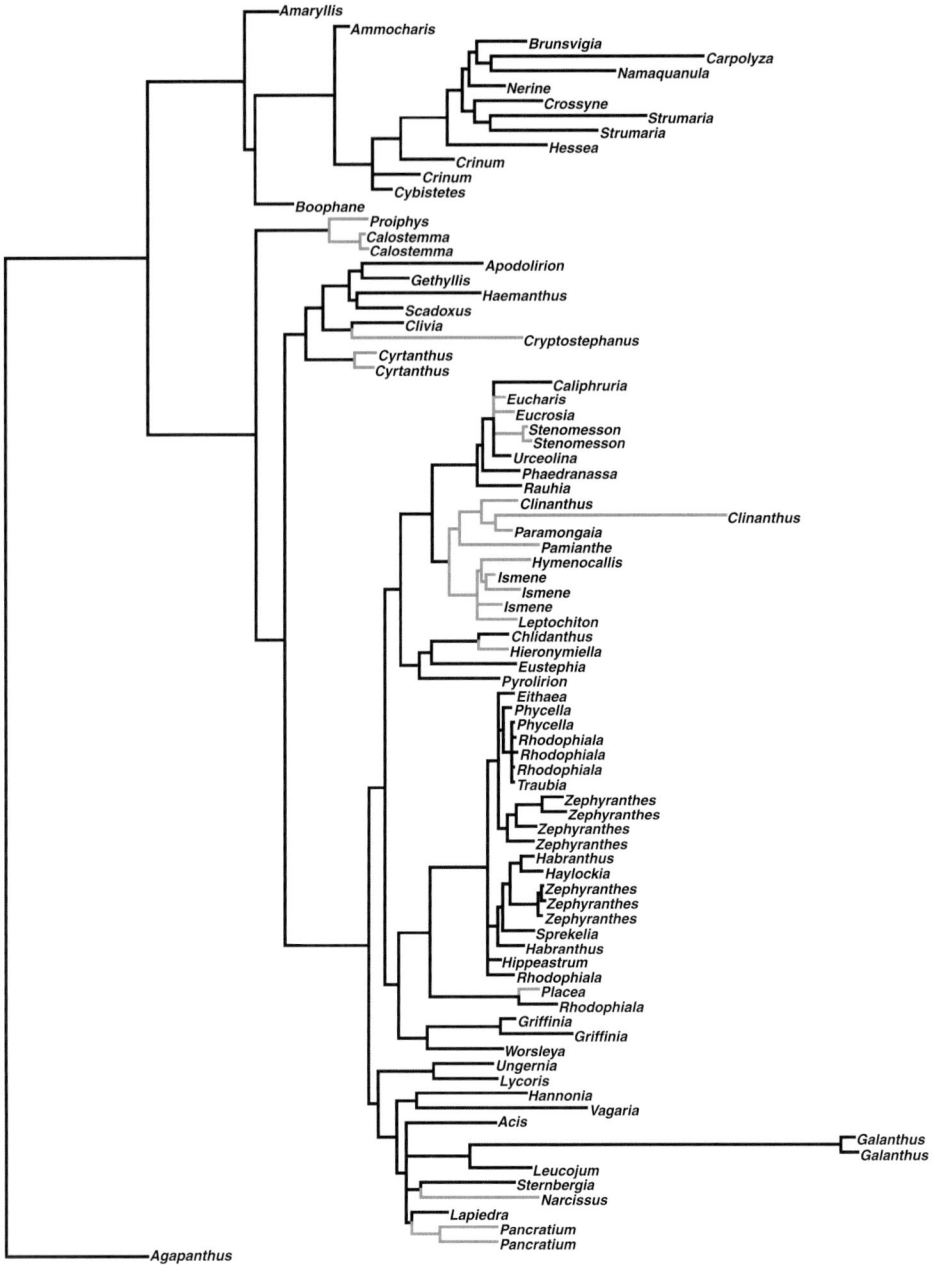

Fig 12.3 Generic-level phylogeny of Amaryllidaceae estimated from available DNA sequences of ITS from GenBank. The tree was inferred by Bayesian methods using the generalized reversible model (nst = 6). Genera with at least some species with a corona are coloured as green terminal branches. The corona has evolved multiple times within Amaryllidaceae. Colour version to be found in colour plate section.

12.2 A new perspective

The ABC model of floral development formulated by Coen and Meyerowitz (1991) was based on two model plant species, *Antirrhinum majus* and *Arabidopsis thaliana*. Similar phenotypic mutants were described in which the identity of floral organs was affected in pairs of adjacent whorls: sepals and petals, petals and stamens, and stamens and carpels. Homeotic transformation was responsible for altering organ identity. Three classes of genes – A, B and C – were postulated to be active in each of two whorls of the meristem: A function in the outer two whorls, B function in the middle two whorls and C function in the inner two whorls. Each floral whorl has a unique combination of gene function to determine the identity of particular organs in that whorl. A function for sepals, A + B function for petals, B + C function for stamens and C function for carpels. After 20 years of research, the ABC model has been modified to accommodate various exceptions and differences from that originally proposed (Litt and Kramer, 2010). Nevertheless the presence and expression of the C-class gene (*Agamous*) is restricted to stamens and carpels and therefore has the potential to contribute some new information as to the homology of the daffodil corona. Therefore this project seeks to clone ABC floral organ identity genes and determine their expression patterns in distinct floral whorls with a view to discovering whether the corona shares similar expression patterns to tepals or to stamens. It is fully appreciated that these observations, in themselves, will not be conclusive, but may, alongside detailed anatomical observations, aid our understanding of the corona.

12.2.1 A few preliminary observations

Figure 12.4a is a scanning electron micrograph of a very young flower of *Narcissus bulbocodium* excised from a bulb in which it developed in the autumn six months before it is visible as an above-ground part in the following spring. The corona, indicated by the arrow, is relatively short at this early stage of floral development, compared to maturity (Fig 12.1a), because it develops after all the other floral organs. Figure 12.4b is a longitudinal section of a young flower at the same stage showing the early development of the corona relative to the staminal filament and tepal. Figure 12.4c is a transverse section of a young flower at a slightly later stage. Figure 12.4d is a schematic flower showing the relative positions of the carpels (blue), stamens (green), corona (yellow) and tepals red. Black lines indicate the position in mature flowers that led Arber (1937) to reach her conclusions that the corona is split off from the tepals. However, the stamens and tepals develop before the corona, thus an important issue is the identity of the area here interpreted as the hypanthium, in which the corona initiates.

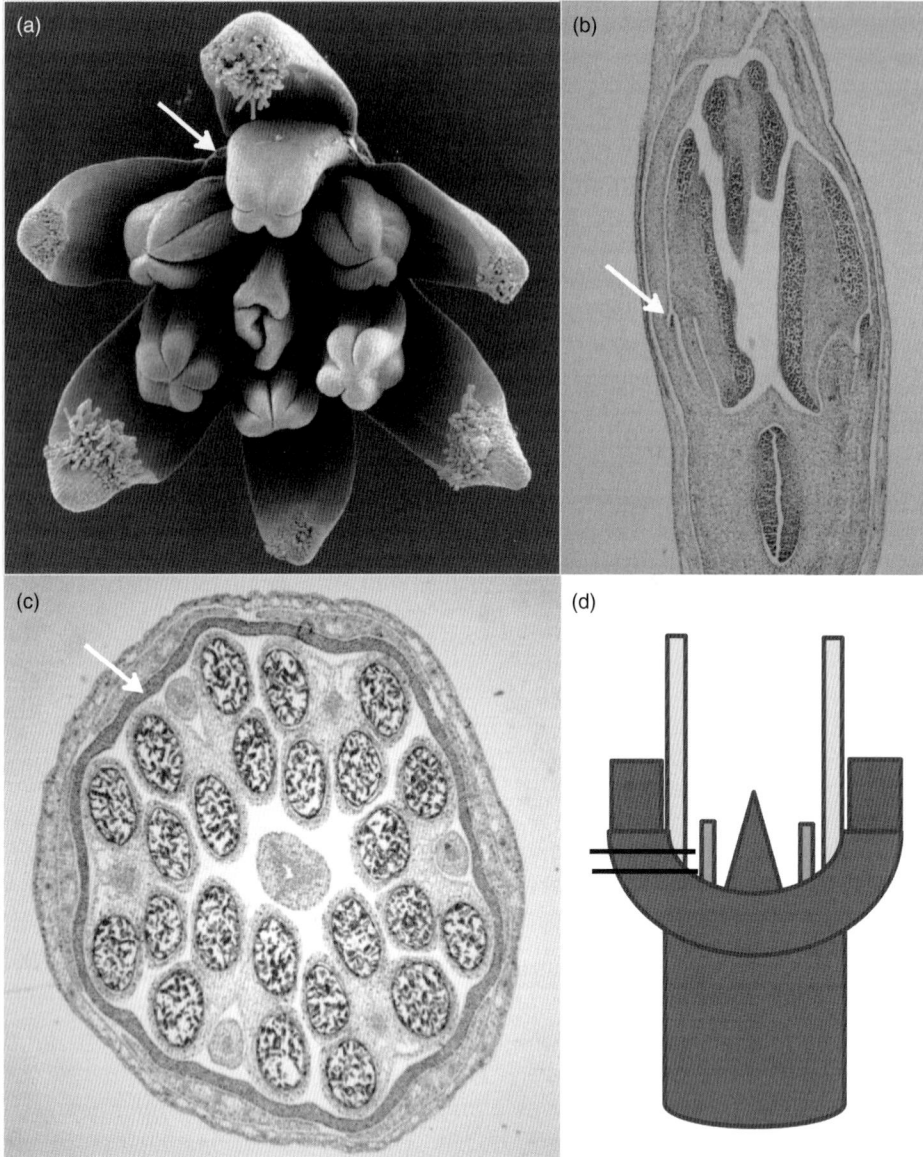

Fig 12.4 a: Very young flower of *Narcissus bulbocodium* with corona initiating late relative to the other floral organs. b: Longitudinal section of young flower of *Narcissus bulbocodium* with initiating corona shown by arrow. c: Transverse section of young flower of *Narcissus bulbocodium* with early stage corona indicated by an arrow. d: Schematic of flower showing carpels (blue), stamens (green), corona (yellow) and tepals (red). The hypanthium region between the stamens and tepals is the point of corona initiation. Black lines indicate the plane of sectioning of Arber (1937). Colour version to be found in colour plate section.

12.3 Conclusion

Developmental genetics is an area of research trying to understand the relationship between phenotype and genotype. Therefore, age-old botanical enigmas like the homology of the daffodil corona can be revisited using floral organ identity genes as a tool to elucidate and understand whether these structures are modified from existing organs or should, more accurately, be considered as novelties (Wagner and Lynch, 2010), perhaps derived from a unique and novel combination of pre-existing developmental programmes (Shubin et al., 2009).

12.4 Acknowledgements

I thank Paul Wilkin for the invitation, two anonymous reviewers for comments and the Gatsby Charitable Foundation for funding.

12.5 References

Arber, A. (1937). Studies in flower structure III. On the 'corona' and androecium in certain Amaryllidaceae. *Annals of Botany*, **1**, 293–304.

Celakovsky, L. J. (1898). Ueber die Bedeutung und den Ursprung der Paracorolle der Narcisseen. *Bulletin international de l'Académie des sciences de Bohême* **8**, 1–15.

Coen, E. S. and Meyerowitz, E. M. (1991). The war of the whorls: genetic interactions controlling flower development. *Nature*, **353**, 31–37.

Doell, J. C. (1857). *Flora des Grossherzogthums, Volume 1*. Carlsruhe: G. Braun.

Eichler, A. W. (1875). *Bluthendiagramme, Volume 1*. Leipzig: W. Engelmann.

Goebel, K. (1933). *Organographie der Pflanzen*, 3rd edn., Part 3. Jena: G. Fischer.

Litt, A. and Kramer, E. M. (2010). The ABC model and the diversification of floral organ identity. *Seminars in Cell and Developmental Biology*, **21**, 129–137.

Masters, M. T. (1865). On the corona of *Narcissus. Seemann's Journal of Botany*, **3**, 105–109.

Pax, F. (1888). Amaryllidaceae. In *Die natürlichen Pflanzenfamilien*, ed. A. Engler and K. Prantl. Leipzig: W. Engelmann, pp. 97–124.

Shubin, N., Tabin, C. and Carroll, S. (2009). Deep homology and the origins of evolutionary novelty. *Nature*, **457**, 818–823.

Singh, V. (1972). Floral morphology of Amaryllidaceae.1. Subfamily Amaryllidioideae. *Canadian Journal of Botany*, **50**, 1555–1565.

Smith, W. G. (1866). The corona of *Narcissus. Seemann's Journal of Botany* **4**, 169–171.

Velenovsky, J. (1907). *Vergleichende Morphologie der Pflanzen, Volume II*. Prague: Fr. Rivnác.

Velenovsky, J. (1910). *Vergleichende Morphologie der Pflanzen, Volume III*. Prague: Fr. Rivnác.

Wagner, G. P. and Lynch, V. (2010). Evolutionary novelties. *Current Biology*, **20**, R48–R52.

13

Anther, ovule and embryological characters in Velloziaceae in relation to the systematics of Pandanales

Maria Das Graças Sajo, Renato Mello-Silva and Paula J. Rudall

13.1 Introduction

Velloziaceae are a monocot family of five genera and *c.* 250 species (Mello-Silva, 2009). The family is now assigned to the order Pandanales, a small but morphologically diverse monocot order of five families: Cyclanthaceae, Pandanaceae, Stemonaceae (including *Pentastemona* Steenis), Triuridaceae and Velloziaceae (APG III, 2009). Within Velloziaceae, three genera occur in South America, of which two are endemic to Brazil (*Barbacenia* Vand., *Vellozia* Vand.) and the third is Andean (*Barbaceniopsis* L.B.Sm.). A fourth genus, *Xerophyta* Juss., grows in tropical Africa and the fifth, *Acanthochlamys* P.C.Kao, is endemic to China.

In this chapter, we present new data on the anther, ovule and embryology of Velloziaceae in the context of ongoing related studies on floral evolution (e.g. Sajo et al., 2010) and the systematics of Pandanales (Furness and Rudall, 2006; Rudall and Bateman, 2006). Earlier studies on ovule and seed development and androecial features were used to delimit the genera of Velloziaceae (Menezes, 1976, 1980). Embryological features of Pandanales are currently poorly known, though some studies suggest that they are potentially systematically useful. For example, Cheah and Stone (1975) reported an unusual megagametophyte in some

Early Events in Monocot Evolution, eds P. Wilkin and S. J. Mayo. Published by Cambridge University Press. © The Systematics Association 2013.

Pandanus species, in which nucellar nuclei migrate into the developing embryo sac, and tapetum type is unusually variable within Pandanales (Furness and Rudall, 2006).

13.2 Material and methods

Most of the material examined in this study was collected from plants in their natural habitats in Minas Gerais and Bahia, Brazil and deposited in the Herbaria CDBI, HUEFS, K, SPF (details listed in Sajo et al., 2010). The following species were examined: *Barbacenia blanchetii* Goethart & Henrard, *B. flava* Mart. ex Schult. & Schult.f., *B. gentianoides* Taub. ex Goethart & Henrard, *B. riparia* (N.L.Menezes & Mello-Silva) Mello-Silva, *B. spiralis* L.B.Sm. & Ayensu, *Vellozia burlemarxii* L.B.Sm. & Ayensu, *V. canelinha* Mello-Silva, *V. declinans* Goethart & Henrard, *V. epidendroides* Mart. ex Schult. & Schult.f., *V. jolyi* L.B.Sm., *V. taxifolia* (Mart. ex Schult. & Schult.f.) Mart. ex Seub., *Vellozia* sp. nov. 1, *Vellozia* sp. nov. 2.

Flowers were fixed in formalin-acetic alcohol (FAA) and stored in 50% ethanol. For light microscopy, flowers were embedded in Paraplast using standard methods (Johansen, 1940) and serially sectioned at *c.* 13 μm thickness using a rotary microtome. For light microscopy (LM), sections were mounted onto microscope slides, stained in safranin and alcian blue, dehydrated through an ethanol series to 100% ethanol, transferred to Histoclear, and mounted in DPX mounting medium (distrene, with dibutyl phthalate and xylene). Slides were examined using a Leica DMLB photomicroscope fitted with a Zeiss Axiocam digital camera. For scanning electron microscope (SEM) examination, fixed flowers were carefully dissected in 70% ethanol and then dehydrated in an ethanol series to 100% ethanol. They were critical-point dried using a Bal-Tec 030 critical-point dryer, mounted onto pin stubs, coated with platinum using an Emitech K550 sputter coater, and examined using a Hitachi cold field emission SEM S-4700 at 2KV.

13.3 Results

13.3.1 Anther and microspore development

Anthers are tetrasporangiate in both *Barbacenia* and *Vellozia*. In *Barbacenia*, the anthers are introrse and dorsifixed in the coronal appendages (Fig 13.1A, B, D, E), except in some species with anthers that are basifixed to stamen filaments. In *Vellozia*, the anthers are latrorse and predominantly basifixed (Fig 13.1G), or dorsifixed in some species (e.g. *V. burlemarxii*, *V. declinans*, *Vellozia* sp. nov.1 and *Vellozia* sp. nov. 2). Each anther is four-lobed at early stages (Fig 13.1B). When the sporogenous cells are differentiated, the anther wall consists of four layers: epidermis, endothecium, one middle layer and a tapetum (Fig 13.1C), indicating

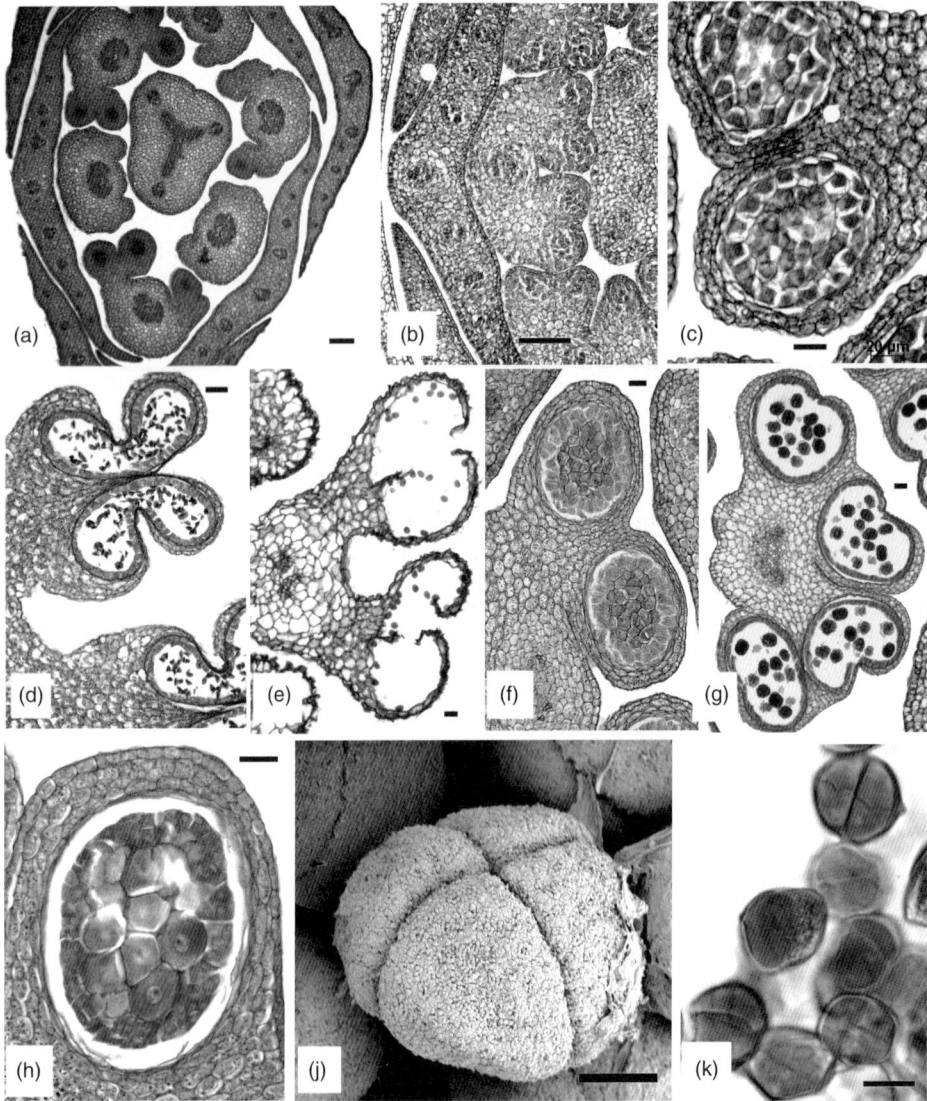

Fig 13.1 Velloziaceae, male structures. A: *Barbacenia spiralis*, TS young flower bud through anthers, above the level of the coronal appendages, which are not visible in this section. B: *B. blanchetii*, TS young anther. C: *B. riparia*, TS young anther with microspore mother cells and anther wall with four layers. D: *B. flava*, TS anther of immediately pre-anthetic flower, showing four microsporangia in each anther (coronal appendages not visible at this level). E: *B. riparia*, TS open (dehisced) anther. F: *B. gentianoides*, TS one side of a young anther (showing two of four microsporangia), with microspore mother cells in anther locules. G: *Vellozia epidendroides*, TS anther of immediately pre-anthetic flower, showing four microsporangia containing tetrads. H: *Vellozia* sp.nov.1, TS anther showing microspore mother cells. J: *V. epidendroides*, SEM pollen tetrad. K: *Barbacenia flava*, TS young anther showing developing microspores at dyad and (tetragonal) tetrad stages. Scales: A = 100 μm, B, D = 50 μm, C, E–H = 20 μm, J, K = 10 μm.

that anther wall development is of the monocotyledonous type. Epidermal cells are undifferentiated or slightly papillate (Fig 13.1C–G). The endothecium lacks thickenings and persists at anthesis (Fig 13.1E). The tapetum is secretory and two-layered in some regions. It consists of uninucleate cells of dense cytoplasm (Fig 13.1C). In each anther locule, the sporogenous tissue produces microsporocytes distinguishable by their large size, dense cytoplasm, and conspicuous nucleus (Fig 13.1C, H). Two successive meiotic divisions give rise to isobilateral tetrads (Fig 13.1K). The anthers dehisce by longitudinal slits (Fig 13.1E). In *Vellozia* the pollen grains are dispersed as tetrads (Fig 13.1J).

13.3.2 Ovule and megaspore development

The inferior ovary consists of three carpels and possesses three septal nectaries (Sajo et al. 2010). In both *Barbacenia* and *Vellozia* the ovules are anatropous and arranged in two rows on an axile placenta in each carpel (Fig 13.2A, B, C, E). Many ovule primordia arise at the same time in each carpel and within each ovule primordium a hypodermal archesporial cell differentiates at the micropylar end of the nucellus (Fig 13.2D). As soon as the two integuments are initiated, differential growth in the ovule primordium leads to ovule curvature (Fig 13.3A, B). The archesporial cell increases in size and gives rise directly to a megasporocyte without cutting off a parietal cell (Fig 13.3C). At this stage, the inner integument is more well-developed, but later both of them extend beyond the nucellus to form the micropyle. Both integuments are two-layered. The megasporocyte undergoes meiosis to form a linear tetrad of megaspores (Fig 13.3D–F, H, I). The chalazal functional megaspore develops into a seven-celled, eight-nucleate (*Polygonum* type) megagametophyte with a globular egg cell, two pear-shaped synergids, three antipodal cells and a central cell with two polar nuclei (Fig 13.3G, J, K). The mature embryo sac is curved and relatively narrow at the chalazal end (Fig 13.3K, 13.4C–D); this curvature becomes more pronounced after fertilization (Fig 13.4E), when the antipodal cells appear conspicuously enlarged (Fig 13.4A, B, E).

Following fertilization (Fig 13.4), the primary nucleus divides earlier than the zygote, and the developing endosperm conforms to the 'nuclear type'. Cellularization of the endosperm occurs after the proembryo achieves the globular stage (Fig 13.4E, F). The pericarp is dry and consists of 15–17 layers of mostly parenchymatous cells. The two outer cell layers are thick walled.

13.4 Discussion

Among the diverse families of Pandanales, flowers of Velloziaceae possess perhaps the greatest number of plesiomorphic floral conditions, including tricarpellate, trilocular ovaries with septal nectaries and axile placentation (Rudall and Bateman,

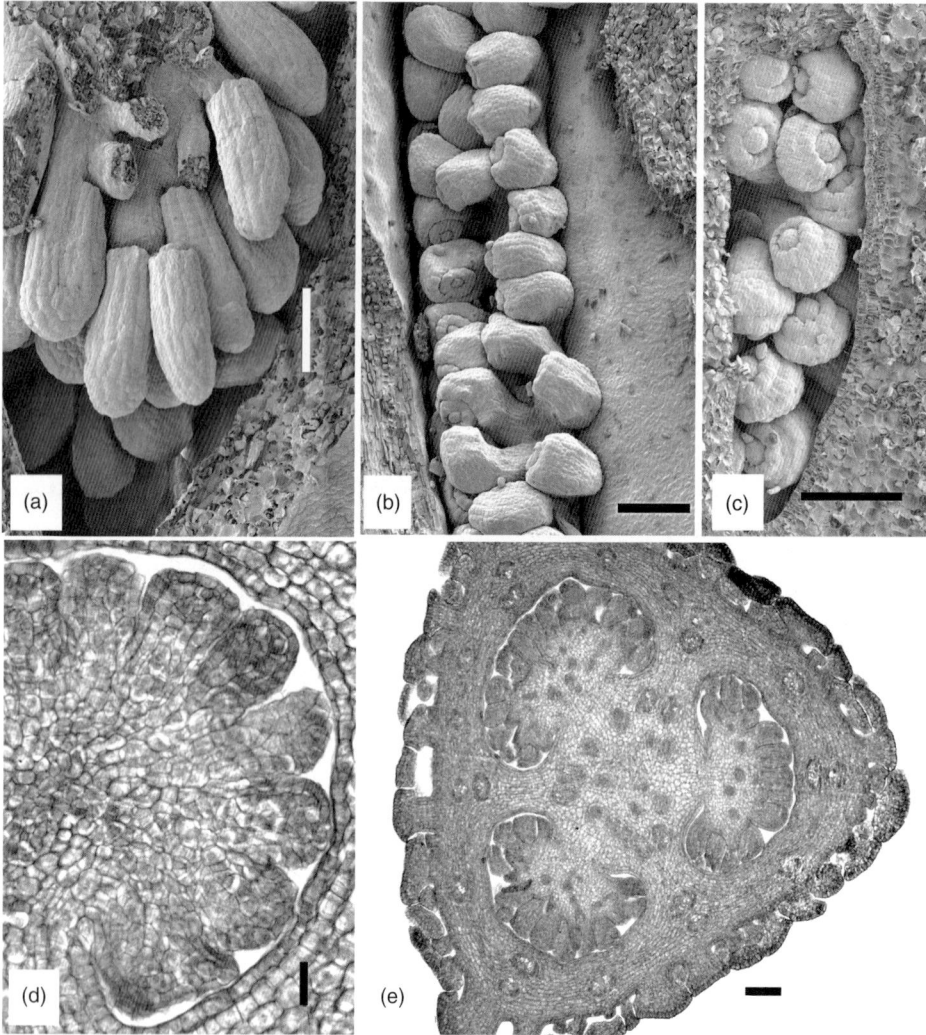

Fig 13.2 Ovule. A–C: SEM ovules. A: *Barbacenia riparia*. B: *Vellozia declinans*. C: *V. jolyi*. D: *Barbacenia blanchetii*, TS ovary showing ovules at archesporial stages. E: *Vellozia burlemarxii*, TS young ovary. Scales: A–C, E = 100 μm, D = 20 μm.

2006; Sajo et al., 2010). This high degree of plesiomorphy is consistent with their putative phylogenetic placement as sister to all other Pandanales (e.g. Chase et al., 2006; Givnish et al., 2006; Graham et al., 2006).

Mature anther characters are relatively variable within Velloziaceae, especially within *Vellozia*. For example, basifixed anther attachment characterizes most Pandanales (Dahlgren and Clifford, 1982). However, within Velloziaceae, the anthers are dorsifixed in most *Barbacenia* species (except *B. pungens*,

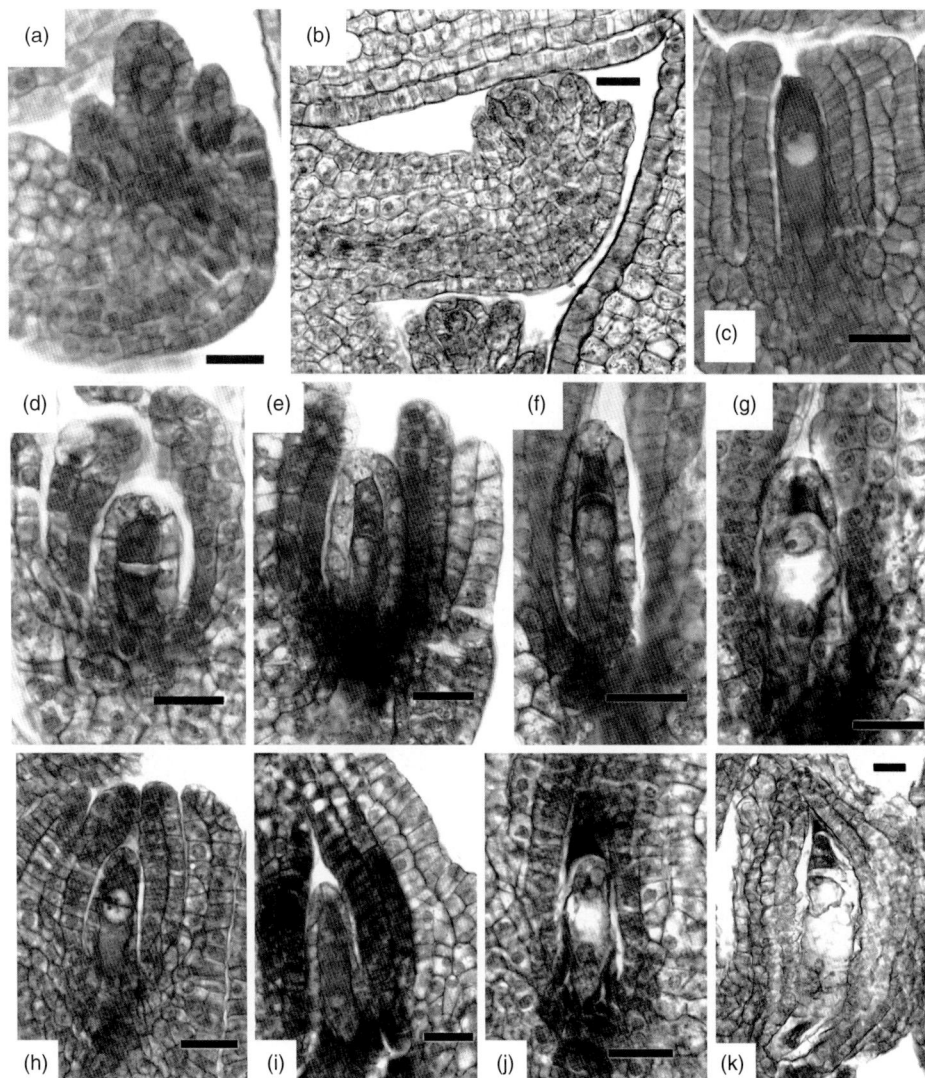

Fig 13.3 A: *Barbacenia flava*, ovule at megaspore mother cell stage. B: *B. gentianoides*, ovule at megaspore mother cell stage. C: *Vellozia epidendroides*, megaspore mother cell. D: *Barbacenia flava*, dyad of megaspores. E: *B. flava*, tetrad of megaspores. F: *B. flava*, tetrad with chalazal megaspore enlarged and micropylar megaspores degenerating. G: *B. flava*, megagametophyte at 2-celled stage, with remains of micropylar megaspores visible. H–J: *Vellozia epidendroides*. H: megaspore mother cell about to divide. I: two-celled stage. J: eight-nucleate megagametophophyte of the *Polygonum* type with remains of chalazal megaspores visible. K: *Vellozia* sp. nov. 1, fertilized ovule with proembryo and early free-nuclear endosperm divisions. Scales = 20 μm.

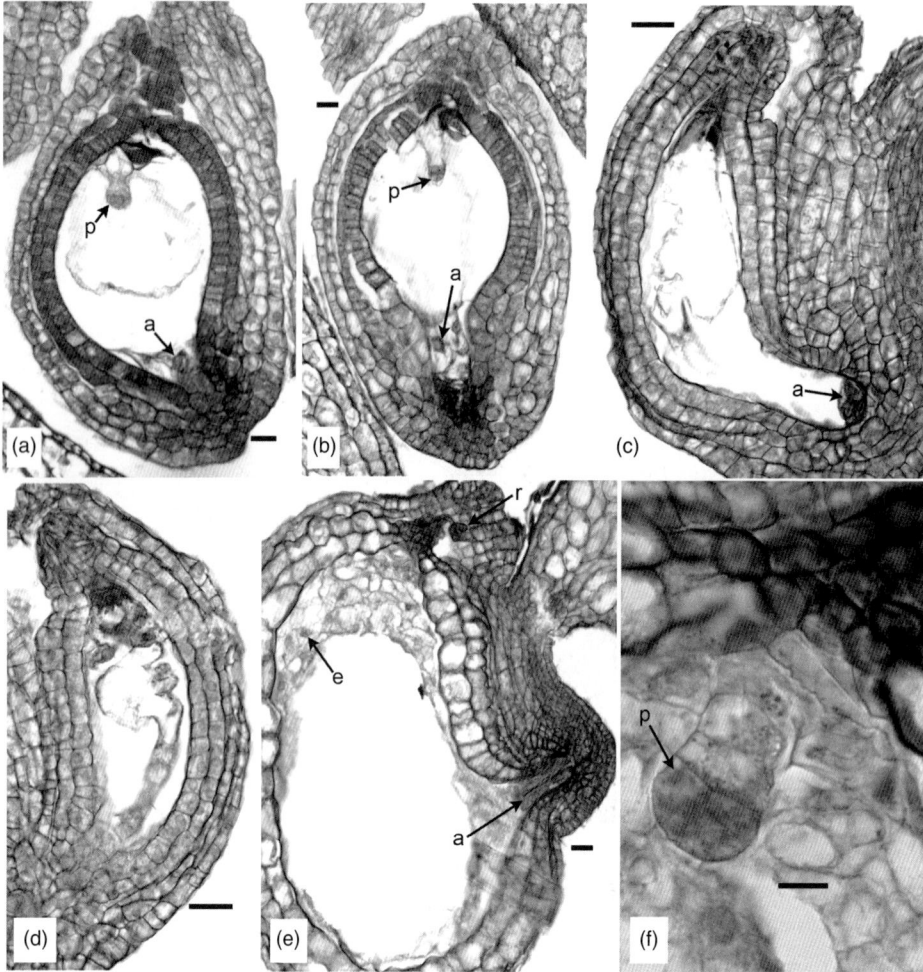

Fig 13.4 A, B: *Barbacenia gentianoides*, fertilized ovules with young proembryo and three persistent antipodals. C–F: *Vellozia* sp. nov.1. C: fertilized ovule with three persistent antipodals. D: fertilized ovule. E: older fertilized ovule with endosperm and three persistent antipodals. F: proembryo. a = antipodals, e = endosperm, p = proembryo, r = remains of pollen tube. Scales = 20 μm, except F = 10 μm.

B. rodriguesii, B. spiralis). Anther attachment in *Vellozia* is normally described as basifixed (Menezes et al., 1994; Mello-Silva, 2000, 2005), but *V. burlemarxii*, *V. declinans* and two related undescribed species possess dorsifixed anthers, a new condition for the genus. The two undescribed *Vellozia* species are assigned to *Vellozia* section *Xerophytoides*, a group with several derived features, but *V. burlemarxii* and *V. declinans* belong to a different section of *Vellozia* (Mello-Silva, 2005), making this character state homoplasious within the genus. The occurrence of dorsifixed anthers in half the species examined here suggests that this character merits more attention. Anther dehiscence is also relatively variable within Pandanales, from latrorse in Cyclanthaceae (Dahlgren and Clifford, 1982) and Triuridaceae (Rudall, 2008) to introrse in Stemonaceae (Rudall et al., 2005). Within Velloziaceae, anthers are introrse in *Barbacenia* and some *Vellozia* species, and latrorse in other *Vellozia* species.

Velloziaceae share many embryological characters with other Pandanales and Dioscoreales, including monocot-type anther wall development, successive microsporogenesis and *Polygonum*-type megagametophyte development, all conditions that are widespread among monocots (Dahlgren and Clifford, 1982; Rudall, 1997; Furness and Rudall, 1998, 1999; Harling et al., 1998). Crassinucellate ovules, the plesiomorphic condition for monocots, occur in both Pandanales and Dioscoreales (e.g. Rudall, 1997), though tenuinucellate ovules have been reported for some Triuridaceae (Maas-van de Kamer and Weustenfeld, 1998) and Burmanniaceae (Maas-van de Kamer, 1998). However, some embryological features – the tapetum type and the development of the embryo sac and endosperm – have systematic potential within Pandanales.

Tapetum type is unusually variable in Pandanales (reviewed by Furness and Rudall, 2006). The secretory (glandular) tapetum type, in which a layer of tapetal cells remains intact around the anther locule, represents the plesiomorphic condition in monocots, as in angiosperms in general (Furness and Rudall, 1998, 2001a, 2001b). Most members of the lilioid orders Asparagales, Dioscoreales and Liliales possess a secretory tapetum (Furness and Rudall, 1998). In contrast, within Pandanales the tapetum is secretory in Stemonaceae and Cyclanthaceae, but a plasmodial tapetum occurs in Pandanaceae and both secretory and intermediate types have been reported in Triuridaceae and Velloziaceae (Furness and Rudall, 1998, 2006). Li et al. (1992) reported a secretory tapetum in *Acanthochlamys*, the Chinese genus that is sister to all other Velloziaceae in some analyses (Chase et al., 1995, 2006; Behnke et al., 2000; Salatino et al., 2001; Mello-Silva, 2005). Furness and Rudall (2006) observed an intermediate tapetum type in *Talbotia* (Velloziaceae), in which the tapetal cell walls degenerate at the tetrad stage and the multinucleate tapetal protoplasts dissociate and extend into the anther locule, though it does not form a true plasmodium. They suggested that this represents an invasive nonsyncytial tapetum, which is unusual in noncommelinid monocots.

Most Velloziaceae possess a 'typical' eight-nucleate *Polygonum*-type embryo sac with three antipodal cells, as in the Chinese genus *Acanthochlamys* (Li et al., 1992) and most other Pandanales, including *Freycinetia* (Pandanaceae: Stromberg, 1956), Cyclanthaceae (Harling, 1946) and Triuridaceae (Rubsamen-Weustenfeld, 1991). However, Menezes (1976) reported that the antipodals proliferate after fertilization in Brazilian Velloziaceae and illustrated a well-defined multinucleate chalazal haustorium. Our observations show that the antipodals persist, but remain undivided, at least at early post-fertilization stages (Fig 13.4B, C). At later stages a chalazal haustorium is present (Fig 13.4E), but it is not clear whether this tissue is derived from the antipodals or the chalazal region of the nucellus. Endosperm is nuclear in Velloziaceae, as in most other Pandanales, though it is reportedly helobial in Cyclanthaceae (Pandanales) and some Dioscoreales (Dahlgren and Clifford, 1982). These results recall a similar controversy within *Pandanus*. Campbell (1909) reported an increased number of nuclei (14 instead of 8) in the embryo sac of *Pandanus* and concluded that the supernumerary nuclei are derived from the antipodal cells. However, Cheah and Stone (1975) subsequently reinterpreted the chalazal haustorium of *Pandanus* as nucellar, produced by enlargement, detachment and proliferation of nucellus cells located between the hypostase and the embryo sac. They observed that the chalazal nucellar cells are conspicuously different from the thicker-walled hypostase cells, even at the 4-nucleate stage. Given this controversy, our observations suggest that this aspect would merit further detailed observation within Velloziaceae.

13.5 References

APG III (Angiosperm Phylogeny Group III). (2009). An update of the Angiosperm Phylogeny Group classification for the orders and families of flowering plants: APG III. *Botanical Journal of the Linnean Society*, **161**, 105–121.

Behnke, H.D., Treutlein, J., Wink, M. et al. (2000). Systematics and evolution of Velloziaceae, with special reference to sieve-element plastids and *rbc*L sequence data. *Botanical Journal of the Linnean Society*, **134**, 93–129.

Campbell, D.H. (1909). The embryo sac of *Pandanus. Bulletin of the Torrey Botanical Club*, **36**, 205–220.

Chase, M.W., Stevenson, D.W., Wilkin, P. and Rudall, P.J. (1995). Monocot systematics: a combined analysis. In *Monocotyledons: Systematics and Evolution*, ed. P.J. Rudall, P.J. Cribb, D.F. Cutler and C.J. Humphries. Kew: Royal Botanic Gardens, pp. 685–730.

Chase, M.W., Fay, M.F., Devey, D.S. et al. (2006). Multigene analyses of monocot relationships: a summary. In *Monocots:*

Comparative Biology and Evolution, ed. J.T. Columbus, E.A. Friar, C.W. Hamilton et al. Claremont, CA: Rancho Santa Ana Botanic Garden, pp. 63–75.

Cheah, C.H. and Stone, B.C. (1975). Embryo sac and microsporangium development in *Pandanus* (Pandanaceae). *Phytomorphology*, **25**, 228–238.

Dahlgren, R.T.M. and Clifford, H.T. (1982). *The Monocotyledons: A Comparative Study*. London: Academic Press

Furness C.A. and Rudall, P.J. (1998). The tapetum and systematics in monocotyledons. *Botanical Review*, **64**, 201–239.

Furness C.A. and Rudall, P.J. (1999). Microsporogenesis in monocotyledons. *Annals of Botany*, **84**, 475–499.

Furness C.A. and Rudall, P.J. (2001a). Pollen and anther characters in monocot systematics. *Grana*, **40**, 17–25.

Furness C.A. and Rudall, P.J. (2001b). The tapetum in basal angiosperms: early diversity. *International Journal of Plant Sciences*, **162**, 375–392.

Furness C.A. and Rudall, P.J. (2006). Comparative structure and development of pollen and tapetum in Pandanales. *International Journal of Plant Sciences*, **167**, 341–348.

Givnish, T.J., Pires, J.C., Graham, S.W. et al. (2006). Phylogenetic relationship of monocots based on the highly informative plastid gene *ndh*F: evidence for widespread concerted convergence. In *Monocots: Comparative Biology and Evolution*, ed. J.T. Columbus, E.A. Friar, C.W. Hamilton et al. Claremont, CA: Rancho Santa Ana Botanic Garden, pp. 28–51.

Graham, S.W., Zgurski, J.M., McPherson, M.A. et al. (2006). Robust inference of monocot deep phylogeny using an expanded multigene plastid data set. In *Monocots: Comparative Biology and Evolution*, ed. J.T. Columbus, E.A. Friar, C.W. Hamilton et al. Claremont, CA: Rancho Santa Ana Botanic Garden, pp. 3–21.

Harling, C. (1946). Studien uber den Blütenbau und die Embryologie der Familie Cyclanthaceae. *Svensk Botanisk Tidskrift*. **40**, 257–272.

Harling G., Wilder G.J. and Eriksson R. (1998). Cyclanthaceae. In *The Families and Genera of Vascular Plants, Volume 3: Flowering Plants, Monocotyledons, Lilianae (except Orchidaceae)*, ed. K. Kubitzki. Berlin: Springer-Verlag, pp. 202–215.

Johansen, D.A. (1940). *Plant Microtechnique*. New York: McGraw-Hill.

Li, P., Gao, B.C., Chen, F. and Luo, H.X. (1992). Studies on morphology and embryology of *Acanthochlamys bracteata*. II. The anther and ovule development. *Bulletin of Botanical Research Northeast Forest University* **12**, 389–398. [Chin.; Eng. summ.]

Maas-van de Kamer. H. (1998). Burmanniaceae. In *The Families and Genera of Vascular Plants, Volume 3: Flowering Plants, Monocotyledons, Lilianae (except Orchidaceae)*, ed. K. Kubitzki. Berlin: Springer-Verlag, pp. 154–163.

Maas-van de Kamer, H. and Weustenfeld, T. (1998). Triuridaceae. In *The Families and Genera of Vascular Plants, Volume 3: Flowering Plants, Monocotyledons, Lilianae (except Orchidaceae)*, ed. K. Kubitzki. Berlin: Springer-Verlag, pp. 452–458.

Mello-Silva, R. (2000). Partial cladistic analysis of *Vellozia* and characters for phylogeny of Velloziaceae. In: *Monocots: Systematics and Evolution*, ed. K.L. Wilson, D.A. Morrison. Collingwood: CSIRO Publishing, pp. 505–522.

Mello-Silva, R. (2005). Morphological analysis, phylogenies and classification in Velloziaceae. *Botanical Journal of the Linnean Society*, **148**, 157–173.

Mello-Silva, R. (2009 onwards). Velloziaceae. In: *Neotropikey*. Version 1, March 2009. Kew: Royal Botanic Gardens. http://www.kew.org/science/tropamerica/neotropikey.htm

Menezes, N.L. (1976). Megasporogênese, Megagametogênese e Embriogênese em Velloziaceae. *Boletim de Botânica da Universidade de São Paulo*, **4**, 41–59.

Menezes, N.L. (1980). Evolution in Velloziaceae with special reference to androecial characters. In *Petaloid Monocotyledons: Horticultural and Botanical Research*, ed. C.D. Brickell, D.F. Cutler, M. Gregory. London: Academic Press, pp. 117–139.

Menezes, N.L., Mayo, S.J. and Mello-Silva, R. (1994). A cladistic analysis of the Velloziaceae. *Kew Bulletin*, **49**, 71–92.

Rübsamen-Weustenfeld, T. (1991). Morphologische, embryologische und systematische Untersuchungen an

Triuridaceae. *Biblioteca Botanica*, **140**, 1–113.

Rudall, P.J. (1997). The nucellus and chalaza in monocotyledons: structure and systematics. *Botanical Review*, **63**, 140–181.

Rudall, P.J. (2008). Fascicles and filamentous structures: comparative ontogeny of morphological novelties in Triuridaceae. *International Journal of Plant Sciences*, **169**, 1023–1037.

Rudall P.J. and Bateman, R.M. (2006). Morphological phylogenetic analysis of Pandanales: testing contrasting hypotheses of floral evolution. *Systematic Botany*, **31**, 223–238.

Rudall, P.J., Cunniff, J., Wilkin, P. and Caddick, L.R. (2005). Evolution of dimery, pentamery and the monocarpellary condition in the monocot family Stemonaceae (Pandanales). *Taxon* **54**, 701–711.

Sajo, M.G., Mello-Silva, R. and Rudall, P.J. (2010). Homologies of floral structures in Velloziaceae, with particular reference to the corona. *International Journal of Plant Sciences*, **171**, 595–606.

Salatino, A., Salatino, M.L.F., Mello-Silva, R. et al. (2001). Phylogenetic inference in Velloziaceae using chloroplast *trn*L-F sequences. *Systematic Botany*, **26**, 92–103.

Stromberg, B. (1956). The embryo sac development of the genus *Freycinetia*. *Svensk Botanisk Tidskrift* **50**: 129–134.

14

Contrasting patterns of support among plastid genes and genomes for major clades of the monocotyledons

Jerrold I. Davis, Joel R. McNeal, Craig F. Barrett,
Mark W. Chase, James I. Cohen, Melvin R. Duvall,
Thomas J. Givnish, Sean W. Graham, Gitte Petersen,
J. Chris Pires, Ole Seberg, Dennis W. Stevenson
and Jim Leebens-Mack

14.1 Introduction

The monocots, representing a major radiation of angiosperms, exhibit a diverse array of life forms (e.g. Kubitzki, 1998a, 1998b; Judziewicz et al., 1999; Dransfield et al., 2008), including perennial geophytes (many families, including Amaryllidaceae and Liliaceae), long-lived, woody textured 'trees' (Arecaceae), genuinely woody plants with a functioning secondary thickening meristem (genera of Asparagaceae subfamily Nolinoideae), emergent aquatics (Cyperaceae, Acoraceae, Eriocaulaceae), floating and submerged aquatics (several families of Alismatales), bamboos and other perennial and annual grasses (Poaceae), rosette epiphytes (Bromeliaceae), succulent epiphytes (Orchidaceae), mycoheterotrophs (Corsiaceae, Burmanniaceae, Triuridaceae) and other forms. The monocots also exhibit a broad range of reproductive features, including (among many other striking cases of extreme modification) the smallest and largest seeds (in Orchidaceae and Arecaceae, respectively). Generation of a comprehensive phylogeny of this group, integrating

Early Events in Monocot Evolution, eds P. Wilkin and S. J. Mayo. Published by Cambridge University Press. © The Systematics Association 2013.

structural and molecular features as sources of evidence, and in turn helping to clarify diversification patterns in these features, is a goal of several working groups at this time, and the present contribution represents a step toward that goal.

Numerous phylogenetic analyses of the monocots conducted over the past 20 years, employing a variety of molecular and morphological characters, have contributed to the development of an increasingly stable taxonomic system for the *c.* 80–100 conventionally recognized families of monocots (e.g. Duvall et al., 1993; Chase et al., 1995, 2000; Stevenson and Loconte, 1995; Neyland and Hennigan, 2003; Davis et al., 2004, 2006; Janssen and Bremer, 2004; Tamura et al., 2004; Givnish et al., 2006; Graham et al., 2006; Xiao-Xian and Zhe-Kun, 2007). Current understandings of relationships among major groups are summarized in the APG III system (APG III, 2009), in which 12 major clades of monocots are recognized, 11 of them at the ordinal level, and the remaining group (the family Dasypogonaceae) unassigned to order, but included in the informally recognized commelinid alliance with Commelinales, Zingiberales, Arecales and Poales. Outside the commelinids, a generally pectinate structure is recognized for these major groups, with six major clades (all but one of them consisting of a single order) diverging in succession from the line that leads to the commelinids, as follows (starting from the basal node of the monocots): Acorales, Alismatales, Petrosaviales, Dioscoreales + Pandanales, Liliales, Asparagales (i.e. the last recognized as sister of the commelinids).

Most molecular phylogenies generated for plants over the past 20 years have been based on one or a few genes of the plastid genome (i.e. the plastome), and this rule holds for monocots as well (see citations above). A recent innovation in plant molecular phylogenetics, facilitated by the availability of new DNA sequencing techniques, has been the use of whole plastid genome sequences, or more specifically, large numbers of genes (60–80 or more; e.g. Goremykin et al., 2003a, 2003b, 2005; Leebens-Mack et al., 2005; Jansen et al., 2006, 2007; Moore et al., 2007, 2010; Guisinger et al., 2008, 2010; Parks et al., 2009; Soltis et al., 2009, 2010; Wang et al., 2009; Givnish et al., 2010; also see Cui et al., 2006). These studies have addressed long-standing questions concerning relationships ranging in scale from those within families to those involving the deepest branches within the angiosperms and beyond. Broad analyses based on plastome sequences generally have employed relatively sparse taxon sampling, for the expense and effort required to generate and annotate such sequences are still substantial. In this respect, these studies differ from previous molecular analyses that employed more extensive taxon sampling, but relied on far fewer nucleotides per taxon. This distinction has been considered by systematists, often in terms of a choice to be made between the inclusion of greater numbers of taxa or greater amounts of data per taxon in an analysis (e.g. Rokas and Carroll, 2005). As sequencing costs continue to drop, and as data-handling methods continue to improve, it is to be expected that the necessity of choosing between more taxa and more data per taxon will diminish. For the present, though, it is possible that the ongoing production of plastid genomic data can complement the existing body of data in which many taxa are each represented by a few genes.

One of the major goals of the Monocot Tree of Life project has been to develop a data matrix for *c.* 600 taxa, comprising morphological characters and sequence data for all of the taxa from a few genes from each of the three major plant genomes, with a substantially greater number of genes scored for a selected subset of the taxa (see http://www.botany.wisc.edu/monatol/). The present contribution represents a step towards achieving that goal.

Here we adopt a hybrid approach – with many taxa sampled for a few genes, and with a subset of taxa sampled for many genes – and we consider questions concerning the relative contributions of the various portions of the data set to the resulting phylogeny. For example, if there is conflict between the phylogenetic structures supported by two data partitions, will one of the two dominate the other in determining the results obtained by the combined matrices? Perhaps more importantly, for a project that is currently in progress, can results obtained at an intermediate stage of completion be used to inform further sampling of data and taxa? In the present contribution we examine areas of conflict and congruence within a matrix that consists entirely of data from the plastid genome, with *rbc*L sampled for all taxa, *mat*K sampled for most, and a set of 60 additional genes sampled for about one-third of the taxon set. Because all data in the present analysis are drawn from the plastid genome, we do not examine intergenomic incongruence; but under these circumstances, does a unified 'plastid signal' emerge? In addressing relationships among monocots, we focus on higher-level structure across this major clade, with specific attention to the commelinids, a major alliance that includes multiple orders, and which at this point has been sampled in greater depth than most of the other monocot lineages.

14.2 Materials and methods

The taxon set comprises 273 terminal taxa (i.e. 'terminals'), including representatives of all 78 monocot families recognized by APG III (APG III, 2009; Haston et al., 2009), except for Corsiaccac and Triuridaceae, which are putatively holomyco-trophic, plus representatives of other major angiosperm lineages as outgroups (Appendix). Each of the 273 terminals represents either a species (i.e. all sequence data collected from a single species) or a genus (i.e. a 'composite terminal', with sequence data for different genes drawn from two or more different species of a genus), and in five cases multiple species of a genus are included as separate terminals (*Acorus, Dioscorea, Eichhornia, Trichopus* and *Xyris*). The taxa fall within

19 major mutually exclusive groups of monocots (here designated 'principal clades'), plus 10 such groups of nonmonocot outgroups. These groups were defined partially on the basis of a priori considerations (including resolution of each group as monophyletic by prior analyses, frequent formal or informal recognition of the groups, and availability for the present analysis of plastid genome data for at least one representative of each group), and partially on the basis of an a posteriori consideration (in particular, resolution of each group as monophyletic by the data matrix assembled for this analysis). All but two of the groups (Amborellales and Ceratophyllales), in addition to including at least one terminal represented by a plastid genome sequence, also include at least two terminals, and therefore are tested for monophyly. All except 5 of the 29 principal clades correspond to groups recognized at the family rank or higher by APG III; the five exceptions (Appendix; Fig 14.1) are Asparagales sens. strict. (comprising all of Asparagales sens. APG III except Orchidaceae) and four subsets of Poales sens. APG III. Poales is sampled here in greater depth than most groups, so relationships among the 16 families placed in this order by APG III can be examined in greater detail, and this set of families therefore is divided provisionally into seven principal clades, three of which correspond to individual families (Bromeliaceae, Typhaceae and Rapateaceae) and four of which, as provisional orders, consist of three to four families each, these being Juncales (Cyperaceae, Juncaceae and Thurniaceae), Xyridales (Eriocaulaceae, Mayacaceae and Xyridaceae), Restionales (Anarthriaceae, Centrolepidaceae and Restionaceae), and Poales sens. strict. (Ecdeiocoleaceae, Flagellariaceae, Joinvilleaceae and Poaceae). These four subdivisions correspond to groups sometimes referred to as the cyperids, xyrids, restiids and graminids, respectively (Linder and Rudall, 2005), except that Hydatellaceae was originally included among the xyrids, and since has been removed from the monocots (Saarela et al., 2007). A third circumscription of Poales, that of Dahlgren et al. (1985), includes the seven families of the combined Poales sens. strict. and Restionales, and thus is intermediate in size.

A major source of data was the data set assembled by Givnish et al. (2010), consisting of 83 complete and draft plastid genome sequences, each comprising sequences of most or all genes from a set of 81 genes (Jansen et al., 2007); throughout the present paper, sets of multiple gene sequences from individual plastid genomes (usually consisting of c. 60 genes, depending on degree of completeness), are referred to as 'plastid genome sequences'. Four additional previously published plastid sequences were added to the set assembled by Givnish et al., and one new plastid sequence was assembled (that of *Japonolirion osense*), yielding a set of 88 taxa represented as plastid genome sequences (Appendix). An *rbc*L sequence is present for each of these 88 taxa, and a *mat*K sequence for 83 of them. In a few cases, sequences of these two genes represent different plant accessions to those that yielded the majority of the

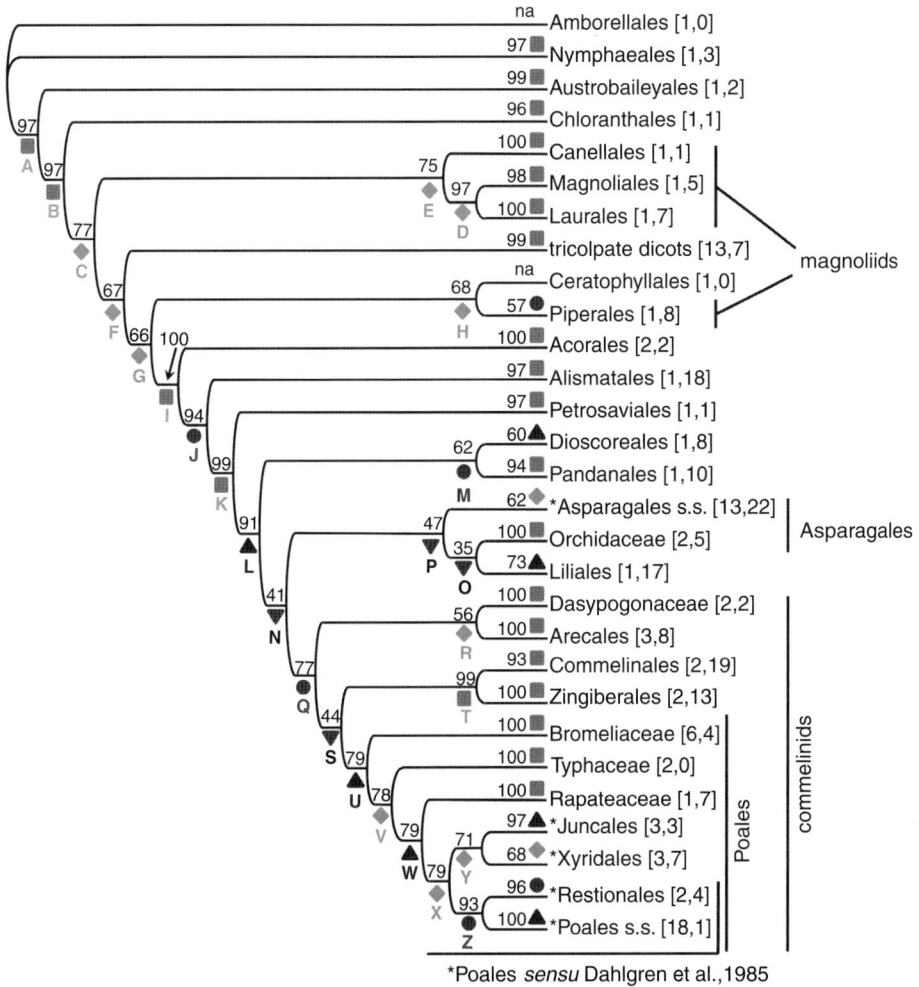

Fig 14.1 Summary consensus tree, depicting principal clades resolved by parsimony analysis of the complete data set, plus two groups represented by one taxon each, and relationships among these 29 groups. Names of groups follow APG III except those marked by asterisks (see text). Following the name of each principal clade, in brackets, is the number of constituent taxa represented by plastid genomes in the data set, and the number represented by *rbc*L and *mat*K sequences (in total, the number of terminals in each of the groups; Appendix). Four higher-level groups are signified by labelled bars at right and bottom. Strict consensus jackknife values are indicated for all principal clades except the two represented by solitary taxa ('na'), and for all relationships among them. Coloured symbols signify five support patterns, as defined in the text and Table 14.3 (green squares = group I; blue triangles, pointed upwards = group II; purple circles = group III; orange diamonds = group IV; grey triangles, pointed downwards = group V). Colour version to be found in colour plate section.

genome sequence, and in one case (*Syngonanthus*) the *rbc*L sequence represents a different species of the genus.

The plastid genome sequence for *Japonolirion* was generated as described by Givnish et al. (2010). In brief, shotgun sequences were generated on an Illumina GS sequencer using the manufacturer's protocol, from a sample estimated to consist of at least 5% plastid DNA, on the basis of RT-PCR results, as compared to those of a sample of a cloned portion of *rbc*L of known concentration. Sequences were subjected to reference-based assembly using the YASRA assembler (Ratan, 2009; download available with documentation at http://www.bx.psu. edu/miller_lab) to layer short reads on a reference genome, while allowing substantial sequence divergence (<85% sequence identity). The final assembly was compiled in Sequencher (Gene Codes Corporation), and genes were annotated and extracted from the resulting contigs using the DOGMA webserver (http:// dogma.ccbb.utexas.edu, Wyman et al., 2004), which identifies genes through BLASTX searches against a database of amino acid sequences extracted from exemplar plastid genomes.

In addition to the set of 88 taxa represented by plastid genomes, 185 taxa are represented by sequences of just *rbc*L and *mat*K (all 185 by *rbc*L sequences, and 141 by *mat*K sequences as well), either generated by the authors or obtained from GenBank (Appendix). The resulting matrix of 273 terminals includes an *rbc*L sequence for every taxon, a *mat*K sequence for 224 of the taxa, and varying numbers of sequences of 60 other genes of the plastid genome for the set of 88 taxa represented by plastid genomes.

For the 88 taxa represented by plastid genomes, sequences of the 62 genes with the greatest numbers of nucleotides in the 81-gene set of Jansen et al. (2007) were assembled into a data matrix (Table 14.1); this set includes all genes of length greater than 200 nucleotides from the 81-gene set, and comprises more than 96% of the total number of nucleotides in the 81-gene set, as determined from gene lengths in *Amborella*. Each gene was aligned with the web-based implementation of MUSCLE (http://www.ebi.ac.uk/Tools/msa/muscle/, Edgar, 2004), and inspected; regions that were regarded as ambiguously aligned were inactivated for analysis. The set of 62 genes was assembled into a matrix with *rbc*L placed first, then *mat*K, and then the remaining genes, in order of decreasing length in *Amborella*, as listed in Table 14.1 (data available from JID). Parsimony analyses of the data were conducted after parsimony-uninformative characters were excluded from the set of active characters, and all reports of tree length, consistency index (CI), and retention index (RI) are based on the active, parsimony-informative subset of the character set (Table 14.2). Parsimony analyses were conducted on the complete set of parsimony-informative characters for 273 taxa and 62 genes, and on various subsets (described below). Maximum Likelihood analyses (ML – see below) were conducted with the complete set of active

Table 14.1 Summary information for 62 genes and gene regions in data matrix, arranged (after *rbc*L and *mat*K) in descending order of length in *Amborella* († = exon/intron structure present in some or all taxa; in these cases the data refer to the sum of two or more exons or to just one exon included in the matrix). Three genes in boldface are those within which the total number of nucleotides in genes other than *rbc*L and *mat*K is divided into quartiles (see text). Numbers in round brackets for *rbc*L and *mat*K refer to the set of 88 taxa for which plastome data are available, for comparison with those for other genes.

Gene or gene region	No. of sites in *Amborella*	No. of active aligned sites in matrix, and % that are parsimony informative	No. of parsimony-informative characters, and running total in matrix	% of total parsimony-informative characters in matrix, and running total
1. *rbc*L	1428	1367, 44.6 (34.0)	610 (466), 610	2.9, 2.9
2. *mat*K	1506	1272, 80.3 (63.7)	1021 (810), 1631	4.9, 7.8
3. *ycf2*	6915	6123, 21.3	1305, 2936	6.3, 14.1
4. *rpoC2*	4110	2788, 46.0	1283, 4219	6.2, 20.3
5. *rpoB*	3219	3062, 42.3	1295, 5514	6.2, 26.5
6. 23S rDNA	2816	2772, 8.6	239, 5753	1.2, 27.7
7. *psaA*	2253	2319, 28.6	664, 6417	3.2, 30.9
8. *ndh*F	**2241**	**2042, 49.0**	**1001, 7418**	**4.8, 35.7**
9. *psaB*	2205	2205, 27.4	605, 8023	2.9, 38.6
10. †*rpoC1*	2043	1942, 41.3	802, 8825	3.9, 42.4
11. †*ndhB*	1533	1547, 14.5	225, 9050	1.1, 43.5
12. *psbB*	1527	1517, 29.9	453, 9503	2.2, 45.7
13. *atpA*	1524	1511, 37.5	566, 10 069	2.7, 48.4
14. *ndhD*	1503	1537, 44.8	689, 10 758	3.3, 51.7
15. *atp*B	**1503**	**1480, 34.5**	**511, 11 269**	**2.5, 54.2**
16. 16S rDNA	1490	1476, 10.4	154, 11 423	0.7, 54.9
17. *psbC*	1422	1458, 27.7	404, 11 827	1.9, 56.9
18. *ndhH*	1182	1180, 38.6	455, 12 282	2.2, 59.1
19. †*ndhA*	1092	1062, 42.9	456, 12 738	2.2, 61.3
20. *psbD*	1062	1078, 25.5	275, 13 013	1.3, 62.6

Table 14.1 (*cont.*)

Gene or gene region	No. of sites in *Amborella*	No. of active aligned sites in matrix, and % that are parsimony informative	No. of parsimony-informative characters, and running total in matrix	% of total parsimony-informative characters in matrix, and running total
21. *psbA*	1053	1063, 24.9	265, 13 278	1.3, 63.9
22. *rpoA*	1005	872, 48.3	421, 13 699	2.0, 65.9
23. *petA*	963	924, 38.0	351, 14 050	1.7, 67.6
24. *ccsA*	942	763, 52.2	398, 14 448	1.9, 69.5
25. †*rpl2*	830	824, 24.6	203, 14 651	1.0, 70.5
26. *atpI*	747	754, 33.4	252, 14 903	1.2, 71.7
27. *rps2*	711	687, 48.3	332, 15 235	1.6, 73.3
28. *ycf4*	708	518, 49.6	257, 15 492	1.2, 74.5
29. *cemA*	690	656, 51.8	340, 15 832	1.6, 76.1
30. *ndhK*	**678**	**657, 37.4**	**246, 16 078**	**1.2, 77.3**
31. *rps3*	657	629, 49.3	310, 16 388	1.5, 78.8
32. †*petB*	648	645, 26.0	168, 16 556	0.8, 79.6
33. †*clpP*	609	544, 45.6	248, 16 804	1.2, 80.8
34. *rps4*	606	524, 41.2	216, 17 020	1.0, 81.9
35. †*atpF*	555	541, 47.9	259, 17 279	1.2, 83.1
36. *ndhI*	543	493, 38.5	190, 17 469	0.9, 84.0
37. *ndhG*	534	521, 49.7	259, 17 728	1.2, 85.3
38. †*ycf3*	507	511, 28.0	143, 17 871	0.7, 85.9
39. *ndhJ*	477	477, 40.0	191, 18 062	0.9, 86.9
40. †*petD*	475	472, 31.1	147, 18 209	0.7, 87.6
41. *rps7*	468	432, 23.6	102, 18 311	0.5, 88.1
42. *rps11*	417	342, 45.9	157, 18 468	0.8, 88.8
43. *atpE*	405	379, 46.7	177, 18 645	0.9, 89.7
44. *rps8*	399	367, 49.0	180, 18 825	0.9, 90.5

Table 14.1 (*cont.*)

Gene or gene region	No. of sites in *Amborella*	No. of active aligned sites in matrix, and % that are parsimony informative	No. of parsimony-informative characters, and running total in matrix	% of total parsimony-informative characters in matrix, and running total
45. †*rpl16*	399	390, 43.8	171, 18 996	0.8, 91.4
46. *rpl22*	375	280, 61.8	173, 19 169	0.8, 92.2
47. *rpl14*	369	366, 44.3	162, 19 331	0.8, 93.0
48. *ndhC*	363	363, 33.3	121, 19 452	0.6, 93.5
49. *rpl20*	360	328, 57.3	188, 19 640	0.9, 94.5
50. *ndhE*	306	298, 43.6	130, 19 770	0.6, 95.1
51. *rps18*	306	168, 29.8	50, 19 820	0.2, 95.3
52. *rps14*	303	286, 46.5	133, 19 953	0.6, 96.0
53. *rpl23*	288	245, 24.9	61, 20 014	0.3, 96.2
54. *rps19*	279	248, 51.2	127, 20 141	0.6, 96.9
55. *rps15*	264	197, 47.2	93, 20 234	0.4, 97.3
56. *psbE*	252	252, 25.8	65, 20 299	0.3, 97.6
57. *psaC*	246	246, 32.1	79, 20 378	0.4, 98.0
58. *atpH*	246	246, 24.8	61, 20 439	0.3, 98.3
59. *infA*	234	202, 45.5	92, 20 531	0.4, 98.7
60. *psbH*	222	207, 39.1	81, 20 612	0.4, 99.1
61. *rpl33*	207	195, 51.8	101, 20 713	0.5, 99.6
62. †*rps16*	197	190, 42.6	81, 20 794	0.4, 100.0
Totals	63 417	59 040	20 794	100.00

characters for all taxa and genes. Data manipulation and preparation for analysis, including assembly of the matrix, marking and removal of inactive and parsimony-uninformative characters, display for examination of tree structures and support values, and preparation of trees for publication, were conducted with WinClada version 1.03 (Nixon, 2002).

Table 14.2 Summary results of cladistic analyses of matrix of 273 taxa scored for plastid sequence data, and eight subsets (Appendix).

	rbcL, 273 taxa	rbcL + matK, 273 taxa	rbcL + matK + one quartile of remaining plastid genome, 273 taxa				All data, 273 taxa	Plastid genome including rbcL and matK, 88 taxa	Plastid genome excluding rbcL and matK, 88 taxa
			+ quartile 1	+ quartile 2	+ quartile 3	+ quartile 4			
Number of informative characters	610	1631	6422	6422	6422	6421	20 794	20 439	19 163
Length and number of most-parsimonious trees	7454; >100 000	19 060; 24	41 145; 768	49 262; 4800	47 885; 6000	46 694; 7680	127 773; 23 040	117 141; 2	108 645; 1
Consistency Index, Retention Index	0.15; 0.62	0.18; 0.62	0.26; 0.62	0.24; 0.61	0.24; 0.61	0.25; 0.62	0.28; 0.61	0.30; 0.61	0.30; 0.61

Parsimony analyses were conducted with TNT version 1.1 (Goloboff et al., 2008) using collapse option 3 (branches considered unsupported if any optimization lacks support). Each analysis began with 1000 search initiations, with each initiation consisting of tree construction using a random taxon addition sequence, followed by TBR branch swapping with up to 20 trees retained, and then by a ratchet sequence (Nixon, 1999) of 100 cycles of swapping while alternating between equal weighting of all characters and random up- and downweighting of characters with each character having a 10% chance of being upweighted and a 5% chance of being downweighted. Most-parsimonious trees from all 1000 search initiations were pooled, and TBR swapping was conducted on these and any additional trees accumulated during this round of swapping, with up to 100 000 trees retained, followed by construction of a strict consensus tree. For each basic search a corresponding 'strict-consensus jackknife' analysis was conducted (Farris et al., 1996; Soreng and Davis, 1998; Davis et al., 2004; Freudenstein and Davis, 2010), involving 1000 jackknife replicates with a 37% deletion probability for each character. Each jackknife replicate search was conducted using the same procedures and settings as used in the basic tree search, except that the initial phase of each search consisted of four initiations, with 16 trees retained, and that 1100 trees were retained during the final round of TBR swapping, prior to construction of the strict consensus tree for that replicate.

Maximum likelihood analyses were conducted on combined alignments using RAxML version 7.2.7 (Stamatakis, 2004), as implemented at the CIPRES Science Gateway (http://www.phylo.org/portal2/). A single GTR substitution process (the standard model in RAxML) was modelled for the entire aligned matrix of active characters, with all data included in a single partition. Among-site variation in substitution rate was modelled using the discrete approximation of the gamma distribution (Yang, 1994) with 25 rate categories. The ML bootstrap analysis was conducted using the rapid bootstrapping option, and consisted of 1000 pseudo-replicates drawn from the combined matrix.

In order to identify different patterns of support for various clades by different subsets of the character set, a series of parsimony analyses was conducted with different character subsets included, and support from these analyses was assessed for clades that had been resolved by parsimony analysis of the complete data set. In addition to the analysis of the complete data set, all 273 terminals were included in six character-subset analyses, i.e. *rbc*L only; *rbc*L + *mat*K; and *rbc*L + *mat*K + each of the four quartiles of remaining plastid genome data. The boundaries of the four quartiles were determined by the sequence in which the 60 genes other than *rbc*L and *mat*K are arranged in the data set (Table 14.1): the first quartile runs from *ycf*2 to a point within *ndh*F, the second from that point to a point within *atp*B; the third to a point within *ndh*K, and the fourth from there to the end of the matrix. Thus, each of the four quartiles comprises 4791 parsimony-informative nucleotides (except the fourth, which includes 4790), but the first consists of five relatively long genes and a portion of another, the fourth consists of 32 shorter genes and a portion of another, and the second and third quartiles consist of intermediate numbers of genes. Two additional analyses also were conducted, each including just the 88 taxa for which plastid genomic data were available. One of these analyses included all data for these taxa, and the other included data from all genes except *rbc*L and *mat*K.

Patterns of support for the 29 principal clades and relationships among them, as determined from the six character-subset analyses that included all 273 taxa, were used to sort clades into five categories. For purposes of this categorization of clades by support pattern, the analysis based on *rbc*L + *mat*K, and the one based on the complete data set (*rbc*L + *mat*K + 60 additional genes for the plastid genome taxon subset), were recognized as endpoints in a range, and those based on *rbc*L + *mat*K + individual quartiles of the plastid genomic data represented intermediate points in this range. Five clade-support patterns were recognized, based on presence/absence of clades in the consensus trees from the six analyses, and levels of jackknife support for these groups. The only taxonomic groups categorized are those that were resolved in the consensus tree of the all-data analysis, so this is a baseline criterion for all five categories. *Support pattern I* applies to clades that are present in the consensus trees of all six reference analyses, with jackknife support ≥ 90% in both endpoint analyses, and with jackknife support in the four intermediate analyses lying no more than 10% beyond the boundaries set by the two endpoints. Because inclusion in this category requires support ≥ 90% by the two endpoints, support by the four intermediate analyses can only lie more than 10% outside the range that they set by having a jackknife score lower than those of both of them. *Support pattern II* applies to clades that are present in the consensus trees of all six analyses, with support ≥ 50% in all six, but <90% in one or both of the two endpoints, and as in support pattern I, with support in the four intermediate analyses lying no more than 10% beyond the boundaries set by the two endpoints. *Support pattern III* applies to clades that are present in the two endpoint analyses, with support ≥ 50% in both, but with the group either absent from the consensus tree for one or more of the intermediate analyses, or having jackknife support <50%, or with support in one or more of the intermediate analyses lying >10% beyond the boundaries set by the two endpoints. *Support pattern IV* applies to clades that are present in the consensus tree of the all-data analysis, with support ≥ 50%, in combination with either the absence of the clade from the consensus tree of the two-gene analysis, or with support <50% in that analysis, or both. Support levels by intermediate analyses are irrelevant to inclusion in this category. *Support pattern V* applies to clades that are present in the consensus tree of the all-data analysis, but with jackknife support <50%; there are no other criteria for inclusion in this category.

14.3 Results

Summary information for the 62 genes in the data matrix is presented in Table 14.1. Because the matrix includes *rbc*L and *mat*K sequences for more taxa than the 88 for which plastid genome sequences are available, corresponding numbers are provided for these two genes for the set of 88 taxa, as well as for the entire taxon set. The genes range in length from 197 to 6915 nucleotides (*rps16* and *ycf2*, respectively), with between 168 and 6123 active, aligned characters (*rps18* and *ycf2*), and the percentage of active, aligned sites that are parsimony informative (for the set of 88 taxa) ranges from 8.6 to 63.7 (23S rDNA and *mat*K), yielding between 50 and 1305 cladistically informative sites (*rps18* and *ycf2*). The total combined sequences of the 62 genes, at 63 417 sites (in *Amborella*), yield about 93% as many active, aligned sites (59 040), and of the latter, about 35% (20 794) are parsimony informative.

Cladistic analysis of nine subsets of the overall matrix (Table 14.2) yield as few as one or two equally parsimonious trees (two analyses of 88 taxa), and as few as 24 trees for subsets that include all 273 taxa (*rbc*L + *mat*K). The number of trees is greatest for the *rbc*L matrix (>100 000), and the next highest number of trees observed is with the entire data set (23 040). CI is lowest with *rbc*L alone (0.15), highest with the two analyses of 88 taxa (0.30), and *c.* 0.24–0.28 for the analyses that include 273 taxa and part or all of the plastid genome data (other than *rbc*L and *mat*K). RI is within a range of 0.61–0.62 across the analyses.

The monocots are resolved as monophyletic by the complete data set, with Ceratophyllales + Piperales placed as their sister group. Within the monocots, five lineages diverge in succession from the line that includes the commelinids (Fig 14.1); the five groups, in their sequence of divergence from the point of origin of the monocots to the sister group of the commelinids, are Acorales, Alismatales, Petrosaviales, Dioscoreales + Pandanales and Liliales + Asparagales. Dioscoreales and Pandanales are sister groups, and within the clade consisting of Liliales + Asparagales, Orchidaceae and Liliales are sister groups, and the sister of this clade is Asparagales sens. strict. (all elements of the order except Orchidaceae); thus, Asparagales sens. APG is not monophyletic. Within the commelinids, Arecales + Dasypogonaceae is the sister of a clade that includes all other members of the group, and within the latter, Commelinales + Zingiberales is sister of a clade that corresponds to Poales sens. APG III. Within the latter group, Bromeliaceae, Typhaceae and Rapateaceae diverge in succession from a clade in which Juncales and Cyperales are sisters, as are Restionales and Poales sens. strict., with these two pairs of groups sisters of each other.

The optimal tree obtained by Maximum Likelihood analysis (Fig 14.2), with log likelihood of –765980.223512, resolves all except one of the 27 principal clades that consist of two or more taxa and that are resolved by the parsimony analysis. The

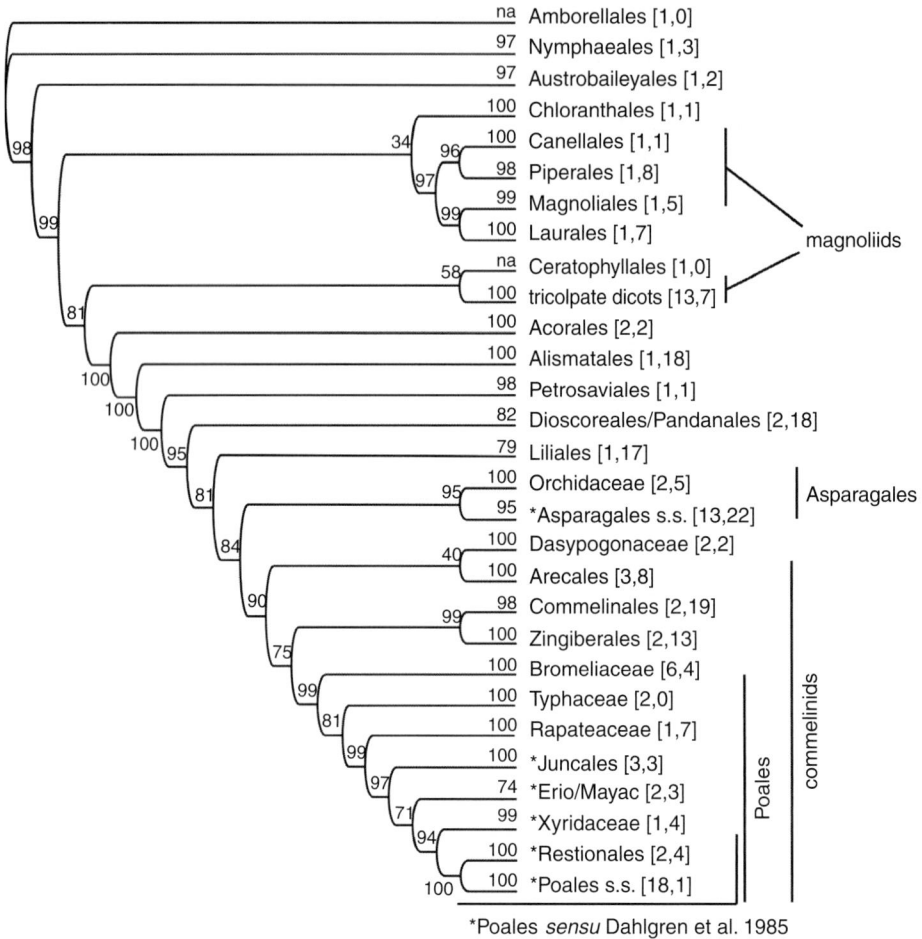

Fig 14.2 Summary consensus tree depicting principal clades resolved by maximum likelihood analysis of the complete data set, plus two groups represented by one taxon each, and relationships among these 29 groups. Bootstrap values are indicated for all principal clades except the two represented by solitary taxa ('na'), and for all relationships among them. Labels are as in Fig 14.1.

exception is Xyridales; rather than resolving a monophyletic grouping of Xyridaceae, Eriocaulaceae and Mayacaceae, the ML analysis places all elements of Xyridaceae (i.e. *Xyris, Abolboda, Aratitiyopea* and *Orectanthe*) in a clade (with 91% bootstrap support) that is the sister of Poales sens. strict. + Restionales, and places *Mayaca* and all elements of Eriocaulaceae (i.e. *Eriocaulon, Syngonanthus, Lachnocaulon* and *Tonina*) in a clade, with 74% support, that is sister to the clade that consists of Xyridaceae, Poales sens. strict. and Restionales. Elsewhere in the ML tree, most relationships among the principal clades are identical to those in the parsimony tree for the complete data set, with five areas that differ, three of them outside the monocots: (1) Piperales and Chloranthales are placed in the clade that includes Canellales, Magnoliales and Laurales, with 34% bootstrap support for this group of five orders. This set of five orders is sister of a clade (with 81% support) that includes all other taxa except Amborellales, Nymphaeales and Austrobaileyales; (2) within this set of five orders, Chloranthales is sister of a clade that includes the other four orders (with 97% support), and among them, Laurales + Magnoliales (99% support) is sister of Canellales + Piperales (96%); (3) Ceratophyllales and the tricolpate dicots are sister groups (with 58% support), and together they are the sister of the monocots (100% support). The other differences between the ML and parsimony trees are inside the monocots, as follows: (1) Liliales (79% support) is sister of a clade (with 84% support) that consists of the commelinids plus a monophyletic Asparagales (i.e. including Orchidaceae) with 95% support; within the latter group, Orchidaceae is sister of a clade (with 100% support) that includes all other elements of Asparagales; and (2) As previously noted, Xyridales is not monophyletic, and Juncales (sister of Xyridales in the parsimony analysis) is sister of the clade with 71% support that consists of Mayacaceae, Eriocaulaceae, Xyridaceae, Restionales and Poales sens. strict.

Of the 27 principal clades represented by two or more taxa, jackknife support in the parsimony analysis of the complete data set is 90% or greater for 22, and between 57 and 73% for the other five. Nineteen of these groups exhibit support pattern I, four exhibit support pattern II, and two each exhibit support patterns III and IV. Among the 26 clades consisting of two or more principal clades, as resolved by the complete data set, jackknife support ranges from 35% to 100%. The number of groups that exhibit support patterns I through V are five, three, four, ten and four, respectively.

Of the 53 clades of interest resolved by parsimony analysis of the complete data set (27 principal clades represented by two or more taxa, and the 26 clades consisting of two or more of them), 22 are not resolved by the *rbc*L analysis, 15 are not resolved by analysis of *rbc*L + *mat*K, and 14, 10, 14 and 9 of these groups are not resolved by analyses of *rbc*L + *mat*K + quartiles 1, 2, 3 and 4 of the plastid genome data, respectively (Table 14.3). Although none of the matrices that include a quartile of the plastid genome data fails to resolve more than 14 of these groups, 18 different groups are not resolved by one or more of these four matrices. In each of the two analyses of 88 terminals represented by plastid genome data, 15 principal clades, rather than two (as in the matrices that include 273 taxa), are represented by just one taxon. Therefore, only 40 of the 53 clades of interest resolved by analysis of the complete data set were tested for monophyly by these two analyses. The matrix of 88 taxa that included *rbc*L and *mat*K resolved all but three of these 40 groups, and the matrix that excluded these two genes resolved all 40.

Table 14.3 Strict-consensus jackknife support for principal clades of monocots and relatives, and clades consisting of two or more of these groups, as determined from a matrix of 273 taxa scored for plastid sequence data, and eight subsets of this matrix (cf. Table 14.1, Appendix). Clades of two or more groups are signified by letters as in Fig 14.1. Bold-facing of jackknife scores signifies presence of a group in the strict consensus of the analysis, and nonbold type signifies absence; 'na' signifies presence of just one representative in the analysis, and therefore lack of test of monophyly or support.

Principal clades	*rbc*L, 273 taxa	*rbc*L + *mat*K, 273 taxa	*rbc*L + *mat*K + quartile 1, 2, 3, or 4 of remaining plastid genome, 273 taxa	All data, 273 taxa	Plastid genome including *rbc*L and *mat*K, 88 taxa	Plastid genome excluding *rbc*L and *mat*K, 88 taxa
Amborellales	na	na	na, na, na, na	na	na	na
Nymphaeales	**100**	**98**	**98, 98, 97, 98**	**97**	na	na
Austrobaileyales	**97**	**97**	**98, 97, 98, 98**	**99**	na	na
Chloranthales	**85**	**96**	**89, 96, 94, 96**	**96**	na	na
Canellales	**100**	**100**	**100, 100, 100, 100**	**100**	na	na
Magnoliales	**94**	**98**	**98, 98, 99, 99**	**98**	na	na
Laurales	**76**	**100**	**100, 100, 100, 100**	**100**	na	na
tricolpate dicots	**90**	**99**	**99, 100, 100, 99**	**99**	**100**	**100**
Ceratophyllales	na	na	na, na, na, na	na	na	na
Piperales (incl. *Lactoris*)	**46**	**95**	**81, 58, 49, 85**	**57**	na	na
Acorales	**100**	**100**	**100, 100, 100, 100**	**100**	**100**	**100**
Alismatales	**87**	**98**	**99, 97, 99, 98**	**97**	na	na
Petrosaviales	**79**	**98**	**96, 94, 98, 98**	**97**	na	na
Dioscoreales	0	**67**	**62, 57, 71, 63**	**60**	na	na
Pandanales	**77**	**96**	**93, 95, 95, 95**	**94**	na	na
Asparagales s.s.	11	17	46, 67, 13, **64**	**62**	**68**	**91**
Orchidaceae	**100**	**100**	**100, 100, 100, 100**	**100**	**100**	**100**
Liliales	**73**	**80**	**83**, 72, **76**, 70	**73**	na	na
Dasypogonaceae	**100**	**100**	**100, 100, 100, 100**	**100**	**100**	**100**

Table 14.3 (*cont.*)

Principal clades	*rbc*L, 273 taxa	*rbc*L + *mat*K, 273 taxa	*rbc*L + *mat*K + quartile 1, 2, 3, or 4 of remaining plastid genome, 273 taxa	All data, 273 taxa	Plastid genome including *rbc*L and *mat*K, 88 taxa	Plastid genome excluding *rbc*L and *mat*K, 88 taxa
Arecales	99	100	100, 100, 100, 100	100	100	100
Commelinales	43	91	95, 95, 92, 91	93	100	100
Zingiberales	100	100	100, 100, 100, 100	100	100	100
Bromeliaceae	100	100	100, 100, 100, 100	100	100	100
Typhaceae	90	100	100, 100, 100, 100	100	100	100
Rapateaceae	99	99	100, 100, 100, 100	100	na	na
Juncales	88	87	96, 98, 98, 97	97	100	100
Xyridales	3	2	13, **66**, 28, 25	68	90	88
Restionales	17	**79**	67, **99**, 39, **63**	96	100	100
Poales s.s.	7	**71**	86, **99**, 100, 82	100	100	100
Clades consisting of two or more principal clades						
A	95	98	98, 98, 97, 98	97	100	100
B	94	98	98, 98, 97, 98	97	100	100
C	6	10	32, 5, **66**, **52**	77	81	67
D	6	38	**95**, **66**, **64**, **95**	97	100	99
E	0	4	19, **51**, 35, 16	75	98	97
F	5	0	9, 15, 23, 9	67	93	91
G	1	0	10, **44**, 18, 9	66	89	95
H	12	1	18, **50**, **55**, 13	68	100	97
I	88	100	100, 98, 100, 100	100	100	100
J	95	99	97, 35, **99**, 96	94	78	87
K	98	99	99, 97, 99, 97	99	100	100

Table 14.3 (*cont.*)

Principal clades	rbcL, 273 taxa	rbcL + matK, 273 taxa	rbcL + matK + quartile 1, 2, 3, or 4 of remaining plastid genome, 273 taxa	All data, 273 taxa	Plastid genome including rbcL and matK, 88 taxa	Plastid genome excluding rbcL and matK, 88 taxa
L	30	60	90, 85, 67, 80	91	99	100
M	0	52	64, 58, 73, 64	62	100	100
N	9	37	2, 22, 27, 43	41	26	74
O	0	0	2, 28, 7, 40	35	27	75
P	2	8	1, 25, 1, 47	47	25	65
Q	30	77	80, 76, 88, 81	77	100	100
R	10	3	15, 38, 34, 1	56	61	63
S	23	4	21, 39, 32, 0	44	61	63
T	68	94	99, 99, 99, 99	99	100	100
U	41	76	80, 77, 89, 82	79	100	100
V	4	1	27, 20, 51, 72	78	99	100
W	5	66	80, 77, 89, 82	79	100	100
X	1	8	75, 74, 32, 74	79	100	100
Y	0	12	25, 74, 2, 26	71	76	80
Z	18	55	39, 99, 7, 48	93	94	95

Among the 40 clades of interest resolved by parsimony analysis of the complete data set, and tested for monophyly by the two matrices of 88 taxa, all 40 were resolved by the matrix of 88 taxa represented by plastid genome data (*rbc*L and *mat*K excluded), and 15 were resolved by the *rbc*L + *mat*K matrix (Table 14.3). These two subsets of the data are mutually exclusive, and together they include all data in the complete data set. Examples of clades resolved by the matrix of plastid genome data, but not by the *rbc*L + *mat*K matrix, are the clade that consists of all elements of Poales *sensu* APG except Bromeliaceae (present with jackknife = 100% in analysis of plastid genome data, absent and with jackknife = 1% in the analysis of *rbc*L + *mat*K); Arecales + Dasypogonaceae (present and 63% vs. absent and 3%), and the placement of Ceratophyllales with Piperales (present and 97% vs. absent and 1%).

14.4 Discussion

The present analysis focusses on the resolution of major groups of monocots, particularly those that are widely recognized at the ordinal level, and on relationships among these groups. A significant difference between the results obtained here via parsimony analysis (Fig 14.1) and the consensus relationships indicated by APG III (2009) lies in the placement of Piperales and Ceratophyllales as sister taxa, with the clade that consists of these two orders placed as sister of the monocots. This placement however, is not supported by the present ML analysis, which places Ceratophyllales with the tricolpate dicots (and these two groups together as sister of the monocots), and places Piperales and Chloranthales with Magnoliales, Laurales and Canellales. In these respects the ML tree is in accord with summary relationships recognized by APG III.

Within the monocots, the parsimony analysis of the complete data set also differs from the APG III summary in supporting a monophyletic grouping of Liliales, Asparagales sens. strict. and Orchidaceae, with Orchidaceae more closely related to Liliales than to Asparagales sens. strict. With respect to these relationships, results of the ML analysis also differ from those of the parsimony analysis, and agree with the APG III summary. Finally, within the commelinids, the APG III summary does not favour a specific relationship among four major clades (Dasypogonaceae, Arecales, Poales sens. APG III and Commelinales + Zingiberales), while the current analysis supports the placement of the first two of these groups as sisters, the latter two as sisters, and these two pairs of groups as sisters. With respect to these four groups, the ML analysis agrees with the parsimony analysis. However, as with relationships among Asparagales sens. strict., Liliales and Orchidaceae, the highest-level relationships within the commelinids are among the most weakly supported elements of the overall monocot structure, and further investigation of all of these relationships is warranted.

It is evident that the relationships resolved by the complete data set are dominated by the plastid genome data. If congruence between the results of the various parsimony analyses is evaluated strictly in terms of presence vs. absence of clades in consensus trees, and if inapplicable instances are excluded from consideration (that is, groups represented by only one terminal), the closest correspondence of results from any of the data subsets examined to those obtained by the entire data set is seen in the analysis of plastome data only, with *rbc*L and *mat*K excluded (Table 14.3). There are, in fact, no instances of disagreement between the results of these two analyses, though there are three cases of disagreement (groups N, O and P) between results of the all-data analysis and those of the plastome-only analysis that includes *rbc*L and *mat*K (Table 14.3, Fig 14.1). If the same comparison is made in terms of jackknife support, results of the plastome-only analysis that includes *rbc*L and *mat*K generally agree more closely with those of the all-data analysis. For example, support levels for groups N and O are closer, numerically, between the all-data analysis and the plastome-only analysis that includes *rbc*L and *mat*K, than between the all-data analysis and the plastome-only analysis that excludes these two genes. The same is true for Asparagales sens. strict. and group C, but there are nontrivial counter-examples, such as groups J and P.

In contrast to the general similarity observed between results of the all-data analysis and the two plastome-only analyses, there are numerous distinctions of substantial magnitude between results of the all-data analysis and those of the *rbc*L and two-gene analyses. Striking cases include Dioscoreales, Asparagales sens. strict., Commelinales, Xyridales, Restionales, Poales sens. strict., groups C-H, L-P, and several other groups. These cases illustrate the extent to which the combined analysis is dominated by the plastome data.

Results of the four analyses that include different quartiles of the plastome data shed light on the roles of various portions of that data partition in determining the results of the combined analysis. To this purpose we review the five support patterns that are defined in the Methods section, and some of the clades that exhibit each pattern. The goal of this classification of support patterns is to elucidate overall patterns of support among six of the analyses (the two-gene analysis, as one endpoint, the all-data analysis, as the other endpoint, and the four analyses that include *rbc*L, *mat*K, and individual quartiles of the plastome data, as intermediate analyses).

Support pattern I specifies groups that are present in the consensus trees of all six of these analyses, with strong support by all six, and little deviation among intermediate analyses from the jackknife support obtained by the two endpoint analyses. These are groups for which support is strong, and for which the addition of plastome data and various subsets to the two-gene portion of the matrix does not alter the results substantially from those that are obtained by the two-gene analysis. As such, these are groups with minimal conflict among the data partitions. Most clades within this group also receive strong support in the analysis of *rbc*L alone, though this is not a criterion for inclusion in this category. Commelinales thus represents an exception to this general rule, for the group is not resolved by *rbc*L alone, though it is resolved with strong support by the two-gene analysis and by all data subsets that include plastome data (Table 14.3). Most of the principal clades exhibit pattern I, the exceptions being Piperales, Dioscoreales, Liliales, Asparagales sens. strict. and four of the seven constituent groups within Poales sens. APG III (Fig 14.1). Among clades consisting of two or more of the 29 principal clades, only a minority exhibit this pattern of support, including the grouping of all monocots, the grouping of all monocots except Acorales and Asparagales (groups I and K, respectively). Thus, the two-gene data set is sufficient to resolve most of the principal clades that are resolved by all of the data, and with strong support, but the plastome data are required to resolve (with strong support) most of the observed relationships among these groups.

Support pattern II specifies groups that are present in the consensus trees of all six analyses, but with support less than 90% in one or both of the endpoint analyses, and little deviation among intermediate analyses from the jackknife support obtained by the endpoint analyses. Hence, these are groups that are resolved by all of the data subsets, but with weaker support than groups with support pattern I. Groups with either of these support patterns are resolved by the two-gene and all-data analyses, and exhibit little conflict between these data partitions. Among groups in this category are Dioscoreales, Liliales, Juncales, Poales sens. strict. and three higher-level groupings, such as the clade that consists of Juncales, Xyridales, Restionales and Poales sens. strict. Although groups with this pattern of support are universally supported by the two-gene matrix, some have quite little support from *rbc*L alone, such as Dioscoreales and Poales sens. strict., which are absent from the consensus tree of that analysis, and have 0% and 7% jackknife support, respectively. With *rbc*L, a group that includes all elements of Dioscoreales other than Nartheciaceae has 94% support, but Nartheciaceae is nearly always placed elsewhere in the tree. Similarly, a group that consists of all elements of Poales sens. strict. except *Flagellaria* has 100% jackknife support in analyses of *rbc*L alone, but as *Flagellaria* is placed elsewhere in the tree, the grouping of all elements of Poales sens. strict. has only 7% support.

Support pattern III specifies groups that are present with jackknife support ≥ 50% in the two-gene and all-data analysis, while results of at least one of the intermediate analyses either do not include the group in the consensus tree, or provide support at less than 50%, or at a level that is more than 10% beyond the range set by the endpoints. For clades in this category, though supported by both data partitions, there is conflict among portions of the plastid genomic data, and thus an indication that different results would have been obtained if only a particular subset of the genomic data had been sampled. Most of the clades that fall in this category do so because of weak support by one or more of the intermediate analyses (such as group J, Table 14.3 and Fig 14.1), but groups M and Q (commelinids and Dioscoreales + Pandanales) fall in this category because one of the analyses that includes a quartile of the plastid genome data provides jackknife support that is more than 10% higher than that obtained by either of the endpoint analyses. If the 10% criterion had been altered by one percentage point, to 11%, these two clades would have exhibited support pattern II. In all four other cases, Support pattern III takes the form of lack of resolution and low jackknife support for the specified group by one or more of the intermediate analyses. A remarkable case involves group Z (Poales sens. Dahlgren et al., 1985), which has 55% jackknife support by the two-gene analysis, and 93% support by the complete data set, yet is unresolved by three of the four analyses that include the two genes plus a quartile of the plastome data, with only 7% support in one of these analyses. Restionales has a similar support pattern, and the support patterns of these two groups are related, for a recurring topology among jackknife trees is one in which Restionales is paraphyletic, with Xyridales and Poales sens. strict. nested within it; this structure is inconsistent with the monophyly of both Restionales and Poales sens. Dahlgren et al. (1985).

Support pattern IV specifies groups that are present in the consensus tree of the all-data analysis, with jackknife support ≥ 50%, but either are absent from the consensus tree of the two-gene analysis, or have support <50%, or both. The results of the intermediate analyses are not part of the definition of this support pattern, which, therefore, simply indicates conflict between the two endpoint analyses. Because the classification system of support patterns is applied only to groups that are present in the consensus tree for the all-data analysis, the converse of this support group – clades that are supported by the two-gene analysis, but not by the all-data analysis – is not considered. All groups that fall within this category are absent from the consensus trees of the two-gene analysis, and several have extremely low levels of support in that analysis (seven with jackknife support <5% from the two-gene analysis). This category therefore comprises clades that would not have been detected had the genomic data not been included. An example of interest is group R (the placement of Arecales and Dasypogonaceae as sister taxa). This group is resolved by the entire data set, though with low jackknife support (56%), and is not resolved by the two-gene matrix, which provides only 3% support for it. A similar pattern is observed for clade V, which includes all taxa of Poales sens. APG III except Bromeliaceae, and for two additional groups within Poales, these being Juncales + Xyridales (group Y), and the clade that includes these two clades plus Restionales and Poales sens. strict. (group X). In all three of these cases, the groups are resolved by the complete data set, with support greater than 70%, but support by the two-gene data set is 12% or less. Notably, both analyses of plastome data alone resolve all three of these groups, and with support greater than that provided by the all-data analysis in all three cases.

Support pattern V specifies groups that are present in the consensus tree for the all-data analysis, but with jackknife support <50%. Four groups fall in this category, and none of the four has strong support in any of the 273-taxon analyses, though all four have jackknife support well above 50% in the 88-taxon analysis of plastid genomes alone (excluding *rbc*L and *mat*K), and one of them has comparable support in the 88-taxon analysis of plastid genomes that includes these two genes. Three of the groups involve relationships among Liliales, Orchidaceae, Asparagales sens. strict. and the placement of these three groups among the monocots. The fourth group includes all commelinids except Arecales and Dasypogonaceae, and thus its support is related to that of the clade that includes just those two groups, since the placement of either Arecales or Dasypogonaceae with any other element of the commelinids would preclude resolution of the group that includes all commelinids except these two.

The placement of Petrosaviales, which includes the species for which new plastome data are provided here (*Japonolirion osense*), is a case in which the plastome data add to support for a relationship that had been detected in previous analyses. Analysis of the entire data set provides 99% support for the placement of this order within a clade that includes all monocots other than Acorales and Alismatales (Fig 14.1), and 91% support for the resolution of the sister group of this order (groups K and L, respectively), thus placing Petrosaviales precisely. As indicated in Table 14.3, group L is resolved by the two-gene analysis, but with only 60% support.

As illustrated by the foregoing cases, the addition of plastome data to the two-gene matrix results in major changes in the groups that are resolved, as well as in support levels for these groups, with certain regions of the overall phylogeny affected in a major way. Of the 19 principal clades within the monocots, all but three exhibit support pattern I or II (Fig 14.1). The exceptions are Asparagales sens. strict., Xyridales and Restionales, and all three of these groups lie in regions of the tree in which higher-level groupings tend to fall within Support Groups III, IV or V. Thus, the placement of Liliales, Orchidaceae and Asparagales sens. strict., relative to each other, and to other monocot lineages, are inter-related matters, as are details of the internal structure of Poales sens. APG III. A third set of inter-related problems involves Arecales, Dasypogonaceae, Commelinales + Zingiberales and Poales sens. APG III, that is, the internal structure of the commelinids. A review of areas of congruence and conflict among previous phylogenetic studies of the monocots is beyond the scope of this paper, and readers are referred to Givnish et al. (2010) and citations within.

Beyond identifying particular areas of conflict among portions of the present data set, a significant result of these analyses is the demonstration of a divergence of signals among the four data subsets that include individual quartiles of the plastid genome data. Each of these data sets includes *rbc*L and *mat*K, plus several additional genes of the plastid genome (more than 30 in the fourth quartile). A striking instance of incongruence is seen in the resolution of Poales sens. Dahlgren et al., and in the support values for this group, which range from 7% (and lack of resolution) to 99% (and resolution) among these four data subsets. This clade exhibits support pattern III, and although it has jackknife support of 93% by the complete analysis, it would seem to be unwise to assume that this group will continue to be resolved as additional plastid sequence data accumulate. Several other lineages exhibit similar support patterns.

In considering the divergence of signals among portions of the plastome data, it should be noted that the quartiles that were examined in this study represent one of a countless number of possible subsets of genes or nucleotides that could have been treated as data partitions. Consequently, the various clades that were examined might have fallen into different support categories had the plastid genome been subdivided differently. As sampling of additional plastid genomes proceeds, it seems likely that this portion of the data will continue to dominate the results of combined matrices like the present one, and that some groups that are supported by one or a few genes will continue to be refuted by the larger gene set, as in the present instance. As indicated by the results of analyses with quartiles of the plastome data, there is also considerable conflict among large subsets of the plastid genome. Had any single quartile of the plastid genome data been sampled alone, it would not have been evident that the results obtained with that quartile differed from those that would have been obtained with others, or with the combination of all four. For example, if group Z is again considered (Restionales plus Poales sens. strict.), a matrix consisting entirely of *rbc*L, *mat*K and the second quartile of the plastome data provides overwhelming support for this group (resolution with 99% jackknife support), while a matrix of identical magnitude, consisting of *rbc*L, *mat*K, and the third quartile of the plastome data, does not resolve this group, and provides only 7% support for it. Thus, even the use of large portions of the plastome can provide strongly conflicting results. In light of the observation that the analysis of different quartiles of plastome sequences can yield strongly conflicting results, it seems unwise to regard the results that eventually are obtained with whole plastome sequences to be definitive.

14.5 Acknowledgements

J. Davis thanks the organisers of the conference on early monocot evolution for the invitation to participate. We acknowledge support by the US NSF Monocot Tree of Life project for JID, TJG, SWG, JCP, DWS, and JL-M (and other PIs), and by the Plant Molecular Biology Center, Northern Illinois University, DeKalb for MRD.

14.6 References

APG III (Angiosperm Phylogeny Group III). (2009). An update of the Angiosperm Phylogeny Group classification for the orders and families of flowering plants: APG III. *Botanical Journal of the Linnean Society*, **161**, 105–121

Asano, T., Tsudzuki, T., Takahashi, S., Shimada, H. and Kadowski, K. (2004). Complete nucleotide sequence of the sugarcane (*Saccharum officinarum*) chloroplast genome: a comparative analysis of four monocot chloroplast genomes. *DNA Research* **11**, 93–99.

Cai, Z.Q., Penaflor, C., Kuehl, J.V. et al. (2006). Complete plastid genome sequences of *Drimys*, *Liriodendron*, and *Piper*: implications for the phylogenetic relationships of magnoliids. *BMC Evolutionary Biology*, **6**, 77.

Chang, C.C., Lin, H.C., Lin, I.P. et al. (2006). The chloroplast genome of *Phalaenopsis aphrodite* (Orchidaceae): comparative analysis of evolutionary rate with that of grasses and its phylogenetic implications. *Molecular Biology and Evolution*, **23**, 279-291.

Chase, M.W., Stevenson, D.W., Wilkin, P. and Rudall, P.J. (1995). Monocot systematics: a combined analysis. In *Monocotyledons: Systematics and Evolution*, ed. P.J. Rudall, P.J. Cribb, D.F. Cutler and C.J. Humphries. Kew: Royal Botanic Gardens, pp. 685-730.

Chase, M.W., Soltis, D.E., Soltis, P.S. et al. (2000). Higher-level systematics of the monocotyledons: an assessment of current knowledge and a new classification. In *Monocots: Systematics and Evolution*, ed. K.L. Wilson and D.A. Morrison. Collingwood: CSIRO, pp. 3-16.

Cui, L., Veeraraghavan, N., Richter, A. et al. (2006). ChloroplastDB: the chloroplast genome database. *Nucleic Acids Research*, **34**, D692-D696.

Dahlgren, R.M.T., Clifford, H.T. and Yeo, P.F. (1985). *The Families of the Monocotyledons*. Berlin: Springer-Verlag.

Daniell, H., Wurdack, K.J., Kanagaraj, A. et al. (2008). The complete nucleotide sequence of the cassava (*Manihot esculenta*) chloroplast genome and the evolution of *atpF* in Malpighiales: RNA editing and multiple losses of a group II intron. *Theoretical and Applied Genetics*, **116**, 723-737.

Davis, J.I., Stevenson, D.W., Petersen, G. et al. (2004). Phylogeny of the monocots, as inferred from *rbcL* and *atpA* sequence variation, and a comparison of methods for calculating jackknife and bootstrap values. *Systematic Botany*, **29**, 467-510.

Davis, J.I., Petersen, G., Seberg, O. et al. (2006). Are mitochondrial genes useful for the analysis of monocot relationships? *Taxon*, **55**, 857-870.

Dransfield, J., Uhl, N.W., Asmussen, C.B. et al. (2008). *Genera Palmarum – The Evolution and Classification of Palms*. Kew: Royal Botanic Gardens.

Duvall, M.R., Clegg, M.T., Chase, M.W. et al. (1993). Phylogenetic hypotheses for the monocotyledons constructed from *rbcL* sequence data. *Annals of the Missouri Botanical Garden*, **80**, 607-619.

Duvall, M.R., Leseberg, C.H., Grennan, C.P. and Morris, L.M. (2010). Molecular evolution and phylogenetics of complete chloroplast genomes in Poaceae. In *Diversity, Phylogeny, and Evolution in the Monocotyledons*, ed. O. Seberg, G. Petersen, A.S. Barfod and J.I. Davis. Aarhus: Aarhus University Press, pp. 438-450.

Edgar, R.C. (2004). MUSCLE: multiple sequence alignment with high accuracy and high throughput. *Nucleic Acids Research*, **32**, 1792-1797.

Farris, J.S., Albert, V.A., Källersjö, M., Lipscomb, D. and Kluge, A.G. (1996). Parsimony jackknifing outperforms neighbor-joining. *Cladistics*, **12**, 99-124.

Freudenstein, J.V. and Davis, J.I. (2010). Branch support via resampling: an empirical study. *Cladistics*, **26**, 1-14.

Givnish, T.J., Pires, J.C., Graham, S.W. et al. (2006). Phylogeny of the monocotyledons based on the highly informative plastid gene *ndhF*: evidence for widespread concerted convergence. In *Monocots: Comparative Biology and Evolution (excluding Poales)*, ed. J.T. Columbus, E.A. Friar, C.W. Hamilton et al. Claremont, CA: Rancho Santa Ana Botanic Garden, pp. 28-51.

Givnish, T.J., Ames, M., McNeal, J.R. et al. (2010). Assembling the tree of the monocotyledons: plastome sequence phylogeny and evolution of Poales. *Annals of the Missouri Botanical Garden*, **97**, 584–616.

Goloboff, P.A, Farris, J.S. and Nixon, K.C. (2008). TNT, a free program for phylogenetic analysis. *Cladistics*, **24**, 774–786.

Goremykin, V.V., Hirsch-Ernst, K.I., Wolfl, S. and Hellwig, F.H. (2003a). Analysis of the *Amborella trichopoda* chloroplast genome sequence suggests that *Amborella* is not a basal angiosperm. *Molecular Biology and Evolution*, **20**, 1499–1505.

Goremykin, V.V., Hirsch-Ernst, K.I., Wolfl, S. and Hellwig, F.H. (2003b). The chloroplast genome of the 'basal' angiosperm *Calycanthus fertilis* – structural and phylogenetic analyses. *Plant Systematics and Evolution*, **242**, 119–135.

Goremykin, V.V., Holland, B., Hirsch-Ernst, K.I. and Hellwig, F.H. (2005). Analysis of *Acorus calamus* chloroplast genome and its phylogenetic implications. *Molecular Biology and Evolution* **22**, 1813–1822.

Graham, S.W., Zgurski, J.M., McPherson, M.A. et al. (2006). Robust inference of monocot deep phylogeny using an expanded multigene plastid data set. In *Monocots: Comparative Biology and Evolution (excluding Poales)*, ed. T.J. Columbus, E.A. Friar, C.W. Hamilton et al. Claremont, CA: Rancho Santa Ana Botanic Garden, pp. 3–20.

Guisinger, M.M., Kuehl, J.V., Boore, J.L. and Jansen, R.K. (2008). Genome-wide analyses of Geraniaceae plastid DNA reveal patterns of increased nonsynonymous substitutions. *Proceedings of the National Academy of Sciences USA*, **105**, 18424–18429.

Guisinger, M.M., Chumley, T.W., Kuehl, J.V., Boore, J.L. and Jansen, R.K. (2010). Implications of the plastid genome sequence of *Typha* (Typhaceae, Poales) for understanding genome evolution in Poaceae. *Journal of Molecular Evolution*, **70**, 149–166.

Hansen, D.R., Dastidar, S.G., Cai, Z.Q. et al. (2007). Phylogenetic and evolutionary implications of complete chloroplast genome sequences of four early-diverging angiosperms: *Buxus* (Buxaceae), *Chloranthus* (Chloranthaceae), *Dioscorea* (Dioscoreaceae), and *Illicium* (Schisandraceae). *Molecular Phylogenetics and Evolution*, **45**, 547–563.

Haston, E., Richardson, J.E., Stevens, P.F., Chase, M.W. and Harris, D.J. (2009). The Linear Angiosperm Phylogeny Group (LAPG) III: a linear sequence of the families in APG III. *Botanical Journal of the Linnean Society*, **161**, 128–131

Hiratsuka, J., Shimada, H., Whittier, R. et al. (1989). The complete sequence of the rice (*Oryza sativa*) chloroplast genome – intermolecular recombination between distinct transfer RNA genes accounts for a major plastid DNA inversion during the evolution of the cereals. *Molecular and General Genetics*, **217**, 185–194.

Jansen, R.K., Kaittanis, C., Saski, C. et al. (2006). Phylogenetic analyses of *Vitis* (Vitaceae) based on complete chloroplast genome sequences: effects of taxon sampling and phylogenetic methods on resolving relationships among rosids. *BMC Evolutionary Biology* **6**, 32.

Jansen, R.K., Cai, Z., Raubeson, L.A. et al. (2007). Analysis of 81 genes from 64 plastid genomes resolves relationships in angiosperms and identifies genome-scale evolutionary patterns. *Proceedings of the National Academy of Sciences USA*, **104**, 19369–19374.

Janssen, T. and K. Bremer. (2004). The age of major monocot groups inferred from 800+ *rbcL* sequences. *Botanical Journal of the Linnean Society*, **146**, 385–398.

Judziewicz, E.J., Clark, L.G., Londoño, X. and Stern, M.J. (1999). *American Bamboos*. Washington: Smithsonian Institution Press.

Kim, K.J. and Lee, H.L. (2004). Complete chloroplast genome sequences from Korean ginseng (*Panax schinseng* Nees) and comparative analysis of sequence evolution among 17 vascular plants. *DNA Research*, **11**, 247–261.

Kubitzki, K., ed. (1998a). *The Families and Genera of Vascular Plants, Volume III : Flowering Plants. Monocotyledons. Lilianae (except Orchidaceae)*. New York: Springer-Verlag.

Kubitzki, K., ed. (1998b). *The Families and Genera of Vascular Plants, Volume IV : Flowering Plants. Monocotyledons. Alismatanae and Commelinanae (except Gramineae)*, New York: Springer-Verlag.

Leebens-Mack, J., Raubeson, L.A., Cui, L. et al. (2005). Identifying the basal angiosperm node in chloroplast genome phylogenies: sampling one's way out of the Felsenstein zone. *Molecular Biology and Evolution*, **22**, 1948–1963.

Leseberg, C.H. and Duvall, M.R. (2009). The complete chloroplast genome of *Coix lachyrma-jobi* and a comparative molecular evolutionary analysis of plastomes in cereals. *Journal of Molecular Evolution*, **69**, 311–318.

Linder, H.P. and Rudall, P.J. (2005). Evolutionary history of Poales. *Annual Review of Ecology and Systematics*, **36**, 107–124.

Maier, R.M., Neckermann, K., Igloi, G.L. and Kossel, H. (1995). Complete sequence of the maize chloroplast genome – gene content, hotspots of divergence and fine-tuning of genetic information by transcript editing. *Journal of Molecular Evolution*, **251**, 614–628.

Mardanov, A.V., Ravin, N.V., Kuznetsov, B.B. et al. (2008). Complete sequence of the duckweed (*Lemna minor*) chloroplast genome: structural organization and phylogenetic relationships to other angiosperms. *Journal of Molecular Evolution*, **66**, 555–564.

Matsushima, R., Hu, Y., Toyoda, K., Sodmergen and Sakamoto, W. (2008). The model plant *Medicago truncatula* exhibits biparental plastid inheritance. *Plant and Cell Physiology*, **49**, 81–91.

Moore, M.J., Dhingra, A., Soltis, P.S. et al. (2006). Rapid and accurate pyrosequencing of angiosperm plastid genomes. *BMC Plant Biology*, **6**, 17.

Moore, M.J., Bell, C.D., Soltis, P.S. and Soltis, D.E. (2007). Using plastid genome-scale data to resolve enigmatic relationships among basal angiosperms. *Proceedings of the National Academy of Sciences USA*, **104**, 19363–19368.

Moore, M.J., Soltis, P.S., Bell, C.D., Burleigh, J.G. and Soltis, D.E. (2010). Phylogenetic analysis of 83 plastid genes further resolves the early diversification of eudicots. *Proceedings of the National Academy of Sciences USA*, **107**, 4623–4628

Morris, L.M. and Duvall, M.R. (2010). The chloroplast genome of *Anomochloa marantoidea* (Anomochlooideae; Poaceae) comprises a mixture of grass-like and unique features. *American Journal of Botany*, **97**, 620–627.

Neyland, R. and Hennigan, M. (2003). A phylogenetic analysis of large-subunit (26S) ribosome DNA sequences suggests that the Corsiaceae are polyphyletic. *New Zealand Journal of Botany*, **41**, 1–11.

Nixon, K.C. (1999). The parsimony ratchet, a new method for rapid parsimony analysis. *Cladistics*, **15**, 407–414.

Nixon, K.C. (2002). *WinClada version 1.03*. Computer program distributed by the author, Cornell University, Ithaca, New York, USA; available at http://www.cladistics.com/

Ogihara, Y., Isono, K., Kojima, T. et al. (2002). Structural features of a wheat plastome as revealed by complete sequencing of chloroplast DNA. *Molecular Genetics and Genomics*, **266**, 740–746.

Okumura, S., Sawada, M., Park, Y.W. et al. (2006). Transformation of poplar (*Populus alba*) plastids and expression of foreign proteins in tree chloroplasts. *Transgenic Research*, **15**, 637–646.

Parks, M., Cronn, R. and Liston, A. (2009). Increasing phylogenetic resolution at low taxonomic levels using massively parallel sequencing of chloroplast genomes. *BMC Biology*, **7**, 84.

Plader, W., Yukawa, Y., Sugiura, M. and Malepszy, S. (2007). The complete structure of the cucumber (*Cucumis sativus* L.) chloroplast genome: its composition and comparative analysis. *Cellular and Molecular Biology Letters*, **12**, 584–594.

Ratan, A. (2009). *Assembly Algorithms for Next-generation Sequence data*. Ph.D. Dissertation, Pennsylvania State University.

Raubeson, L.A., Peery, R., Chumley, T.W. et al. (2007). Comparative chloroplast genomics: analyses including new sequences from the angiosperms *Nuphar advena* and *Ranunculus macranthus*. *BMC Genomics*, **8**, 174.

Rokas, A. and Carroll, S.B. (2005). More genes or more taxa? The relative contribution of gene number and taxon number to phylogenetic accuracy. *Molecular Biology and Evolution*, **22**, 1337–1344.

Saarela, J.M., Rai, H.S., Doyle, J.A. et al. (2007). Hydatellaceae identified as a new branch near the base of the angiosperm phylogenetic tree. *Nature*, **446**, 312–315.

Samson, N., Bausher, M.G., Lee, S.-B., Jansen, R.K. and Daniell, H. (2007). The complete nucleotide sequence of the coffee (*Coffea arabica* L.) chloroplast genome: organization and implications for biotechnology and phylogenetic relationships amongst angiosperms. *Plant Biotechnology Journal*, **5**, 339–353.

Saski, C., Lee, S.B., Fjellheim, S. et al. (2007). Complete chloroplast genome sequences of *Hordeum vulgare*, *Sorghum bicolor* and *Agrostis stolonifera*, and comparative analyses with other grass genomes. *Theoretical and Applied Genetics*, **115**, 571–590.

Schmitz-Linneweber, C., Maier, R.M., Alcaraz, J.P. et al. (2001). The plastid chromosome of spinach (*Spinacia oleracea*): complete nucleotide sequence and gene organization. *Plant Molecular Biology*, **45**, 307–315.

Soltis, D.E., Moore, M.J., Burleigh, J.G. and Soltis, P.S. (2009). Molecular markers and concepts of plant evolutionary relationships: progress, promise, and future prospects. *Critical Reviews in Plant Sciences*, **28**, 1–15.

Soltis, D.E., Moore, M.J., Burleigh, J.G., Bell, C.D. and Soltis, P.S. (2010). Assembling the angiosperm tree of life: progress and future prospects. *Annals of the Missouri Botanical Garden*, **97**, 514–526.

Soreng, R.J. and Davis, J.I. (1998). Phylogenetics and character evolution in the grass family (Poaceae): simultaneous analysis of morphological and chloroplast DNA restriction site character sets. *Botanical Review*, **64**, 1–85.

Stamatakis, A. (2004). RAxML-VI-HPC: Maximum likelihood-based phylogenetic analyses with thousands of taxa and mixed models. *Bioinformatics*, **22**, 2688–2690.

Stevenson, D.W. and Loconte, H. (1995). Cladistic analysis of monocot families. In *Monocotyledons: Systematics and Evolution*, ed. P.J. Rudall, P.J. Cribb, D. F. Cutler, and C.J. Humphries. Kew: Royal Botanic Gardens, pp. 543–578.

Tamura, M.N., Yamashita, J., Fuse, S. and Haraguchi, M. (2004). Molecular phylogeny of monocotyledons inferred from combined analysis of plastid *matK* and *rbcL* gene sequences. *Journal of Plant Research*, **117**, 109–120.

Wang, H., Moore, M.J., Soltis, P.S. et al. (2009). Rosid radiation and the rapid rise of angiosperm-dominated forests. *Proceedings of the National Academy of Sciences USA*, **106**, 3853–3858.

Wu, F.-H., Kan, D.-P., Lee, S.-B. et al. (2009). Complete nucleotide sequence of *Dendrocalamus latiflorus* and *Bambusa oldhamii* chloroplast genomes. *Tree Physiology*, **29**, 847–856.

Wyman, S.K., Jansen, R.K. and Boore, J.L. (2004). Automatic annotation of organellar genomes with DOGMA. *Bioinformatics*, **20**, 3252–3255.

Xiao-Xian, L. and Zhe-Kun, Z. (2007). The higher-level phylogeny of monocots based on *matK*, *rbcL* and 18S rDNA sequences. *Acta Phytotaxonomica Sinica*, **45**, 113–133.

Yang, Z.H. (1994). Maximum likelihood phylogenetic estimation from DNA sequences with variable rates over sites. *Journal of Molecular Evolution*, **39**, 306–314.

14.7 Appendix

Sources of DNA sequences for 273 taxa included in phylogenetic analyses. Gen-Bank accession numbers are provided for all taxa, and literature citations also are provided for 87 taxa (in boldface) represented by previously published plastid genomes (▲ = Givnish et al. (2010); GenBank numbers HQ180399-HQ183709); plastid genome data for *Japonolirion osense* are published here for the first time, as are two new *rbc*L sequences (*Acanthochlamys* and *Haemodorum*, for which voucher information is provided). Taxa are sorted into 29 mutually exclusive principal clades recognized by APG III at the ordinal rank and below, except as indicated by asterisks (see text), and are listed alphabetically by genus and species within each group, regardless of family assignment. Within square brackets after the name of each group are the number of taxa represented by plastid genomes, and the number represented only by *rbc*L and *mat*K sequences. When two Gen-Bank voucher numbers are present within a single pair of round brackets, the first is for an *rbc*L sequence, and the second for a *mat*K sequence.

AMBORELLALES [1,0]. *Amborella trichopoda* **Baill.; NC_005086; Goremykin et al. (2003a)**.

NYMPHAEALES [1,3]. ***Nuphar advena* Ait.; NC_008788; Raubeson et al. (2007).** *Nymphaea odorata* Aiton (*rbc*L M77034); *Nymphaea alba* L. (*mat*K AJ627251). *Trithuria submersa* Hook. f. (*rbc*L DQ915188). *Victoria cruziana* Orb. (*rbc*L M77036); *Victoria amazonica* (Poepp.) J.C. Sowerby (*mat*K AF092991).

AUSTROBAILEYALES [1,2]. *Austrobaileya scandens* C.T. White (L12632, AF465286). ***Illicium oligandrum* Merr. & Chun; NC_009600; Hansen et al. (2007).** *Schisandra sphenanthera* Rehder & E.H. Wilson (*rbc*L L12665); *Schisandra chinensis* (Turcz.) Baill. (*mat*K DQ185526).

CHLORANTHALES [1,1]. ***Chloranthus spicatus* (Thunb.) Makino; NC_009598; Hansen et al. (2007).** *Hedyosmum orientale* (*rbc*L AY236848); *Hedyosmum arborescens* Sw. (*mat*K AF465296).

CANELLALES [1,1]. ***Drimys granadensis* L. f.; NC_008456; Cai et al. (2006).** *Tasmannia lanceolata* (Poir.) A.C. Sm. (AY298851, DQ882241).

MAGNOLIALES [1,5]. *Annona muricata* L. (L12629, AF543722). *Eupomatia bennettii* F. Muell. (L12644, DQ401341). ***Liriodendron tulipifera* L.; NC_008326; Cai et al. (2006).** *Magnolia grandiflora* L. (AY298837, AF548640). *Michelia figo* (Lour.) Spreng. (L12659, AF123467). *Myristica fragrans* Houtt. (*rbc*L AY298839); *Myristica maingayi* Hook. f. (*mat*K AY220452).

LAURALES [1,7]. ***Calycanthus floridus* L.; NC_004993; Goremykin et al. (2003b).** *Chimonanthus praecox* (L.) Link; (L12639, AY525340). *Doryphora aromatica* (F.M. Bailey) L.S. Sm. (*rbc*L L77211); *Doryphora sassafras* Endl. (*mat*K AF542568). *Gyrocarpus* sp. (*rbc*L L12647); *Gyrocarpus americanus* Jacq. (*mat*K AF465295). *Hedycarya arborea* J.R. Forst. & G. Forst. (L12648, AM396509). *Idiospermum australiense* (Diels) S.T. Blake; (L12651, AY525342). *Neolitsea cassia* (L.) Kosterm. (*rbc*L AY298841); *Neolitsea levinei* Merr. (*mat*K AF244393). *Peumus boldus* Molina (AF040664, AJ247183).

TRICOLPATE DICOTS (i.e. 'EUDICOTS') [13,7]. *Akebia quinata* (Houtt.) (L12627, AB069851). ***Anethum graveolens* L.; EU016721-EU016801; Jansen et al. (2007). *Buxus microphylla* Siebold & Zucc.; NC_009599; Hansen et al. (2007).** *Cercidiphyllum japonicum* Siebold & Zucc. (L11673, AM396508). ***Coffea arabica* L.; NC_008535; Samson et al. (2007). *Cucumis sativus* L.; NC_007144; Plader et al. (2007).** *Decaisnea fargesii* Franch. (D85692, GU266595). *Epimedium koreanum* Nakai (L75869, AB069837). ***Helianthus annuus* L.; NC_007977; Jansen et al. (2007).** *Mahonia bealei* (Fortune) Carrière (*rbc*L L75871); *Mahonia japonica* (Thunb.) DC. (*mat*K AB038184). ***Manihot esculenta* Crantz; NC_010433; Daniell et al. (2008). *Medicago truncatula* Gaertn.; NC_003119; Matsushima et al. (2008). *Nandina domestica* Thunb.; NC_008336; Raubeson et al. (2007).** *Nelumbo lutea* Willd. (*rbc*L M77032); *Nelumbo nucifera* Gaertn. (*mat*K AF543740). ***Panax ginseng* C.A. Mey.; NC_006290; Kim & Lee (2004). *Platanus occidentalis* L.; NC_008335; Moore et al. (2006). *Populus alba* Torr. & A. Gray; NC_008235; Okumura et al. (2006).** *Sargentodoxa cuneata* (Oliv.) Rehder & E.H.

Wilson (FJ626605, DQ401351). ***Spinacia oleracea* L.; NC_002202; Schmitz-Linneweber et al. (2001).** *Vitis vinifera* L.; NC_007957; Jansen et al. (2006).

CERATOPHYLLALES [1,0]. ***Ceratophyllum demersum* L.; NC_009962; Moore et al. (2007).**

PIPERALES [1,8]. *Aristolochia macrophylla* Lam. (*rbc*L L12630); *Aristolochia gigantea* Mart. & Zucc. (*mat*K AB060794). *Asarum canadense* L. (*rbc*L L14290); *Asarum caudigerum* Hance (*mat*K AY952420). *Houttuynia cordata* Thunb. (L08762, AF543737). *Lactoris fernandeziana* Phil. (L08763, AF543739). *Macropiper excelsum* (Forster f.) Miq. (*rbc*L AY298836; *Macropiper hooglandii* I. Hutton & P.S. Green (*mat*K DQ882228). *Peperomia pellucida* (L.) Kunth (EF450306, DQ212738). ***Piper coenoclatum* Diels; NC_008457; Cai et al. (2006).** *Saruma henryi* Oliver (L12664, AF543748). *Saururus cernuus* L. (L14294, AF543749).

ACORALES [2,2]. ***Acorus americanus* (Raf.) Raf.; DQ069337-DQ069702, EU016701-EUO16720; Leebens-Mack et al. (2005). *Acorus calamus* L.; NC_007407; Goremykin et al. (2005).** *Acorus gramineus* Aiton (D28866, AB040155). *Acorus tatarinowii* Schott (AY298815, DQ008866).

ALISMATALES [1,18]. *Alisma plantago-aquatica* L. (*rbc*L L08759); *Alisma canaliculatum* A. Braun & Bouché (*mat*K AB040179). *Aponogeton fenestralis* (Pers.) Hook. f. (AB088808, AB088779). *Arisaema triphyllum* (L.) Schott (*rbc*L AY298817); *Arisaema tortuosum* (Wall.) Schott (*mat*K AF387428). *Butomus umbellatus* L., (U80685, AY952416). *Caldesia oligococca* (F. Von Mueller) Buche (AY277799, AY952427). *Gymnostachys anceps* R. Br. (M91629, AB040177). *Halodule uninervis* (Forssk.) Asch. (AY952436, AY952424). ***Lemna minor* L.; NC_010109; Mardanov et al. (2008).** *Orontium aquaticum* L. (AJ005632, AF543744). *Ottelia alismoides* (L.) Pers. (*rbc*L U80707); *Ottelia acuminata* (Gagnep.) Dandy (*mat*K AY952432). *Pleea tenuifolia* Michx. (AJ131774, AB183407). *Posidonia oceanica* (L.) Delile (*rbc*L U80719). *Potamogeton richardsonii* (A. Benn.) Rydb. (*rbc*L U03730); *Potamogeton perfoliatus* L. (*mat*K AY952425). *Sagittaria latifolia* Willd. (*rbc*L L08767); *Sagittaria sprucei* Micheli (*mat*K EF088104). *Scheuchzeria palustris* L. (U03728, GQ452338). *Symplocarpus foetidus* (L.) W. Barton (L10247, AM920551). *Tofieldia pusilla* (Michx.) Pers. (*rbc*L AJ286562); *Tofieldia calyculata* (L.) Wheldon (*mat*K AB183403). *Triglochin maritima* L. (U80714, AB088782). *Zostera capensis* Setch. (AM235166, AB096165).

PETROSAVIALES [1,1]. *Japonolirion osense* T. Nakai; *M. Chase 3000*, deposited at K; GenBank accessions (JQ068951-JQ069028). *Petrosavia stellaris* Becc. (*rbc*L AF206806); *Petrosavia sakuraii* (Makino) J.J. Sm. ex Steenis (*mat*K AB040156).

DIOSCOREALES [1,8]. *Aletris spicata* (Thunb.) Franch. (AB088834, AB040174). *Burmannia longifolia* Becc. (*rbc*L AF307484); *Burmannia biflora* L. (*mat*K AJ581397). *Dioscorea communis* (L.) Caddick & Wilkin (AF307474, AF465303). ***Dioscorea elephantipes* (L'Hér.) Engl.; NC_009601; Hansen et al. (2007).** *Narthecium ossifragum* (L.) Huds. (*rbc*L AJ286560); *Narthecium asiaticum* Maxim.

(*mat*K AB040162). *Tacca chantrieri* André (AJ235810, AY973837). *Thismia rodwayi* (*rbc*L AY939892). *Trichopus sempervirens* (H. Perrier) Caddick & Wilkin (AY298818, AY973844). *Trichopus zeylanicus* Gaertn. (AF307477, AY973845).

PANDANALES [1,10]. *Acanthochlamys bracteata* P.C. Kao (HQ845619 [*Kao 1993*, K], AY952413). *Barbaceniopsis* sp. (*rbc*L AY298819). *Carludovica palmata* Ruiz & Pav. (AF197596, AB088793). *Chorigyne cylindrica* R. Erikss. (*rbc*L AY298823). *Croomia pauciflora* (Nutt.) Torr. (AY298827, AY437815). *Cyclanthus bipartitus* Poit. ex A. Rich. (*rbc*L AY007660). *Freycinetia scandens* Gaudich. (*rbc*L AF206770); *Freycinetia formosana* Hemsl. (*mat*K AB040209). **Pandanus utilis Bory;** ▲. *Sphaeradenia pendula* Hammel (*rbc*L AJ235808). *Stemona japonica* (Blume) Miq. (AJ131948, AB040210). *Talbotia elegans* Balf. (AY149358, AY491664).

*ASPARAGALES sens. strict. [13,22]. **Agapanthus praecox Willd.;** ▲. *Alania endlicheri* Kunth (*rbc*L Y14982). **Albuca kirkii (Baker) Brenan;** ▲. *Anthericum liliago* L. (Z69225, AJ511426). *Arthropodium cirratum* (G. Frost.) R. Br. (*rbc*L Z69233). **Asparagus officinalis L.;** ▲. *Astelia pumila* (G. Forst.) R. Br. (*rbc*L AF307906); *Astelia alpina* Banks & Solander ex R. Br. (*mat*K AY368372). *Blandfordia punicea* (Labill.) Sweet (Z73694, AY557206). *Borya septentrionalis* F. Muell. (*rbc*L Y14985); *Borya laciniata* Churchill (*mat*K AY368373). **Chlorophytum rhizopendulum Bjorå & Hemp;** ▲. **Curculigo capitulata (Lourt.) Kunze;** ▲. *Dianella ensifolia* (L.) DC. (M96960, AB088787). *Doryanthes excelsa* Corrêa (Z73697, AB088785). *Eustrephus latifolius* R. Br. (*rbc*L AY298831). *Geitonoplesium cymosum* (R. Br.) A. Cunn. ex Hook. (*rbc*L AY298833). *Hemerocallis fulva* (L.) L. (L05036, AB017318). **Hesperaloe parviflora (Torr.) J.M. Coult;** ▲. **Hosta ventricosa (Salis.) Stearn;** ▲. *Hypoxis glabella* R. Br. (*rbc*L Y14989); *Hypoxis leptocarpa* (Engelm. & A. Gray) Small (*mat*K AY368375). **Iris virginica L.;** ▲. *Ixiolirion tataricum* (Pall.) Herb. (Z73704, AB017327). *Johnsonia pubescens* Lindl. (*rbc*L Z77304). *Lanaria lanata* (L.) T. Durand & Schinz (Z77313, AY368376). **Lomandra longifolia Labill.;** ▲. **Neoastelia spectabilis J.B. Williams;** ▲. *Neomarica northiana* (Schneev.) Sprague (AY298842, AJ579972). **Nolina atopocarpa Bartlett;** ▲. **Phormium tenax Forst.;** ▲. **Sisyrinchium micranthum** Cav. (Z77290, AJ579982). *Sowerbaea juncea* Andrews (*rbc*L Z69234). *Tecophilaea cyanocrocus* Leyb. (*rbc*L Z73709). *Thysanotus spiniger* Brittan (*rbc*L Z69236). *Xanthorrhoea hastilis* R. Br. (*rbc*L Z73710); *Xanthorrhoea quadrangulata* F. Muell. (*mat*K DQ401345). *Xeronema callistemon* W.R.B. Oliv. (*rbc*L Z69235). **Yucca schidigera Ortgies; DQ069337-DQ069702, EU016681-EU016700; Leebens-Mack et al. (2005)**.

ORCHIDACEAE [2,5]. **Apostasia wallichii R. Br.;** ▲. *Calopogon tuberosus* (L.) Britton, Sterns & Poggenb. (AF074119, AF263635). *Cypripedium passerinum* Richardson (*rbc*L AF074142); *Cypripedium calceolus* L. var. *pubescens* (Willd.) Correll (*mat*K AY557208). *Epipactis helleborine* (L.) Crantz (Z73707, AF263659). *Isotria verticillata* (Muhl. ex Willd.) Raf. (*rbc*L AF074180). *Neuwiedia veratrifolia*

Blume (AF074200, AY557211). ***Phalaenopsis aphrodite* Rchb. f.; NC_007499; Chang et al. (2006)**.

LILIALES [1,17]. *Alstroemeria* sp. (Z77254, AY624481). *Amianthium muscaetoxicum* (Walter) A. Gray (*rbc*L AJ417895). *Burchardia umbellata* R. Br. (*rbc*L Z77266). *Calochortus minimus* Ownbey (*rbc*L Z77263); *Calochortus uniflorus* Hook. & Arn. (*mat*K AY624478). *Campynema lineare* Labill. (*rbc*L Z77264). *Chamaelirium luteum* (L.) A. Gray (AJ276347, AB040196). *Clintonia borealis* (Aiton) Raf. (D17372, AB024542). *Colchicum speciosum* Steven (L12673, AB040181). *Disporum sessile* D. Don (D17376, AB040182). *Drymophila moorei* Baker (AB088812, AB040180). ***Lilium superbum* L.; ▲**. *Petermannia cirrosa* F. Muell. (*rbc*L AY298844). *Philesia buxifolia* J.F. Gmel. (Z77302, AY624479). *Rhipogonum elseyanum* F. Muell. (*rbc*L Z77309). *Schelhammera multiflora* R. Br. (*rbc*L AY298849). *Smilax glauca* Walter (*rbc*L Z77310); *Smilax china* L. (*mat*K AB040204). *Trillium grandiflorum* (Michx.) Salisb. (D28164, AB017392). *Veratrum album* L. (*rbc*L D28168); *Veratrum maackii* Regel (*mat*K AB040183).

DASYPOGONACEAE [2,2]. *Baxteria australis* R. Br. ex Hook. (AY123230, DQ888764). *Calectasia cyanea* R. Br. (AY123231, DQ888765). ***Dasypogon bromeliiformis* R. Br.; ▲. *Kingia australis* R. Br.; ▲**.

ARECALES [3,8]. *Areca triandra* Roxb. ex Buch.-Ham. (AJ404819, AY952428). *Calamus hollrungii* Becc. (*rbc*L AJ404775); *Calamus aruensis* Becc. (*mat*K AM114551). ***Chamaedorea seifrizii* Burret; ▲. *Elaeis oleifera* (Kunth) Cortés; EU016883-EU016962; Leebens-Mack et al. (2005)**. *Euterpe oleracea* Mart. (AY298832, AM114647). *Nypa fruticans* Wurmb (M81813, AM114552). *Phoenix reclinata* Jacq. (*rbc*L M81814); *Phoenix dactylifera* L. (*mat*K AB040211). *Phytelephas aequatorialis* Spruce (AY298846, AM114613). *Plectocomia elongata* Mart. ex. Blume (*rbc*L AY298848); *Plectocomia mulleri* Blume (*mat*K AM114550). ***Ravenea hildebrandtii* C.D. Bouché; ▲**. *Trithrinax acanthocoma* Drude (*rbc*L AY298852); *Trithrinax campestris* (Burmeist.) Drude & Griseb. (*mat*K AM114556).

COMMELINALES [2,19]. *Anigozanthos flavidus* DC. in Redouté (AJ404843, AB088796). ***Belosynapsis ciliata* (Blume) R.S. Rao; ▲**. *Callisia warszewicziana* (Kunth & Bouché) D.R. Hunt (*rbc*L AY298821); *Callisia fragrans* (Lindl.) Woodson (*mat*K GU135072). *Cochliostema odoratissimum* Lem. (*rbc*L AY298824). *Commelina communis* L. (*rbc*L AY298825); *Commelina purpurea* C.B. Clarke (*mat*K GQ248103). *Dichorisandra thyrsiflora* J.C. Mikan (*rbc*L AY298828). *Eichhornia crassipes* (Mart.) Solms (U41574, AB040212). *Eichhornia paniculata* (Spreng.) Solms (*rbc*L U41578). *Haemodorum laxum* R. Br. (*rbc*L HQ845620 [*Chase 2977*, K]); *Haemodorum discolor* T.D. Macfarl. (*mat*K EU499279). *Hanguana malayana* Merr. (AJ417896, AB088800). *Helmholtzia glaberrima* (Hook. f.) Caruel (AY298834, EU499296). *Heteranthera rotundifolia* (Kunth) Griseb. (*rbc*L U41585). *Hydrothrix gardneri* Hook. f. (*rbc*L U41582). *Monochoria korsakowii* Regel & Maack (U41590, AB088795). *Murdannia clarkeana* Brenan (*rbc*L AF312256). *Palisota bracteosa* C.B.

Clarke (*rbc*L AY298843). *Philydrella pygmaea* (R. Br.) Caruel (AY298845, AF434870). *Philydrum lanuginosum* Banks & Sol. ex Gaertn. (U41596, AY952429). *Pontederia cordata* L. (U41592, AF434872). **Tradescantia ohiensis Raf.; ▲**. *Xiphidium caeruleum* Aubl. (AY149359, EU499294).

ZINGIBERALES [2,13]. *Alpinia purpurata* (Vieill.) K. Schum. (*rbc*L AY298816); *Alpinia intermedia* Gagnep. (*mat*K AB040213). *Calathea loeseneri* J.F. Macbr. (AF243842, AY140273). *Canna indica* L. (*rbc*L AF378763); *Canna flaccida* Salisb. (*mat*K AF478906). *Costus lateriflorus* Baker (*rbc*L AY298826); *Costus pulverulentus* C. Presl (*mat*K AF478907). *Dimerocostus argenteus* (Ruiz & Pav.) Maas (*rbc*L AY298829). *Globba atrosanguinea* Teijsm. & Binn. (AF378777, AF478852). *Heliconia indica* Lam. (*rbc*L AF378765); *Heliconia irrasa* Lane ex R.R. Sm. (*mat*K AF478908). *Maranta bicolor* Ker Gawl. (*rbc*L AF378768); *Maranta leuconeura* E. Morren (*mat*K AY140303). *Monocostus uniflorus* (Poepp. ex Petersen) Maas (*rbc*L AF243839). **Musa acuminata Colla; EU016983-EU017063; Leebens-Mack et al. (2005)**. *Orchidantha fimbriata* Holttum (AF243841, AY952417). *Ravenala madagascariensis* Sonn. (L20138, AF434873). **Renealmia alpinia (Rottb.) Maas; ▲**. *Strelitzia nicolai* Regel & Körn. (*rbc*L AF243846); *Strelitzia alba* Skeels (*mat*K AF434874). *Tapeinochilos ananassae* K. Schum. (*rbc*L AF243840).

BROMELIACEAE [6,4]. *Ananas comosus* (L.) Merr. (*rbc*L L19977); *Ananas ananassoides* (Baker) L.B. Sm. (*mat*K AF162227). **Brocchinia micrantha (Baker) Mes; ▲**. *Catopsis montana* L.B. Sm. (*rbc*L L19976); *Catopsis nutans* (Sw.) Griseb. var. *nutans* (*mat*K AY614026). **Fosterella caulescens Rauh; ▲**. *Hechtia montana* Brand. (*rbc*L L19974); *Hechtia carlsoniae* Burt-Utley & Utley (*mat*K AY614020). **Navia saxicola L. B. Sm.; ▲. Neoregelia carolinae (Beer) L.B. Sm. cv. argentea; ▲. Pitcairnia feliciana (A. Chev.) Harms & Mildbr.; ▲. Puya laxa L.B. Sm.; ▲**. *Tillandsia elizabethiae* Rauh (*rbc*L L19971); *Tillandsia usneoides* (L.) L. (*mat*K AY614122).

TYPHACEAE [2,0]. **Sparganium eurycarpum Engelm.; ▲. Typha latifolia L.; NC_013823; Guisinger et al. (2010)**.

RAPATEACEAE [1,7]. *Cephalostemon flavus* (Link) Steyerm. (*rbc*L AY298822). *Epidryos allenii* (Steyerm.) Maguire (AY298830, AF162225). *Kunhardtia radiata* Maguire & Steyerm. (*rbc*L AF036883). **Potarophytum riparium Sandwith; ▲**. *Rapatea xiphoides* Sandwith (*rbc*L AF460969); *Rapatea* sp. (*mat*K AF539958). *Schoenocephalium cucullatum* Maguire (*rbc*L AF460970). *Spathanthus bicolor* Ducke (*rbc*L AF460971). *Stegolepis parvipetala* Steyerm. (AY123242, AY614014).

*JUNCALES [3,3]. *Carex monostachya* A. Rich. (*rbc*L Y12998); *Carex saximontana* (*mat*K FJ597250). **Cyperus alternifolius L.; ▲. Juncus effusus L.; ▲**. *Luzula multiflora* (Ehrh.) Lej. (*rbc*L AJ419945); *Luzula wahlenbergii* Rupr. (*mat*K AY973518). *Prionium serratum* (L. f.) Drège ex E. Mey. (*rbc*L U49223). **Thurnia sphaerocephala Hook. f.; ▲**.

*XYRIDALES [3,7]. **Abolboda macrostachya Spruce ex Malme; ▲**. *Aratitiyopea lopezii* (L.B. Sm.) Steyerm. & P.E. Berry (*rbc*L AF461418). *Eriocaulon humboldtii*

Kunth (*rbc*L AY123236); *Eriocaulon septangulare* With. (*mat*K AY952430). *Lachnocaulon anceps* (Walter) Morong (*rbc*L AY298835). **Mayaca fluviatilis Aubl.; ▲**. *Orectanthe sceptrum* (Oliv. ex Thurn) Maguire (*rbc*L AY123241). **Syngonanthus chrysanthus Ruhland; ▲ plus Syngonanthus flavidulus (Michx.) Ruhland (*rbc*L AY298850)**. *Tonina fluviatilis* Aubl. (*rbc*L AY123237). *Xyris bicephala* Gleason (*rbc*L AY123243). *Xyris jupicai* Rich. (*rbc*L AY298854).

*RESTIONALES [2,4]. *Anarthria polyphylla* Nees (*rbc*L AF148760); *Anarthria prolifera* R. Br. (*mat*K DQ257499). *Baloskion tetraphyllum* (Labill.) B.G. Briggs & L.A.S. Johnson (AF148761, AF164379). **Centrolepis monogyna (Hook. f.) Benth.; ▲ plus Centrolepis monogyna (Hook. f.) Benth. (*mat*K DQ257505)**. *Elegia fenestrata* Pillans (*rbc*L AY123238); *Elegia cuspidata* Mast. (*mat*K DQ257512). *Lepyrodia glauca* (Nees) F. Muell. (AF148785, DQ257521). **Thamnochortus insignis Mast.; ▲**.

*POALES sens. strict. [18,1]. **Agrostis stolonifera L.; NC_008591; Saski et al. (2007). Anomochloa marantoidea Brongn.; NC_014062; Morris & Duvall (2010). Bambusa oldhamii Munro; NC_012927; Wu et al. (2009). Dendrocalamus latiflorus Munro; NC_013088; Wu et al. (2009). Ecdeiocolea monostachya F. Muell.; ▲. Eleusine coracana (L.) Gaertn.; ▲. Flagellaria indica L.; ▲ plus Flagellaria indica L. (*rbc*L DQ307445). Georgeantha hexandra B.G. Briggs & L. A.S. Johnson; ▲. Hordeum vulgare L.; NC_008590; Saski et al. (2007). Joinvillea ascendens Gaudich. ex Brongn. & Gris; ▲. Joinvillea plicata (Hook. f.) Newell & B.C. Stone; FJ486219-FJ486269; Leseberg & Duvall (2009). Oryza sativa L.; NC_001320; Hiratsuka et al. (1989). Pharus latifolius L. (AY357724, AF164388). Puelia olyriformis (Franch.) Clayton; HQ603991-HQ604067; ▲ and Duvall et al. (2010). Saccharum officinarum L.; NC_006084; Asano et al. (2004). Sorghum bicolor (L.) Moench.; NC_008602; Saski et al. (2007). Streptochaeta angustifolia Soderst.; ▲. Triticum aestivum L.; NC_002762; Ogihara et al. (2002). Zea mays L.; NC_001666; Maier et al. (1995).**

Taxonomic index

All taxonomic names corresponding to orders, family, genera and intermediate ranks between them are indexed. Names at the rank of species and below are omitted.

Subject index

Systematics Association Publications

[†] Published by the Systematics Association (out of print)
[*] Published by the Palaeontological Association in conjunction with the Systematics Association
[‡] Published by Oxford University Press for the Systematics Association

Systematics Association Special Volumes

---

[a] Published by Clarendon Press for the Systematics Association
[*] Published by Academic Press for the Systematics Association
[‡] Published by Oxford University Press for the Systematics Association
[**] Published by Chapman & Hall for the Systematics Association
[‡‡] Published by CRC Press for the Systematics Association

DATE DUE

TN: 4447692

Pieces: 1

ILL: 127882739

FHM 08/29/14

PRINTED IN U.S.A.